AF595319

# Nanobody

# Nanobody

Editors

**Ulrich Rothbauer**
**Patrick Chames**

MDPI • Basel • Beijing • Wuhan • Barcelona • Belgrade • Manchester • Tokyo • Cluj • Tianjin

*Editors*
Ulrich Rothbauer
Eberhard Karls University
Tuebingen

Patrick Chames
Aix Marseille University
France

*Editorial Office*
MDPI
St. Alban-Anlage 66
4052 Basel, Switzerland

This is a reprint of articles from the Special Issue published online in the open access journal *Antibodies* (ISSN 2073-4468) (available at: https://www.mdpi.com/journal/antibodies/special_issues/nanobody).

For citation purposes, cite each article independently as indicated on the article page online and as indicated below:

LastName, A.A.; LastName, B.B.; LastName, C.C. Article Title. *Journal Name* **Year**, *Volume Number*, Page Range.

**ISBN 978-3-0365-0378-3 (Hbk)**
**ISBN 978-3-0365-0379-0 (PDF)**

# Contents

# About the Editors

**Ulrich Rothbauer** is Professor for pharmaceutical Biotechnology at the University of Tuebingen and Head of the Pharma and Biotechnology Department of the Natural and Medical Sciences Institute at the University of Tuebingen (NMI). He received his PhD at Ludwig-Maximilians University (LMU), Munich, in the group of Prof. Dr. Walter Neupert revealing the pathomechanism of a mitochondrial disease. He started his work on nanobodies in 2004 as a postdoc in the lab of Prof. Dr. Heinrich Leonhardt at the LMU Biocenter in Munich. In 2006 he became an independent GO-Bio group leader focusing on the development of nanobody-derived tools for protein purification, proteomics and cellular diagnostics. 2008 he founded the Biotech company ChromoTek, which becomes the leading provider of innovative research reagents and technologies based on the nano-/chromobody-technology.

**Patrick Chames** obtained his Ph.D. at the "University de la Méditerranée", Marseille, France in 1997 in the field of antibody engineering. From 1997 to 2001, he worked in the laboratory of phage display pioneer Hennie R. Hoogenboom, Maastricht, NL where he isolated by phage display the first human antibody fragment specifically binding to a cancer related class I MHC complex (TCR-like antibodies). From 2001 to 2005, he worked for a french start-up company (Cellectis SA, Paris) in the field of genome engineering where he significantly contributed to the set up of an in vivo method, leading to the isolation of homing endonucleases capable of performing specific double strand break in a whole genome. In 2005 he accepted a permanent position for the French National center for research (CNRS). Since 2009, he is working for the French institute for health and medical research (INSERM, Marseille, France) in the field of therapeutic bispecific antibodies and blocking antibodies and is specialized in the use of single domain antibodies. Since 2012, he is leader of the "Antibody therapeutics and Immunotargeting team" of the Cancer Research Center of Marseille. He authored above 109 publications including 62 peer reviewed papers, 11 book chapters and 15 patents.

Editorial

# Special Issue: Nanobody

**Patrick Chames [1,*] and Ulrich Rothbauer [2,3,*]**

1 Aix Marseille University, CNRS, INSERM, Institute Paoli-Calmettes, CRCM, 13009 Marseille, France
2 Pharmaceutical Biotechnology, Eberhard Karls University, 72076 Tuebingen, Germany
3 Natural and Medical Sciences Institute at the University of Tuebingen, Markwiesenstr. 55, 72770 Reutlingen, Germany
* Correspondence: patrick.chames@inserm.fr (P.C.); ulrich.rothbauer@uni-tuebingen.de (U.R.); Tel.: +33-491-828-833 (P.C.); +49-7121-51530-415 (U.R.)

Received: 13 February 2020; Accepted: 4 March 2020; Published: 6 March 2020

Since their first description in 1993 [1], single-domain antibody fragments derived from heavy-chain-only antibodies of camelids have received increasing attention as highly versatile binding molecules in the fields of biotechnology and medicine. The term "nanobody"—originally introduced as a trademark of the company Ablynx in 2003—became the general label for those proteins, perfectly reflecting their small dimensional size (2.5 nm × 4 nm; ~13 kDa). Since the expiration of main patent claims in 2013, there has been an emerging tendency in commercializing nanobodies as research, diagnostic and therapy agents.

Nanobodies can be efficiently selected from large (semi-) synthetic/naive or immunized cDNA-libraries using well established display technologies like phage- or yeast-display [2,3]. The simple and single-gene format enables the production of purified nanobodies in the mg–g range per liter of culture, thereby offering an unlimited supply of consistent binding molecules. Additionally, nanobodies can be easily genetically or chemically engineered. Nanobodies are characterized by high affinities and specificities, robust structures, including stable and soluble behaviors in hydrophilic environments and superior cryptic cleft accessibility, low-off target accumulation, and deep tissue penetration [4]. To date, many nanobodies have been evolved into versatile research and diagnostic tools and the list of therapeutic nanobodies applied in clinical trials is constantly growing [5]. Nanobody-derived formats comprise the nanobody itself, homo- or heteromultimers, nanobody-coated nanoparticles or matrixes, nanobody-displayed bacteriophages or enzymatic-, fluorescent- or radionuclide-labeled nanobodies. All these formats were successfully applied in basic biomedical research, cellular and molecular imaging, diagnosis or targeted drug delivery and therapy. With caplacizumab from Sanofi, the first therapeutic active nanobody, was approved by the FDA in February 2019 [6].

This Special Issue on "Nanobodies" includes original manuscripts and reviews covering various aspects related to the discovery, characterization, engineering and application of nanobodies for biomedical research, diagnostics and therapy.

Starting a series of original articles, Longhin et al. selected a set of six novel nanobodies from an immunized library directed against the zinc-transporting $P_{IB}$-ATPase ZntA from *Shigella sonnei* (SsZntA). Further exploiting their ability of bind to cavities and active sites of the target protein, with Nb9, the authors identified a highly selective inhibitor of the ATPase activity of SsZntA. These nanobodies provide a versatile toolset for structural and functional studies of this subset of ATPases [7]. Focusing on more therapeutic application, nanobodies can be a rich source of neutralizing anti-viral reagents. Liu et al. selected a panel of high affinity nanobodies against the E2/E3E2 envelope protein of the Western equine encephalitis virus (WEEV) and demonstrated their potential as detection reagents. The intrinsic modularity and stability of such nanobodies might also be exploited to create stable neutralizing molecules adapted to storage in resource-limited areas [8]. Similarly, Ramage et al. used alpacas immunized with recombinant hemagglutinin from two representative Influenza B viruses to generate nanobodies with both cross-reactive and lineage-specific binding, and carefully analyzed their specificities over a large panel of viruses. The broadly reactive nanobodies might have interesting applications in Influenza B virus diagnostics, vaccine potency testing and possibly as neutralizing immunotherapeutics with potential for intranasal delivery [9]. Exploiting a similar concept, Strokappe et al. generated a panel of neutralizing nanobodies targeting the HIV gp41 and gp120 envelope proteins, thereby describing three new epitopes on these targets. Interestingly, using detailed biophysical and structural characterization, the author took advantage of the modularity of nanobodies to successfully design bispecific constructs with up to 1400-fold higher neutralization potencies than the mixture of the individual nanobodies, thus endowed with a high therapeutic or microbicide potential [10]. Nanobodies also have therapeutic potential beyond virology. In this issue, Heukers et al. took advantage of the small size of nanobodies to generate a new generation of biopharmaceuticals with nanomolar potency by combining anti-hepatocyte growth factor receptor nanobodies to a photosensitizer, thus allowing efficient targeted photodynamic therapy upon local illumination [11]. A detailed epitope mapping is extremely helpful for downstream applications of nanobodies. In their study, Angalakurthi and colleagues used hydrogen exchange-mass spectrometry (HX-MS) to identify the epitopes of 21 nanobodies directed against the ribosome-inactivating subunit (RTA) of ricin toxin. Modelling these epitopes on the surface of RTA not only showed the potential of HX-MS to identify three dimensional epitopes but also supports the generation of a comprehensive B-cell epitope map of ricin toxin [12]. One of the most important features of nanobodies is that they can be genetically engineered for their desired downstream application. In this context, Anderson et al. demonstrated the potential of nanobodies fused to Beta-galactosidase to detect antigens in immunoassays. Using the example of a nanobody specific for the Bacillus collagen-like protein of anthracis (BclA), the authors highlight the potential to engineer nanobodies as highly sensitive reagents for one-step detection of antigen spores in sandwich immunoassays [13].

To generate an intracellular biosensor which monitors the activation of RHO-GTPases, Laura Keller et al. selected a nanobody (RH57) specifically for the GTP-bound version of RHO-GTPase from a synthetic library. When expressed as a fluorescent fusion protein (chromobody), it visualizes the localization of activated endogenous RHO at the plasma membrane without interfering with signaling. As a BRET-based biosensor, the RH57 nanobody was able to monitor RHO spatio-temporal resolved activation in living cells [14]. To optimize the expression of such chromobodies for antigen visualization in living cells, Bettina Keller and colleagues presented a strategy to stabilize biosensors introduced into various cell lines. By site-directed integration of antigen sensitive chromobodies into the AAVS1 safe harbor locus of human cells using CRISPR/Cas9 gene editing, they generated stable chromobody cell lines which not only visualize the localization of the endogenous antigen but can also be used to monitor changes in antigen concentration by quantitative imaging [15]. Nanobodies fused to fluorescent proteins can also be applied for preclinical in vivo imaging. In this context, Gorshkova et al. generated and produced two previously reported TNF-$\alpha$ specific nanobodies fused to the far-red fluorescent protein Katushka. They evaluated the ability of both fluorescently labeled nanobodies to bind and neutralize TNF-$\alpha$ in vitro and to serve as fluorescent probes for in vitro and non-invasive molecular in vivo imaging. In addition to the visualization of local expression of TNF-$\alpha$, they demonstrated that in vivo fluorescence of the engineered nanobodies correlates with TNF levels in living mice [16].

This set of original work is further complemented by a series of reviews highlighting the emerging potential of nanobodies in biomedical research, diagnostics and therapy. Aguilar and colleagues, the pioneers in the field, summarized recent developments on how intracellularly functional nanobodies combined with functional or structural units can be used to study and manipulate protein function in multicellular organisms and developmental biology [17]. As exemplified by several studies in this Special Issue, nanobodies open new avenues for the treatment of viral infections. De Vlieger et al. presented here an overview of the literature covering the use of nanobodies and derived formats to combat viruses including influenza viruses, human immunodeficiency virus-1, and human respiratory syncytial virus [18]. Jank et al. described another field of applications of nanobodies, namely their use as diagnostic and therapeutic reagents against stroke. They covered the advantages of nanobodies over conventional antibody-based therapeutics in the context of brain ischemia and described several innovative nanobody-based treatment protocols aiming at improving stroke diagnostic and therapy [19]. Exploring another very promising and new therapeutic field afforded by the peculiar nature of nanobodies, Bélanger et al. presented the most recent advances in the development of nanobodies as potential therapeutics across brain barriers, including their use for the delivery of biologics across the blood–brain and blood–cerebrospinal fluid barriers, the treatment of neurodegenerative diseases and the molecular imaging of brain targets [20]. Highlighting the unique potential and increasing applications of nanobodies for in vivo imaging, Pieterjan Debie and colleagues presented a comprehensive overview on the current state of the art on how to generate, functionalize and apply nanobodies as molecular tracers for nuclear imaging and image-guided surgery [21]. Finally, Chanier and Chames provided an in-depth coverage of the use of nanobodies as innovative building blocks providing new solutions for the detection and imaging of cancer cells, as well as the development of next-generation cancer immunotherapy approaches, including multispecific constructs for effector cell retargeting, cytokine and immune checkpoint blockade, cargo delivery or the design of optimized CAR T cells [22].

We are convinced that this collection of articles will provide novel insights and information which are valuable to many readers working on different aspects of nanobodies. The editors would like to thank all the contributors for their excellent submissions to this Special Issue, as well as the reviewers and the editorial office of MDPI Antibodies, namely Arya Zou and Nathan Li, for their outstanding support.

**Conflicts of Interest:** The author declares no conflict of interest of interest.

## References

1. Hamers-Casterman, C.; Atarhouch, T.; Muyldermans, S.; Robinson, G.; Hamers, C.; Songa, E.B.; Bendahman, N.; Hamers, R. Naturally occurring antibodies devoid of light chains. *Nature* **1993**, *363*, 446–448. [CrossRef]
2. Moutel, S.; Bery, N.; Bernard, V.; Keller, L.; Lemesre, E.; de Marco, A.; Ligat, L.; Rain, J.C.; Favre, G.; Olichon, A., et al. Nali-h1: A universal synthetic library of humanized nanobodies providing highly functional antibodies and intrabodies. *Elife* **2016**, *5*, e16228. [CrossRef]
3. Pardon, E.; Laeremans, T.; Triest, S.; Rasmussen, S.G.; Wohlkonig, A.; Ruf, A.; Muyldermans, S.; Hol, W.G.; Kobilka, B.K.; Steyaert, J. A general protocol for the generation of nanobodies for structural biology. *Nat. Protoc.* **2014**, *9*, 674–693. [CrossRef]
4. Muyldermans, S. Nanobodies: Natural single-domain antibodies. *Annu. Rev. Biochem.* **2013**, *82*, 775–797. [CrossRef]
5. Steeland, S.; Vandenbroucke, R.E.; Libert, C. Nanobodies as therapeutics: Big opportunities for small antibodies. *Drug Discov. Today* **2016**, *21*, 1076–1113. [CrossRef]
6. Morrison, C. Nanobody approval gives domain antibodies a boost. *Nat. Rev. Drug Discov.* **2019**, *18*, 485–487. [CrossRef]
7. Longhin, E.; Gronberg, C.; Hu, Q.; Duelli, A.S.; Andersen, K.R.; Laursen, N.S.; Gourdon, P. Isolation and characterization of nanobodies against a zinc-transporting p-type atpase. *Antibodies (Basel)* **2018**, *7*, 39. [CrossRef]
8. Liu, J.L.; Shriver-Lake, L.C.; Zabetakis, D.; Goldman, E.R.; Anderson, G.P. Selection of single-domain antibodies towards western equine encephalitis virus. *Antibodies (Basel)* **2018**, *7*, 44. [CrossRef]
9. Ramage, W.; Gaiotto, T.; Ball, C.; Risley, P.; Carnell, G.W.; Temperton, N.; Cheung, C.Y.; Engelhardt, O.G.; Hufton, S.E. Cross-reactive and lineage-specific single domain antibodies against influenza b hemagglutinin. *Antibodies (Basel)* **2019**, *8*, 14. [CrossRef]
10. Strokappe, N.M.; Hock, M.; Rutten, L.; McCoy, L.E.; Back, J.W.; Caillat, C.; Haffke, M.; Weiss, R.A.; Weissenhorn, W.; Verrips, T. Super potent bispecific llama vhh antibodies neutralize hiv via a combination of gp41 and gp120 epitopes. *Antibodies (Basel)* **2019**, *8*, 38. [CrossRef]

11. Heukers, R.; Mashayekhi, V.; Ramirez-Escudero, M.; de Haard, H.; Verrips, T.C.; van Bergen En Henegouwen, P.M.P.; Oliveira, S. Vhh-photosensitizer conjugates for targeted photodynamic therapy of met-overexpressing tumor cells. *Antibodies (Basel)* **2019**, *8*, 26. [CrossRef]
12. Angalakurthi, S.K.; Vance, D.J.; Rong, Y.; Nguyen, C.M.T.; Rudolph, M.J.; Volkin, D.; Middaugh, C.R.; Weis, D.D.; Mantis, N.J. A collection of single-domain antibodies that crowd ricin toxin's active site. *Antibodies (Basel)* **2018**, *7*, 45. [CrossRef]
13. Anderson, G.P.; Shriver-Lake, L.C.; Walper, S.A.; Ashford, L.; Zabetakis, D.; Liu, J.L.; Breger, J.C.; Brozozog Lee, P.A.; Goldman, E.R. Genetic fusion of an anti-bcla single-domain antibody with beta galactosidase. *Antibodies (Basel)* **2018**, *7*, 36. [CrossRef]
14. Keller, L.; Bery, N.; Tardy, C.; Ligat, L.; Favre, G.; Rabbitts, T.H.; Olichon, A. Selection and characterization of a nanobody biosensor of gtp-bound rho activities. *Antibodies (Basel)* **2019**, *8*, 8. [CrossRef]
15. Keller, B.M.; Maier, J.; Weldle, M.; Segan, S.; Traenkle, B.; Rothbauer, U. A strategy to optimize the generation of stable chromobody cell lines for visualization and quantification of endogenous proteins in living cells. *Antibodies (Basel)* **2019**, *8*, 10. [CrossRef]
16. Gorshkova, E.N.; Efimov, G.A.; Ermakova, K.D.; Vasilenko, E.A.; Yuzhakova, D.V.; Shirmanova, M.V.; Mokhonov, V.V.; Tillib, S.V.; Nedospasov, S.A.; Astrakhantseva, I.V. Properties of fluorescent far-red anti-tnf nanobodies. *Antibodies (Basel)* **2018**, *7*, 43. [CrossRef]
17. Aguilar, G.; Matsuda, S.; Vigano, M.A.; Affolter, M. Using nanobodies to study protein function in developing organisms. *Antibodies (Basel)* **2019**, *8*, 16. [CrossRef]
18. De Vlieger, D.; Ballegeer, M.; Rossey, I.; Schepens, B.; Saelens, X. Single-domain antibodies and their formatting to combat viral infections. *Antibodies (Basel)* **2018**, *8*, 1. [CrossRef]
19. Jank, L.; Pinto-Espinoza, C.; Duan, Y.; Koch-Nolte, F.; Magnus, T.; Rissiek, B. Current approaches and future perspectives for nanobodies in stroke diagnostic and therapy. *Antibodies (Basel)* **2019**, *8*, 5. [CrossRef]
20. Belanger, K.; Iqbal, U.; Tanha, J.; MacKenzie, R.; Moreno, M.; Stanimirovic, D. Single-domain antibodies as therapeutic and imaging agents for the treatment of cns diseases. *Antibodies (Basel)* **2019**, *8*, 27. [CrossRef]
21. Debie, P.; Devoogdt, N.; Hernot, S. Targeted nanobody-based molecular tracers for nuclear imaging and image-guided surgery. *Antibodies (Basel)* **2019**, *8*, 12. [CrossRef] [PubMed]
22. Chanier, T.; Chames, P. Nanobody engineering: Toward next generation immunotherapies and immunoimaging of cancer. *Antibodies (Basel)* **2019**, *8*, 13. [CrossRef] [PubMed]

*Article*

# Isolation and Characterization of Nanobodies against a Zinc-Transporting P-Type ATPase

**Elena Longhin [1], Christina Grønberg [1,†], Qiaoxia Hu [1,†], Annette Susanne Duelli [1], Kasper Røjkjær Andersen [2,*], Nick Stub Laursen [2] and Pontus Gourdon [1,3,*]**

[1] Department of Biomedical Sciences, University of Copenhagen, Blegdamsvej 3B, DK-2200 Copenhagen, Denmark; elonghin@sund.ku.dk (E.L.); christina.groenberg@sund.ku.dk (C.G.); qiaoxia@sund.ku.dk (Q.H.); duelli@sund.ku.dk (A.S.D.)

[2] Department of Molecular Biology and Genetics, Aarhus University, Gustav Wieds Vej 10c, DK-8000 Aarhus C, Denmark; nsl@mbg.au.dk

[3] Department of Experimental Medical Science, Lund University, Sölvegatan 19, SE-221 84 Lund, Sweden

* Correspondence: kra@mbg.au.dk (K.R.A.); pontus@sund.ku.dk (P.G.); Tel.: +45-2095-5917 (K.R.A.); +45-5033-9990 (P.G.)

† These authors contributed equally to this work.

Received: 17 October 2018; Accepted: 4 November 2018; Published: 7 November 2018

**Abstract:** P-type ATPases form a large and ubiquitous superfamily of ion and lipid transporters that use ATP (adenosine triphosphate) to carry out their function. The IB subclass ($P_{IB}$-ATPases) allows flux of heavy metals and are key players in metal detoxification, critical for human health, crops, and survival of pathogens. Nevertheless, $P_{IB}$-ATPases remain poorly understood at a molecular level. In this study, nanobodies (Nbs) are selected against the zinc-transporting $P_{IB}$-ATPase ZntA from *Shigella sonnei* (SsZntA), aiming at developing tools to assist the characterization of the structure and function of this class of transporters. We identify six different Nbs that bind detergent stabilized SsZntA. We further assess the effect of the Nbs on the catalytic function of SsZntA, and find that five nanobodies associate without affecting the function, while one nanobody significantly reduces the ATPase activity. This study paves the way for more refined mechanistical and structural studies of zinc-transporting $P_{IB}$-ATPases.

**Keywords:** P-type ATPase; nanobody; llama; Zinc-transport; Zinc-transporting P-ATPase; ZntA

---

## 1. Introduction

The protein superfamily of P-type ATPases is formed by phylogenetically related pumps that actively transport ions and lipids across biological membranes of prokaryotes and eukaryotes [1] at the expense of adenosine triphosphate (ATP). They are divided in five subfamilies ($P_I$-$P_V$) based on sequence similarity and transport specificity [2]. $P_I$-ATPases transport cations, with the $P_{IB}$-subclass being specific for heavy metals such copper and zinc. Noteworthy members of the other subfamilies include the calcium and sodium-potassium ATPases of $P_{II}$ and the proton ATPase of $P_{III}$. The focus here is on class 2 $P_{IB}$-ATPases, $P_{IB\text{-}2}$-ATPases, which comprises zinc-transporting P-type ATPases. These ATPases are relatively poorly characterized from a mechanistic and functional point of view, and only E2 states (metal-free) have been resolved structurally [3]. One reason is that metals such as zinc render these targets unstable, and another that there are no identified compounds that can bind specifically and exclusively to several specific states (including metal bound E1 conformations) of $P_{IB}$-ATPases. The overall structural architecture is conserved in all P-type ATPases, with four domains [4]: The soluble domains, P (phosphorylation), N (nucleotide binding), and A (actuator), and the M domain in the transmembrane region. The P domain contains the highly conserved aspartic acid—lysine—threonine—glycine—threonine (DKTGT) motif with the catalytic aspartate that is targeted by ATP stimulated autophosphorylation. The N domain is responsible for orienting the ATP towards the P domain. The A domain comprises the conserved threonine—glycine—glutamic acid (TGE) loop, which allows for dephosphorylation of the catalytic aspartate in the P-domain and the M-domain is composed by a variable number of helices that enclose membranous ion-binding site(s) that are critical for transport. In addition, zinc transporting $P_{IB-2}$-ATPases possess one or more soluble subfamily-specific domains known as heavy metal-binding domains (HMBDs), whose function remains unclear [5]. These domains work in a tightly coupled manner in order to achieve transport, and the reaction cycle is summarized in the so called Post-Albers scheme [6–8] (Figure 1).

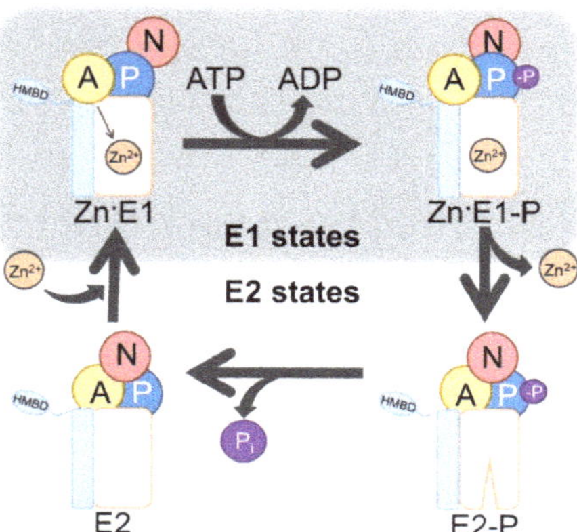


**Figure 1.** Post-Albers scheme of $P_{IB-2}$-ATPases. The E1 (high zinc affinity) and E2 (low zinc affinity) states of the enzyme alternate, and couple ATP (adenosine triphosphate) hydrolysis to the export of zinc. The E1 state accepts one zinc ($Zn^{2+}$) ion and ATP from the intracellular side, which promotes autophosphorylation, reaching the zinc occluded Zn·E1-P state and releasing ADP (adenosine diphosphate). Completion of phosphorylation triggers considerable conformational changes that opens the pump towards the outside, allowing release of zinc in the E2-P state. Metal discharge is associated with auto dephosphorylation, liberation of inorganic phosphate ($P_i$), and allows the enzyme to reach the E2 conformation. The domains are represented as follows: The actuator (A) domain in yellow, the phosphorylation (P) domain in blue, the nucleotide-binding (N) domain in red, the transmembrane domain in light orange. Features specific for $P_{IB}$-ATPases are shown in light blue, and includes two transmembrane helices and heavy-metal binding domain(s) (HMBD).

Antibodies, or immunoglobulins, are large plasma proteins that play a fundamental role in protection against pathogens, such as microorganisms, and are used for numerous basic and applied science applications. Immunoglobulin gamma 1 (IgG1), which is the most abundant immunoglobulin, comprises four polypeptide chains: Two heavy chains, each formed by a variable domain ($V_H$) and three constant domains ($C_H1$, $C_H2$, and $C_H3$), and two light chains, composed by a variable ($V_L$) and a constant ($C_L$) domain. The paratope (antigen binding-site) is formed by the $V_L$ and $V_H$ domains and mediates the interaction with the antigen [9]. However, heavy-chain only antibodies are present in certain species [10]: They are smaller (about 75 kDa) than other antibody isotypes and are formed by two heavy chains, each containing a $V_{HH}$, $C_H2$, and $C_H3$ domain. Their paratope permits antigen-recognition despite being formed by a single $V_{HH}$ domain only, paving the way for the development of single-domain antibodies also called nanobodies. These engineered antibodies are derived from such heavy-chain only antibodies and consist of a single polypeptide chain (about 13 kDa) folding into a variable domain ($V_{HH}$). They can be obtained by immunization of camelids (e.g., llamas) with the target antigen, followed by generation of phage display libraries and screening for antigen binding [11].

The aim of this work is to isolate nanobodies (Nbs) that selectively associate with the zinc-transporter, ZntA, from *Shigella sonnei* (SsZntA), the target employed previously for structural characterization of $P_{IB-2}$-ATPases [3], to develop inhibitors for further structural and functional studies. We successfully raise and purify Nbs against SsZntA and perform experiments to assess binding and inhibition capacities. Notably, we identify six Nbs, which bind specifically to SsZntA, including one that exhibits an inhibitory effect on the ATPase activity.

## 2. Materials and Methods

### 2.1. SsZntA Production

In the following text we refer to the manufacturers Sigma-Aldrich with location in Schnelldorf, Germany; and to VWR with location in Søborg, Denmark, unless other is stated. The gene for ZntA from the bacterium *Shigella sonnei* (UniProtID Q3YW59) was cloned in the vector, pET22 (Merck, Novagen®, Darmstadt, Germany), containing an amino-terminus tag of eight histidine residues (HisTag) for downstream affinity chromatography purification and a cleavage site for TEV protease (TEVp) to allow removal of the HisTag. The construct, pET22-HisTag-SsZntA, was transformed into the *E. coli* C43(DE3) expression strain and cells were grown in Terrific-Broth medium (12 % peptone (Sigma-Aldrich), 24% yeast extract (Sigma-Aldrich), 4% glycerol (VWR), 50 mM Phosphate buffer pH 7 (VWR)) at 37 °C until OD600 reached 1. Then, protein production was induced with 1 mM isopropyl-β-D-thiogalactoside (IPTG) (Biosynth AG, Staad, Switzerland) at 18 °C for 24 h. Cells were harvested at 8000× *g* for 15 min, and resuspended at a concentration of 5 mL/g wet cells in buffer containing 50 mM Tris-HCl pH 8 (Sigma-Aldrich), 200 mM KCl (VWR), 20% *v*/*v* glycerol (VWR), 5 mM β-mercaptoethanol (BME) (VWR), 1 SIGMAFAST™ protease inhibitor tablet (Sigma-Aldrich) per 6 L culture, and then stored at −20 °C. To the thawed cells, a final concentration of 1 mM $MgCl_2$ (VWR), 2 µg/mL DNase I (Sigma-Aldrich) and 1 mM phenylmethanesulphonyl fluoride (PMSF) (Sigma-Aldrich) were added before lysis. The solution was passed through a Constant Systems cell disruptor (Constant Systems Limited, Daventry, UK) twice at 25 kpsi, large cell debris were spun down at 20,000× *g* for 40 min, and membranes were isolated by ultracentrifugation at 190,000× *g* for 3 h. The membrane pellet was resuspended in 20 mM Tris-HCl pH 7.5, 200 mM KCl, 20% *v*/*v* glycerol, 1 mM $MgCl_2$, 5 mM BME, 1 mM PMSF at 3 mg (total protein) per mL (buffer), and solubilized in 1% *w*/*v* n-Dodecyl-β-D-maltoside (DDM) (Anatrace, Maumee, OH, USA) for 1.5 h, followed by ultracentrifugation at 190,000× *g* for 45 min to remove insolubilized material. The supernatant from 6 L culture was adjusted to 50 mM imidazole (Sigma-Aldrich) and 500 mM KCl prior to loading on a 5 mL HisTrap HP (GE Healthcare, Life Sciences, Uppsala, Sweden) equilibrated in buffer containing 20 mM Tris-HCl pH 7.5, 200 mM KCl, 20% *v*/*v* glycerol, 1 mM $MgCl_2$, 0.015% *w*/*v* octaethylene glycol monododecyl ether ($C_{12}E_8$) (Nikko Chemicals Co., Ltd., Tokyo, Japan), 5 mM BME, using an Äkta pure chromatographic system (GE Healthcare, Life Sciences, Uppsala, Sweden), and was eluted with the same buffer with 500 mM imidazole added. Protein containing fractions were pooled and treated with TEVp to remove the HisTag while dialyzing to diminish the excess of imidazole. The cleaved sample was loaded on the HisTrap again (Reverse-affinity chromatography or R-IMAC) to separate uncleaved (HisTagged) protein and the TEVp; the flow through was collected and tested by Western-blot using a conjugated antibody against 6× HisTag (6× His mAb-HRP conjugated by Takara® Bio Europe AB, Göteborg, Sweden) to assess cleavage. The cleaved sample was concentrated to 12 mg/mL and run on a 24 mL size-exclusion chromatography (SEC) column with Superose6 beads (GE Healthcare, Life Sciences, Uppsala, Sweden) equilibrated in *SEC buffer* (20 mM MOPS (3-(N-morpholino)propanesulfonic acid) (VWR) pH 6.8, 80 mM KCl, 20% *v*/*v* glycerol, 3 mM $MgCl_2$, 0.03% *w*/*v* DDM or 0.015% *w*/*v* $C_{12}E_8$, 5 mM BME). The fractions corresponding to the main peak were collected, assessed for purity by SDS-PAGE (sodium dodecyl sulfate-polyacrylamide gel electrophoresis) (Thermo Fisher Scientific, Roskilde, Denmark), concentrated to 10 mg/mL, and stored at −80 °C.

*2.2. LpCopA and MmCadA Production*

CopA from *Legionella pneumophila* (LpCopA, UniProtID Q5ZWR1) and CadA from *Mesorhizobium metallidurans* (MmCadA, UniProtID I4IY19) were produced with the same buffers as SsZntA, but with somewhat different approaches. LpCopA was cloned in the pET22 vector, without any affinity chromatography tag nor cleavage site, and was purified by $Ni^{2+}$-affinity chromatography exploiting the endogenous histidine rich amino-terminus (no engineered HisTag). MmCadA was cloned in the pET52 vector (Merck, Novagen®, Darmstadt, Germany) that includes an N-terminal Strep-tag II with a HRV 3C cleavage site and was purified by StrepTactin® Superflow® (IBA GmbH, Göttingen, Germany) [12] affinity chromatography at pH 7.8, followed by SEC at pH 7.4; this tag does not bind metal ions and therefore does not need to be removed.

*2.3. Llama Immunization and Nanobodies Identification*

Llama immunization and library generation was performed as previously described, now using a mixture of proteins including purified SsZntA for immunization [13]. Briefly, SsZntA solubilized in 0.03% *w*/*v* DDM were injected four times (100 µg/injection) during a period of 12 weeks. The immunization was performed under the permit of Capralogics Inc., which provides a healthy housing environment for all animals and adheres strictly to the United States Department of Agriculture Animal Welfare Act regulations for Animal Care and Use. Peripheral blood mononuclear cells (PMBCs) were isolated with Ficoll paque plus (GE healthcare, Life Sciences, Uppsala, Sweden), and total RNA were extracted using a RNeasy plus kit (Qiagen, Hilden, Germany). cDNA was generated with Superscript III first strand (Invitrogen) and amplified using primers specific for the VHH genes. PCR products were cloned into a phagemid vector designed to express Nbs as pIII fusions and with a C-terminal E-detection tag. VCSM13 helper phages were used for generation of a M13 phage-display library. For selection, 20 µg biotinylated SsZntA (solubilized in 0.015% *w*/*v* $C_{12}E_8$) bound to streptavidin beads were blocked in *SEC buffer* (containing 0.015% *w*/*v* $C_{12}E_8$ and supplemented with 2% *w*/*v* bovine serum albumin (BSA) (Sigma-Aldrich) for 30 min. $5 \times 10^{13}$ M13 phage particles were incubated with the protein for 1 h before the beads were washed 15 times with *SEC buffer* (with 0.015% *w*/*v* $C_{12}E_8$). Elution of the bound phage particles were achieved by addition of 500 µL of 0.2 M glycine (Sigma-Aldrich) pH 2.2 for 10 min, which were added to 75 µL of 1 M Tris pH 9.1 for neutralization before being added to *E. coli* ER2738 cells. Cells were incubated for 1 h at 37 °C and plated on agar plates with 2% *w*/*v* glucose (Sigma-Aldrich). The enriched library was amplified and used in a second round of phage display performed as the first round, but with 1 µg SsZntA and $2 \times 10^{12}$ M13 phage particles. For ELISA, single colonies were transferred to a 96-well plate format and grown for 4 h in LB medium, before Nb production was induced by addition of IPTG to 0.8 mM. The plate was incubated by shaking overnight at 30 °C. Next, the plate was centrifuged and 50 µL of the supernatant was transferred to an ELISA plate coated with a total of 50 µg SsZntA in *SEC buffer* (with 0.015% *w*/*v* $C_{12}E_8$) blocked with 2% *w*/*v* BSA. After incubation for 1 h, the plate was washed four times with *SEC buffer* (with 0.015% *w*/*v* $C_{12}E_8$, without BME), and the anti-E-tag-HPR antibody (Bethyl Laboratories Inc., Montgomery, TX, USA) was added. The plate was then washed four times in *SEC buffer* (with 0.015% *w*/*v* $C_{12}E_8$, without BME) followed by the addition of 50 µL 3,3′,5,5′-tetramethyl-benzidine (Sigma-Aldrich). The reaction was quenched by addition of 50 µL of 1 M HCl and absorbance was measured at 450 nm. Positive phagemids were sequenced and subcloned in pET22, containing a PelB signal at the amino-terminus for periplasmic secretion and a 6xHisTag at the carboxyl-terminus.

*2.4. Nanobodies Sequence Analysis*

ELISA-identified Nb hits were sequenced and the obtained sequences were aligned using the SeaView software (version 4.7, PRABI-Doua, Lyon, France) [14], and a phylogenetic tree was built using the BLOSUM62 matrix. The aligned sequences were visualized using the Sequence Manipulation

Suite© (Multiple Align Show) (version 2, Paul Stothard, University of Alberta, Edmonton, NA, Canada) [15] to highlight the level of identity with a threshold of 90%, which allowed selection of non-redundant Nbs for downstream efforts. The predicted secondary structure was obtained combining the predictions from ABody Builder Antibody Modelling software (Oxford Protein Informatics Group, Oxford, UK) [16] and iCAN analysis platform for nanobodies (Southeast University, Nanjing, Jiangsu Province, China) [17].

### 2.5. Nanobodies Production

The vectors containing 11 selected Nbs (Nb1 to Nb11) were transformed in *E. coli* BL21 strain and grown in 2 L Terrific Broth medium each, at 37 °C until OD600 of 0.6. The expression was induced with 1 mM IPTG at 18 °C over-night. Cells were harvested at 8000× *g* for 15 min and solubilized in 30 mL *Nb buffer* (20 mM Tris-HCl pH 7.5, 400 mM KCl, 20 mM imidazole) prior to sonication and centrifugation at 25,000× *g* for 20min. Nb proteins were purified from the clarified supernatant by incubation with 1 mL $Ni^{2+}$ beads (Ni Sepharose™ 6 Fast Flow) (GE Healthcare, Life Sciences, Uppsala, Sweden) for 1 h at 4 °C, followed by 3 washes in *Nb buffer* and elution in 15 mL *Nb buffer* with 400 mM imidazole added. Next, the eluted Nbs were concentrated to 10–15 mg/mL and run on a SEC column with *SEC buffer* (20 mM MOPS pH 6.8, 80 mM KCl, 20% *v/v* glycerol, 3 mM $MgCl_2$, 0.015% *w/v* $C_{12}E_8$) without reducing agent. The main peak was collected, concentrated to 5 mg/mL (370 μM), and stored at −80 °C.

### 2.6. Functional Assay

#### 2.6.1. SEC Co-Elution

Samples of SsZntA (40 μM) were incubated with Nb from 1 to 11 (40 μM) in 40 mM MOPS pH 6.8, 80 mM KCl, 20% *v/v* glycerol, 1 mM $MgCl_2$, 1 mM TCEP (tris(2-carboxyethyl) phosphine) (Sigma-Aldrich), 0.15 mg/mL $C_{12}E_8$ for 30 min at room temperature in a final volume of 50 μL and run on analytic-scale SEC (column volume of 2.4 mL, flow 0.04 mL/min) equilibrated with the same buffer, to separate SsZntA-Nb complex from unbound proteins. Then, fractions were run on a denaturing SDS-PAGE to visualize the presence of SsZntA and Nb. For the Nbs that showed complex formation, the assay was scaled-up to obtain more reproducible and stable signals: 40 μM SsZntA was incubated with 100 μM Nb in 50 μL and then diluted to 500 μL prior to running in a medium-scale SEC (column volume of 24 mL, flow 0.05 mL/min).

#### 2.6.2. Ni-NTA Co-Elution

In this assay, 13 nmol of SsZntA were incubated with 44 nmol of one of Nb1-2-4-5-8-9 in 20 mM MOPS pH 7.5, 200 mM KCl, 20% *v/v* glycerol, 1 mM $MgCl_2$, 0.015% *w/v* $C_{12}E_8$, 5 mM BME, for 1 h on ice in a final volume of 500 μL. The samples were then adjusted to 400 mM KCl and 50 mM imidazole, and run on a 1 mL HisTrap HP (GE Healthcare, Life Sciences, Uppsala, Sweden) equilibrated with the same buffer at 0.3 mL/min. Fractions were collected and run on a SDS-PAGE gel, then stained with InVision™ HisTag In-gel Stain (Thermo Fisher Scientific, Roskilde, Denmark), according to the manual, to assess the presence of HisTagged proteins followed by Coomassie-staining for total protein detection.

#### 2.6.3. ATPase Activity Assay

The Baginski assay was used to measure ATPases activity and was conducted as previously described [3]. Briefly, 0.8 μM SsZntA was incubated with no Nb, 1.6 μM Nb 1 to 9, 1.6 μM $AlF_3$, 2 mM $AlF_3$, and 0.2, 0.4, 0.8, 1.2, 1.6, 2, 2.4, 2.8, 3.2, 6.4 μM Nb9 in 40 mM MOPS pH 6.8, 150 mM NaCl, 5 mM $MgCl_2$, 20 mM $(NH_4)_2SO_4$ (VWR), 20 mM L-cysteine (VWR), 5 mM $NaN_3$ (Sigma-Aldrich), 0.25 mM $Na_2MoO_4$ (Sigma-Aldrich), 1.2 mg/mL soybean lipids (Sigma-Aldrich), 0.3% *w/v* $C_{12}E_8$, 0.5 mM $ZnSO_4$ (Alfa Aesar by Thermo Fisher Scientific, Karlsruhe, Germany), in a total volume of 50 μL for 1 h at room temperature and the assay was started by addition of 5 mM ATP. The reaction was stopped after 15 min and after 30 min by adding 50 µL Stop solution (2.5% *w*/*v* ascorbic acid (Sigma-Aldrich), 0.4 M HCl, 0.48% *w*/*v* $(NH_4)_6Mo_7O_{24}$ (Fluka Analytical, Bucharest, Romania), 0.8% *w*/*v* SDS (Sigma-Aldrich)) and then 75 µL of arsenic solution (2% *w*/*v* arsenite (Sigma-Aldrich, 2% *v*/*v* acetic acid (VWR), 3.5% *w*/*v* sodium citrate (VWR)), respectively. The absorbance was measured in a microplate reader at 860 nm and the signal normalized with a sample without protein as phosphate background. When testing Nb9 with LpCopA and MmCadA, the same conditions were applied, except the use of $CuSO_4$ as a metal ion with LpCopA.

## 3. Results

### 3.1. Isolation of Native SsZntA

SsZntA was produced in *E. coli* C43 cells and initially purified by immobilized metal affinity chromatography (IMAC). The 8× histidine tag (HisTag) was removed using the TEVp cleavage site, which is important since the HisTag can bind zinc and other metals, potentially interfering with activity and binding assays. Detergent solubilized SsZntA was purified by HisTag-based affinity chromatography and treated with TEVp to remove the HisTag and subjected to affinity purification again, achieving complete separation of cleaved SsZntA, uncleaved (HisTagged) SsZntA, and HisTagged TEVp (Figure 2). The flow-through of the second affinity chromatography (R-IMAC) contained completely cleaved and pure SsZntA, while uncleaved SsZntA and TEVp were present in fractions containing 250 mM and 500 mM imidazole, respectively. SsZntA was further purified using size-exclusion chromatography (SEC), achieving a high degree of purity, as assessed by SDS-PAGE, and homogeneity, as assessed by SEC (Figure 3). The final yield of SsZntA is 5 to 10 mg per liter of culture.

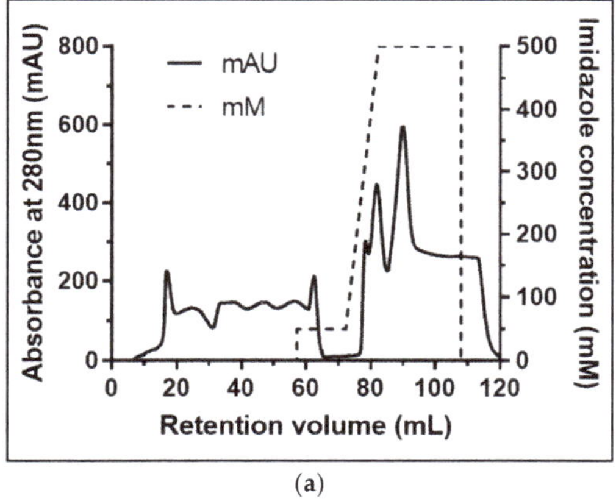


(a)

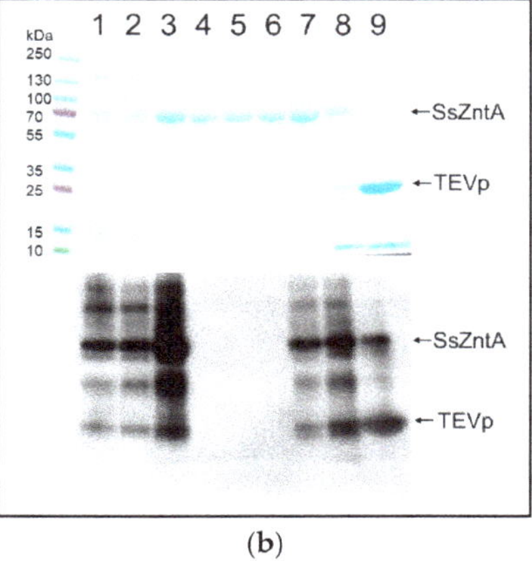


(b)

**Figure 2.** Purificaiton of the zinc-transporter SsZntA from *Shigella sonnei*: (Reverse-affinity chromatography) R-IMAC. (**a**) Reverse (second) affinity chromatography profile: The first 60 mL represent the flow-through containing cleaved SsZntA (without HisTag); at 80 mL, corresponding to 250 mM imidazole, uncleaved SsZntA is eluted; while at 90 mL, at 500 mM imidazole, HisTagged TEVp is eluted. (**b**) On top, Coomassie stained sodium dodecyl sulfate-polyacrylamide gel electrophoresis (SDS-PAGE), and on the bottom, Western blot anti-HisTag; lane 1: Solubilized membranes, lane 2: Clarified solubilized membranes, lane 3: Affinity-chromatography purified SsZntA (with HisTag), lanes 4 to 6: Flow through of reverse affinity chromatography (SsZntA without HisTag) corresponding to retention volume from 0 mL to 60 mL, lanes 7 to 9: Eluted fractions corresponding to retention volume of 80, 85, and 90 mL, respectively.

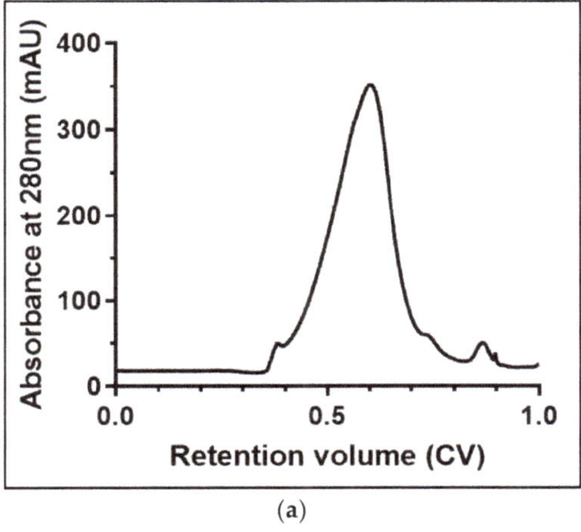


(a)

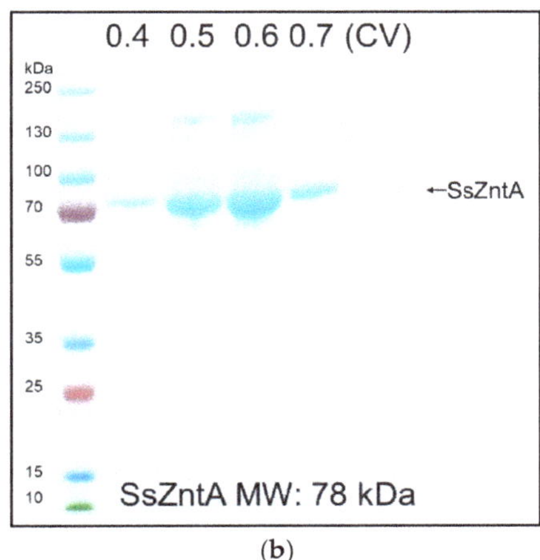


(b)

**Figure 3.** SsZntA purification: SEC. (**a**) Elution profile from a Superose6 size-exclusion column with a column volume (CV) = 24 mL: The main peak elutes at 0.6 CV, followed by minor contaminants. (**b**) Coomassie-stained SDS-PAGE containing the SEC fractions collected at 0.4, 0.5, 0.6, 0.7 CV; upper band at 150 kDa is possibly SsZntA dimers (due to SDS artifacts).

### *3.2. Isolation of Nanobodies*

The purified SsZntA sample (in 0.03% *w*/*v* DDM) was used for llama immunization with multiple injections during a period of 12 weeks. We generated an Nb-library and, following two rounds of phage-display followed by enzyme-linked immunosorbent assay ELISA analysis, we found about 100 positive clones. These were reduced for highly redundant ones, resulting in 45 different Nb sequences that were grouped in three main families (Figure 4a). We decided to investigate 11 Nbs covering most of the overall sequence variability: Nb1, Nb2, and Nb3 belong to the first family; Nb4 and Nb5 to the second family; Nb6 and Nb7 to the third family; Nb8, Nb9, Nb10, and Nb11 did not fit in the previous families and are therefore considered outliers. Figure 4b displays the primary structures of the selected 11 Nbs, where the portions corresponding to the three complementarity-determining regions (CDRs), as well as the predicted conserved secondary structure, are indicated. We then expressed and purified the selected 11 Nbs (with HisTag) and tested them for SsZntA binding and inhibition. Expression in *E. coli* BL21 yielded 4–5 g wet cell weight, which resulted in 9–13 mg affinity-purified Nb, per liter of culture. Nb3 precipitated following affinity-chromatography and further work with it was, therefore, discontinued. The purified Nbs were subjected to SEC, achieving relatively pure Nb samples as analyzed using SDS-PAGE (Figure 5c). Purified SsZntA was tested with three different assays to assess binding (size-exclusion and nickel-affinity chromatography) and inhibitory (ATPase activity) properties of the purified Nbs.

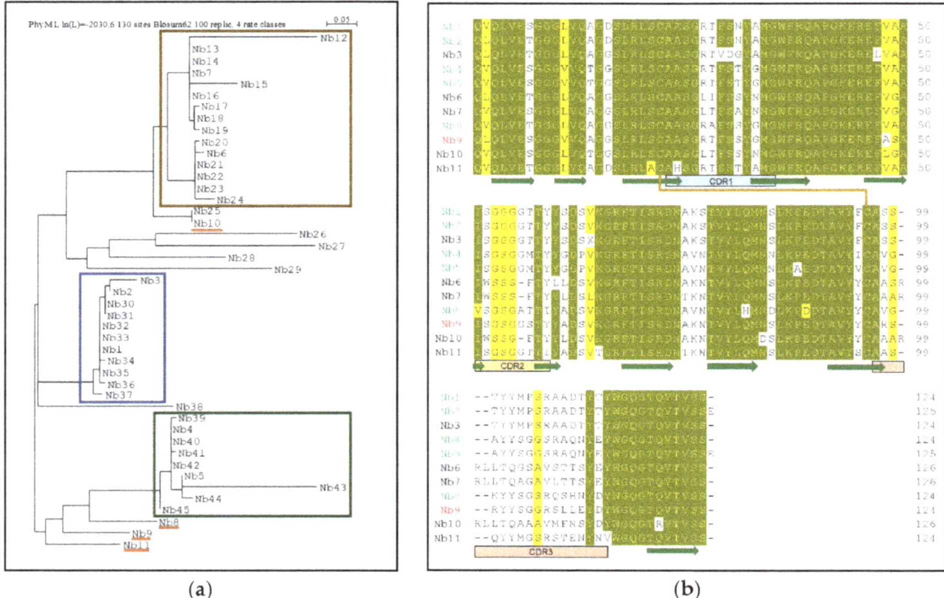

**Figure 4.** Sequence analysis of the identified Nbs. (**a**) Phylogenetic tree of 45 Nb sequences: The three main families of Nbs are highlighted by colored boxes: The first family (blue box) includes Nb1-2-3, the second family (green) includes Nb4-5, the third family (ochre) is composed of two subfamilies, which include Nb6 and Nb7, while Nb8-9-10-11 were sampled as outliers. (**b**) Sequence alignment of the 11 selected Nbs containing the predicted secondary structure (β-sheets are shown as green arrows), the complementarity-determining regions (CDR1 to 3), and the conserved disulphide bond (orange line). Identical (90% threshold) residues are in dark yellow while similar amino acids in light yellow. Nbs that bind to SsZntA, but do not inhibit activity, are represented with a green name (Nb1-2-4-5-8), while in red, the Nb that significantly inhibits ATPase activity (Nb9). The Nbs that precipitated during purification (Nb3) or that caused precipitation of SsZntA (Nb6-7-10-11) are represented in black.

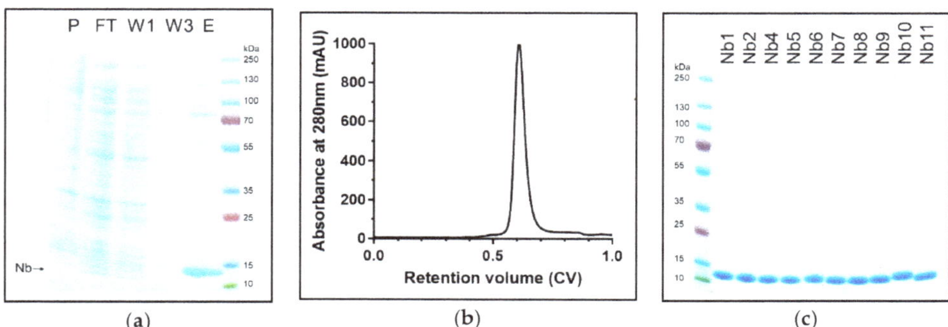


**Figure 5.** Nanobody purification. Samples from purification of Nb1 are representative of all the assessed Nbs. (**a**) Samples from insoluble fraction (pellet, P), flow-through of the IMAC (FT), first and third wash of $Ni^{2+}$-beads (W1 and W3, respectively), and eluted fraction containing Nb1 loaded on Coomassie-stained SDS-PAGE. (**b**) SEC (24mL, Superose75) elution profile: The main peak elutes at 0.6 CV with only minor contaminants. (**c**) Coomassie-stained SDS-PAGE of SEC-purified Nbs: Nbs have a molecular weight of approximately 13.5 kDa.

### 3.3. Size-Exclusion Chromatography (SEC) Co-Elution

Nbs were incubated with SsZntA and subjected to SEC to assess complex formation. We tested the binding in a small-scale size-exclusion column (2.4 mL) and in a medium-scale column (24 mL). The latter was used only on Nbs, which proved to bind to SsZntA in order to obtain a more stable signal and improve the signal-to-noise ratio. The chromatogram reported in Figure 6a is acquired from the medium scale column obtained from Nb1 and is representative of those obtained from Nb2-4-5-8-9 (results not shown), and shows the first peak at 0.6 CV (14.5 mL) containing the SsZntA-Nb complex and the second peak at 0.85 CV (20.5 mL) containing unbound Nb. The overlaying dashed and dotted lines represent the chromatograms obtained from Nb1 and SsZntA, respectively, under the same conditions. As summarized in Table 1, six Nbs (Nb1-2-4-5-8-9) displayed co-elution, suggesting SsZntA-Nb interaction. Nb6-7-10-11 caused precipitation of SsZntA and hence these Nbs were not tested in subsequent assays.

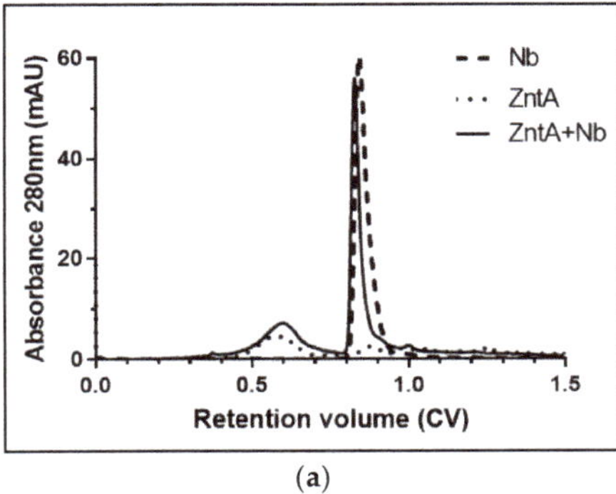


(**a**)

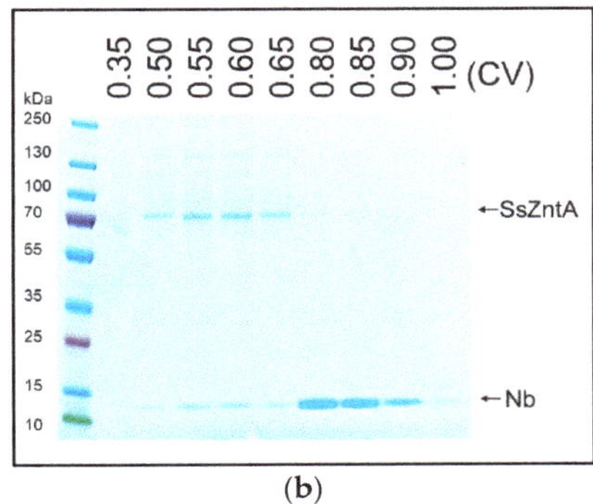


(**b**)

**Figure 6.** SsZntA-Nb complex analysis using size-exclusion chromatography. (**a**) Black line: Chromatogram of SsZntA (40 μM) incubated with Nb1 (100 μM) in a final volume of 50 μL (then diluted to 500 μM) and run through a size-exclusion column (Superose6, 1 CV = 24 mL) showing a peak at 0.6 CV, which contains the SsZntA-Nb complex, and a peak at 0.85 CV, which contains unbound Nb. The dashed line represents the chromatogram of Nb1 alone, while the dotted line the one from SsZntA alone (under the same condition). (**b**) Coomassie-stained SDS-PAGE gel of the fractions collected at 0.35, 0.5, 0.55, 0.6, 0.65, 0.8, 0.85, 0.9, 1 CV.

**Table 1.** Size-exclusion chromatography co-elution results.

| Tested Nb | Binding | Precipitation |
|---|---|---|
| Nb1 | + | - |
| Nb2 | + | - |
| Nb4 | + | - |
| Nb5 | + | - |
| Nb6 | N.D. | + |
| Nb7 | N.D. | + |
| Nb8 | + | - |
| Nb9 | + | - |
| Nb10 | N.D. | + |
| Nb11 | N.D. | + |

Not determined (N.D.): Due to sample precipitation, binding could not be assessed.

### 3.4. Affinity Purification Co-Elution

The engineered HisTag of the Nbs was used to assess binding, as illustrated in Figure 7. Nb1-2-4-5-8-9 were incubated with HisTag-free SsZntA and subjected to HisTag-based affinity chromatography. The samples were then subjected to SDS-PAGE and stained with InVision® HisTag

In-gel staining to evaluate the presence of the HisTag in the protein bands. The results in Figure 8 represent control experiments. Panel (**a**) confirms that the HisTag-free SsZntA appears in the flow-through and is therefore lacking $Ni^{2+}$-binding capacity; panel (**b**) shows that HisTagged SsZntA (lane +) is stained by the InVision® dye, resulting in a clear band. Figure 9a reveals that binding of Nb1 to SsZntA allows for retention of the latter in the $Ni^{2+}$-immobilized resin, proving that an Nb-SsZntA complex forms; Nb2-4-5-8-9 behaved similarly to Nb1 (results not shown). In Figure 9b, it is verified that the bands corresponding to SsZntA do not present a HisTag, while the Nbs do.

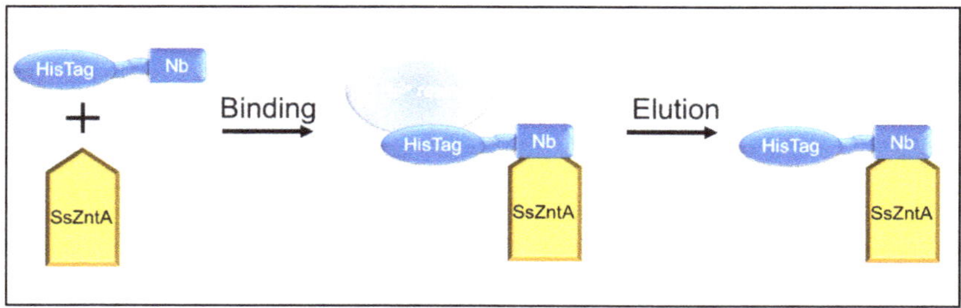


**Figure 7.** Affinity chromatography binding assay. SsZntA is treated with TEVp to remove the HisTag, in this way it cannot bind to a $Ni^{2+}$ beads, while the HisTagged nanobody can. The binding was assessed by incubating HisTag-free SsZntA with HisTagged Nb and loading them to an Ni-NTA column. When SsZntA forms a complex with the Nb, it indirectly binds to the $Ni^{2+}$ beads (via the Nb's HisTag) and can be eluted from the $Ni^{2+}$-immobilized resin. N.B. All the representations are out of scale for clarity.

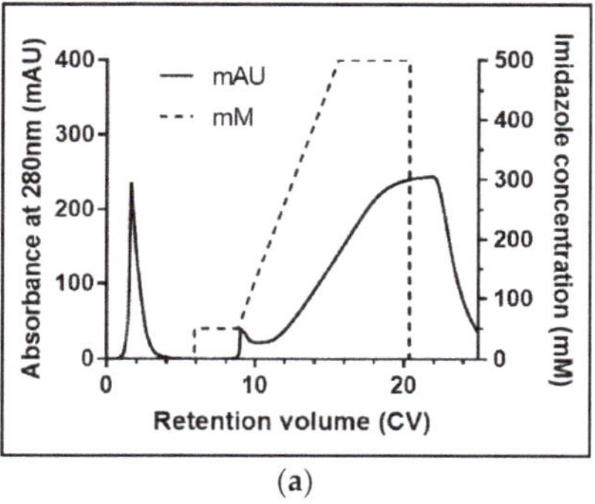


(**a**)

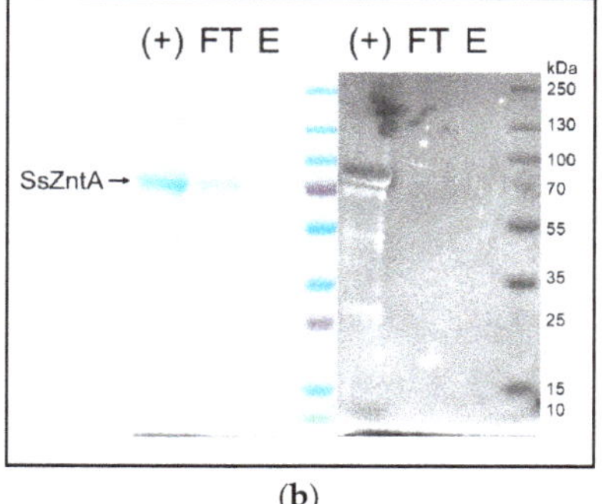


(**b**)

**Figure 8.** SsZntA-Nb complex analysis using affinity chromatography: SsZntA only. (**a**) Chromatogram of HisTag-free SsZntA loaded on an $Ni^{2+}$-immobilized column. Without Nb, the ATPase elutes in the flow through (control). (**b**) Coomassie (left) and InVision® (right) stained SDS-PAGE of HisTagged SsZntA as positive control for the staining (+), flow through (FT), and eluted fraction (E) of HisTag free SsZntA loaded on the $Ni^{2+}$-immobilized resin as negative control.

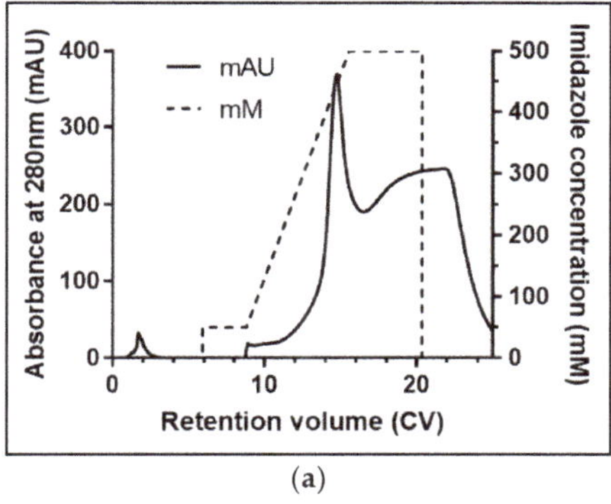


(a)

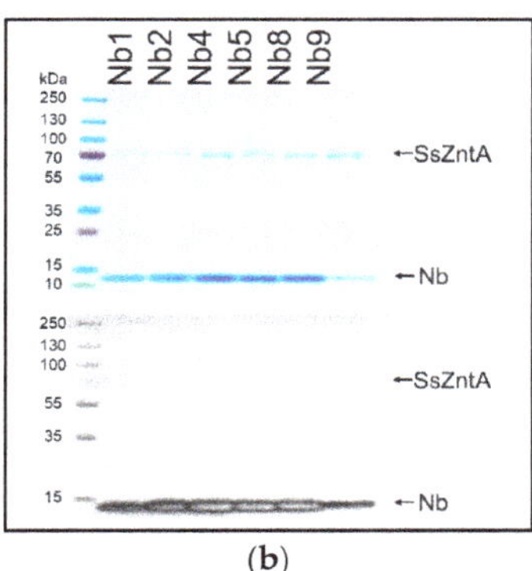


(b)

**Figure 9.** SsZntA-Nb complex analysis using affinity chromatography: SsZntA + Nb. (**a**) Chromatogram of HisTag-free SsZntA incubated with HisTagged Nb1. The complex elutes at 400 mM imidazole and is here reported as representative of the other Nbs. (**b**) Coomassie (top) and InVision® (bottom) stained SDS-PAGE of the eluted fractions at 400 mM imidazole of HisTag-free SsZntA in complex with, respectively, Nb1-2-4-5-8-9.

### 3.5. Nb Effect on SsZntA Functionality

We used the Baginski assay to access Nb-induced inhibition of SsZntA, which estimates the amount of released inorganic phosphate ($P_i$) associated with the catalytical cycle (ATP turn-over, 1 mole $P_i$ equals 1 mole of transported zinc). Different batches of Nbs were tested for inhibitory activity at a molar ratio of 1 (SsZntA) to 2 (Nb), corresponding to 8 μM SsZntA and 16 μM Nbs, yielding the results shown in Figure 10a: Wild-type SsZntA without added Nb displays a specific activity of $885 \pm 87$ nmol $P_i$ mg$^{-1}$ min$^{-1}$, which is comparable to the activity previously reported [18]. Nb9 reduces the ATPase activity to around 50% of wild-type, showing a significant inhibition with a $p$ value < 0.0001 (Dunn's multiple comparison test). The inhibition on the ATPase activity by the other Nbs is non-significant at the concentrations tested. We included the phosphate analog, aluminum fluoride ($AlF_3$), as a control of the ATPase inhibition assay at two different concentrations: 1.6 μM (as the Nbs) and 2 mM (previously reported [3] working concentration). $AlF_3$ does not affect ATPase activity at 1.6 μM, showing that the affinity for SsZntA of Nb9 is higher than that of $AlF_3$. In Figure 10b, a titration experiment of Nb9 with different molar ratios is reported: Nb9 inhibits ATPase activity of SsZntA when present at least at a molar ratio 1:1 (SsZntA:Nb9) and has the maximum inhibitory effect (at about 50%) at the molar ratio of 1:3 (SsZntA:Nb9). To assess Nb9 specificity, it was further tested against CopA from *Legionella pneumophila* (LpCopA): A member of the same subfamily as SsZntA ($P_{IB}$-type, 34% identity to SsZntA), but transporting $Cu^+$ instead of $Zn^{2+}$, and against CadA from *Mesorhizobium metallidurans* (MmCadA); a $Cd^{2+}$ and $Zn^{2+}$-transporting ATPase from the same subclass as SsZntA ($P_{IB-2}$-type, 53% identity). LpCopA and MmCadA were purified similarly to SsZntA and yield highly pure protein (Figure 11a). As shown in Figure 11b, Nb9 did not inhibit ATPase activity for any of the two proteins, suggesting that Nb9 is specific for SsZntA.

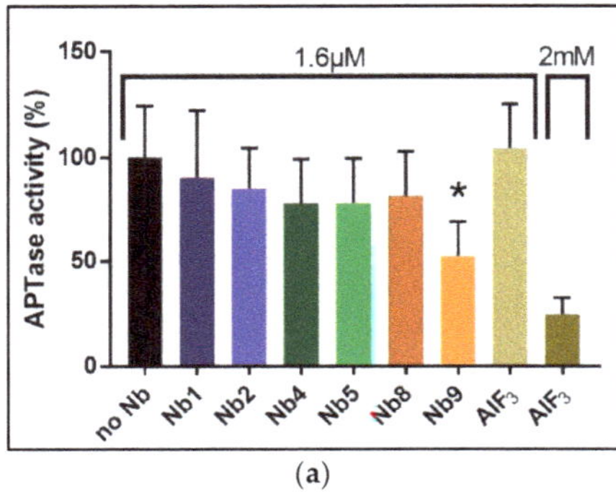

(a)

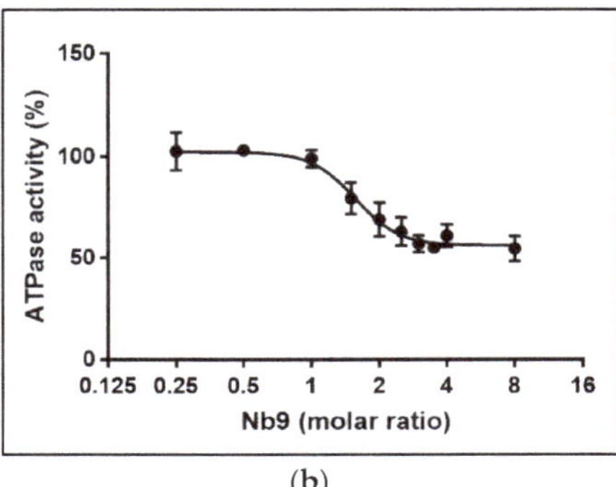

(b)

**Figure 10.** Nanobody effect on the catalytical function. (**a**) SsZntA was incubated with each Nb at a 1:2 molar ratio (i.e., 1.6 μM) and activity was assessed. Nb9 displayed an inhibitory effect of about 50% compared to untreated (no Nb) SsZntA. The difference between no Nb and Nb9 has a *p* value of $<0.0001$ (*), as assessed by a Dunn's multiple comparison test. The ATPase inhibitor aluminum fluoride ($AlF_3$) is shown as a positive control, at 1.6 μM (corresponding to the concentration of the tested Nbs) and at 2 mM. (**b**) The inhibition stoichiometry was assessed as titration of SsZntA with different molar ratios of Nb9. The inhibitory effect appears when Nb9 is present at a molar ratio of 1:1 (ZntA:Nb9), and reaches the inhibitory plateau at a molar ratio of 1:3 (ZntA:Nb9), with an $IC_{50}$ = 1.57 molar ratio.

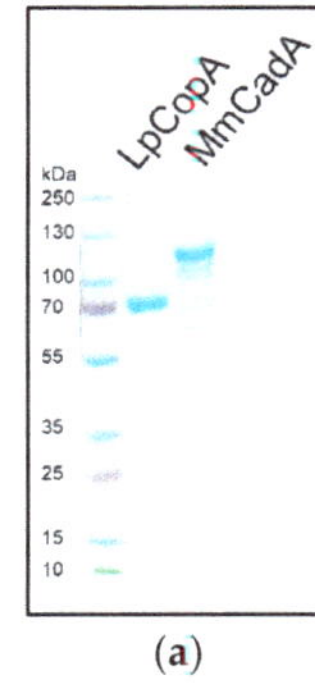

(a)

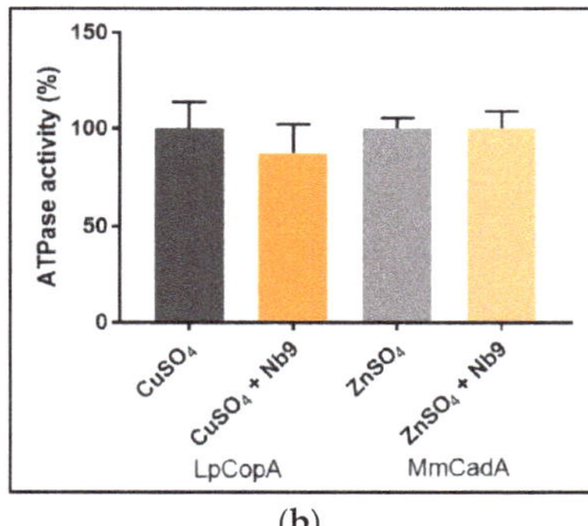

(b)

**Figure 11.** Nb9 specificity. (**a**) SDS-PAGE of purified $Cu^{+}$-transporting $P_{IB-1}$-ATPase from *Legionella pneumophila* (LpCopA) and $Cd^{2+}$ and $Zn^{2+}$-transporting $P_{IB-2}$-ATPase from *Mesorhizobium metallidurans* (MmCadA) (**b**) Nb9 was tested for specificity against LpCopA and against MmCadA, in the presence of $Cu^{+}$ or $Zn^{2+}$, respectively. The results are normalized for the background in presence of the metal ion chelator EDTA.

## 4. Discussion

P-type ATPases are characterized by a transmembrane domain, embedding the ion binding site(s), and a large soluble portion, which includes catalytic and regulatory domains. Each domain has a particular function that requires an overall maintained structural integrity: The relative arrangements of these domains, and their ability to undergo conformational changes, is of crucial importance for the ATPase function and the ion translocation activity [19]. Hence, ATP and phosphate analogs (e.g., AMPPCP (Adenylylmethylenediphosphonate disodium salt) [20] or $AlF_3$ [21]) that occupy the nucleotide binding or phosphorylation site in the soluble portion, preventing hydrolysis, inhibit also translocation of transported ion(s) through the transmembrane domain. In this study, Nbs were selected as tool compounds, exploiting their ability to bind to cavities and active sites of proteins due to a combination of the small size and convex paratope [22]. We identified Nbs that bind specifically to the model zinc-transporting $P_{IB}$-type ATPase SsZntA to widen the toolset available for structural and functional studies of this subclass of ATPases.

Isolated membranes containing SsZntA were solubilized in DDM, followed by detergent exchange to $C_{12}E_8$ using affinity chromatography. For the samples exploited for immunization, but not those used for selection, an additional detergent exchange to DDM was performed during SEC. DDM is frequently the detergent of choice for membrane proteins, providing efficient solubilization, stabilization, and preventing non-specific interaction for many membrane protein targets. This means that DDM is suitable if the aim is to reduce interaction and aggregation with other soluble and membrane proteins, as in the case of llama immunization. The SsZntA samples used for Nb selection were instead purified in 0.015% *w/v* $C_{12}E_8$: This detergent provides a smaller micelle size and, therefore, leaves more of the protein exposed for interactions. Detergents that give a small micelle size are preferred for structural biology studies, increasing the possibility for protein-protein interactions, which are essential for crystal-lattice formation [23]. In this work, the final aim was to provide new tools for SsZntA crystallization, and thus the identified Nbs need to bind and/or inhibit in the presence of the detergent used for crystallization studies of many P-type ATPases: $C_{12}E_8$ [3,24–26]. Therefore, the selection was performed using $C_{12}E_8$ solubilized SsZntA, but we anticipate similar results in the presence of other mild detergents, including DDM.

Following screening of the initial Nb hits, we selected 11 for further studies, attempting to address variability in the sequences. As shown in Figure 4 (**b**), the selected Nbs display distinct differences in the complementarity determining regions (CDR), while the conserved motifs and predicted scaffold structure are maintained. This observation is in line with previous findings on Nbs, since the CDRs are constituted by residues responsible for antigen binding, and different residue combinations potentially bind to different antigens. Among the selected 11 Nbs, one (Nb3) could not be purified and four (Nb6, Nb7, Nb10, Nb11) caused precipitation of the sample when mixed with SsZntA, and could thus not be tested in further assays. We believe that the observed precipitation can relate to the presence of detergent, perhaps in combination with Nb binding, which apparently compromises SsZntA further in this environment (e.g., through interaction with unstable intermediates or at sensitive regions of the proteins, such as the membrane interface). We successfully identified six Nbs (Nb1, Nb2, Nb4, Nb5, Nb8, Nb9) which bind to SsZntA, while only one (Nb9) displays significant (<50%) inhibition of the ATPase activity. Nb9 was tested against two other members of the heavy metal-transporting subfamily in order to assess its specificity: It does not inhibit the ATPase activity of neither a copper- nor a zinc-transporting ATPase, denoting a high degree of specificity and a tight interaction with specific residues and/or conformations of SsZntA, which make it an ideal tool for studying in detail the transport cycle of this protein.

Further studies are needed to assess whether Nb9 inhibits the activity by binding between domains (as allosteric inhibitors), or if it occupies the ATP or the metal binding site, thus inhibiting the activity as competitive ligands, and if it blocks the reaction cycle by binding to a specific reaction cycle state. The latter is of particular interest, since it could be the first E1-specific inhibitor to be found for a metal-transporting P-type ATPase and would thus have extensive applications in both structural and functional characterization efforts. The fact that five out of six Nbs bind, but do not inhibit, the activity may hint that these Nbs associate to, for example, the more exposed portions of the soluble domains, and not to domain-domain interfaces. In any case, they can still be exploited as tools for structural biology studies: Co-crystallization of the Nbs with SsZntA is likely to alter crystal contacts/crystal packing by, for instance, generating a wider hydrophilic surface for lattice contacts. If stronger crystal contacts are obtained, it may result in an increased resolution of the diffraction data generated from these crystals. Moreover, the identified Nbs may serve as chaperones for cryo-EM experiments, with the aim of facilitating images' orientation, domain identification, and to render particles larger. In conclusion, this work provides the foundation for further usage of Nbs in structural studies of $P_{IB}$-type ATPases, aiming at expanding the available toolbox to achieve a greater understanding of this important, yet elusive, subfamily of enzymes.

**Author Contributions:** Conceptualization, K.R.A., N.S.L. and P.G.; Formal analysis, E.L., C.G. and Q.H.; Funding acquisition, P.G.; Investigation, E.L., C.G., Q.H. and A.S.D.; Methodology, K.R.A.; Project administration, P.G.; Resources, K.R.A., N.S.L. and P.G.; Supervision, C.G., K.R.A., N.S.L. and P.G.; Validation, E.L., C.G., Q.H. and A.S.D.; Visualization, E.L.; Writing—original draft, E.L.; Writing—review & editing, E.L., C.G., K.R.A., N.S.L. and P.G.

**Funding:** E.L. was funded by Lundbeck Foundation grant number R218-2016-1548. P.G. is supported by the following Foundations: Lundbeck, Knut and Alice Wallenberg, Carlsberg, Novo-Nordisk, Brødrene Hartmann, Agnes og Poul Friis, Augustinus, Crafoord as well as The Per-Eric and Ulla Schyberg. Funding is also obtained from The Independent Research Fund Denmark, the Swedish Research Council and through a Michaelsen scholarship. N.S.L. was funded by The Lundbeck Foundation.

**Conflicts of Interest:** The authors declare no conflict of interest.

## References

1. Bublitz, M.; Morth, J.P.; Nissen, P. P-type ATPases at a glance. *J. Cell Sci.* **2011**, *124*, 2515–2519. [CrossRef] [PubMed]
2. Axelsen, K.B.; Palmgren, M.G. Evolution of substrate specificities in the P-type ATPase superfamily. *J. Mol. Evol.* **1998**, *46*, 84–101. [CrossRef] [PubMed]
3. Wang, K.; Sitsel, O.; Meloni, G.; Autzen, H.E.; Andersson, M.; Klymchuk, T.; Nielsen, A.M.; Rees, D.C.; Nissen, P.; Gourdon, P. Structure and mechanism of Zn2+-transporting P-type ATPases. *Nature* **2014**, *514*, 518. [CrossRef] [PubMed]
4. Kühlbrandt, W. Biology, structure and mechanism of P-type ATPases. *Nat. Rev. Mol. Cell. Biol.* **2004**, *5*, 282. [CrossRef] [PubMed]
5. Sitsel, O.; Duelli, A.; Gourdon, P. Zinc-Transporting P-Type ATPases. In *Encyclopedia of Inorganic and Bioinorganic Chemistry*; Scott, R.A., Ed.; John Wiley & Sons, Ltd.: Hoboken, NJ, USA, 2016.
6. Post, R.L.; Sen, A.K.; Rosenthal, A.S. A phosphorylated intermediate in adenosine triphosphate-dependent sodium and potassium transport across kidney membranes. *J. Biol. Chem.* **1965**, *240*, 1437–1445. [PubMed]
7. Albers, R.W. Biochemical Aspects of Active Transport. *Annu. Rev. Biochem.* **1967**, *36*, 727–756. [CrossRef] [PubMed]
8. Apell, H.J. How do P-type ATPases transport ions? *Bioelectrochemistry* **2004**, *63*, 149–156. [CrossRef] [PubMed]
9. De Meyer, T.; Muyldermans, S.; Depicker, A. Nanobody-based products as research and diagnostic tools. *Trends Biotechnol.* **2014**, *32*, 263–270. [CrossRef] [PubMed]
10. Hamers-Casterman, C.; Atarhouch, T.; Muyldermans, S.; Robinson, G.; Hamers, C.; Songa, E.B.; Bendahman, N.; Hamers, R. Naturally occurring antibodies devoid of light chains. *Nature* **1993**, *363*, 446–448. [CrossRef] [PubMed]
11. Arbabi Ghahroudi, M.; Desmyter, A.; Wyns, L.; Hamers, R.; Muyldermans, S. Selection and identification of single domain antibody fragments from camel heavy-chain antibodies. *FEBS Lett.* **1997**, *414*, 521–526. [CrossRef]
12. Schmidt, T.G.M.; Skerra, A. The Strep-tag system for one-step purification and high-affinity detection or capturing of proteins. *Nat. Protoc.* **2007**, *2*, 1528. [CrossRef] [PubMed]
13. Hansen, S.B.; Laursen, N.S.; Andersen, G.R.; Andersen, K.R. Introducing site-specific cysteines into nanobodies for mercury labelling allows de novo phasing of their crystal structures. *Acta Crystallogr. D Struct. Biol.* **2017**, *73*, 804–813. [CrossRef] [PubMed]
14. Gouy, M.; Guindon, S.; Gascuel, O. SeaView Version 4: A Multiplatform Graphical User Interface for Sequence Alignment and Phylogenetic Tree Building. *Mol. Biol. Evol.* **2010**, *27*, 221–224. [CrossRef] [PubMed]
15. Stothard, P. The Sequence Manipulation Suite: JavaScript Programs for Analyzing and Formatting Protein and DNA Sequences. *BioTechniques* **2000**, *28*, 1102–1104. [CrossRef] [PubMed]
16. Leem, J.; Dunbar, J.; Georges, G.; Shi, J.; Deane, C.M. ABodyBuilder: Automated antibody structure prediction with data-driven accuracy estimation. *mAbs* **2016**, *8*, 1259–1268. [CrossRef] [PubMed]
17. Zuo, J.; Li, J.; Zhang, R.; Xu, L.; Chen, H.; Jia, X.; Su, Z.; Zhao, L.; Huang, X.; Xie, W. Institute collection and analysis of Nanobodies (iCAN): A comprehensive database and analysis platform for nanobodies. *BMC Genomics* **2017**, *18*, 797. [CrossRef] [PubMed]
18. Sharma, R.; Rensing, C.; Rosen, B.P.; Mitra, B. The ATP Hydrolytic Activity of Purified ZntA, a Pb(II)/Cd(II)/Zn(II)-translocating ATPase from Escherichia coli. *J. Biol. Chem.* **2000**, *275*, 3873–3878. [CrossRef] [PubMed]

19. Meng, D.; Bruschweiler-Li, L.; Zhang, F.; Brüschweiler, R. Modulation and Functional Role of the Orientations of the N- and P-Domains of Cu+-Transporting ATPase along the Ion Transport Cycle. *Biochemistry* **2015**, *54*, 5095–5102. [CrossRef] [PubMed]
20. Krasteva, M.; Barth, A. Structures of the Ca2+—ATPase complexes with ATP, AMPPCP and AMPPNP. An FTIR study. *Biochim. Biophys. Acta.* **2007**, *1767*, 114–123. [CrossRef] [PubMed]
21. Cho, H.; Wang, W.; Kim, R.; Yokota, H.; Damo, S.; Kim, S.-H.; Wemmer, D.; Kustu, S.; Yan, D. BeF3- acts as a phosphate analog in proteins phosphorylated on aspartate: Structure of a BeF3-complex with phosphoserine phosphatase. *Proc. Nati. Acad. Sci. USA* **2001**, *98*, 8525–8530. [CrossRef] [PubMed]
22. Beghein, E.; Gettemans, J. Nanobody Technology: A Versatile Toolkit for Microscopic Imaging, Protein–Protein Interaction Analysis, and Protein Function Exploration. *Front. Immunol.* **2017**, *8*. [CrossRef] [PubMed]
23. Gutmann, D.A.P.; Mizohata, E.; Newstead, S.; Ferrandon, S.; Henderson, P.J.F.; van Veen, H.W.; Byrne, B. A high-throughput method for membrane protein solubility screening: The ultracentrifugation dispersity sedimentation assay. *Protein Sci.* **2007**, *16*, 1422–1428. [CrossRef] [PubMed]
24. Andersson, M.; Mattle, D.; Sitsel, O.; Klymchuk, T.; Nielsen, A.M.; Møller, L.B.; White, S.H.; Nissen, P.; Gourdon, P. Copper-transporting P-type ATPases use a unique ion-release pathway. *Nat. Struct. Mol. Biol.* **2014**, *21*, 43–48. [CrossRef] [PubMed]
25. Olesen, C.; Picard, M.; Winther, A.-M.L.; Gyrup, C.; Morth, J.P.; Oxvig, C.; Møller, J.V.; Nissen, P. The structural basis of calcium transport by the calcium pump. *Nature* **2007**, *450*, 1036. [CrossRef] [PubMed]
26. Morth, J.P.; Pedersen, B.P.; Toustrup-Jensen, M.S.; Sørensen, T.L.M.; Petersen, J.; Andersen, J.P.; Vilsen, B.; Nissen, P. Crystal structure of the sodium–potassium pump. *Nature* **2007**, *450*, 1043. [CrossRef] [PubMed]

*Article*

# Selection of Single-Domain Antibodies towards Western Equine Encephalitis Virus

**Jinny L. Liu, Lisa C. Shriver-Lake, Dan Zabetakis, Ellen R. Goldman and George P. Anderson ***

Naval Research Laboratory, Center for Biomolecular Science and Engineering, Washington, DC 20375, USA; jinny.liu@nrl.navy.mil (J.L.L.); lisa.shriverlake@nrl.navy.mil (L.C.S.-L.); daniel.zabetakis@nrl.navy.mil (D.Z.); ellen.goldman@nrl.navy.mil (E.R.G.)

* Correspondence: george.anderson@nrl.navy.mil; Tel.: +1-202-404-6033

Received: 26 November 2018; Accepted: 12 December 2018; Published: 15 December 2018

**Abstract:** In this work, we describe the selection and characterization of single-domain antibodies (sdAb) towards the E2/E3E2 envelope protein of the Western equine encephalitis virus (WEEV). Our purpose was to identify novel recognition elements which could be used for the detection, diagnosis, and perhaps treatment of western equine encephalitis (WEE). To achieve this goal, we prepared an immune phage display library derived from the peripheral blood lymphocytes of a llama that had been immunized with an equine vaccine that includes killed WEEV (West Nile Innovator + VEWT). This library was panned against recombinant envelope (E2/E3E2) protein from WEEV, and seven representative sdAb from the five identified sequence families were characterized. The specificity, affinity, and melting point of each sdAb was determined, and their ability to detect the recombinant protein in a MagPlex sandwich immunoassay was confirmed. Thus, these new binders represent novel recognition elements for the E2/E3E2 proteins of WEEV that are available to the research community for further investigation into their applicability for use in the diagnosis or treatment of WEE.

**Keywords:** single-domain antibody; Western equine encephalitis virus; MagPlex

## 1. Introduction

Western equine encephalitis virus (WEEV), an alphavirus in the Togaviridae family, is an arbovirus transmitted to people and horses by mosquitoes and is the causative agent of western equine encephalitis (WEE). WEEV originated when Eastern equine encephalitis virus (EEEV), a new world virus, and Sindbis virus, an old world virus, recombined [1–3]. WEEV, EEEV and the related Venezuelan equine encephalitis virus (VEEV) can spread to the central nervous system, causing symptoms ranging from mild febrile reactions to encephalitis, often resulting in permanent neurological damage that can lead to death. In North America, WEE is seen primarily in U.S. States and Canadian provinces west of the Mississippi River [4]. The disease is also seen in South and Central American countries. WEEV causes serious disease in horses, with a case-fatality rate of 20–30% [5]. Although human cases are relatively rare, in 1941 an outbreak in Canada caused more than 1000 human infections [6]. For humans, the WEEV case-fatality rate has been estimated at 3 to 7%, with 15 to 30% of convalescent patients developing secondary neurological damage [7,8]. In addition to central nervous system impairment, demyelination is a known sequela of this disease. Additional complications can include mental retardation, behavioral changes, paralysis, permanent focal neurologic deficits, seizure disorders, cerebellar damage, and choreoathetosis.

Alphaviruses can be produced in large quantities, are easy to disseminate and are highly infectious aerosols [9,10]; thus, WEEV, as well as VEEV and EEEV, are considered to be potential biological weapons [11–13] and are classified as category B bioterrorism agents/diseases by the US Centers for Disease Control and Prevention (https://emergency.cdc.gov/agent/agentlist-category.asp).

WEEV, like other alphaviruses, is an enveloped, positive-stranded RNA virus. Two envelope proteins, glycoproteins E1 and E2, associate as trimers of E1-E2 dimers on the viral surface, and make up 80 spikes on the viral surface. Structural studies of alphaviruses including cryo-EM of WEEV have been reported [14]. The E3 protein binds the E1-E2 spike, and protects it from the low pH of the secretory pathway [15].

There are available equine vaccines for WEEV which are recommended for the majority of horses. The vaccines typically contain inactivated virus and an adjuvant. There is currently no vaccine to protect humans from WEEV, however various vaccination strategies have been investigated. Due to the virus's broad geographic distribution but low levels of infection accompanied with serious complications, an active human vaccination program is unlikely even if an effective vaccine were available. Nonetheless, rapid and inexpensive detection methodologies are still needed, and once identified, post-exposure therapies to reduce the risk of complications would be beneficial. A number of antibodies have been developed, including two antibody fragments (scFv) generated from murine IgGs for diagnostic purposes [16–18]. To date, the most extensively studied antibodies are scFv-Fc recombinant antibodies selected via phage display from two macaques immunized with inactivated WEEV. Several of these antibodies were found to be neutralizing [19] and in a later study were found to protect mice from an aerosol challenge [20]. While much progress has been made, there is clearly a need for additional reagents that possess properties different than those currently being investigated.

Single-domain antibodies (sdAb) are small and stable binding domains derived from the variable domain of the heavy-chain-only antibodies (termed VHH) found in camelids including camels, llamas, and alpacas. SdAb combine the sensitivity and specificity of conventional antibodies with advantages that come from being comprised of only a single domain, such as high physical-chemical stability including heat-resistance, the ability to refold after denaturation, excellent solubility in water, and the capacity to be produced using recombinant technology in good yield [21–26]. SdAb that are produced using recombinant technology, most often in *Escherichia coli*, are amenable to the formation of fusion constructs to tailor their integration into a variety of assay formats and sensor systems [27–32]. They can also be modified to improve their biophysical properties; mutagenesis has led to variants with improved protein production and stability, as assessed by the protein's melting point [33–37]. In addition, one can take what is already a rugged and reliable immunoreagent and create even more robust versions for detection applications in resource-limited areas that lack refrigeration. Challenges remain for transitioning sdAb as therapeutics including rapid clearance from circulation due to their small size [38], and the presence of anti-domain antibodies in many individuals [39]. In this work, we describe the selection and characterization of seven new sdAb towards the E2/E3E2 envelope protein of WEEV.

## 2. Materials and Methods

### 2.1. Reagents

Unless otherwise specified, chemical reagents were from Sigma Aldrich (St. Louis, MO, USA), Thermo Fisher Scientific (Waltham, MA, USA), or VWR International (Radnor, PA, USA). Restriction endonucleases and ligation reagents were from New England Biolabs (Ipswich, MA, USA). Recombinant WEEV glycoprotein (E2/E3E2) was purchased from IBT Bioservices (Rockville, MD, USA). The data in the product insert indicated that the majority of the preparation is E2, however E3E2 was also observed by both gel electrophoresis and Western blotting. For simplicity, we refer to this product as E3E2. The West Nile innovator + VEWT equine vaccine is produced by Zoetis (Parsippany-Troy Hills, NJ, USA) and is available from numerous US veterinary supply outlets.

A number of recombinantly produced proteins and virus-like particles (VLPs) were used to assess the specificity of selected binders. Chikungunya virus (CHIKV) E2 protein, E1 protein, and VLPs were from The Native Antigen Company (UK); EEEV E2/E3E2 was from IBT Bioservices; West Nile virus envelope protein was from Prospec (Israel); Lassa virus VLPs were from Zalgen (Germantown, MD, USA).

### 2.2. Library Construction, Panning, and Production of sdAb

Llama immunizations were through Triple J Farms, Bellingham, WA. Peripheral blood lymphocytes were isolated from a llama immunized with West Nile Innovator + VEWT, an equine vaccine that includes killed WEEV. Starting from these cells, we prepared total RNA, produced cDNA, amplified the coding sequence from the variable heavy domains, and constructed a phage display library as described previously [40]. Three rounds of panning, using the E3E2 protein adsorbed to wells of 96-well plates, were carried out essentially as previously described [40]. Positive clones were identified by a combination of monoclonal phage ELISA and monoclonal phage MagPlex assay after the second and third rounds [41].

The coding sequences for the sdAb were each mobilized from the pecan21 [42] phage display vector into pET22b as *NcoI-NotI* fragments as described previously [43]. In two cases, the sequence of the identified clone had an amber stop codon. In these cases, oligos were purchased to revert the amber stop codon to glutamine [44], and the clone is indicated by the suffix "f". The sdAb expression plasmids were transformed into Tuner (DE3) for protein production. Freshly transformed colonies were used to start overnight cultures in 50 mL terrific broth (TB) containing ampicillin (100 μg/mL) at 25 °C. The next day, the overnight cultures were poured into 450 mL of TB with ampicillin and grown for 2 h at 25 °C prior to induction with isopropyl-D-1 thiogalactoside (IPTG, 0.5 mM) and a further 2 h growth.

Purification of sdAb expressed from pET22b, the periplasmic expression vector, was carried out through an osmotic shock protocol as described previously [45], followed by immobilized metal affinity chromatography (IMAC) resin (Ni Sepharose High Performance, GE Healthcare, Marlborough, MA, USA) eluted with 0.25 M imidazole prior to purification by fast protein liquid chromatography (FPLC) on a Enrich SEC 70 (10 × 300 mm) column (Bio-Rad, Hercules, CA, USA) equilibrated with phosphate buffered saline (PBS) with 0.02% sodium azide. Typical FPLC results are shown in Appendix A Figure A1. All sdAb eluted nearly entirely as monomers; only minor contaminants were left to be removed by gel filtration following IMAC chromatography. Only the monomeric fraction was used for further characterization. SdAb concentration was determined by UV absorption and stored at 4 °C or at −80 °C for long-term storage.

### 2.3. Surface Plasmon Resonance

Affinity and kinetics measurements were performed using the ProteOn XPR36 (Bio-Rad, Hercules, CA, USA). Two lanes of a general layer compact (GLC) chip were individually coated with WEEV E3E2 His tagged recombinant protein from IBT Bioservices diluted to 20 μg/mL in 10 mM sodium acetate pH 5.0, as described previously [46]. The other four lanes were left uncoated for this work. The binding affinity of each sdAb was determined by doing what is referred to as "One-Shot kinetics", wherein each sdAb is flowed over the chip at a range of five concentrations and a blank to rapidly provide an array of binding curves. The chip is then regenerated by brief exposure to 0.085% phosphoric acid, then the next sdAb is analyzed. Data analysis was performed with ProteOn Manager 2.1 software, corrected by subtraction of the zero-antibody concentration column as well as interspot correction. The standard error of the fits was less than 10%. Binding constants were determined using the Langmuir model or the Langmuir with Mass Transfer built into the analysis software with the average of two determinations reported.

### 2.4. Determining Melting Temperature by Fluorescent Dye Melt Assay

The fluorescent dye melt assay which allows us to assess the inherent thermal stability of each sdAb was performed as described previously [47]. Each sdAb was first diluted to a concentration of 500 μg/mL in a final volume of 20 μL PBS. Next, a 1:1000 dilution of Sypro Orange dye (Sigma Aldrich) was added to each sample. Samples were measured in triplicate using a Step One Real-Time polymerase chain reaction (PCR) machine (Applied Biosystems, Foster City, CA, USA). The heating program was run in continuous mode from 25 °C–99 °C at a heating rate of 1% (~2 °C per minute), and data was recorded using the ROX filter. The melting point was determined to be the peak of the first derivative of the fluorescence intensity.

### *2.5. MagPlex Direct Binding and Sandwich Assays*

Specificity was evaluated via direct binding to WEEV E3E2 recombinant protein immobilized on MagPlex magnetic microspheres (Luminex, Austin, TX, USA). The WEEV E3E2 along with a number of other viral proteins and VLPs in PBS (pH 7.2), as listed in 2.1, were immobilized to different sets of MagPlex microspheres using the standard two-step 1-ethyl-3-(3-dimethylaminopropyl)carbodiimide hydrochloride /N-hydroxysulfosuccinimide (Thermo Fisher Scientific) immobilization protocol recommended by the manufacture. Each of the sdAb was biotinylated using a 10-fold excess of EZ-Link NHS-LC-LC-Biotin (Thermo Fisher Scientific) for 30 min and then the excess biotin was removed using Zeba spin columns (Thermo Fisher Scientific) with the sdAb concentration determined by absorbance at 280 nM. Dilutions of each biotinylated sdAb (Bt-sdAb) in PBSTB (PBS + 0.05% Tween + 0.1% BSA) were prepared in round bottom polypropylene microtiter plates (VWR). To each was added the mixture of antigen-coated MagPlex microspheres sufficient to provide counts of least 50 for each set per well. After an incubation of 30 min, the plate was washed twice with PBST and then incubated with 5 µg/mL streptavidin-conjugated phycoerythrin (Thermo Fisher Scientific) for 30 min, washed, and binding evaluated on the MAGPIX instrument (Luminex).

Sandwich format MagPlex bead assays were performed in order to demonstrate the ability of the sdAb to act as both the capture and recognition reagent for the detection of WEEV E3E2 protein. For this assay, each sdAb was immobilized to a set of MagPlex microspheres as described above, and added to dilutions of WEEV E3E2 in PBSTB also as above. Then, each of the Bt-sdAb was utilized as the recognition molecule.

## 3. Results

Following immunization of a llama with the West Nile Innovator + VEWT equine vaccine, a phage display library was prepared that contained as much of the VHH immune repertoire of the animal as possible. After three rounds of panning on WEEV E3E2 recombinant protein immobilized on microtiter plates followed by evaluation of monoclonal phage from the second and third rounds by ELISA and MagPlex, 24 potential binding clones were identified. Four positive clones were identified from round 2 out of 32 screened, while 20 out of 64 round 3 clones had signal at least twice the background and were considered positive. Following sequencing of the identified potential binding clones, it was determined that the isolated clones segregated into five different sequence families based on homology of their CDR 3 sequence. Every sequence family except the one typified by WF4 had at least two different members. At least one representative clone was chosen from each family for further characterization (Figure 1).

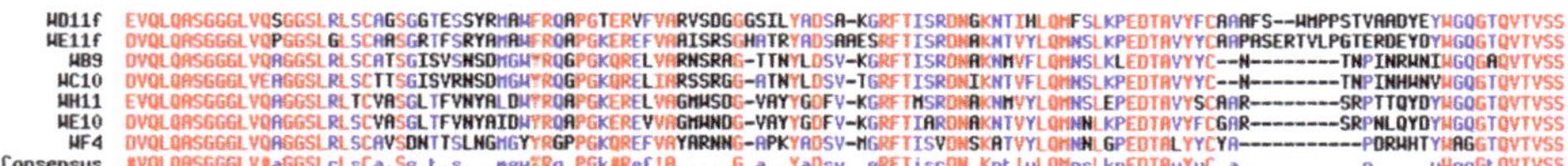

**Figure 1.** Deduced protein sequences of the seven single-domain antibodies (sdAb) that were evaluated. Sequences are given in single letter amino acid code. Alignment was performed using Multalin; high homology positions are shown in red, where lower homology is in blue [48].

Each of the selected sequences was then cloned into an expression vector (pET22b) and produced via *E. coli* in 0.5 liter-scale shake flasks and purified. Protein yields are shown in Table 1. The majority of produced protein was monomeric with very little potential aggregation observed; a typical FPLC chromatogram is shown in Appendix A Figure A1. With the exception of WF4, all produced more than 10 mg/L; WE10 had the best yield at ~55 mg/L. The melting temperature (Tm) of each clone was measured using a fluorescent dye melt assay to evaluate their innate utility for high temperature applications, Table 1. Clone WF4 was least stable with a Tm of 43 °C, while WE10 was the most stable, melting at 75 °C. In general, however, no correlation between yield and Tm was observed.

**Table 1.** Anti-WEEV sdAb yields, melting points, and binding affinities.

| Clone | Yield (mg/L) | Tm °C | ka (1/Ms) | kd (1/s) | KD (M) |
|---|---|---|---|---|---|
| WC10 | 48 | 51 | $1.0 \times 10^{8}$ | $8.5 \times 10^{-1}$ | $8.5 \times 10^{-9}$ |
| WD11f | 18 | 50 | $7.3 \times 10^{7}$ | $5.3 \times 10^{-1}$ | $7.2 \times 10^{-9}$ |
| WE10 | 55 | 75 | $7.0 \times 10^{4}$ | $8.3 \times 10^{-4}$ | $1.2 \times 10^{-8}$ |
| WE11f | 38 | 60 | nbo | nbo | nbo |
| WH11 | 15 | 72 | $1.5 \times 10^{5}$ | $3.3 \times 10^{-3}$ | $2.2 \times 10^{-8}$ |
| WB9 | 21 | 54 | $1.1 \times 10^{6}$ | $2.9 \times 10^{-3}$ | $2.6 \times 10^{-9}$ |
| WF4 | 3 | 43 | $6.3 \times 10^{4}$ | $1.5 \times 10^{-3}$ | $2.4 \times 10^{-8}$ |

nbo indicates no binding observed.

The affinity of each of the sdAb for the WEEV E3E2 recombinant protein was determined by SPR; results also shown in Table 1. Representative SPR plots are shown in Appendix A Figure A2. The sdAb showed a wide range of on rates as well as off rates, however they resulted in dissociation constants that varied between 2.6 and 24 nM. Only one of the sdAb, WE11f, failed to show binding via SPR. While it could be that its epitope was obscured during immobilization of the E3E2 recombinant protein to the sensor chip, based on the direct binding results presented below, it is clear that it was at best a relatively weak binder.

The MagPlex direct binding results, shown in Figure 2, do not show as much correlation with the binding affinities determined by SPR as one might expect. The three best binders in the MagPlex assay were WC10, WD11f, and WE10. While one did not expect much of WE11f, as it failed to bind via the SPR, the other three possess affinities as good as or better than the three which displayed the best binding in this assay. The two assays, SPR and MagPlex, have important differences, which can easily lead to a lack of correlation; the SPR sensor chips and MagPlex microspheres both use a similar chemistry for protein immobilization, but the reaction is done at different pHs, pH 5.0 for SPR and pH 7.2 for MagPlex, which can impact the orientation of antigen attachment. Even more critically, the sdAb is biotinylated for the MagPlex assays which can negatively impact the activity of the binder. Nonetheless, the most important aspect of this assay was that all the selected binding sdAb showed good specificity, having virtually no binding to any other antigen evaluated. The anti-CHIKV sdAb CC3, tested primarily to demonstrate the activity of at least one of the specificity controls, showed better binding activity than the selected WEEV binders likely due to its superior affinity for CHIKV VLP (ka: $5.2 \times 10^{5}$; kd: $3.3 \times 10^{-4}$, KD: $6.5 \times 10^{-10}$) [49].

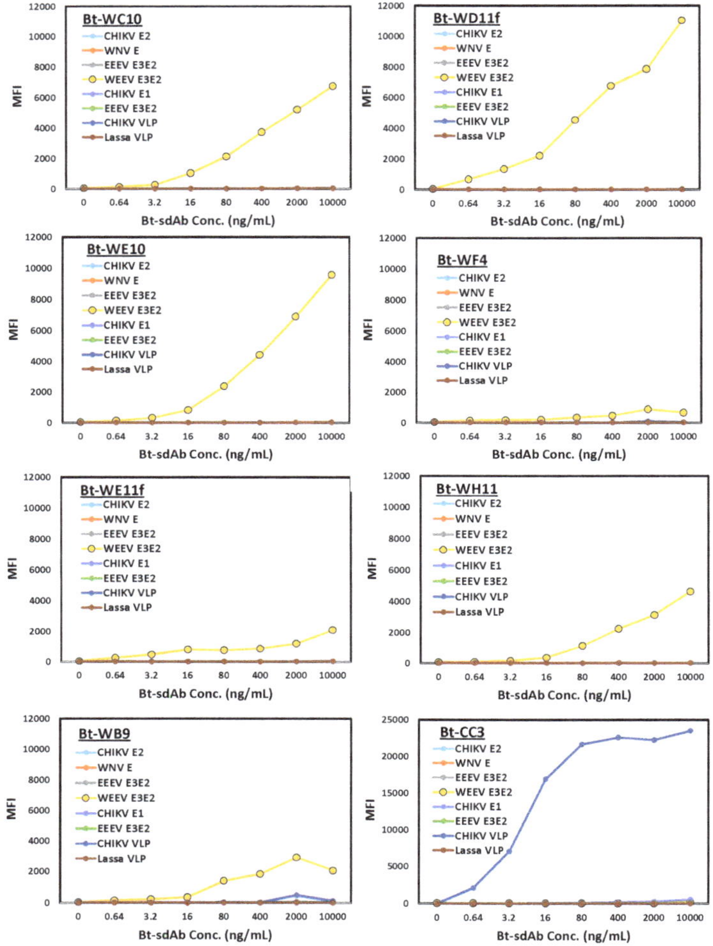


**Figure 2.** MagPlex direct binding assay to evaluate the specificity of the putative anti-WEEV E3E2 binding sdAb. Each sdAb was biotin labeled and mixed with MagPlex microspheres coated with various antigens at a range of concentrations. Biotinylated sdAb are denoted by Bt-clone name. Median fluorescence intensity versus Bt-sdAb concentration is plotted. The anti CHIKV sdAb CC3 was also tested as a positive control for the activity of the other sets of microspheres [49].

The final experiment looked at using the anti-WEEV sdAb as both capture and biotinylated recognition reagents in a MagPlex sandwich fluoroimmunoassay. The four best biotinylated recognition elements paired with each of the capture bead sets are shown in Figure 3. Not surprisingly, three of the four best biotinylated recognition sdAb were WC10, WD11f, and WE10, as those three also performed best in the MagPlex direct binding assay. WB9 also performed well in this assay. While not doing well in the direct binding assay, WB9 has a sequence very similar to WC10 and had the lowest dissociation constant (2.6 nM), thus it was not surprising that it did well. These same four sdAb were also the four best capture sdAb. Since the same sdAb could function as both the capture and biotinylated recognition reagent, it can be assumed that the E3E2 protein is at least partially aggregated, however the best capture for each biotinylated recognition sdAb was one of the other three, so it would appear a sufficient amount of the WEEV E3E2 protein was monomeric to make the use of separate epitopes for the capture and recognition reagents beneficial.

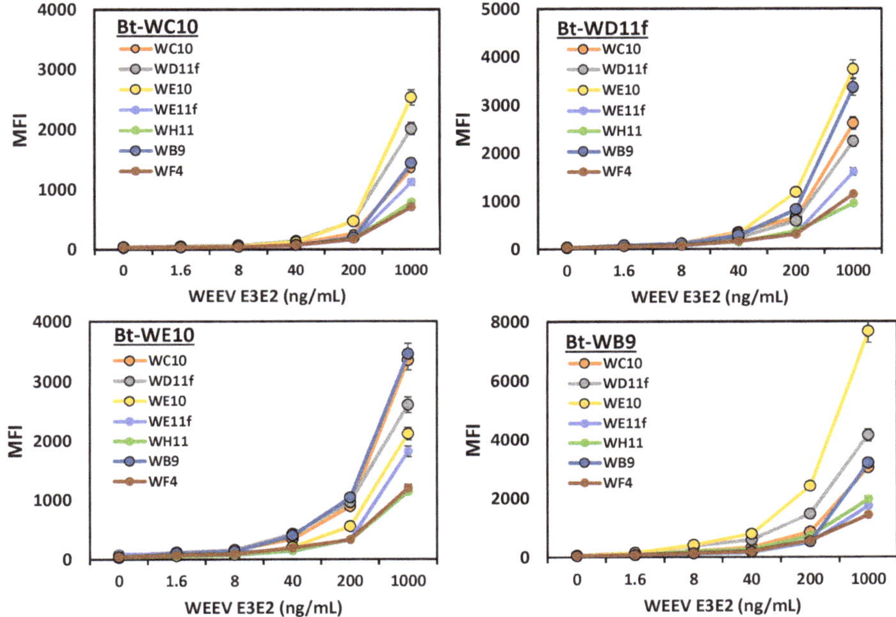


**Figure 3.** MagPlex sandwich fluoroimmunoassay to evaluate the utility of the sdAb to function as both a capture and recognition reagent for the detection of WEEV E3E2 protein. Only the four best Bt-sdAb recognition reagents are shown. Median fluorescence intensity versus WEEV E3E2 concentration is plotted; error bars represent the SEM.

## 4. Discussion

This work describes the initial isolation and characterization of anti-WEEV E2/E3E2 sdAb. Several antibodies have now been described that are useful for the detection and perhaps the treatment of WEEV [19,20]. We selected sdAb for binding to WEEV envelope protein, specifically E2 and E3E2. This is a promising target, as several neutralizing epitopes have been found on the E2 of the related VEEV [50]. There is also evidence that suggests antibodies directed against E3 can provide protection against alphaviruses in mice [51].

SdAb offer an alternative to conventional antibodies and their derivatives that can be easily tailored to possess additional functionalities as desired. For instance, the sdAb isolated here all have affinities between ~2.6 and 24 nM, whereas to obtain adequate sensitivity or activity one may desire sub-nM apparent affinities. This can easily be achieved for sdAb by the formation of multimer constructs that can take advantage of avidity. For viral targets that are naturally multimeric, this is a valid approach to increase the apparent binding affinity. These types of constructs have been demonstrated by many groups including ourselves [52–54]. Additionally, several strategies have been described for increasing the stability of sdAb [37], and have been shown to increase melting temperatures by as much as 20 °C [43]. Similarly, sdAb can be engineered to tailor them for specific assay formats [28,32,55].

There are many examples of viral neutralizing sdAb [56,57]; expressing the sdAb as genetic fusions has led to improved neutralization [58–60]. In one example, viral neutralizing sdAb have been generated through selections against the trimeric envelope proteins of Respiratory Syncytial

Virus, Rabies virus and H5N1 Influenza. In that work, the researchers identified neutralizing sdAb recognizing different epitopes in the receptor binding sites on the spikes with affinities in the low nanomolar range. Multimeric constructs, in which the sdAb were genetically linked improved neutralization potencies up to 4000-fold for RSV, 1500-fold for Rabies virus and 75-fold for Influenza H5N1 and had potencies similar to or better than the best performing monoclonal antibodies [59]. The trivalent sdAb construct (ALX-0171) that inhibits RSV has also been successfully delivered by inhalation [61], a route that may also prove valuable for treatment of alphaviruses that are transmitted via aerosols.

It is probable that additional binders for WEEV E2 and E3E2 exist in our constructed library. This was only our first look at this library and if additional binders are desired, a greater number of clones can be evaluated or additional rounds of panning can easily be performed. If these anti-WEEV binding sdAb are proven to be neutralizing and protective, they may make for attractive alternatives to the larger antibody fragments, as their robust nature may allow them to be stored without refrigeration making them ideal for use in resource-limited areas that lack the power grid infrastructure and where the mosquito transmission of these arbovirus is widespread.

**Author Contributions:** Conceptualization: G.P.A., J.L.L. and E.R.G.; Investigation: G.P.A., J.L.L., L.C.S.-L., D.Z. and E.R.G.; Methodology: E.R.G., G.P.A. and L.C.S.-L.; Project administration: G.P.A. and E.R.G.; Resources: D.Z.; Supervision: G.P.A. and E.R.G.; Visualization: G.P.A., L.C.S.-L., D.Z. and E.R.G.; Writing—original draft: G.P.A. and E.R.G.; Writing—review & editing: G.P.A., L.C.S.-L., J.L.L., D.Z. and E.R.G.

**Funding:** This work was supported by Naval Research Laboratory base funds.

**Conflicts of Interest:** The authors declare no conflict of interest. The funding sponsors had no role in the design of the study; in the collection, analyses, or interpretation of data; in the writing of the manuscript, and in the decision to publish the results.

## Appendix A

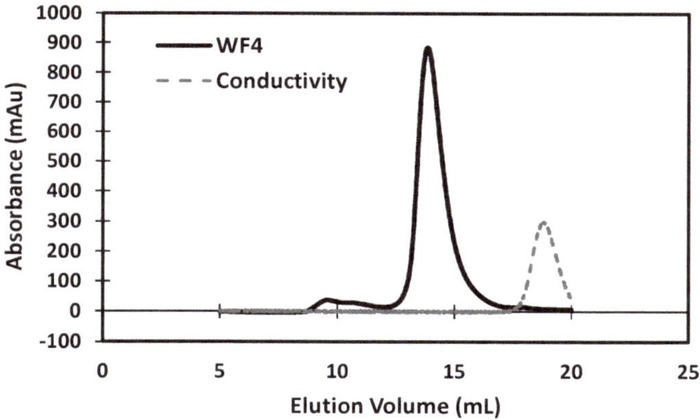


**Figure A1.** Representative gel-filtration chromatography trace.

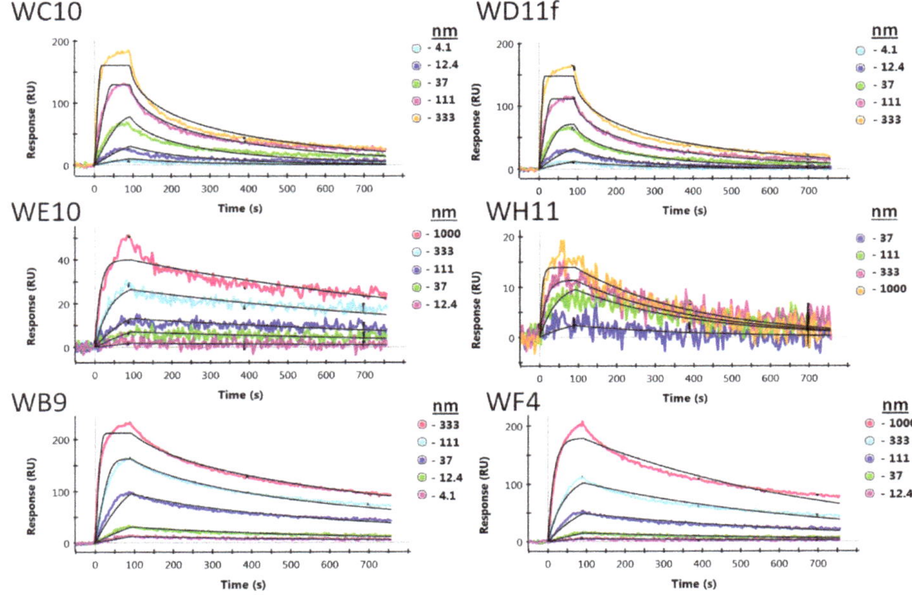


**Figure A2.** Representative surface plasmon resonance data.

## References

1. Hahn, C.S.; Lustig, S.; Strauss, E.G.; Strauss, J.H. Western equine encephalitis virus is a recombinant virus. *Proc. Natl. Acad. Sci. USA* **1988**, *85*, 5997–6001. [CrossRef] [PubMed]
2. Powers, A.M.; Brault, A.C.; Shirako, Y.; Strauss, E.G.; Kang, W.; Strauss, J.H.; Weaver, S.C. Evolutionary relationships and systematics of the alphaviruses. *J. Virol.* **2001**, *75*, 10118–10131. [CrossRef] [PubMed]
3. Strauss, J.H.; Strauss, E.G. The alphaviruses: Gene expression, replication, and evolution. *Microb. Rev.* **1994**, *58*, 491–562.
4. Goh, L.Y.; Hobson-Peters, J.; Prow, N.A.; Gardner, J.; Bielefeldt-Ohmann, H.; Pyke, A.T.; Suhrbier, A.; Hall, R.A. Neutralizing monoclonal antibodies to the e2 protein of chikungunya virus protects against disease in a mouse model. *Clin. Immunol.* **2013**, *149*, 487–497. [CrossRef] [PubMed]
5. Eastern, Western and Venezuelan Equine Encephalomyelitis. Available online: http://www.cfsph.iastate.edu/Factsheets/pdfs/easter_wester_venezuelan_equine_encephalomyelitis.pdf (accessed on 25 October 2018).
6. Reisen, W.; Monath, T. Western equine encephalitis. In *The Arboviruses: Epidemiology and Ecology*; Monath, T.P., Ed.; CRC Press: Boca Raton, FL, USA, 1989; Volume V, pp. 89–137.
7. Zacks, M.A.; Paessler, S. Encephalitic alphaviruses. *Vet. Microbiol.* **2010**, *140*, 281–286. [CrossRef] [PubMed]
8. Reeves, W.C.; Hutson, G.A.; Bellamy, R.E.; Scrivani, R.P. Chronic latent infections of birds with western equine encephalomyelitis virus. *Proc. Soc. Exp. Biol. Med.* **1958**, *97*, 733–736. [CrossRef] [PubMed]
9. Phillpotts, R.J. Venezuelan equine encephalitis virus complex-specific monoclonal antibody provides broad protection, in murine models, against airborne challenge with viruses from serogroups i, ii and iii. *Virus Res.* **2006**, *120*, 107–112. [CrossRef]
10. Reed, D.S.; Larsen, T.; Sullivan, L.J.; Lind, C.M.; Lackemeyer, M.G.; Pratt, W.D.; Parker, M.D. Aerosol exposure to western equine encephalitis virus causes fever and encephalitis in cynomolgus macaques. *J. Infect. Dis.* **2005**, *192*, 1173–1182. [CrossRef]
11. Hawley, R.J.; Eitzen, E.M., Jr. Biological weapons—A primer for microbiologists. *Annu. Rev. Microb.* **2001**, *55*, 235–253. [CrossRef]
12. Sidwell, R.W.; Smee, D.F. Viruses of the bunya- and togaviridae families: Potential as bioterrorism agents and means of control. *Antiv. Res.* **2003**, *57*, 101–111. [CrossRef]

13. Nagata, L.P.; Wong, J.P.; Hu, W.-G.; Wu, J.Q. Vaccines and therapeutics for the encephalitic alphaviruses. *Future Virol.* **2013**, *8*, 661–674. [CrossRef]
14. Sherman, M.B.; Weaver, S.C. Structure of the recombinant alphavirus western equine encephalitis virus revealed by cryoelectron microscopy. *J. Virol.* **2010**, *84*, 9775–9782. [CrossRef]
15. Uchime, O.; Fields, W.; Kielian, M. The role of e3 in ph protection during alphavirus assembly and exit. *J. Virol.* **2013**, *87*, 10255–10262. [CrossRef] [PubMed]
16. Das, D.; Kriangkum, J.; Nagata, L.P.; Fulton, R.E.; Suresh, M.R. Development of a biotin mimic tagged scfv antibody against western equine encephalitis virus: Bacterial expression and refolding. *J. Virol. Methods* **2004**, *117*, 169–177. [CrossRef] [PubMed]
17. Xu, B.; Kriangkum, J.; Nagata, L.P.; Fulton, R.E.; Suresh, M.R. A single chain fv specific against western equine encephalitis virus. *Hybridoma* **1999**, *18*, 315–323. [CrossRef]
18. Long, M.C.; Jager, S.; Mah, D.C.W.; Jebailey, L.; Mah, M.A.; Masri, S.A.; Nagata, L.P. Construction and characterization of a novel recombinant single-chain variable fragment antibody against western equine encephalitis virus. *Hybridoma* **2000**, *19*, 1–13. [CrossRef] [PubMed]
19. Hülseweh, B.; Rülker, T.; Pelat, T.; Langermann, C.; Frenzel, A.; Schirrmann, T.; Dübel, S.; Thullier, P.; Hust, M. Human-like antibodies neutralizing western equine encephalitis virus. *mAbs* **2014**, *6*, 717–726. [CrossRef] [PubMed]
20. Burke, C.W.; Froude, J.W.; Miethe, S.; Hülseweh, B.; Hust, M.; Glass, P.J. Human-like neutralizing antibodies protect mice from aerosol exposure with western equine encephalitis virus. *Viruses* **2018**, *10*, 147. [CrossRef]
21. Hamers-Casterman, C.; Atarhouch, T.; Muyldermans, S.; Robinson, G.; Hamers, C.; Songa, E.B.; Bendahman, N.; Hamers, R. Naturally occurring antibodies devoid of light chains. *Nature* **1993**, *363*, 446–448. [CrossRef]
22. Ghahroudi, M.A.; Desmyter, A.; Wyns, L.; Hamers, R.; Muyldermans, S. Selection and identification of single domain antibody fragments from camel heavy-chain antibodies. *FEBS Lett.* **1997**, *414*, 521–526. [CrossRef]
23. De Marco, A. Biotechnological applications of recombinant single-domain antibody fragments. *Microb. Cell Fact.* **2011**, *10*, 44. [CrossRef] [PubMed]
24. Muyldermans, S. Nanobodies: Natural single-domain antibodies. *Annu. Rev. Biochem.* **2013**, *82*, 775–797. [CrossRef] [PubMed]
25. Wesolowski, J.; Alzogaray, V.; Reyelt, J.; Unger, M.; Juarez, K.; Urrutia, M.; Cauerhff, A.; Danquah, W.; Rissiek, B.; Scheuplein, F.; et al. Single domain antibodies: Promising experimental and therapeutic tools in infection and immunity. *Med. Microbiol. Immunol.* **2009**, *198*, 157–174. [CrossRef] [PubMed]
26. Eyer, L.; Hruska, K. Single-domain antibody fragments derived from heavy-chain antibodies: A review. *Vet. Med.* **2012**, *57*, 439–513. [CrossRef]
27. Hussack, G.; Hirama, T.; Ding, W.; MacKenzie, R.; Tanha, J. Engineered single-domain antibodies with high protease resistance and thermal stability. *PLoS ONE* **2011**, *6*. [CrossRef] [PubMed]
28. Liu, J.L.; Walper, S.A.; Turner, K.B.; Brozozog-Lee, A.; Medintz, I.L.; Susumu, K.; Oh, E.; Zabetakis, D.; Goldman, E.R.; Anderson, G.P. Conjugation of biotin-coated luminescent quantum dots with single domain antibody-rhizavidin fusions. *Biotechnol. Rep.* **2016**, *10*, 56–65. [CrossRef] [PubMed]
29. Liu, J.L.; Zabetakis, D.; Brozozog Lee, P.A.; Goldman, E.R.; Anderson, G.P. Single domain antibody alkaline phosphatase fusion proteins for antigen detection—Analysis of affinity and thermal stability of single domain antibody. *J. Immunol. Methods* **2013**, *393*, 1–7. [CrossRef]
30. Pleschberger, M.; Saerens, D.; Weigert, S.; Sleytr, U.B.; Muyldermans, S.; Sara, M.; Egelseer, E.M. An s-layer heavy chain camel antibody fusion protein for generation of a nanopatterned sensing layer to detect the prostate-specific antigen by surface plasmon resonance technology. *Bioconjug. Chem.* **2004**, *15*, 664–671. [CrossRef]
31. Raphael, M.P.; Christodoulides, J.A.; Byers, J.M.; Anderson, G.P.; Liu, J.L.; Turner, K.B.; Goldman, E.R.; Delehanty, J.B. Optimizing nanoplasmonic biosensor sensitivity with orientated single domain antibodies. *Plasmonics* **2015**, *10*, 1649–1655. [CrossRef]
32. Sherwood, L.J.; Hayhurst, A. Hapten mediated display and pairing of recombinant antibodies accelerates assay assembly for biothreat countermeasures. *Sci. Rep.* **2012**, *2*, 807. [CrossRef]
33. Liu, J.L.; Goldman, E.R.; Zabetakis, D.; Walper, S.A.; Turner, K.B.; Shriver-Lake, L.C.; Anderson, G.P. Enhanced production of a single domain antibody with an engineered stabilizing extra disulfide bond. *Microb. Cell Fact.* **2015**, *14*, 158. [CrossRef] [PubMed]

34. Hagihara, Y.; Mine, S.; Uegaki, K. Stabilization of an immunoglobulin fold domain by an engineered disulfide bond at the buried hydrophobic region. *J. Biol. Chem.* **2007**, *282*, 36489–36495. [CrossRef] [PubMed]
35. Saerens, D.; Conrath, K.; Govaert, J.; Muyldermans, S. Disulfide bond introduction for general stabilization of immunoglobulin heavy-chain variable domains. *J. Mol. Biol.* **2008**, *377*, 478–488. [CrossRef] [PubMed]
36. Turner, K.B.; Liu, J.L.; Zabetakis, D.; Lee, A.B.; Anderson, G.P.; Goldman, E.R. Improving the biophysical properties of anti-ricin single-domain antibodies. *Biotechnol. Rep.* **2015**, *6*, 27–35. [CrossRef] [PubMed]
37. Goldman, E.R.; Liu, J.L.; Zabetakis, D.; Anderson, G.P. Enhancing stability of camelid and shark single domain antibodies: An overview. *Front. Immunol.* **2017**, *8*, 865. [CrossRef] [PubMed]
38. Hu, Y.; Liu, C.; Muyldermans, S. Nanobody-based delivery systems for diagnosis and targeted tumor therapy. *Front. Immunol.* **2017**, *8*, 1442. [CrossRef] [PubMed]
39. Cordy, J.C.; Morley, P.J.; Wright, T.J.; Birchler, M.A.; Lewis, A.P.; Emmins, R.; Chen, Y.Z.; Powley, W.M.; Bareille, P.J.; Wilson, R.; et al. Specificity of human anti-variable heavy (vh) chain autoantibodies and impact on the design and clinical testing of a vh domain antibody antagonist of tumour necrosis factor-alpha receptor 1. *Clin. Exp. Immunol.* **2015**, *182*, 139–148. [CrossRef]
40. Liu, J.L.; Shriver-Lake, L.C.; Anderson, G.P.; Zabetakis, D.; Goldman, E.R. Selection, characterization, and thermal stabilization of llama single domain antibodies towards ebola virus glycoprotein. *Microb. Cell Fact.* **2017**, *16*, 223. [CrossRef]
41. Anderson, G.; Matney, R.; Liu, J.; Hayhurst, A.; Goldman, E. Multiplexed fluid array screening of phage displayed anti-ricin single domain antibodies for rapid assessment of specificity. *Biotechniques* **2007**, *43*, 806–811. [CrossRef]
42. Goldman, E.; Anderson, G.; Liu, J.; Delehanty, J.; Sherwood, L.; Osborn, L.; Cummins, L.; Hayhurst, A. Facile generation of heat-stable antiviral and antitoxin single domain antibodies from a semisynthetic llama library. *Analyt. Chem.* **2006**, *78*, 8245–8255. [CrossRef]
43. Walper, S.A.; Liu, J.L.; Zabetakis, D.; Anderson, G.P.; Goldman, E.R. Development and evaluation of single domain antibodies for vaccinia and the l1 antigen. *PLoS ONE* **2014**, *9*, e106263. [CrossRef] [PubMed]
44. Marcus, W.D.; Lindsay, S.M.; Sierks, M.R. Identification and repair of positive binding antibodies containing randomly generated amber codons from synthetic phage display libraries. *Biotechnol. Prog.* **2006**, *22*, 919–922. [CrossRef] [PubMed]
45. Shriver-Lake, L.C.; Zabetakis, D.; Goldman, E.R.; Anderson, G.P. Evaluation of anti-botulinum neurotoxin single domain antibodies with additional optimization for improved production and stability. *Toxicon* **2017**, *135*, 51–58. [CrossRef] [PubMed]
46. Walper, S.A.; Lee, P.A.B.; Anderson, G.P.; Goldman, E.R. Selection and characterization of single domain antibodies specific for bacillus anthracis spore proteins. *Antibodies* **2013**, *2*, 152–167. [CrossRef]
47. Liu, J.L.; Zabetakis, D.; Goldman, E.R.; Anderson, G.P. Selection and evaluation of single domain antibodies toward ms2 phage and coat protein. *Mol. Immunol.* **2013**, *53*, 118–125. [CrossRef]
48. Corpet, F. Multiple sequence alignment with hierarchical-clustering. *Nucleic. Acids Res.* **1988**, *16*, 10881–10890. [CrossRef] [PubMed]
49. Liu, J.L.; Shriver-Lake, L.C.; Zabetakis, D.; Anderson, G.P.; Goldman, E.R. Selection and characterization of protective anti-chikungunya virus single domain antibodies. *Mol. Immunol.* **2019**, *105*, 190–197. [CrossRef]
50. Hunt, A.R.; Frederickson, S.; Maruyama, T.; Roehrig, J.T.; Blair, C.D. The first human epitope map of the alphaviral e1 and e2 proteins reveals a new e2 epitope with significant virus neutralizing activity. *PLOS Negl. Trop. Dis.* **2010**, *4*, e739. [CrossRef]
51. Parker, M.D.; Buckley, M.J.; Melanson, V.R.; Glass, P.J.; Norwood, D.; Hart, M.K. Antibody to the e3 glycoprotein protects mice against lethal venezuelan equine encephalitis virus infection. *J. Virol.* **2010**, *84*, 12683–12690. [CrossRef]
52. Liu, J.L.; Zabetakis, D.; Walper, S.A.; Goldman, E.R.; Anderson, G.P. Bioconjugates of rhizavidin with single domain antibodies as bifunctional immunoreagents. *J. Immunol. Methods* **2014**, *411*, 37–42. [CrossRef]
53. Walper, S.A.; Brozozog Lee, P.A.; Goldman, E.R.; Anderson, G.P. Comparison of single domain antibody immobilization strategies evaluated by surface plasmon resonance. *J. Immunol. Methods* **2013**, *388*, 68–77. [CrossRef] [PubMed]
54. Zhang, J.; Tanha, J.; Hirama, T.; Khieu, N.H.; To, R.; Tong-Sevinc, H.; Stone, E.; Brisson, J.-R.; Roger MacKenzie, C. Pentamerization of single-domain antibodies from phage libraries: A novel strategy for the rapid generation of high-avidity antibody reagents. *J. Mol. Biol.* **2004**, *335*, 49–56. [CrossRef] [PubMed]

55. Anderson, G.; Shriver-Lake, L.; Walper, S.; Ashford, L.; Zabetakis, D.; Liu, J.; Breger, J.; Brozozog Lee, P.; Goldman, E. Genetic fusion of an anti-bcla single-domain antibody with beta galactosidase. *Antibodies* **2018**, *7*, 36. [CrossRef]
56. Vanlandschoot, P.; Stortelers, C.; Beirnaert, E.; Ibanez, L.I.; Schepens, B.; Depla, E.; Saelens, X. Nanobodies(r): New ammunition to battle viruses. *Antiviral. Res.* **2011**, *92*, 389–407. [CrossRef] [PubMed]
57. Wu, Y.; Jiang, S.; Ying, T. Single-domain antibodies as therapeutics against human viral diseases. *Front. Immunol.* **2017**, *8*, 1802. [CrossRef] [PubMed]
58. Boruah, B.M.; Liu, D.; Ye, D.; Gu, T.-J.; Jiang, C.-L.; Qu, M.; Wright, E.; Wang, W.; He, W.; Liu, C.; et al. Single domain antibody multimers confer protection against rabies infection. *PLoS ONE* **2013**, *8*, e71383. [CrossRef]
59. Hultberg, A.; Temperton, N.J.; Rosseels, V.; Koenders, M.; Gonzalez-Pajuelo, M.; Schepens, B.; Ibañez, L.I.; Vanlandschoot, P.; Schillemans, J.; Saunders, M.; et al. Llama-derived single domain antibodies to build multivalent, superpotent and broadened neutralizing anti-viral molecules. *PLoS ONE* **2011**, *6*, e17665. [CrossRef] [PubMed]
60. Matz, J.; Kessler, P.; Bouchet, J.; Combes, O.; Ramos, O.H.; Barin, F.; Baty, D.; Martin, L.; Benichou, S.; Chames, P. Straightforward selection of broadly neutralizing single-domain antibodies targeting the conserved cd4 and coreceptor binding sites of hiv-1 gp120. *J. Virol.* **2013**, *87*, 1137–1149. [CrossRef]
61. Larios Mora, A.; Detalle, L.; Gallup, J.M.; Van Geelen, A.; Stohr, T.; Duprez, L.; Ackermann, M.R. Delivery of alx-0171 by inhalation greatly reduces respiratory syncytial virus disease in newborn lambs. *mAbs* **2018**, *10*, 778–795. [CrossRef]

*Article*

# Cross-Reactive and Lineage-Specific Single Domain Antibodies against Influenza B Hemagglutinin

**Walter Ramage [1], Tiziano Gaiotto [1], Christina Ball [1], Paul Risley [1], George W. Carnell [2], Nigel Temperton [2], Chung Y. Cheung [3], Othmar G. Engelhardt [3] and Simon E. Hufton [1,*]**

1 Biotherapeutics Division, National Institute for Biological Standards and Control, Blanche Lane, South Mimms, Potters Bar, Hertfordshire EN6 3QG, UK; walter.ramage@nibsc.org (W.R.); Tiziano.Gaiotto@gmail.com (T.G.); chris.ball@nibsc.org (C.B.); paul.risley@nibsc.org (P.R.)

2 Infectious Diseases and Allergy Group, School of Pharmacy, University of Kent, Kent ME4 4TB, UK; g.carnell@gmail.com (G.W.C.); n.temperton@kent.ac.uk (N.T.)

3 Division of Virology, National Institute for Biological Standards and Control, Blanche Lane, South Mimms, Potters Bar, Hertfordshire EN6 3QG, UK; chung.cheung@hli.ubc.ca (C.Y.C.); othmar.engelhardt@nibsc.org (O.G.E.)

* Correspondence: simon.hufton@nibsc.org

Received: 11 January 2019; Accepted: 1 February 2019; Published: 10 February 2019

**Abstract:** Influenza B virus (IBV) circulates in the human population and causes considerable disease burden worldwide, each year. Current IBV vaccines can struggle to mount an effective cross-reactive immune response, as strains become mismatched, due to constant antigenic changes. Additional strategies which use monoclonal antibodies, with broad reactivity, are of considerable interest, both, as diagnostics and as immunotherapeutics. Alternatives to conventional monoclonal antibodies, such as single domain antibodies (Nanobodies$^{TM}$) with well-documented advantages for applications in infectious disease, have been emerging. In this study we have isolated single domain antibodies (sdAbs), specific to IBV, using alpacas immunised with recombinant hemagglutinin (HA) from two representative viruses, B/Florida/04/2006 (B/Yamagata lineage) and B/Brisbane/60/2008 (B/Victoria lineage). Using phage display, we have isolated a panel of single domain antibodies (sdAbs), with both cross-reactive and lineage-specific binding. Several sdAbs recognise whole virus antigens, corresponding to influenza B strains included in vaccines spanning over 20 years, and were capable of neutralising IBV pseudotypes corresponding to prototype strains from both lineages. Lineage-specific sdAbs recognised the head domain, whereas, sdAbs identified as cross-reactive could be classified as either head binding or stem binding. Using yeast display, we were able to correlate lineage specificity with naturally occurring sequence divergence, at residue 122 in the highly variable 120 loop of the HA1 domain. The single domain antibodies described, might have applications in IBV diagnostics, vaccine potency testing and as immunotherapeutics.

**Keywords:** influenza; influenza B virus; hemagglutinin; single domain antibody; Nanobody$^{TM}$; phage display; yeast display; epitope mapping

## 1. Introduction

Seasonal influenza caused by the influenza A virus (IAV) and the influenza B virus (IBV) remains a considerable public health challenge [1]. IBV has received less attention than IAV, largely because it does not cause pandemics as there are no naturally occurring, non-human reservoirs to facilitate the extensive antigenic changes, characteristic of a pandemic. However, there has been a recent increase in the rate of IBV infection, globally, with between 20–30% of the total influenza burden now due to influenza B [2]. Current vaccines can struggle to induce sufficient levels of cross-protective immunity against IBV and strains often become mismatched, due to the constant antigenic changes in the influenza virus [3]. The virus also co-circulates as two antigenically distinct lineages and, therefore, the dominant circulating lineage must be predicted for inclusion in the trivalent vaccine. This highlights the need to develop additional approaches to both treat and protect from the influenza B virus infection. Passive immunotherapy, using broadly neutralising antibodies against the major viral coat protein hemagglutinin (HA), is one approach that has shown great promise for the treatment of IAV infection [4]. The precursor HA protein (HA0) is cleaved by the host proteases into a form that comprises a highly variable distal head domain, HA1, and a more conserved proximal stem region, HA2 [5]. Most broadly, neutralising anti-HA monoclonal antibodies that have been described to date, target the HA stem of IAV [6–8]; however, monoclonal antibodies with broad reactivity have been described, which bind the more variable head domain, using a single CDR3 loop for antigenic recognition [9]. The more recently reported broadly neutralising monoclonal antibodies against IBV have also shown the presence of conserved epitopes, in both the stem region and the head domain [10–12]. A feature common between the broadly neutralising human monoclonal antibodies against both IBV and IAV, is the low levels of somatic hypermutation and 'heavy-chain-only' binding [7,10]. 'Heavy-chain-only' binding has been suggested to be the preferred mode of binding to influenza HA, as has been described for HIV [13]. These observations highlight that the VL domain might not be required for binding occluded epitopes onto the HA stem. Guided by these observations, we have isolated naturally occurring 'heavy-chain-only' antibodies from camelids (also known as Nanobodies™), believing their small size, single domain structure and long CDR3 loops make this unique antibody format well-equipped to access similar HA stem epitopes [14–16].

In the 1980's, the influenza B virus diverged to give two distinct lineages, termed the B/Yamagata/16/88 and B/Victoria/2/87 lineages [17]. Although the HA stem region remains highly conserved, the head domain has diverged to become sufficiently antigenically and genetically distinct, to warrant an independent inclusion in the current quadrivalent influenza vaccines [18]. The inclusion of seasonal strains from both IBV lineages, in turn, presents a challenge for both vaccine production and potency determination. Separate monoclonal antibodies, with either B-Victoria or B-Yamagata lineage-specific recognition could find applications in determining the individual antigenic content of each IBV lineage strain included in the seasonal vaccines [19]. However, the challenge in identifying lineage-specific binding antibodies is that, such epitopes are expected to be in the more variable head domain of the HA and, as such, are susceptible to antigenic change with associated loss of reagent binding. The degree of antigenic divergence, tolerated by a given monoclonal antibody binding to the variable head domain, which is subject to constant selective pressure to change, to some extent, is determined by the size of the epitope footprint. It is interesting to speculate that a small epitope footprint could be an advantage, as there may be less amino acid mutations that can impact antibody reagent binding. Single domain antibodies isolated from camelids, have preferences for smaller crevices and clefts on protein surfaces, as compared to the conventional two-chain monoclonal antibodies, which generally bind to larger flatter epitopes [20,21]. As such, single domain antibodies (sdAbs) may have a smaller epitope footprint and, therefore, might be able to retain binding to a constantly changing virus, for longer. In addition, human monoclonal antibodies to the HA head domain derived from either naturally infected or vaccinated patients might have already contributed to the antigenic pathways of viral escape and might be pre-ordained to the loss of binding. For these reasons, we have chosen to isolate the IBV lineage-specific sdAbs, rather than the conventional MAbs (Monoclonal Antibodies), believing that this unique antibody format could also form the basis of a robust and durable immunoassays for potency determination of the IBV containing vaccines [19].

By immunising alpacas with recombinant HA and the selection of phage display libraries, we have isolated sdAbs with both cross-reactivity and lineage-specific recognition. Through combining conventional phage display selections and next-generation sequencing, we have isolated rare sdAbs with either B-Victoria or B-Yamagata lineage specificity. We have grouped sdAbs, on the basis of head or stem binding, and used yeast display, to further de-lineate an epitope associated with

B-Victoria lineage-specific recognition. The potential applications of these different classes of sdAbs for applications in the IBV vaccine potency assays and as immunotherapeutics, are also discussed.

## 2. Materials and Methods

### 2.1. Influenza Antigens and Immunisation of the Alpacas

The virus antigen standards used in this study were B/Brisbane/60/2008 (National Institute for Biological Standards and Control (NIBSC) 14/146), B/Malaysia/2506/2004 (NIBSC 08/184), B/Shangdong/9/97 (NIBSC 04/128), B/Victoria/2/87 (NIBSC 87/774) of the B-Victoria lineage; the B/Yamagata/16/88 (NIBSC 92/628), B/Yamanashi/166/98 (NIBSC 99/586), B/Harbin/7/94 (NIBSC 97/748), B/Jiangsu/10/2003 (NIBSC 04/202), B/Florida/4/2006 (NIBSC 08/140), B/Massachusetts/02/2012 (NIBSC 13/106), B/Phuket/3037/2013 (NIBSC 14/252), B/Utah/9/2014 (NIBSC 15/100) of the B-Yamagata lineage; and the pre-lineage split strain B/HongKong/8/73 (NIBSC 79/568). Seasonal H1N1 strain A/Brisbane/10/2007 (NIBSC 08/278) was used as a negative control. Purified recombinant hemagglutinins used in this study were B/Brisbane/60/2008 (B-Victoria) and B/Florida/04/2006 (B-Yamagata) (Protein Sciences$^{TM}$, Meriden, CT, USA). Purified HA1 head domains from B/Brisbane/60/2008 (B-Victoria) and B/Florida/04/2006 (B-Yamagata) were also used (Sino Biologicals$^{TM}$, Beijing, China). Juvenile male alpacas were purchased through the Royal Veterinary College, Hertfordshire, UK. All experiments were reviewed by a local ethics committee and performed under a U.K Home Office licence PPL 80/2581. A blood sample, prior to immunization, was obtained from the external jugular vein and this was followed by four intramuscular injections on day 0 (primary immunisation), 21, 43 and 71, with injections being administered to the rear legs (thigh region) on days 0 and 43, and to the front legs (thigh region) on days 21 and 71. The primary immunisation consisted of 50 µg of recombinant HA B/Brisbane/60/2008 (B-Victoria) or B/Florida/04/2006 (B-Yamagata) (ProteinSciences$^{TM}$) in 400 µL of sterile PBS (phosphate buffered saline), which was emulsified with 800 µL of Freund's complete adjuvant (Sigma Aldrich, St. Louis, MO, USA), just prior to immunisation. Similarly, the three booster injections of 50 µg of the same recombinant HA as the primary immunisation in Freund's incomplete adjuvant (Sigma), were administered. Approximately four days after each injection, a 10 mL blood sample was collected from the external jugular vein, from which the serum was prepared, after allowing the blood to clot overnight at 4 °C.

### 2.2. Construction and Selection of the Phage-Displayed Libraries

For the antibody library construction, approximately 10 mL blood samples were collected into the heparinised tubes, from an immunised alpaca. Peripheral blood mononuclear cells were purified using a Ficol Hypaque centrifugation procedure (Sigma) and RNA was extracted using a RiboPure$^{TM}$ RNA extraction kit (Novagen, Thermo Fisher Scientific, Waltham, MA, USA) following the manufacturer's instructions. First strand cDNA synthesis was performed using Superscript III$^{TM}$ reverse transcriptase (Invitrogen, Thermo Fisher Scientific, Waltham, MA, USA) and oligo-dT primer with 200 ng of total RNA per reaction. Primary PCR and secondary PCR to recover the alpaca *VHH* genes (DNA encoding the variable region of the heavy-chain-only antibodies) appended with *Sfi*1 and *Not*1 restriction sites were carried out, as previously described [14]. Approximately 5 µg of the VHH antibody DNA was digested by the *Not*1 and *Sfi*I restriction enzymes (New England BioLabs, Ipswich, MA, USA) and then ligated into a phage-display vector, pNIBS-1, which contains a His tag for sdAb purification, a Myc tag for detection and a suppressible stop codon for soluble expression [14]. After purification (Qiagen, Hilden, Germany), the ligation mix was transformed into the TG1 electro-competent cells (Agilent, Santa Clara, CA, USA), using electroporation. Phage antibody library selections were performed using immunotubes (Nunc, Thermo Fisher Scientific, Waltham, MA, USA) coated overnight, at 4 °C, with 1 mL of the 10 µg/mL recombinant HA (Protein Sciences$^{TM}$) in PBS or the whole virus antigen standards reconstituted in PBS [14].

### 2.3. Antibody Expression and Screening

Individual colonies from each round of selection were picked and grown overnight, at 37 °C, in 2 × TY supplemented with carbenicillin (100 µg/mL) and 2.0% (*w*/*v*) glucose and plasmid DNA was isolated (Qiagen, Hilden, Germany). Cloned VHH genes were sequenced, aligned and grouped, according to the CDR3 length and homology. CDR3 sequences of high homology and identical length were assumed to have been derived from clonally related B cells and were likely to bind to the same, or overlapping, epitope on HA. Representative members from each proposed epitope group were expressed and purified. For single domain antibody production, pNIBS-1 clones were transformed into the non-suppressor *Escherichia coli* strain WK6 (New England Biolabs). Soluble antibody expression was induced with the addition of IPTG to the 1 mM final concentration, followed by a further incubation, overnight, at 30 °C. Periplasmic extracts were prepared [14] and purified by immobilised metal chelate chromatography (IMAC), using Ni-NTA spin columns (Qiagen) or TALON$^{TM}$ resin (Clontech, Takara Bio Inc., Mountain View, CA, USA), according to manufacturer's instructions, depending on the scale. Purified samples were then dialysed, using Slide-A-Lyzer cassettes with a 3.5 kDa molecular weight cut-off (Pierce, Thermo Fisher Scientific, Waltham, MA, USA) into the PBS and the size and purity assessed by the SDS-PAGE. Purified single domain antibodies were screened for binding to the recombinant HAs and to the influenza virus antigen standards. Influenza virus antigen standards (National Institute for Biological Standards and Control, NIBSC) were reconstituted in 1 mL sterile water and then diluted 1/20 in PBS (prior to incubation) overnight, at 4 °C, in a 96-well plate (Nunc), followed by ELISA using an HRP (horseradish peroxidase) conjugated anti-c-Myc secondary reagent and TMB (3,3′,5,5′ tetramethylbenzidine) detection at $OD_{450}$nm [14].

### 2.4. Analysis Using Surface Plasmon Resonance

For binding and affinity ranking, a BIAcore T100 machine (GE Healthcare, Marlborough, MA, USA) was used, in combination with a single-cycle kinetics procedure [22]. In brief, the purified recombinant hemagglutinins from different Influenza B viruses were immobilised onto a BIAcore$^{TM}$ CM5 chip in 10 mM sodium acetate pH 5.5, using an amine coupling kit (GE Healthcare), to approximately 3000 RU. A concentration series from 1–100 nM of purified sdAbs were run over the different antigen surfaces. A reference surface was subtracted, prior to evaluation of the sensorgrams, using the single-cycle kinetics procedure of the BIAevaluation$^{TM}$ software (GE Healthcare) and a 1:1 fitting model. Binding the full length HA0 or the head domain, HA1, of hemagglutinin was evaluated using the recombinant B-Victoria HA0, B/Brisbane/60/2008, and B-Yamagata HA0, B/Florida/04/2006, (Protein Sciences$^{TM}$) or B-Victoria HA1, B/Brisbane/60/2008, and B-Yamagata HA1, B/Florida/04/2006 (Sino Biological Inc., Beijing, China).

### 2.5. Next-Generation-Sequence-Assisted Single Domain Antibody Discovery

Plasmid DNA was extracted from *E. coli* cultures grown from pre- and post-selection libraries (Figure 1), to obtain template DNA for the next-generation sequencing (NGS). A primary PCR reaction was performed using Phusion Hot Start II High Fidelity Polymerase (Thermo Fisher Scientific) and the primers: NGS_Alp_Fr1_Q (5′-tcgtcggcagcgtcagatgtgtataagagacagCAGCCGGCCATGGCACAG-3′) and NGS_FR4_Rev_AD (5′-gtctcgtgggctcggagatgtgtataagagacagTGAGGAGACGGTGACCTG-3′), which encoded the VHH gene flanking plasmid sequence (upper case) and adaptor sequences for the Nextera XT indexing (lower case), resulting in PCR products between 450 bp and 550 bp. The PCR products were purified (Qiagen) and used as a template for a secondary low-cycle number indexing PCR, using a Nextera XT indexing kit (Illumina, San Diego, CA, USA). Resulting PCR products were purified (Qiagen), quantified using the Qubit 2.0 fluorimeter (Thermo Fisher Scientific) and the DNA1000 Kit (Agilent Technologies, Santa Clara, CA, USA), followed by quality checking, using the Bioanalyzer 2100 (Agilent Technologies). Samples were prepared and multiplexed for use with the MiSeq reagent kit v3 (2 × 300 cycle) and run on the Illumina MiSeq, according to the manufacturer's protocol. The de-multiplexing and trimming of sequences were performed on the Illumina MiSeq instrument. Copy numbers of the individual CDR3′s were obtained from the reverse read FastQ files, using the "antibody mining toolbox" [23]. Our intention at this stage was not a full-length sdAb repertoire analysis but to identify the unique high frequency CDR3′s, which were highly enriched by the selection of Influenza B HA1 domain, compared to their frequency in the original phage library. Relative frequency of each CDR3, as a percentage of the total, was determined by dividing the individual CDR3 copy number by a total CDR3 copy number ×100 (%RF) and the enrichment level (fold) was determined by the CDR3 %RF, after selection, divided by the CDR3 %RF, in the absence of any selection (the original phage library). Candidate CDR3′s from the selections on recombinant HA1 that showed high levels of enrichment, were compared with the enrichment levels of the same CDR3′s in the selections of full-length HA0. Full-length DNA sequence for four sdAbs containing these highly enriched candidate CDR3′s were then obtained, using the DNA sequence of the CDR3 of interest, to search the reverse read FastQ file using the Geneious 7.1.2 (https://www.geneious.com), combining the sequences found with the corresponding forward reads from the forward FastQ file. Sequences were synthesised (IDT), the sdAbs were expressed and purified, as before, and the binding specificities were determined by ELISA.

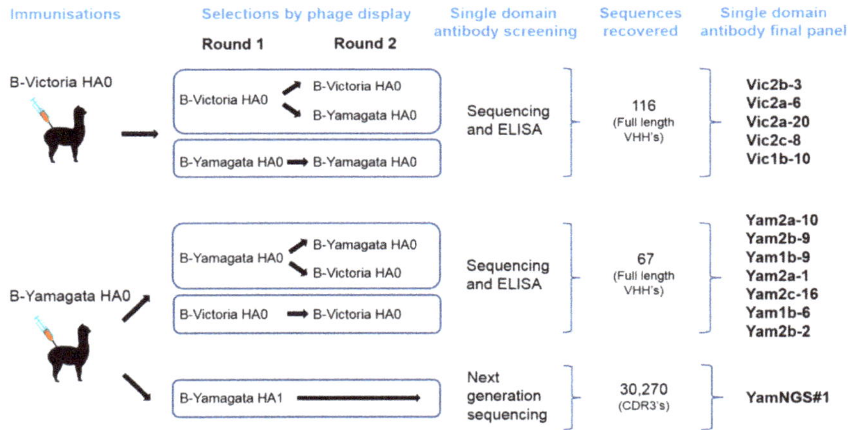


**Figure 1.** Selection strategies to isolate cross-reactive and lineage-specific single domain antibodies (sdAbs) against influenza B hemagglutinin. The figure shows the immunisation approaches, selection and screening strategies. The number of VHH CDR3 sequences isolated from each approach is indicated.

### 2.6. Lentiviral Pseudotype Assays

Lentiviral pseudotypes were produced by transient co-transfection of HEK293T/17 cells, using polyethylenimine [24,25]. Plasmid p8.91 encodes the structural (gag, pol) genes, whereas, pCSFLW represents the genome incorporated into the pseudotypes, bearing the firefly luciferase reporter. Influenza B hemagglutinin genes, in the expression plasmid pI.18, were also added to this mix, alongside the Human Airway Trypsin (HAT) expression plasmid, pCAGGS-HAT to allow for HA0 to HA1/2 maturation. The pseudotype-based microneutralisation assay (pMN) was carried out in Nunc™ F96 microplates (Thermo Fisher Scientific) [24,25]; 1:2 serial dilutions were performed with the relevant sdAbs, across the 96-well plate in a total of 50 µL DMEM (10% foetal bovine serum, 1% penicillin/streptomycin). HIV-1-derived lentiviral pseudotypes, bearing the influenza B HA of choice, was then added to yield a relative luminescence unit (RLU) input of $1.5 \times 10^6$ per well, in a total volume of 50 µL [26]. The plates were then incubated in a humidified incubator at 37 °C, 5% $CO_2$, for one hour, after which $1 \times 10^4$ HEK293T/17 cells were added, per well, in a total volume of 50 µL. After 48 h, the supernatants were removed and a 50:50 mix of PBS and Bright-Glo™ (Promega, Madison,

WI, USA) was added to each well. Plates were incubated at room temperature, for five minutes, and then luminescence was read using a Glomax® luminometer (Promega). Results were normalised to cell and virus only controls, representing 100% and 0% neutralization, respectively, and $IC_{50}$ values were calculated by non-linear regression, using GraphPad 5 Prism.

### 2.7. Construction and Screening of a Randomly Mutagenised HA Library Using Yeast Display

The coding sequence of the Influenza B virus hemagglutinin (HA0) of the strain B/Brisbane/60/2008 (GenBank FJ766840.1, minus the N-terminal signal sequence and the C-terminal transmembrane domain, comprising amino acids D1-I534) was codon-optimised for expression in yeast and synthesised, with 5′ *Sfi*I and 3′ *Not*I restriction sites (IDT). The construct was cloned as a *Sfi*I/*Not*I restriction fragment into the yeast display vector pNIBS-5 [16], which carried an SV5 tag, and transformed into *Saccharomyces cerevisiae* EBY100 (Invitrogen), using a yeast transformation kit (Sigma).

A library of the HA0 mutants was generated by error-prone PCR, at a low error rate, to give approximately, 1 mutation per HA gene [16]. Approximately 20 µg of the HA0 error-prone PCR product was co-transformed with 20 µg of the *Sfi*1/*Not*1 digested pNIBS-5 vector, into the EBY100 competent cells, for a recombination between the PCR products and the yeast vector to take place in the yeast cells [16]. Standard procedures and recipes for growth, induction, yeast cell labelling, media and buffer preparation were used [16]. Staining for the sdAb binding to HA0 was performed by incubating yeast cells with 200 nM of purified sdAbs (c-Myc tagged), followed by a chicken anti c-Myc antibody (Bethyl Laboratories, Montgomery, TX, USA), and then by a goat-anti-chicken IgG AlexaFluor647-labelled secondary antibody (Jackson ImmunoResearch Europe, Ely, UK). Staining of yeast cells for HA0 display was performed by incubating yeast cells with mouse monoclonal anti-SV5 antibody (AbSerotec, Bio-Rad, Hercules, CA, USA), followed by a goat anti-mouse IgG AlexaFluor488-labelled secondary antibody (Thermo Fisher scientific). Staining was performed simultaneously for the sdAb binding and the HA0 display. The mutant yeast library was grown in a selective medium for induction of the HA display and, approximately, $10^6$ cells were co-stained with purified sdAbs at 200 nM, followed by anti-SV5 and anti-c-Myc detection antibodies, and then by fluorescent secondary antibodies, as above [16]. Flow cytometric cell sorting was performed using the BDAria III and FACSDiva software v8.0.1 (Becton Dickenson, Franklin Lakes, NJ, USA). A sorting gate was chosen to sort cells for the display of HA0 (by virtue of the anti-SV5 signal), but also for the absence of sdAb binding. A second round was then performed using the same conditions. Yeast DNA minipreps were made from the final round of sorting and transformed into the TG1 electro-competent *Escherichia coli* cells (Agilent). Then, plasmids from single bacterial colonies were sequenced. Resulting sequences were aligned to wt HA0 (wild-type B/Brisbane/60/2008 HA0), using Geneious 7.1.2 (https//www.geneious.com) to identify candidate mutations. Plasmids containing single residue candidate mutations were re-created, where required, using the QuikChangeII site directed mutagenesis kit (Agilent) and re-transformed into yeast cells. Individual yeast clones containing either a mutant HA0 or a yeast containing the wt HA0, were grown in a selective medium, for induction of the HA display and cells stained for display and sdAb binding, as above. Binding of each sdAb to each mutant HA0 and wt HA0 yeast clone was determined by flow cytometry on a FACS CantoII flow cytometer (Becton Dickinson). Data collection and analysis was performed using the FACSDiva and FlowJo v10, (LLC, Ashland, OR, USA) software.

## 3. Results

### 3.1. Isolation and Characterisation of Cross-Reactive and Lineage-Specific Single Domain Antibodies against the Influenza B Hemagglutinin

Two phage displayed sdAb libraries were constructed from purified peripheral blood mononuclear cells of alpacas immunised with recombinant HA0 from the representative B-Victoria (B/Brisbane/60/2008) or B-Yamagata (B/Florida/04/2006) lineage viruses. The size of the B-Victoria library was $9.4 \times 10^7$ independent clones and the B-Yamagata library was $3.7 \times 10^7$ independent clones. Each phage library was selected for two rounds on the HA antigen used for immunization, for an unbiased recovery of all sdAbs binding to the immunogen. To bias for the selection of cross-reactive sdAbs, a second strategy was also performed, by alternating selections between the B-Victoria lineage and the B-Yamagata lineage antigens (Figure 1).

After selection, sdAbs were sequenced and grouped into clonally related families, based on the CDR3 sequence length and similarity. Single domain antibodies with identical CDR3 length and high sequence similarity were assumed to be clonally related and, thus, binding to the same or overlapping epitopes, so a single representative clone from each family was taken forward for further analysis.

Our ELISA screening identified sdAbs with a pre-dominantly cross-lineage reactivity (Figure 2), which was not surprising, given the high sequence homology between the B-Yamagata and the B-Victoria strains.

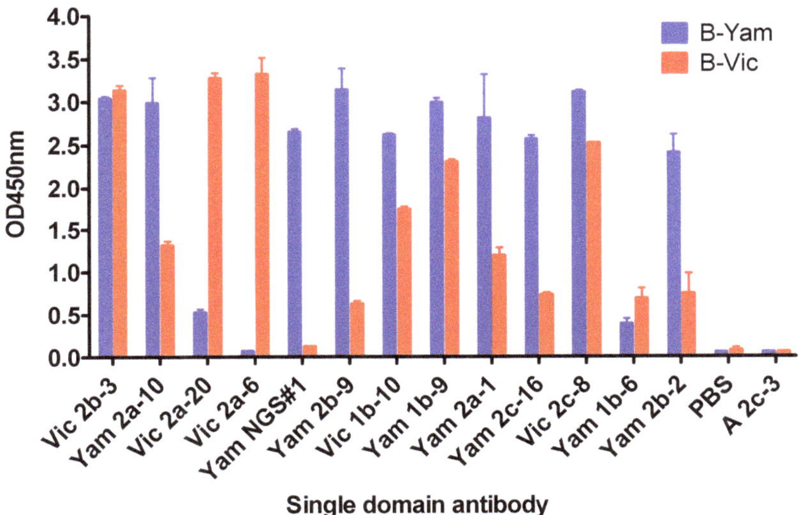


**Figure 2.** ELISA showing sdAb binding to the recombinant hemagglutinin (HA0). sdAb binding to the recombinant HA0 from Influenza B-virus lineage representative strains used in immunisations and phage display library selections. B-Vic is B/Brisbane/60/2008, B-Yam is B/Florida/04/2006. Control sdAb A2c-3 is specific for influenza A hemagglutinin, PBS is a no sdAb negative control. ELISA signal is an average of 2 assays.

In addition to cross-lineage reactive sdAbs, we were able to identify Vic2a-6 with B-Victoria lineage-specific recognition, however, we were unable to isolate any B-Yamagata lineage-specific sdAbs, using the conventional screening. We reasoned that epitopes defining B-Yamagata lineage specificity would be rare and likely to bind to the more variable HA1 head domain [5]. Therefore, we selected the B-Yamagata phage display library on a purified recombinant HA1 domain of B/Florida/4/2006, to bias against the stem reactive sdAbs, which had dominated previous selection strategies, and then used next-generation sequencing to identify new unique CDR3 sequences [27]. Highly enriched CDR3 sequences were identified, relative to the unselected CDR3 pool, and compared to those sequences already recovered, using conventional phage display selections. This led to the identification of several VHH CDR3 sequences which had not previously been identified (Figure S1). Full-length VHH sequences were synthesised and four candidate sdAbs were expressed and tested for B-Yamagata lineage-specific recognition, resulting in one unique B-Yamagata lineage-specific sdAb, YamNGS#1. A final panel of thirteen unique sdAbs were taken forward, which included eleven cross-lineage reactive sdAbs, one B-Victoria lineage, and one B-Yamagata lineage-specific sdAb (Table 1) (Figure 1).

All sdAbs were shown to have the characteristic camelid heavy-chain-only antibody framework 2 "hallmark" residue substitutions [28]. CDR3 lengths varied from 5 to 20 residues (Table 1) and all sdAbs were expressed at yields between 3 mg/L to 27 mg/L.

**Table 1.** VHH CDR3 sequences of specific single domain antibodies.

| Nanobody | CDR3 Sequence | B-Yamagata | B-Victoria |
|---|---|---|---|
| **Vic2b-3** | AADAVVAGPDDEYDY | + | + |
| **Yam2a-10** | NVGFNY | + | + |
| **Vic2a-20** | ASKGTDYIDGIYYISYQFNS | + | + |
| **Vic2a-6** | VASPFSTGRATLPYQYPY | − | + |
| **YamNGS#1** | AAASLCSFSSNDYFY | + | − |
| **Yam2b-9** | AAGSGGRYDY | + | + |
| **Vic1b-10** | NAPTYSN | + | + |
| **Yam1b-9** | ALGDFTGLTNLRQAFYDF | + | + |
| **Yam2a-1** | NFPRSSS | + | + |
| **Yam2c-16** | NTHDY | + | + |
| **Vic2c-8** | ALGDFSGSLWAYEYDF | + | + |
| **Yam1b-6** | AAAKGGGAYSMISAYTY | + | + |
| **Yam2b-2** | RLDHWLVSGY | + | + |

Binding to recombinant HA0 from the B-Victoria strain (B/Brisbane/60/2008) and the B-Yamagata strain (B/Florida/04/2006) in ELISA, indicated as binding (+) or not binding (−).

We tested the extent of cross-reactivity on a panel of whole virus antigen standards, corresponding to the strains included in seasonal vaccines between 1973 and 2014 (Table 2). Most of the sdAbs initially identified as cross-reactive with the B lineage representative strains B/Brisbane/60/2008 (B-Victoria) and B/Florida/04/2006 (B-Yamagata), used for library construction (Table 1), showed extensive cross-reactivity going, both, forward and backwards in time, including the pre-lineage split strain (B/HongKong/08/73).

Single domain antibodies Yam2b-9, Vic1b-10, Yam 1b-9, Yam2a-1, Yam2c-16, Vic2c-8, and Yam1b-6 recognised all antigen standards tested from the pre-lineage split virus B/HongKong/8/73, up to more recent vaccine viruses B/Brisbane/60/2008 (B-Victoria lineage) and B/Utah/9/2014 (B-Yamagata lineage). The B-Victoria lineage-specific sdAb Vic2a-6 showed reactivity to all B-Victoria strains tested between 1987–2008, but also bound the earliest B-Yamagata strain (B/Yamagata/16/1988), just post-split. YamNGS#1 was identified as having B-Yamagata lineage-specific binding and was able to recognize the Yamagata strains between 1994 and 2012, covering a duration of eighteen years (Table 2).

**Table 2.** Single domain antibody recognition of the whole virus antigen standards by ELISA.

| Antigen Standard | Year | Vic2b-3 | Yam2a-10 | Vic2a20 | Vic2a-6 | YamNGS#1 | Yam2b-9 | Vic1b-10 | Yam1b-9 | Yam2a-1 | Yam2c-16 | Vic2c-8 | Yam1b-6 | Yam2b-2 |
|---|---|---|---|---|---|---|---|---|---|---|---|---|---|---|
| B/Brisbane/60/2008 | 2008 | 0.34 | 0.16 | 2.30 | 2.42 | 0.08 | 0.45 | 0.79 | 0.30 | 0.82 | 0.45 | 0.29 | 0.50 | 0.07 |
| B/Malaysia/2506/2004 | 2004 | 0.07 | 0.08 | 3.54 | 1.93 | 0.09 | 0.68 | 1.39 | 0.40 | 1.43 | 0.72 | 0.40 | 0.59 | 0.07 |
| B/Shangdong/7/97 | 1997 | 0.39 | 0.14 | 2.58 | 0.32 | 0.06 | 0.69 | 0.96 | 0.30 | 0.92 | 0.51 | 0.28 | 0.31 | 0.08 |
| B/Victoria/2/87 | 1987 | 0.43 | 0.20 | 3.31 | 0.86 | 0.09 | 0.56 | 1.27 | 0.83 | 1.35 | 1.15 | 0.77 | 1.04 | 0.11 |
| B/HongKong/8/73 | 1973 | 0.46 | 0.39 | 0.06 | 0.07 | 0.13 | 0.34 | 1.48 | 0.66 | 1.41 | 0.72 | 0.56 | 0.29 | 0.09 |
| B/Yamagata/16/88 | 1988 | 0.83 | 1.76 | 2.41 | 1.07 | 0.09 | 1.59 | 1.85 | 0.98 | 1.79 | 1.01 | 0.87 | 0.71 | 0.12 |
| B/Yamanashi/166/98 | 1994 | 0.74 | 1.08 | 0.11 | 0.07 | 0.07 | 1.70 | 2.16 | 1.34 | 2.41 | 1.35 | 1.24 | 1.70 | 0.14 |
| B/Harbin/7/94 | 1994 | 0.28 | 0.26 | 0.14 | 0.09 | 0.44 | 0.38 | 0.57 | 0.27 | 0.52 | 0.36 | 0.27 | 0.30 | 0.06 |
| B/Jiangsu/10/2003 | 2003 | 0.46 | 0.71 | 1.16 | 0.08 | 0.59 | 0.66 | 1.32 | 0.53 | 1.33 | 0.46 | 0.53 | 0.23 | 0.06 |
| B/Florida/4/2006 | 2006 | 0.42 | 0.40 | 0.16 | 0.07 | 0.83 | 0.62 | 0.64 | 0.20 | 0.68 | 0.31 | 0.18 | 0.32 | 0.07 |
| B/Mass'/02/2012 | 2012 | 0.53 | 0.82 | 0.79 | 0.05 | 1.00 | 3.03 | 1.82 | 0.94 | 1.96 | 0.85 | 0.84 | 1.57 | 0.10 |
| B/Phuket/3073/2013 | 2013 | 0.58 | 1.22 | 0.15 | 0.06 | 0.11 | 1.01 | 0.94 | 0.55 | 0.96 | 0.45 | 0.52 | 0.32 | 0.08 |
| B/Utah/9/2014 | 2014 | 0.90 | 1.36 | 0.14 | 0.07 | 0.10 | 0.31 | 1.13 | 0.52 | 1.20 | 0.43 | 0.50 | 0.21 | 0.05 |
| A/Brisbane/10/2007 | 2007 | 0.05 | 0.06 | 0.07 | 0.07 | 0.07 | 0.06 | 0.06 | 0.07 | 0.06 | 0.07 | 0.06 | 0.07 | 0.07 |

B-Victoria strains ranging from 1987–2008 and B-Yamagata strains from 1988–2014. Pre-lineage split strain highlighted in red. Influenza B strains used as immunogens for the sdAb library generation are highlighted in blue. Influenza A strain control highlighted in green. Table arranged to show strains moving away from pre-lineage split in time, with most recent strains being the farthest away. Values shown are the average of two independent assays. sdAb binding was tested at a single concentration of 10 µg/mL. $OD_{450} \geq 1.0$, $OD_{450} \geq 0.5$, $OD_{450} \geq 2 \times$ negative control for each sdAb. B/Mass'/02/2012 is B/Massachusetts/02/2012.

### 3.2. Grouping sdAbs on the Basis of HA1 Head Domain or Stem Domain-Specific Binding

Single domain antibodies were tested for binding to the head domain, or the HA stem region, using surface plasmon resonance on both full-length recombinant HA0 and HA1 (head domain) from representative B-Yamagata and B-Victoria strains (Table 3, Supplementary Figure S2).

**Table 3.** Epitope classification of full-length (HA0) and head domain (HA1) using surface plasmon resonance (SPR).

| Single Domain Antibody | B Yam HA1 [1] B/Florida /4/2006 | B Yam HA0 [2] B/Florida /4/2006 | B Vic HA1 [1] B/Brisbane /60/2008 | B Vic HA0 [2] B/Brisbane /60/2008 | Domain | Specificity |
|---|---|---|---|---|---|---|
| **Vic2b-3** | 0.27 nM | 1.57 nM | 0.06 nM | 0.11 nM | Head | Cross reactive |
| **Yam2a-10** | 3.49 nM | 6.70 nM | + [3] | 3.30 nM | Head | Cross reactive |
| **Vic2a-20** | 173 nM | - | 0.79 nM | 0.48 nM | Head | Cross reactive |
| **Vic2a-6** | - | - | 0.07 nM | 0.08 nM | Head | B-Vic lineage |
| **YamNGS#1** | 11.8 nM | 2.39 nM | - | - | Head | B-Yam lineage |
| **Yam2b-9** | - | 2.13 nM | - | 10.0 nM | Stem | Cross reactive |
| **Vic1b-10** | - | 0.28 nM | - | 1.1 nM | Stem | Cross reactive |
| **Yam1b-9** | - | 0.10 nM | - | 0.08 nM | Stem | Cross reactive |
| **Yam2a-1** | - | 0.14 nM | - | 7.40 nM | Stem | Cross reactive |
| **Yam2c-16** | - | 1.46 nM | - | 0.76 nM | Stem | Cross reactive |
| **Vic2c-8** | - | 0.13 nM | - | 0.37 nM | Stem | Cross reactive |
| **Yam1b-6** | - | 5.43 nM | - | 0.60 nM | Stem | Cross reactive |
| **Yam2b-2** | - | 10.0 nM | - | 1.36 nM | Stem | Cross reactive |

[1] HA1 is purified hemagglutinin head domain (Residues Met1-Arg361 of B/Florida/04/2006 and residues Met1-Arg362 of B/Brisbane/60/2008). [2] HA0 is purified full-length hemagglutinin. [3] Binding could be seen but could not be resolved using the BIAevaluation™ software.

Five sdAbs were shown to be specific to the HA1 domain and eight sdAbs only showed recognition of the full-length HA0. Most of the cross-reactive sdAbs were classified as stem-specific, however, three sdAbs were predicted to bind cross-reactive epitopes in the head domain. The B-Victoria lineage-specific sdAb Vic2a-6, bound to the head domain of the B/Brisbane/60/2008 (B Victoria), but did not bind to the B/Florida/04/2006 head domain or the full-length HA (B-Yamagata), which was consistent with recognition of a B-Victoria lineage-specific epitope in the head domain. YamNGS#1 bound to the head domain of the B/Florida/04/2006 (B-Yamagata), but showed no binding to the head domain of the B/Brisbane 60/2008 (B-Victoria), indicating that it binds to a B-Yamagata lineage-specific epitope in the head domain. The sdAbs had a high affinity down to <100 pM, reflecting the affinity maturation possible using the prime and boost immunisation of alpacas. The panel of sdAbs were grouped as cross-reactive head binding (Vic2b-3, Yam2a-10, Vic2a-20), cross-reactive stem binding (Yam 2b-9, Vic1b-10, Yam1b-9, Yam2a-1, Yam2c-16, Vic2c-8, Yam1b-6, Yam2b-2), B Victoria lineage-specific head binding (Vic2a-6) or B-Yamagata lineage-specific head binding (YamNGS#1) (Table 3).

### 3.3. In Vitro Neutralisation Activity of Single Domain Antibodies

Purified sdAbs were tested for their ability to neutralise influenza virus pseudotypes, corresponding to the different IBV lineages, using micro-neutralisation assays [24]. Six of the eight sdAbs identified as cross-reactive and binding to the epitopes in the HA stem region, were shown to be capable of neutralising pseudoviruses of both lineages with the most potent sdAb, Vic1b-10, having $IC_{50}$ values of 0.2 nM and 15.5 nM on B-Vic and B-Yam lineage viruses, respectively (Table 4). Yam2b-9 and Yam2b-2, although cross-reactive on the recombinant antigen (Figure 2, Table 3), showed only a lineage-specific neutralisation activity. There was generally a good correlation between affinity and neutralisation activity, except for Vic2c-8, which was shown to be the least potent in neutralization, despite having an affinity comparable to the other cross-reactive, sdAbs (Tables 3 and 4).

**Table 4.** Neutralisation of the influenza B pseudotype viruses.

| Single Domain Antibody | B/Yamagata /16/1988 (Yamagata) $IC_{50}$ [nM] | B/Victoria /2/1987 (Victoria) $IC_{50}$ [nM] | B/Florida /4/2006 (Yamagata) $IC_{50}$ [nM] | B/Brisbane /60/2008 (Victoria) $IC_{50}$ [nM] |
|---|---|---|---|---|
| **Vic2b-3** | 104 | 81 | 246 | 240 |
| **Yam2a-10** | - | - | - | - |
| **Vic2a-20** | - | 1.1 | - | 112 |
| **Vic2a-6** | 315 | 159 | - | 10 |
| **YamNGS1#1** | - | - | NT | NT |
| **Yam2b-9** | 30.5 | - | NT | NT |
| **Vic1b-10** | 0.2 | 15.5 | NT | NT |
| **Yam1b-9** | 31.7 | 81.8 | NT | NT |
| **Yam2a-1** | 81.8 | 6 | NT | NT |
| **Yam2c-16** | NT | NT | NT | NT |
| **Vic2c-8** | 202.5 | 228 | NT | NT |
| **Yam1b-6** | 26.6 | 12.6 | NT | NT |
| **Yam2b-2** | - | 113.5 | NT | NT |

NT—Not Tested; $IC_{50}$ is the sdAb concentration in nM that gives 50% inhibition; and (-) indicates no neutralisation activity.

Both, Vic2a-6 and Vic2a-20 showed a B-Vic lineage-specific neutralisation activity with Vic2a-6 also showing a low level of neutralisation activity on the earliest post-split B/Yamagata/16/88 strain. This was consistent with Vic2a-6 recognising an antigen standard corresponding to the same strain (Table 2).

### 3.4. *Identification of the B Victoria Lineage-Specific Epitope*

For a more precise epitope mapping, we used the yeast surface display which we had previously used to epitope map the sdAb binding to HA from A(H1N1)pdm09 [16]. The B/Brisbane/60/2008 HA0 precursor gene (B-Victoria lineage) (corresponding to the residues D1-I534 of the mature protein) was sub-cloned into a yeast display vector, where it was fused to a SV5 epitope tag, enabling the detection of a full-length expression and display of the folded HA0 on the yeast surface (Figure 3). Yeast cells were simultaneously labelled with an anti-SV5 antibody to determine the HA0 display levels and an anti c-Myc antibody to show binding of the individual sdAbs. The head binding sdAbs Vic2b-3, Yam2a-10, Vic2a-20, and the Vic2a-6 bound yeast displayed the B-Victoria lineage HA0, whilst YamNGS#1 did not show any binding, which was consistent with its B-Yamagata lineage specificity (Table 3) (Figure 4). The stem binding sdAbs did not show any binding which suggested that their epitope was not accessible on the yeast cell surface.

A library of the B-Victoria lineage HA0 mutants was generated using a low rate error-prone PCR to give, approximately, 1 mutation per HA gene [16]. The library was then selected for two rounds, for the loss of binding to the B-Victoria lineage-specific sdAb Vic2a-6 (Figure 3). Random clones from the sorted population of yeast cells were sequenced to identify HA mutations that were enriched, relative to the unselected library, and so were predicted to be residues directly involved in the binding epitope. Clones with mutations introducing/replacing cysteine or proline residues were discarded, as they were predicted to have indirect effects on the sdAb binding. Several mutations were identified relative to the unselected library, which were all mapped within the 120 loop of the head domain between residues 116–137 (Figure 3), which has previously been identified as an antigenic site on the IBV-HA [29,30]. The panel of mutant HA's was tested on the other head binding sdAbs, and mutation G133D was shown to interfere with the binding of all head-specific sdAbs, equally. Amino acid 133 sits within a highly structured region 'PGGP', which is consistent with having a more destabilising pleiotropic effect. The V124D mutation also interfered with all head binding sdAbs, equally, and was also classified as having a pleiotropic effect. Both of these mutations could not be associated with lineage-specific binding, so were not investigated further. The I125T mutation was also ruled out as a direct contact residue for the Vic2a-6, as it also abolished binding to all head binding sdAbs, equally. It is interesting to note that this threonine mutation was also present as a 'naturally' occurring variant in the early pre-lineage split strain B/HongKong/08/73, to which two of our head binding sdAbs (Vic2a-6 and Vic2a-20) did not bind (Table 2). The remaining residues 122 and 129, on the other hand, were predicted to correlate with the B-Victoria lineage binding and the enrichment of two separate mutations at position 122 (H122L and H122Y), suggested the importance of histidine 122 for the binding of Vic2a-6.

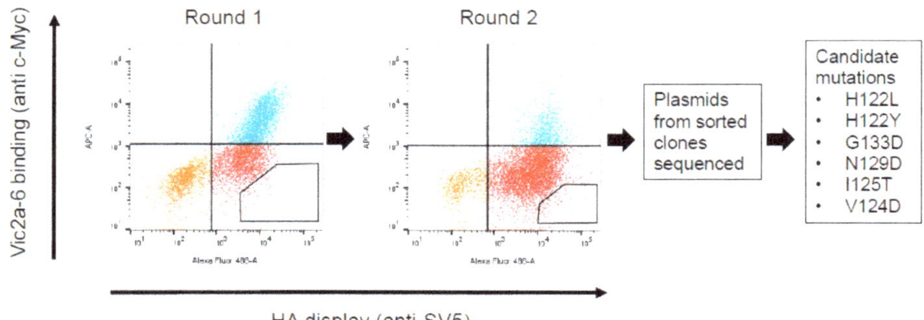


**Figure 3.** Cell sorting and isolation of candidate mutations interfering with the Vic2a-6 sdAb binding. Flow cytometry plots for two rounds of negative sorting for loss of binding to Vic2a-6, by gating cells in the lower right quadrant of the FACS dot plot, as indicated. Plasmids from the sorted yeast clones were sequenced and mutations in the HA which interfered with sdAb binding to HA, were identified. Residue numbering was according to B/HongKong/8/73 [29].

Sequence alignment of the epitope region 116–137 of the B-Victoria and the B-Yamagata strains tested in ELISA (Table 2), showed that histidine 122 and asparagine 129, spanned regions of divergence between the two lineages (Figure 5). We, therefore, re-created three single residue mutants in a wild-type HA0 background, H122L, N129D, and a naturally occurring substitution, H122Q, where the B-Vic preferred residue (histidine) was substituted with the B-Yam preferred residue (glutamine) (Figure 5). Flow cytometry experiments of the sdAb binding to the mutant HA0s (Figure 4), showed that the H122L mutation completely abolished the binding of both Vic2a-6 and Vic2a-20, whereas, the H122Q mutation reduced the binding of both sdAbs to between 14% and 24% of binding to the wt HA0, confirming the importance of this residue in the epitope footprint of both sdAbs (Figure 4). The N129D mutation only interfered with the Vic2a-6 binding, with no effect on the Vic2a-20, suggesting that, although these two sdAbs recognized the overlapping epitopes, they were not identical. This was consistent with their having unique VHH CDR3 sequences which defined their respective paratopes and interactions with HA (Table 2). None of the above three mutations had any effect on the binding of Vic2b-3, which bound to a different epitope on the HA0 head domain of the IBV.

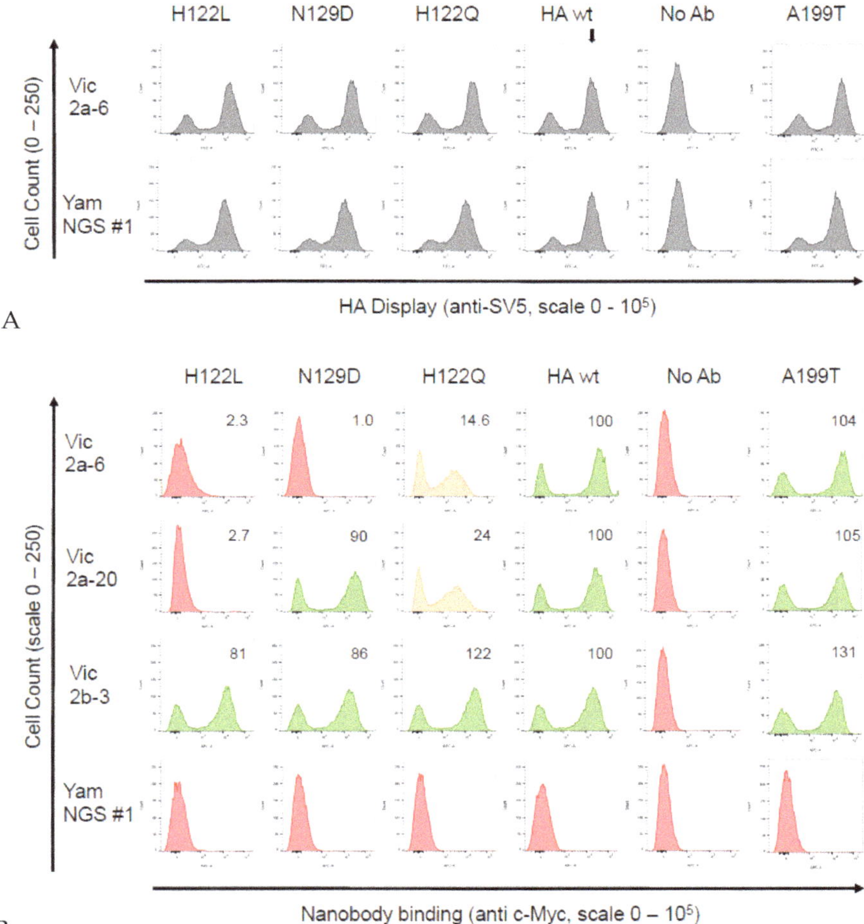


**Figure 4.** Binding of the sdAbs to the B-Victoria single residue mutant HA0's. Candidate single residue mutations predicted to correlate with lineage-specific binding and a naturally occurring substitution H122Q, which differentiated between the B-Victoria and B-Yamagata strains, were generated and binding was tested using yeast display and flow cytometry. (**A**) Detection of the HA0 display level, using detection of the SV5 epitope tag attached to the HA0, showing that the mutations had no effect on the HA0 display level relative to the wild-type (wt) HA0 (indicated by vertical arrow). (**B**) Flow cytometry histograms showing binding of the Vic2a-6, Vic2a-20, Vic2b-3 and the YamNGS#1 to the wild-type HA (B/Brisbane/60/2008 precursor HA0) and selected mutants (the positive population is the right peak, whereas, the left peak shows the unbound and unstained populations). Mutations that completely eliminated sdAb binding are shown in red, those that partially affected binding are shown in orange and those that had no effect are shown in green. We determined the extent of sdAb binding by dividing the geometric mean fluorescence intensity (MFI) of each sdAb mutant pair by the value of the MFI of the wild-type B-Vic HA sdAb interaction and the resulting ratio, normalised to a percentage value of the wild-type interaction. Relative binding of the sdAbs was categorised as 'no binding' (in red)—< 5%; 'intermediate binding' (in orange)—between 20–40%; and 'strong binding' (in green)—> 40%; shown in the upper right-hand quadrant of the flow cytometry histograms. Graphs shown are a representative of at least two independent experiments.

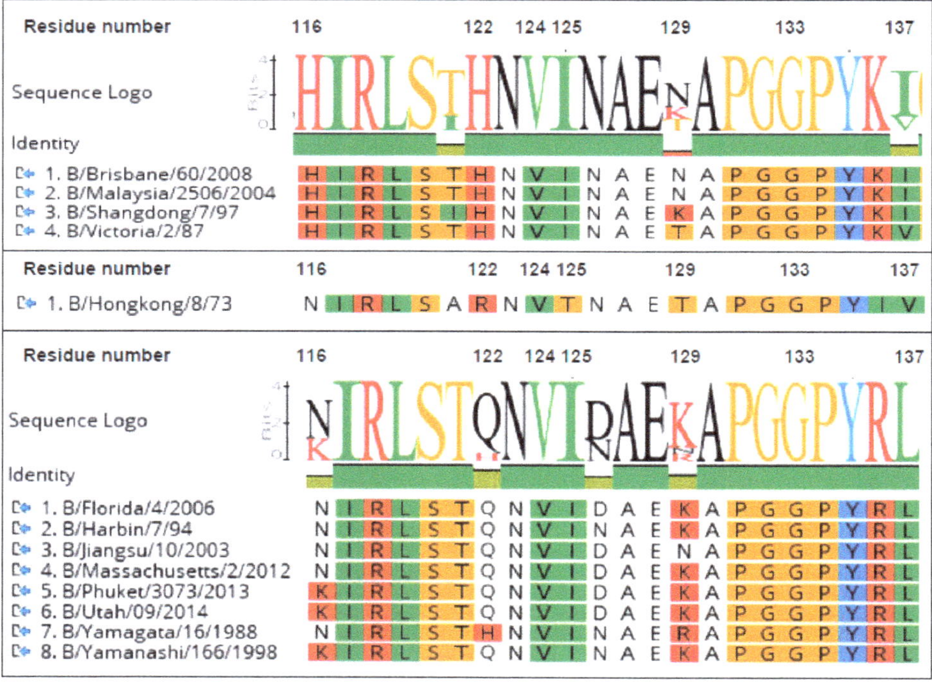


**Figure 5.** Sequence logo of antigen standards spanning predicted the epitope for lineage-specific, head-binding, single domain antibodies. Candidate residues, predicted to have a role in the binding epitope footprint of the Vic2a-6 are numbered, according to the B/HongKong/8/73 [29] within 116–137. Alignment of the B-Victoria (top) pre-lineage split strain (middle) and the B-Yamagata strains (bottom) are shown for the strains used to determine the sdAb-specificity profile (Table 2). Protein sequences used were translated from the DNA sequences: B/Brisbane/60/2008 (gb:ACN29380.1), B/Malaysia/2506/2004 (gb:ACR15732.1), B/Shangdong/9/97 (gb:AAM12546.1), B/Victoria/2/87 (gb:M58428.1) of the B-Victoria lineage; B/Yamagata/16/88 (gb:ABL77255.1), B/Yamanashi/166/98 (gb: CY019531.1), B/Harbin/7/94 (gb:AF060003.1), B/Jiangsu/10/2003 (gb:ACF54202.1), B/Florida/4/2006 (gb:ACF54246.1, B/Massachusetts/02/2012 (gb:KC892118.1), B/Phuket/3037/2013 (GISAID, EPI529345), B/Utah/9/2014 (gb | AMB72003.1) of the B-Yamagata lineage; and the pre-lineage split strain B/HongKong/8/73 (gb: AAA43717.1).

## 4. Discussion

Monoclonal antibodies are generally seen as requiring a paired light and heavy-chain to recognise a target antigen. However, there are now several examples of stem binding human monoclonal antibodies to both IAV and IBV, which only use their heavy-chain for binding, with no contacts being made by the light chain [7,8,10]. Guided by this observation, we have previously used alpacas as a route to high-affinity, cross-neutralising, single domain antibodies (Nanobodies$^{TM}$), naturally devoid of a paired LC, which bind to equivalent epitopes in the HA stem of the IAV [14,16]. This unique antibody format has several advantages over the conventional MAbs and, in particular, their small oblate structure and extended CDR3 loop makes them naturally equipped to access grooves and clefts, such as those on viral surfaces [21,31]. In addition, the potential for a smaller epitope footprint of the sdAbs, compared with conventional antibodies, which have evolved to bind to larger, flatter surfaces, might translate into a higher genetic barrier for an escaping virus and retention of binding for a longer period, in evolutionary time.

Within this study we have identified the sdAbs, specific to the IBV-HA, which have broad cross-reactivity and lineage-specific recognition. Single domain antibodies Vic1b-10, Yam1b-9, and the Yam2a-1 were shown to have a broad cross-reactivity, covering strains spanning over 20 years. This clearly shows that these sdAbs bind to the epitopes resistant to antigenic change and could be expected to maintain recognition of the influenza B HA, over a significant period of time (Table 2). In addition, these sdAbs were capable of neutralizing the influenza lentiviral pseudotypes, corresponding to the two prototype viruses representing the two divergent lineages, B/Victoria/2/87 and B/Yamagata/16/88, suggesting that they might have a potential in immunotherapeutics. Single domain antibodies, because of their simple molecular structure, can be easily re-formatted for half-life extension [32,33]. This could give immediate, short-term immunity, independent of influenza vaccination or the need for a functioning immune system. Longer term passive immunity, using cross-neutralising antibodies, can be achieved using a viral vector-mediated gene delivery [34,35]. In addition, the high stability and ability to withstand nebulisation are distinct advantages of the sdAbs, over human antibody formats, and mean intranasal delivery is possible, allowing deep penetration into the respiratory tract [36]. The potential applications of the cross-reactive sdAbs against IBV, extend beyond immune prophylaxis [10], and include a diagnosis of circulating IBV strains, or vaccine potency assays [19,37], for quantitating the HA content of vaccines.

Given the current requirements of the quadrivalent vaccines we sought to prove that we could isolate lineage-specific sdAbs, in a short timeframe, which would be able to maintain reactivity for a significant length of time, ideally over several seasonal vaccine changes. The identification of sdAbs, which bind to lineage-specific epitopes, is more challenging, given the high sequence and antigenic similarity between the lineages. However, using phage display and next-generation sequencing, we were able to identify, both, a B-Victoria (Vic2a-6) and a B-Yamagata (YamNGS#1) lineage-specific sdAb. Initial attempts using conventional screening to identify B-Yam lineage-specific sdAbs were unsuccessful. We reasoned that this was likely due to the low sampling of the selected clones and the dominance of the cross-reactive sdAbs during selection. Using NGS, we were able to overcome these difficulties and identify a B-Yam lineage-specific sdAb, demonstrating that NGS-assisted screening is a useful addition to the phage display process, particularly, in identifying rare sdAb specificities. The B-Victoria lineage-specific sdAb Vic2a-6 was shown to neutralise B/Brisbane/60/2008 (B-Victoria) but not the B/Florida/04/2006 (B-Yamagata), suggesting that it binds to a lineage-specific epitope in the head domain (Table 3). Another sdAb, Vic2a-20, showed preferential binding to the immobilised B-Victoria strains, using surface plasmon resonance (Table 3), with a greater than 100-fold higher affinity for the B-Victoria than the B-Yamagata HA, which was consistent with the B-Victoria lineage-specific neutralisation observed (Table 4). This suggests that the Vic2a-20 has a preferred B-Victoria lineage-specificity rather than the absolute-specificity shown by the Vic2a-6. The sdAb YamNGS#1 showed an absolute specificity for the B-Yamagata strains, with no binding to any of the B-Victoria strains tested (Table 2). The immunogen used to generate the YamNGS#1 was the 2006 B-Yamagata vaccine strain (B/Florida/04/2006) and lineage-specific binding was maintained for a period of 6 years, up to the 2012 strain (B/Massachusetts/02/2012).

In order to understand the structural basis of lineage-specific binding of the sdAb Vic2a-6, we used yeast display and mutational scanning [16,38]. Yeast display has emerged as a powerful tool for epitope mapping, as its eukaryotic translation machinery acts as a quality control for functional, folded, protein variants [39]. In addition, simultaneous selection for, both, display and sdAb binding, using flow cytometric cell sorting means, each protein variant can be selected on the basis of multiple parameters. Selection of a HA0 mutant library on Vic2a-6 identified several candidate mutations, predicted to specifically interfere with binding. All these mutations lie within the 120 loop spanning residues 116–137, which is a region of high antigenic diversity [29,30]. The I125T mutation was selected from our B-Victoria HA0 mutant library, which is also a naturally occurring substitution (Figure 3). Analysis of naturally occurring IBV-HA sequences identified threonine as being present in the very early pre-lineage split strain (B/Hongkong/8/73) (Figure 5), which mutated to an isoleucine in later IBV strains. The ability to relate the I125T mutation to the absence of binding of Vic2a-6 and Vic2a-20, highlighted the potential of yeast display and mutational scanning to correlate sdAb binding and strain-specificity profiles to the natural sequence divergence of IBV-HA. Mutational scanning predicted that residues 122 and 129 were key residues in the epitope footprint of Vic2a-6. The H122L mutation was shown to completely abolish binding of the Vic2a-6 and Vic2a-20, whereas N129D completely abolished binding of the sdAb Vic2a-6, with no effect on the binding of Vic2a-20. This demonstrated that although these sdAbs' epitopes were overlapping, they were also distinct, which was consistent with the Vic2a-20 only having a 'preferred' B-Vic lineage-specific binding (Table 2). The naturally occurring H122Q mutation associated with the sequence divergence of IBV (Figure 5) gave a significant reduction in binding to between 14% and 24% of the wild-type HA interaction but did not completely abolish binding. This suggests that this conservative polar substitution was associated with B-Victoria lineage-specific binding, but the N129 residue was a more significant determinant of the lineage-specific binding of the Vic2a-6. The natural sequence diversity at residue 122, in the B-Victoria strains (97% His: 1% Gln), compared to the B-Yamagata strains (94% Gln: 5% His) indicated that the preferences for either amino acid, although significant, was not complete [30]. The close correlation of binding specificity in ELISA with the identity of residue 122, was highlighted by the binding of the Vic2a-6 to B/Yamagata/16/88 (Table 2) and the neutralisation of a pseudotype corresponding to this same strain (Table 4). Although this was the earliest B-Yamagata strain, following the divergence of the two lineages, it maintained a histidine at position 122, and retained binding to Vic2a-6, which was only lost in the later B-Yamagata strains with the substitution of a glutamine. Correlating the mutations which interfere with binding of the Vic2a-6 and Vic2a-20, with the structure of HA, showed that the epitope footprint was adjacent to the receptor binding site and within the 120 loop (Figure 6). The 120 loop was one of the four main regions on the influenza B HA1 head domain, identified as being dominant in the antigenic evolution of the most recent strains [29,30,40]. The identification of lineage-specific sdAbs, which are able to retain binding to an epitope, over such a long period of time (20 years in the case of Vic2a-6 and 18 years in the case of YamNGS#1), despite binding to the hyper-variable head domain, was somewhat surprising. It is interesting to speculate that this was due to the well-documented ability of sdAbs to bind to small grooves and pockets on protein surfaces, which might represent a higher genetic barrier for escape than conventional antibodies which bind to larger flatter surfaces [16].

We have shown that it is possible to isolate high-affinity, cross-reactive and lineage-specific sdAbs from alpacas immunised with a single seasonal vaccine strain, which can maintain binding and resistance to natural antigenic changes, over a significant period of time. We have also highlighted that next-generation sequencing analysis of phage displayed libraries can be useful in identifying sdAbs that may have been missed using more limited conventional ELISA-based screening. In addition, using yeast display and mutational scanning we have been able to correlate the lineage-specific binding with the structure of HA and have related natural antigenic diversity within this epitope with the sdAb (Nanobody$^{TM}$) reagent binding. This suggests that yeast display epitope mapping could be adapted to give a comprehensive analysis of the epitope of lineage-specific sdAbs, which could be used for quadrivalent vaccine potency testing, and in addition, it could predict when a sdAb might lose its binding and needs updating [37].

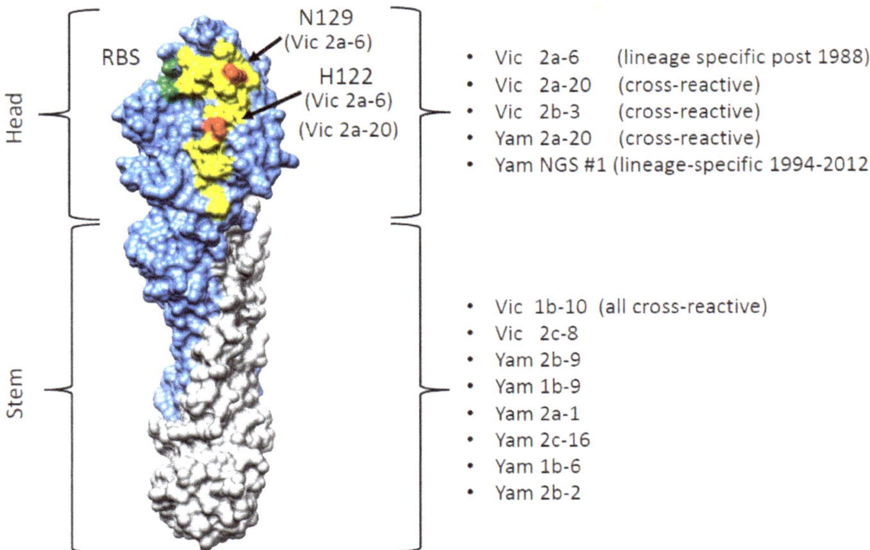


**Figure 6.** Correlation of the sdAb epitopes with the Hemagglutinin structure. The structure of the HA0 monomer (B/Brisbane/60/2008 PDB structure 4FQM) is shown with the HA1 domain in blue, the stem region in grey, the 120 loop in yellow and the receptor binding site (RBS) in green. The residue of H122 and N129 associated with the B-Victoria lineage-specific binding of the sdAb Vic2a-6 or preferred binding of the Vic2a-20 are shown in red. The binding specificity of the sdAb panel are shown in relation to the HA head and stem region.

**Supplementary Materials:** The following are available online at http://www.mdpi.com/2073-4468/8/1/14/s1, Figure S1: Isolation of the B-Yamagata lineage-specific nanobodies, using NGS. High copy number CDR3's using antibody mining toolbox (A) were normalised for each sequencing run and presented as % Relative Frequency (B,C). Fold increase in CDR3 frequencies were then calculated from a CDR3 frequency, in the unselected phage library and the same CDR3's frequency after selection of the HA1 domain, or of the full-length HA0 (D,E). Figure S2: Grouping as head- or stem-specific binding using surface plasmon resonance (SPR). The kinetic binding constants (*kon* and *koff*) of the panel of nanobodies were determined using SPR and single-cycle kinetics. Data are presented as rate plots with iso-affinity diagonals (RAPID) where the diagonals (dotted lines) are connecting the points of equal dissociation constant (KD). Affinity on (A) the B-Yamagata head domain (HA1) of the B/Florida/04/2006, (B) the full-length HA0 of the B/Florida/04/2006, (C) the B-Victoria head domain (HA1) of the B/Brisbane/60/2008, and (D) the full length HA0 of the B/Brisbane/60/2008, is shown. Fitting was with single-cycle kinetics and a 1:1 Langmuir fitting model, using the BIAevaluation$^{TM}$ software (GE Healthcare, Marlborough, MA, USA). Equilibrium dissociation constants are given in Table 3.

**Author Contributions:** Methodology, W.R., T.G., C.B., P.R., G.W.C., C.Y.C. and S.E.H.; Investigation, W.R., T.G., C.B., P.R., C.Y.C., G.W.C. and S.E.H.; Writing—Original Draft Preparation, W.R.; Writing-Review & Editing, W.R., S.E.H., C.B., O.G.E., G.W.C. and N.T.; Supervision, O.G.E. and S.E.H.; Funding Acquisition, O.G.E. and S.E.H.

**Funding:** This project has been funded in whole or in part by the Federal funds from the U.S. Department of Health and Human Services, Office of the Assistant Secretary for Preparedness and Response, Biomedical Advanced Research and Development Authority, under Contract No. HHSO100201300005C.

**Acknowledgments:** The authors would like to thank Edward Mee for helpful discussions and assistance with Next-Generation Sequencing.

**Conflicts of Interest:** The authors declare no conflict of interest.

## References


1. Zhou, H.; Thompson, W.W.; Viboud, C.G.; Ringholz, C.M.; Cheng, P.Y.; Steiner, C.; Abedi, G.R.; Anderson, L.J.; Brammer, L.; Shay, D.K. Hospitalizations associated with influenza and respiratory syncytial virus in the United States, 1993–2008. *Clin. Infect. Dis.* **2012**, *54*, 1427–1436. [CrossRef] [PubMed]
2. Paul, G.W.; Schmier, J.K.; Kuehn, C.M.; Ryan, K.J.; Oxford, J. The burden of influenza B: A structured literature review. *Am. J. Public Health* **2013**, *103*, e43–e51. [CrossRef] [PubMed]
3. Heikkinen, T.; Ikonen, N.; Ziegler, T. Impact of influenza B lineage-level mismatch between trivalent seasonal influenza vaccines and circulating viruses, 1999–2012. *Clin. Infect. Dis.* **2014**, *59*, 1519–1524. [CrossRef] [PubMed]
4. Corti, D.; Cameroni, E.; Guarino, B.; Kallewaard, N.L.; Zhu, Q.; Lanzavecchia, A. Tackling influenza with broadly neutralizing antibodies. *Curr. Opin. Virol.* **2017**, *24*, 60–69. [CrossRef] [PubMed]
5. Skehel, J.J.; Wiley, D.C. Receptor binding and membrane fusion in virus entry: The influenza hemagglutinin. *Annu. Rev. Biochem.* **2000**, *69*, 531–569. [CrossRef] [PubMed]
6. Corti, D.; Voss, J.; Gamblin, S.J.; Codoni, G.; Macagno, A.; Jarrossay, D.; Vachieri, S.G.; Pinna, D.; Minola, A.; Vanzetta, F.; et al. A neutralizing antibody selected from plasma cells that binds to group 1 and group 2 influenza A hemagglutinins. *Science* **2011**, *333*, 850–856. [CrossRef] [PubMed]
7. Ekiert, D.C.; Bhabha, G.; Elsliger, M.A.; Friesen, R.H.; Jongeneelen, M.; Throsby, M.; Goudsmit, J.; Wilson, I.A. Antibody recognition of a highly conserved influenza virus epitope. *Science* **2009**, *324*, 246–251. [CrossRef] [PubMed]
8. Sui, J.; Hwang, W.C.; Perez, S.; Wei, G.; Aird, D.; Chen, L.M.; Santelli, E.; Stec, B.; Cadwell, G.; Ali, M.; et al. Structural and functional bases for broad-spectrum neutralization of avian and human influenza A viruses. *Nat. Struct. Mol. Biol.* **2009**, *16*, 265–273. [CrossRef] [PubMed]
9. Ekiert, D.C.; Kashyap, A.K.; Steel, J.; Rubrum, A.; Bhabha, G.; Khayat, R.; Lee, J.H.; Dillon, M.A.; O'Neil, R.E.; Faynboym, A.M.; et al. Cross-neutralization of influenza A viruses mediated by a single antibody loop. *Nature* **2012**, *489*, 526–532. [CrossRef] [PubMed]
10. Dreyfus, C.; Laursen, N.S.; Kwaks, T.; Zuijdgeest, D.; Khayat, R.; Ekiert, D.C.; Lee, J.H.; Metlagel, Z.; Bujny, M.V.; Jongeneelen, M.; et al. Highly conserved protective epitopes on influenza B viruses. *Science* **2012**, *337*, 1343–1348. [CrossRef] [PubMed]
11. Yasugi, M.; Kubota-Koketsu, R.; Yamashita, A.; Kawashita, N.; Du, A.; Sasaki, T.; Nishimura, M.; Misaki, R.; Kuhara, M.; Boonsathorn, N.; et al. Human monoclonal antibodies broadly neutralizing against influenza B virus. *PLoS Pathog.* **2013**, *9*, e1003150. [CrossRef] [PubMed]
12. Chai, N.; Swem, L.R.; Park, S.; Nakamura, G.; Chiang, N.; Estevez, A.; Fong, R.; Kamen, L.; Kho, E.; Reichelt, M.; et al. A broadly protective therapeutic antibody against influenza B virus with two mechanisms of action. *Nat. Commun.* **2017**, *8*, 14234. [CrossRef] [PubMed]
13. Zhou, T.; Xu, L.; Dey, B.; Hessell, A.J.; Van, R.D.; Xiang, S.H.; Yang, X.; Zhang, M.Y.; Zwick, M.B.; Arthos, J.; et al. Structural definition of a conserved neutralization epitope on HIV-1 gp120. *Nature* **2007**, *445*, 732–737. [CrossRef] [PubMed]
14. Hufton, S.E.; Risley, P.; Ball, C.R.; Major, D.; Engelhardt, O.G.; Poole, S. The breadth of cross sub-type neutralisation activity of a single domain antibody to influenza hemagglutinin can be increased by antibody valency. *PLoS ONE* **2014**, *9*, e103294. [CrossRef] [PubMed]
15. Wu, Y.; Jiang, S.; Ying, T. Single-Domain Antibodies as Therapeutics against Human Viral Diseases. *Front. Immunol.* **2017**, *8*, 1802. [CrossRef] [PubMed]
16. Gaiotto, T.; Hufton, S.E. Cross-Neutralising Nanobodies Bind to a Conserved Pocket in the Hemagglutinin Stem Region Identified Using Yeast Display and Deep Mutational Scanning. *PLoS ONE* **2016**, *11*, e0164296. [CrossRef] [PubMed]
17. Rota, P.A.; Wallis, T.R.; Harmon, M.W.; Rota, J.S.; Kendal, A.P.; Nerome, K. Cocirculation of two distinct evolutionary lineages of influenza type B virus since 1983. *Virology* **1990**, *175*, 59–68. [CrossRef]
18. Weir, J.P.; Gruber, M.F. An overview of the regulation of influenza vaccines in the United States. *Influenza Other Respir. Viruses* **2016**, *10*, 354–360. [CrossRef] [PubMed]
19. Minor, P.D. Assaying the Potency of Influenza Vaccines. *Vaccines* **2015**, *3*, 90–104. [CrossRef] [PubMed]
20. Gonzalez-Sapienza, G.; Rossotti, M.A.; Tabares-da, R.S. Single-Domain Antibodies as Versatile Affinity Reagents for Analytical and Diagnostic Applications. *Front. Immunol.* **2017**, *8*, 977. [CrossRef] [PubMed]

21. De Genst, E.; Silence, K.; Decanniere, K.; Conrath, K.; Loris, R.; Kinne, J.; Muyldermans, S.; Wyns, L. Molecular basis for the preferential cleft recognition by dromedary heavy-chain antibodies. *Proc. Natl. Acad. Sci. USA* **2006**, *103*, 4586–4591. [CrossRef] [PubMed]
22. Karlsson, R.; Katsamba, P.S.; Nordin, H.; Pol, E.; Myszka, D.G. Analyzing a kinetic titration series using affinity biosensors. *Anal. Biochem.* **2006**, *349*, 136–147. [CrossRef] [PubMed]
23. D'Angelo, S.; Glanville, J.; Ferrara, F.; Naranjo, L.; Gleasner, C.D.; Shen, X.; Bradbury, A.R.; Kiss, C. The antibody mining toolbox: An open source tool for the rapid analysis of antibody repertoires. *MAbs* **2014**, *6*, 160–172. [CrossRef] [PubMed]
24. Carnell, G.W.; Ferrara, F.; Grehan, K.; Thompson, C.P.; Temperton, N.J. Pseudotype-based neutralization assays for influenza: A systematic analysis. *Front. Immunol.* **2015**, *6*, 161. [CrossRef] [PubMed]
25. Temperton, N.J.; Hoschler, K.; Major, D.; Nicolson, C.; Manvell, R.; Hien, V.M.; Ha, D.Q.; de Jong, M.; Zambon, M.; Takeuchi, Y.; et al. A sensitive retroviral pseudotype assay for influenza H5N1-neutralizing antibodies. *Influenza Other Respir. Viruses* **2007**, *1*, 105–112. [CrossRef] [PubMed]
26. Ferrara, F.; Carnell, G.; Kinsley, R.; Bottcher-Friebertshauser, E.; Pohlmann, S.; Scott, S.; Fereidouni, S.; Corti, D.; Kellam, P.; Gilbert, S.; et al. Development and use of lentiviral Vectors Pseudotyped with Influenza B Haemagglutinins: Application to vaccine immunogenicity, mAb potency and sero-surveillance studies. *bioRxiv* **2018**. [CrossRef]
27. Ravn, U.; Didelot, G.; Venet, S.; Ng, K.T.; Gueneau, F.; Rousseau, F.; Calloud, S.; Kosco-Vilbois, M.; Fischer, N. Deep sequencing of phage display libraries to support antibody discovery. *Methods* **2013**, *60*, 99–110. [CrossRef] [PubMed]
28. Harmsen, M.M.; Ruuls, R.C.; Nijman, I.J.; Niewold, T.A.; Frenken, L.G.; de Geus, B. Llama heavy-chain V regions consist of at least four distinct subfamilies revealing novel sequence features. *Mol. Immunol.* **2000**, *37*, 579–590. [CrossRef]
29. Wang, Q.; Cheng, F.; Lu, M.; Tian, X.; Ma, J. Crystal structure of unliganded influenza B virus hemagglutinin. *J. Virol.* **2008**, *82*, 3011–3020. [CrossRef] [PubMed]
30. Ni, F.; Kondrashkina, E.; Wang, Q. Structural basis for the divergent evolution of influenza B virus hemagglutinin. *Virology* **2013**, *446*, 112–122. [CrossRef] [PubMed]
31. Stijlemans, B.; Conrath, K.; Cortez-Retamozo, V.; Van, X.H.; Wyns, L.; Senter, P.; Revets, H.; De Baetselier, P.; Muyldermans, S.; Magez, S. Efficient targeting of conserved cryptic epitopes of infectious agents by single domain antibodies. African trypanosomes as paradigm. *J. Biol. Chem.* **2004**, *279*, 1256–1261. [CrossRef] [PubMed]
32. Shen, J.; Vil, M.D.; Jimenez, X.; Iacolina, M.; Zhang, H.; Zhu, Z. Single variable domain-IgG fusion. A novel recombinant approach to Fc domain-containing bispecific antibodies. *J. Biol. Chem.* **2006**, *281*, 10706–10714. [CrossRef] [PubMed]
33. Coppieters, K.; Dreier, T.; Silence, K.; de Haard, H.; Lauwereys, M.; Casteels, P.; Beirnaert, E.; Jonckheere, H.; Van de Wiele, C.; Staelens, L.; et al. Formatted anti-tumor necrosis factor alpha VHH proteins derived from camelids show superior potency and targeting to inflamed joints in a murine model of collagen-induced arthritis. *Arthritis Rheum.* **2006**, *54*, 1856–1866. [CrossRef] [PubMed]
34. Balazs, A.B.; Bloom, J.D.; Hong, C.M.; Rao, D.S.; Baltimore, D. Broad protection against influenza infection by vectored immunoprophylaxis in mice. *Nat. Biotechnol.* **2013**, *31*, 647–652. [CrossRef] [PubMed]
35. Deal, C.E.; Balazs, A.B. Engineering humoral immunity as prophylaxis or therapy. *Curr. Opin. Immunol.* **2015**, *35*, 113–122. [CrossRef] [PubMed]
36. Ibanez, L.I.; De Filette, M.; Hultberg, A.; Verrips, T.; Temperton, N.; Weiss, R.A.; Vandevelde, W.; Schepens, B.; Vanlandschoot, P.; Saelens, X. Nanobodies with in vitro neutralizing activity protect mice against H5N1 influenza virus infection. *J. Infect. Dis.* **2011**, *203*, 1063–1072. [CrossRef] [PubMed]
37. Verma, S.; Soto, J.; Vasudevan, A.; Schmeisser, F.; Alvarado-Facundo, E.; Wang, W.; Weiss, C.D.; Weir, J.P. Determination of influenza B identity and potency in quadrivalent inactivated influenza vaccines using lineage-specific monoclonal antibodies. *PLoS ONE* **2017**, *12*, e0175733. [CrossRef] [PubMed]
38. Chao, G.; Cochran, J.R.; Wittrup, K.D. Fine epitope mapping of anti-epidermal growth factor receptor antibodies through random mutagenesis and yeast surface display. *J. Mol. Biol.* **2004**, *342*, 539–550. [CrossRef] [PubMed]

39. Angelini, A.; Chen, T.F.; de Picciotto, S.; Yang, N.J.; Tzeng, A.; Santos, M.S.; Van Deventer, J.A.; Traxlmayr, M.W.; Wittrup, K.D. Protein Engineering and Selection Using Yeast Surface Display. *Methods Mol. Biol.* **2015**, *1319*, 3–36. [CrossRef] [PubMed]
40. Nakagawa, T.; Higashi, N.; Nakagawa, N. Detection of antigenic variants of the influenza B virus by melting curve analysis with LCGreen. *J. Virol. Methods* **2008**, *148*, 296–299. [CrossRef] [PubMed]

*Article*

# Super Potent Bispecific Llama VHH Antibodies Neutralize HIV via a Combination of gp41 and gp120 Epitopes

**Nika M. Strokappe [1,2,†], Miriam Hock [3,4,†], Lucy Rutten [1,2,†], Laura E. Mccoy [5], Jaap W. Back [6], Christophe Caillat [3], Matthias Haffke [7,8], Robin A. Weiss [5], Winfried Weissenhorn [3] and Theo Verrips [1,2,*]**

1. Department of Biology, Faculty of Sciences, Utrecht University, 3584 CH Utrecht, The Netherlands; n.m.strokappe@gmail.com (N.M.S.); lucy.rutten@gmail.com (L.R.)
2. QVQ Holding bv, Yalelaan 1, 3584 CL Utrecht, The Netherlands
3. Institute de Biologie Structurale (IBS), CNRS, CEA, Université Grenoble Alpes, F-38000 Grenoble, France; Miriam.hock@googlemail.com (M.H.); christophe.caillat@ibs.fr (C.C.); winfried.weissenhorn@ibs.fr (W.W.)
4. Immunocore Ltd., 101 Park Drive, Milto, Abingdon OX14 4RY, UK
5. Division of Infection and Immunity, University College London, London WC1E 6BT, UK; l.mccoy@ucl.ac.uk (L.E.M.); r.weiss@ucl.ac.uk (R.A.W.)
6. Pepscan B.V., Zuidersluisweg 2, 8243 RC Lelystad, The Netherlands; jjaapwillemback@eurofins.com
7. European Molecular Biology Laboratory, Grenoble Outstation, 6 rue Jules Horowitz, 38042 Grenoble, France; Matthias.haffke@novartis.com
8. Global Discovery Chemistry, Novartis Institutes for BioMedical Research, Novartis Pharma AG, Novartis Campus, 4002 Basel, Switzerland

* Correspondence: theo.verrips@outlook.com

† These authors contributed equally.

Received: 17 March 2019; Accepted: 30 May 2019; Published: 18 June 2019

**Abstract:** Broad and potent neutralizing llama single domain antibodies (VHH) against HIV-1 targeting the CD4 binding site (CD4bs) have previously been isolated upon llama immunization. Here we describe the epitopes of three additional VHH groups selected from phage libraries. The 2E7 group binds to a new linear epitope in the first heptad repeat of gp41 that is only exposed in the fusion-intermediate conformation. The 1B5 group competes with co-receptor binding and the 1F10 group interacts with the crown of the gp120 V3 loop, occluded in native Env. We present biophysical and structural details on the 2E7 interaction with gp41. In order to further increase breadth and potency, we constructed bi-specific VHH. The combination of CD4bs VHH (J3/3E3) with 2E7 group VHH enhanced strain-specific neutralization with potencies up to 1400-fold higher than the mixture of the individual VHHs. Thus, these new bivalent VHH are potent new tools to develop therapeutic approaches or microbicide intervention.

**Keywords:** Aids; HIV; Llama Antibodies; bi-specific VHH; pepscan; competition studies; epitope mapping; co-crystallisation

## 1. Introduction

AIDS remains one of the largest global health problems and annually an estimated 1.8 million people die an AIDS related death. While antiretroviral therapy and careful clinical management can render HIV a chronic disease, the treatment is expensive and has many adverse effects [1]. Thus in the absence of a vaccine, prophylactic therapies to prevent HIV-1 infection are urgently needed. The encouraging results of an antiretroviral-based gel microbicide [2] suggest that microbicidal prevention methods merit investigation. Recently, many broad and potent HIV neutralizing monoclonal antibodies (mAb) have been discovered [3–11]. However, while monoclonal antibody (mAb)-based microbicides have been evaluated before [12], the application of this is limited by their expensive production and the need for cold distribution system. The variable domains of the heavy chain of heavy chain only antibodies (VHH) derived from llamas or other *Camelidae* may be better alternatives, as they can be produced relatively cheaply in microorganisms like bacteria or yeast [13] and are often stable at high temperatures [14]. This is predominantly related to their small size, which is a 10-fold smaller than that of a conventional antibody. Moreover, their small size and more than average length of CDR 3 allows them to bind to recessed epitopes, like the CD4 binding site (CD4bs) of HIV-1.

Previous immunizations of llamas (*Lama glama*) have generated anti-HIV-1 VHH specific for the CD4bs on gp120 or gp140 [15–17] with the most potent and broadest VHH, J3, neutralizing 96% of the HIV-1 strains tested from subtypes A, B, C, D, G and circulating recombinant forms AC, ACD, AE, AG, and BC [16]. A mix of these VHH together with J3 was tested in vitro and showed to be neutralizing as potently as any of the individual VHH, and a 100% coverage for the panel of 60 viruses tested was predicted. Similarly, combinations of antibodies that bind independent epitopes have also been shown to neutralize in vitro as efficiently mixed together as when used in isolation [18].

A number of broadly neutralizing VHH, with breadths up to 82%, targeting epitopes on the HIV envelope glycoprotein (Env) other than the CD4bs, e.g., the co-receptor binding site or gp41, have been isolated as well, but their epitopes had not been defined [17,19,20]. Some of these VHH are able to neutralize the few strains that are not neutralized by J3 or by 3E3, another VHH obtained from our immunized llamas, which neutralizes 80% of virus strains [21]. By using a mix of VHH targeting the CD4bs and VHH targeting other epitopes, broader neutralization is expected. An additional advantage of the use of a mix of VHH is that it reduces the chance of the emergence of escape mutants. This has been observed in studies in which a combination of some broadly neutralizing conventional human antibodies targeting independent epitopes were passively transferred to humanized mice [3,4]. Using covalently linked VHH often has an advantage over using monovalent VHH or a mix of VHH, as the bivalent VHH may have higher potencies than the constituent monovalent VHH, due to enhanced avidity [22,23]. Furthermore, HIV-1 neutralization potency can be enhanced by heteroligation of two distinct single-chain Fv (scFv), into bi-specific molecules [24]. In contrast, bi-specific VHH targeting the CD4bs as well as the co-receptor binding site did not show any increase in potency, but an extended breadth was confirmed. Nevertheless bi-specific VHH may have an advantage when used in gene therapy in a vectored-immunoprophylaxis (VIP) adeno-associated virus (AAV) [25], because the size of a bi-specific VHH genes do not exceed the maximal allowed size of the transgene insert.

Therefore, we aimed to design and produce bi-specific anti-HIV-1 VHH, which have neutralization abilities superior to those of the two best VHH, J3 and 3E3, regarding breadth and potency, for various applications. In order to design these molecules, i.e., to make the best combinations, we first determined the epitopes of several VHH. VHH that bound to epitopes other than the CD4bs were linked to J3 and 3E3 and most of these bi-specific VHH have improved breadth and potency against certain viral strains, compared to the equimolar mix of the constituent VHH.

## 2. Results

### 2.1. Competition-Based Determination of 4 Different Epitope Groups

Previously, we have selected over 100 different HIV-1 neutralizing VHH starting with various immune libraries (Most important data summarized in Figures 1 and 2) [15–17,21,26]. To be able to make bi-specific VHH, two VHH that target independent epitopes need to be linked together. Preliminary evidence indicated that a number of the selected VHH do not compete with J3 or 3E3 [19]. To determine the epitopes of the preselected VHH we performed a (cross-) competition assay in which seven of these VHH (3E3, J3, 1B5, 2E7, 1F10, 11F1F, 1H9) were tested for their competition against each other and against six other VHH that neutralize HIV-1 (Figure 3A). In the assay, Env was immobilized and a total of 13 neutralizing and an irrelevant VHH were added in excess to the plate.

Following incubation, seven biotinylated VHH of interest were added and their binding was detected via peroxidase-conjugated streptavidin (Figure 3A), showing the competition against gp140UG37 (and Table S1 showing the competition against gp120IIIB and gp140CN54). These competition assays enabled the clear division into four groups of VHH targeting different epitopes. The first group is comprised of phylogenetically diverse VHH (Figure 2) targeting the CD4bs, which includes J3, 3E3, A12, D7 and C8. The second group includes phylogenetically related 2E7 and 11F1F and unrelated VHH 11F1B and 1E2. The third group of VHH, that probably block the co-receptor binding site, is composed of phylogenetically related 1B5, 1H9 and 2B4F and the unrelated 4D4. The fourth group contains one VHH, 1F10, which does not fully compete with any of the other VHH, indicating that it binds to a separate epitope. To obtain more insight into the location of the different epitopes on Env, a competition assay was performed with these VHH versus b12 (a bnAb targeting the CD4 binding site), 17b (a bnAb targeting the co-receptor binding site), sCD4 and anti-gp41 MPER bnAbs (2F5, 4E10) were performed (Figure 3B). This revealed that group I targets the CD4 binding site, group II targets gp41 independent of the MPER epitope, group III targets the co-receptor binding site and group IV also seems to target the co-receptor site (Figure 3B). Since this competition was only performed in one way, differences in affinity between the VHH and the mAb may have led to false negatives. The epitopes recognized by the VHH were subsequently further characterized by pepscan analyses and structural studies.

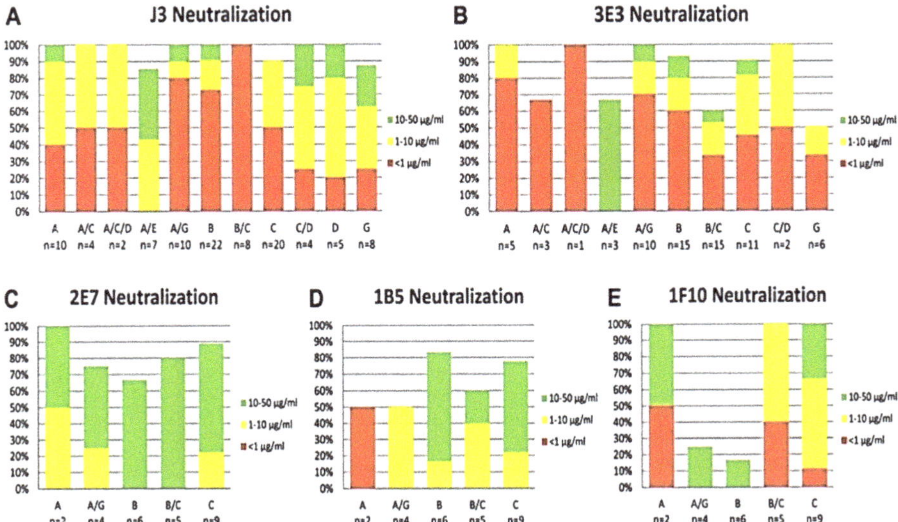


**Figure 1.** Clade Specific Neutralization of VHH. (**A**) J3, (**B**) 3E3, (**C**) 2E7, (**D**) 1B5 and (**E**) 1F10. The total neutralization per clade is shown by the height of the bar in the graph and the neutralization potency by the colors of the bar. Red indicates an IC50 < 1 µg/mL, yellow between 1 and 10 µg/mL and green between 10 and 50 µg/mL.

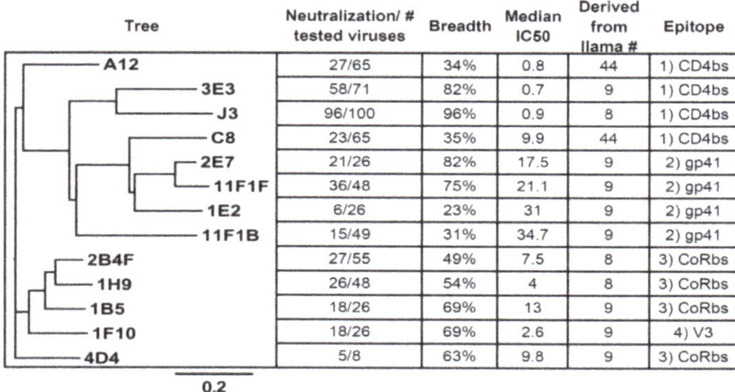

| Tree | Neutralization/ # tested viruses | Breadth | Median IC50 | Derived from llama # | Epitope |
|---|---|---|---|---|---|
| A12 | 27/65 | 34% | 0.8 | 44 | 1) CD4bs |
| 3E3 | 58/71 | 82% | 0.7 | 9 | 1) CD4bs |
| J3 | 96/100 | 96% | 0.9 | 8 | 1) CD4bs |
| C8 | 23/65 | 35% | 9.9 | 44 | 1) CD4bs |
| 2E7 | 21/26 | 82% | 17.5 | 9 | 2) gp41 |
| 11F1F | 36/48 | 75% | 21.1 | 9 | 2) gp41 |
| 1E2 | 6/26 | 23% | 31 | 9 | 2) gp41 |
| 11F1B | 15/49 | 31% | 34.7 | 9 | 2) gp41 |
| 2B4F | 27/55 | 49% | 7.5 | 8 | 3) CoRbs |
| 1H9 | 26/48 | 54% | 4 | 8 | 3) CoRbs |
| 1B5 | 18/26 | 69% | 13 | 9 | 3) CoRbs |
| 1F10 | 18/26 | 69% | 2.6 | 9 | 4) V3 |
| 4D4 | 5/8 | 63% | 9.8 | 9 | 3) CoRbs |



**Figure 2.** Phylogenetic Tree and Characteristics of the Selected VHH. VHH A12 (and D7) and C8 originate from an immunization with gp120 in llama 44, where the other VHH orginate from immunizations of llamas 8 and 9 with gp140 of UG037 and CN54. The neutralization (breadth), Medium IC50 values, epitopes (this study).

A

| | J3 | 3E3 | A12 | C | 2E | 11F1F | 11F1B | 1E | 1B5 | 1H9 | 2B4F | 4D4 | 1F1 | Irr. |
|---|---|---|---|---|---|---|---|---|---|---|---|---|---|---|
| J3 | 0 | -1 | 6 | 28 | 94 | 47 | 66 | 81 | 87 | 92 | 93 | 81 | 96 | 100 |
| 3E3 | 14 | 0 | 19 | 41 | 80 | 48 | 75 | 90 | 95 | 86 | 89 | 77 | 84 | 100 |
| 2E7 | 110 | 104 | 114 | 118 | 0 | 4 | 6 | 1 | 98 | 108 | 110 | 102 | 47 | 100 |
| 11F1F | 83 | 85 | 88 | 95 | 0 | 0 | 0 | 1 | 63 | 64 | 67 | 84 | 24 | 100 |
| 1B5 | 76 | 69 | 84 | 74 | 63 | 48 | 56 | 72 | 0 | 8 | -4 | -3 | 41 | 100 |
| 1H9 | 102 | 79 | 65 | 78 | 67 | 62 | 66 | 34 | 1 | 0 | 0 | 1 | 30 | 100 |
| 1F10 | 83 | 74 | 99 | 86 | 72 | 59 | 57 | 74 | 61 | 60 | 53 | 69 | 0 | 100 |

B

| | Binding | | | Competition | | | | |
|---|---|---|---|---|---|---|---|---|
| | UG37 gp140 | IIIB gp120 | IIIB gp41 | b12 | 17b | sCD4 | 2F5 | 4E10 |
| J3 | 1.3 | 5.4 | nb | 8 | 295 | 0 | nt | nt |
| 3E3 | 3.2 | 8.5 | nb | 2 | 430 | 0 | nt | nt |
| A12 | 6.2 | 5.0 | nb | 1 | 4 | 0 | nt | nt |
| C8 | nb | 3.7 | nb | 36 | 10 | 5 | nt | nt |
| 2E7 | 0.9 | nb | 3.2 | 87 | 79 | 112 | 74 | 85 |
| 11F1F | 1.9 | nb | 1.1 | 98 | 127 | 155 | 124 | 75 |
| 11F1B | 4.6 | nb | 4.8 | 95 | 131 | 148 | 121 | 73 |
| 1E2 | 14.4 | nb | 2.5 | 100 | 84 | 150 | 64 | 80 |
| 1B5 | 3.5 | 29.7 | nb | 122 | 4 | 127 | nt | nt |
| 1H9 | 51.4 | 152.0 | nb | 115 | 5 | 135 | nt | nt |
| 2B4F | 50.0 | 74.5 | nb | 118 | 5 | 176 | nt | nt |
| 4D4 | 100.6 | 58.4 | nb | 84 | 8 | 130 | nt | nt |
| 1F10 | 7.3 | nb | nb | 101 | 21 | 129 | nt | nt |
| Irr. | nb | nb | nb | 95 | 103 | 95 | 98 | 83 |

C

| | Binding | | | Competition | | | | |
|---|---|---|---|---|---|---|---|---|
| | UG37 gp140 | IIIB gp120 | IIIB gp41 | b12 | 17b | sCD4 | 2F5 | 4E10 |
| J3-15-2E7 | 0.8 | 7.2 | 1.6 | 7 | 126 | 0 | 79 | 83 |
| J3-15-11F1F | 2.5 | 11.5 | 4.5 | 5 | 203 | 0 | 80 | 84 |
| 3E3-15-2E7 | 4.4 | 39.0 | 4.9 | 1 | 405 | 0 | 99 | 96 |
| 3E3-15-11F1F | 0.8 | 13.3 | 2.4 | 2 | 427 | 0 | 75 | 84 |
| J3-15-1F10 | 16.1 | 14.3 | nb | 6 | 43 | 0 | nt | nt |
| 3E3-15-1F10 | 14.3 | 57.9 | nb | 2 | 194 | 0 | nt | nt |
| J3-15-1H9 | 9.2 | 14.6 | nb | 2 | 5 | 0 | nt | nt |
| 3E3-15-1H9 | 60.2 | 68.4 | nb | 1 | 41 | 0 | nt | nt |
| 2E7-15-1H9 | 17.9 | 130.1 | 36.4 | 87 | 9 | 85 | 95 | 82 |
| 11F1F-15-1H9 | 9.2 | 321.7 | 17.0 | 91 | 5 | 105 | 83 | 86 |
| Irr. | nb | nb | nb | 81 | 96 | 80 | 103 | 86 |

| Binding IC50 | Competition |
|---|---|
| <1 nM | > 125% |
| 1-10 nM | 75-125% |
| 10-100 nM | 25-75% |
| >100 nM | <25% |

**Figure 3.** Interaction of the selected VHH with HIV-1 Env proteins. (**A**) Is showing the VHH competition among themselves for binding to gp140UG37, this is presented as a percentage. Gp140UG37 trimers used in this study are non-native trimers that contain a substantial amount of "open" Env structures. The signal observed for VHH competing with themselves was defined as 0%, signal during competition with an irrelevant VHH was defined as 100%. The VHH represented in the columns are the competing VHH, which were present in large excess. The VHH represented in the rows are the detected VHH. (**B**,**C**) are showing the binding of VHH (**B**) and bi-valent VHH (**C**) to gp140UG37, gp120IIIB and gp41, and their ability to compete with several mAbs and sCD4. Binding is expressed as the IC50 value in nM, competition as a percentage of the signal obtained where no competing VHH was present.

### 2.1.1. Epitope Group I, VHH That Bind to CD4bs

This group consists of J3, 3E3, A12, C8, D7 and their family members. The VHH were obtained directly from the phage library or selective elution with sCD4, of phages carrying these VHH. Moreover, previous work on competition experiments of these VHH, with sCD4 and b12, pointed out that this group of VHH is targeting the CD4bs [19]. During these studies it became clear that J3 as well as 3E3 enhance binding of 17b by approximately 3- or 4-fold, respectively. Although J3 and 3E3 are phylogenetically not related and were selected from different llamas, they have several features in common. For example, they share the very rare feature that they lack three residues in the CDR2 relative to their respective germ line [27]. The reinsertion of these three residues in J3 or 3E3 abrogates the binding abilities of either VHH (data not shown and McCoy et al. [16]). Unlike e.g., VRC01, which also binds to the CD4bs, J3 and 3E3 do not bind to RSC3, a resurfaced gp120 molecule in which there are many alterations outside the structurally invariant part of the CD4bs, including in the bridging sheets [9]. These data suggest that that J3 and 3E3, like CD4, bind to the bridging sheets of gp120. Moreover, J3 in particular shows unusual degrees of maturation outside of its paratope, although less than bnAb selected from humans. The foot print of J3 and 3E3 were determined by modeling [27]. The most characteristic feature of the binding of J3 to gp120 is that its foot print is very similar to the foot print of CD4. It was predicted by the model that $Tyr99_{J3}$ occupies a position similar to $Phe43_{CD4}$ resulting in a similar hydrophobic interaction with gp120. Recently these interactions were confirmed by co-crystallization [manuscript in preparation]. The competition assay shows that sCD4 competes with J3, 3E3, A12 and C8 for binding to Env. However, A12 and C8 also compete with 17b, indicating that the interaction surface of A12 is also outside the CD4bs, which is less conserved than the CD4bs itself. This may be an explanation for the relatively low breadth of A12, which neutralized 42% of the viruses tested.

### 2.1.2. Epitope Group II Consisting of 2E7, 11F1F, 1E2 and 11F1B and Their Family Members

2E7 and 11F1F share 91% DNA and 89% protein sequence identity and neutralize 21 out of 26 (81%) and 36 out of 45 (76%) viruses tested, respectively. While the potencies of 2E7 and 11F1F (median IC50 of 19 and 21 μg/mL respectively) are much weaker than those of the CD4bs VHH 3E3 and J3, these VHH can neutralize some viral strains (e.g., Du172.17 and CAP45.2.00.G3) that are resistant to either 3E3 or J3 (Table S1). Based on the binding and competition experiments described above it is clear that 2E7, 11F1F, 1E2 and 11F1B bind to gp41 outside the MPER epitope as they don't compete with bnAbs 2F5 and 4E10. Furthermore, no competition was observed with gp41 Heptad Repeat 1 (HR1) Abs HK20 and 3D6, that target amino acids 535–581 [8] and the immunodominant area (amino acids 579–613) [28] of gp41, respectively. Inhibition of 2E7 and 11F1F binding was seen in the presence of antibody 246-D (epitope sequence 590-QQLLGIWG-597 with the epitope core being LLGI), suggesting that these VHH bind to the C-terminal part of the HR1 region of gp41, but not to the epitope of 3D6, which is 599-SGKLICTTA-607.

This region of gp41 has been characterized as being immunodominant, thus it is not surprising that immunization of llamas with recombinant gp140 yielded antibodies targeting this region. However, it is novel that the VHH antibodies elicited can not only neutralize HIV, but do so with a breadth ranging across many subtypes. Therefore, the epitopes of 2E7, 11F1F, 1E2 and 11F1B were investigated by measuring their binding to overlapping linear and cyclic 15-mer peptides covering gp160 proteins derived from subtype A, B, C and CRF BC viruses (strains UG037, HXB2, SF162, ZM96 and CN54). All VHH showed binding to a peptide containing the sequence AVERYLKDQQLLGIWG (corresponding to residues 582–597 in HXB2 numbering, data not shown). Therefore, we focused on the broadest VHH, 2E7 in the following analysis. A peptide containing the epitope of 2E7 was used as a seed for a library in which each amino acid was uniquely substituted with Ala, Arg, Glu, Phe, Gly or Trp to obtain a limited substitution mutagenesis scanning. Clear binding peaks were observed corresponding to the peptides indicated in Figure 4, and were largely consistent for all tested subtypes represented in the peptide sets. The consensus sequence for the 2E7 epitope is (I/V)ERYL(R/K)DQQL (583–592).

The limited substitution mutagenesis screening was in good agreement with the results obtained in the initial gp160 screening, showing that 2E7 is particularly sensitive to mutations in residues K/R588, D589, and L592.

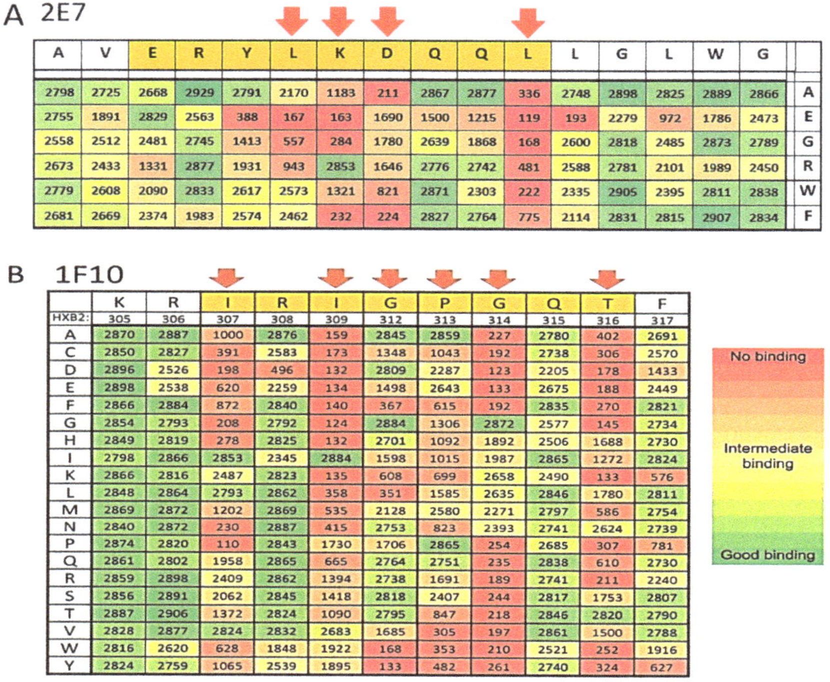


| A | V | E | R | Y | L | K | D | Q | Q | L | L | G | L | W | G | |
|---|---|---|---|---|---|---|---|---|---|---|---|---|---|---|---|---|
| 2798 | 2725 | 2668 | 2929 | 2791 | 2170 | 1183 | 211 | 2867 | 2877 | 336 | 2748 | 2898 | 2825 | 2889 | 2866 | A |
| 2755 | 1891 | 2829 | 2563 | 388 | 167 | 163 | 1690 | 1500 | 1215 | 119 | 193 | 2279 | 972 | 1786 | 2473 | E |
| 2558 | 2512 | 2481 | 2745 | 1413 | 557 | 284 | 1780 | 2639 | 1868 | 168 | 2600 | 2818 | 2485 | 2873 | 2789 | G |
| 2673 | 2433 | 1331 | 2877 | 1931 | 943 | 2853 | 1646 | 2776 | 2742 | 481 | 2588 | 2781 | 2101 | 1989 | 2450 | R |
| 2779 | 2608 | 2090 | 2833 | 2617 | 2573 | 1321 | 821 | 2871 | 2303 | 222 | 2335 | 2905 | 2395 | 2811 | 2838 | W |
| 2681 | 2669 | 2374 | 1983 | 2574 | 2462 | 232 | 224 | 2827 | 2764 | 775 | 2114 | 2831 | 2815 | 2907 | 2834 | F |

| | K | R | I | R | I | G | P | G | Q | T | F |
|---|---|---|---|---|---|---|---|---|---|---|---|
| HXB2: | 305 | 306 | 307 | 308 | 309 | 312 | 313 | 314 | 315 | 316 | 317 |
| A | 2870 | 2887 | 1000 | 2876 | 159 | 2845 | 2859 | 227 | 2780 | 402 | 2691 |
| C | 2850 | 2827 | 391 | 2583 | 173 | 1348 | 1043 | 192 | 2738 | 306 | 2570 |
| D | 2896 | 2526 | 198 | 496 | 132 | 2809 | 2287 | 123 | 2205 | 178 | 1433 |
| E | 2898 | 2538 | 620 | 2259 | 134 | 1498 | 2643 | 133 | 2675 | 188 | 2449 |
| F | 2866 | 2884 | 872 | 2840 | 140 | 367 | 615 | 192 | 2835 | 270 | 2821 |
| G | 2854 | 2793 | 208 | 2792 | 124 | 2884 | 1306 | 2872 | 2577 | 145 | 2734 |
| H | 2849 | 2819 | 278 | 2825 | 132 | 2701 | 1092 | 1892 | 2506 | 1688 | 2730 |
| I | 2798 | 2866 | 2853 | 2345 | 2884 | 1598 | 1015 | 1987 | 2865 | 1272 | 2824 |
| K | 2866 | 2816 | 2487 | 2823 | 135 | 608 | 699 | 2658 | 2490 | 133 | 576 |
| L | 2848 | 2864 | 2793 | 2862 | 358 | 351 | 1585 | 2635 | 2846 | 1780 | 2811 |
| M | 2869 | 2872 | 1202 | 2869 | 535 | 2128 | 2580 | 2271 | 2797 | 586 | 2754 |
| N | 2840 | 2872 | 230 | 2887 | 415 | 2753 | 823 | 2393 | 2741 | 2624 | 2739 |
| P | 2874 | 2820 | 110 | 2843 | 1730 | 1706 | 2865 | 254 | 2685 | 307 | 781 |
| Q | 2861 | 2802 | 1958 | 2865 | 665 | 2764 | 2751 | 235 | 2838 | 610 | 2730 |
| R | 2859 | 2898 | 2409 | 2862 | 1394 | 2738 | 1691 | 189 | 2741 | 211 | 2240 |
| S | 2856 | 2891 | 2062 | 2845 | 1418 | 2818 | 2407 | 244 | 2817 | 1753 | 2807 |
| T | 2887 | 2906 | 1372 | 2824 | 1090 | 2795 | 847 | 218 | 2846 | 2820 | 2790 |
| V | 2828 | 2877 | 2824 | 2832 | 2683 | 1685 | 305 | 197 | 2861 | 1500 | 2788 |
| W | 2816 | 2620 | 628 | 1848 | 1922 | 168 | 353 | 210 | 2521 | 252 | 1916 |
| Y | 2824 | 2759 | 1065 | 2539 | 1895 | 133 | 482 | 261 | 2740 | 324 | 627 |

**Figure 4.** Epitopes Determination of the VHH 2E7, 11F1F and 1F10. After initial pepscan analyses, the amino acids in the gp140 envelop protein that interact with these VHH have been fine mapped by limited substitution of the amino acids of the epitopes of 2E7 (**A**) and full substitution for 1F10 (**B**). The key residues of the interaction are indicated by red arrows.

Gp41 residues 582 to 596 adopt a helical conformation in the native gp140 structure and form the C-terminal part of the HR1 triple stranded coiled coil, which is hidden within the trimer interface [29]. We therefore tested binding of 2E7 to a fusion intermediate conformation of gp41 ($gp41_{int}$) that contains part of HR1 including the 2E7 epitope fused in frame to the pIIGCN4 triple stranded coiled coil [30]. SPR measurements revealed a $K_D$ of 0.592 nM (k(on): $9.52 \times 10^5$ (1/Ms); k(off): $5.64 \times 10^4$ (1/s) (Figure 5A) corroborating the interaction of 2E7 with activated gp41 that has the HR1 coiled coil exposed similar to the mode of action described for HR1 antibodies D5 and HK20 [31,32]. We next solved the crystal structure of 2E7 in complex with the peptide 582-AVERYLKDQQLLGIW-596 to a resolution of 2.9 Å (Table 1). The structure shows that the VHH interacts with the gp41 peptide in an unusual way. The gp41 helix packs lateral to one side of the VHH beta sheet (Figure 5B). The major contacts are hydrogen bonds between gp41 D589 and the hydroxyl and backbone amide of the CDR 2 residue T53 as well as gp41 W596 and framework residue T50. A salt bridge from gp41 K588 to CDR1 D32 (Figure 5B) and hydrophobic contacts of gp41 L592 to CDR1 A33 and CDR2 I51, and gp41 I595 and W596 to framework P47 further stabilize the interaction. Cα super-positioning of the 2E7-gp41 peptide structure and the native gp140 structure confirms that 2E7 would not be able to access the HR1 epitope which is hidden in the native Env conformation (Figure 6). We conclude that 2E7 targets the

HR1 coiled coil, which is temporarily exposed during membrane fusion before refolding into the six helical bundle post fusion conformation [33].

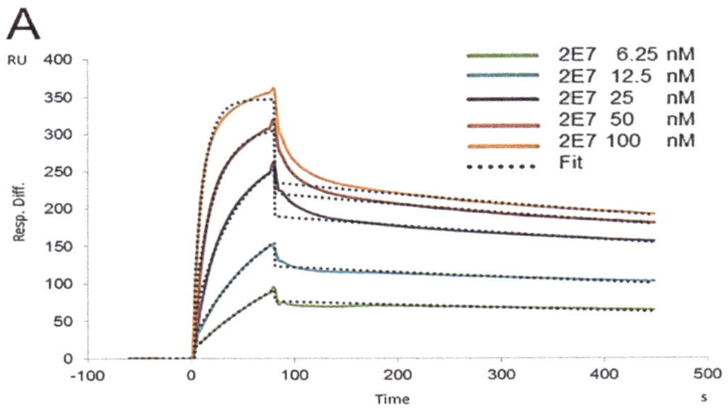


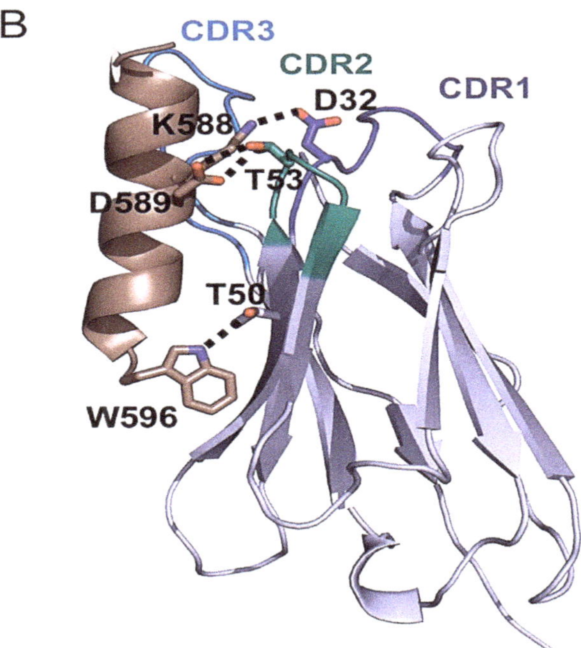


**Figure 5.** Structural analysis of 2E7 in complex with its gp41 HR1 epitope. (**A**) SPR analysis of VHH 2E7 with gp41int. 2E7 concentrations ranging from 6.25 to 100 nM were tested for binding to gp41int. Kinetic constants were derived by global fitting the data corresponding to the five indicated concentrations to a 1:1 Langmuir model (dotted lines) using local Rmax parameters. (**B**)Ribbon diagram of the crystal structure of the 2E7 VHH in complex with a gp41 HR1 peptide. The gp41 peptide is shown in wheat and the VHH CDR regions are indicated by different colors. Polar interactions are shown as dashed lines.

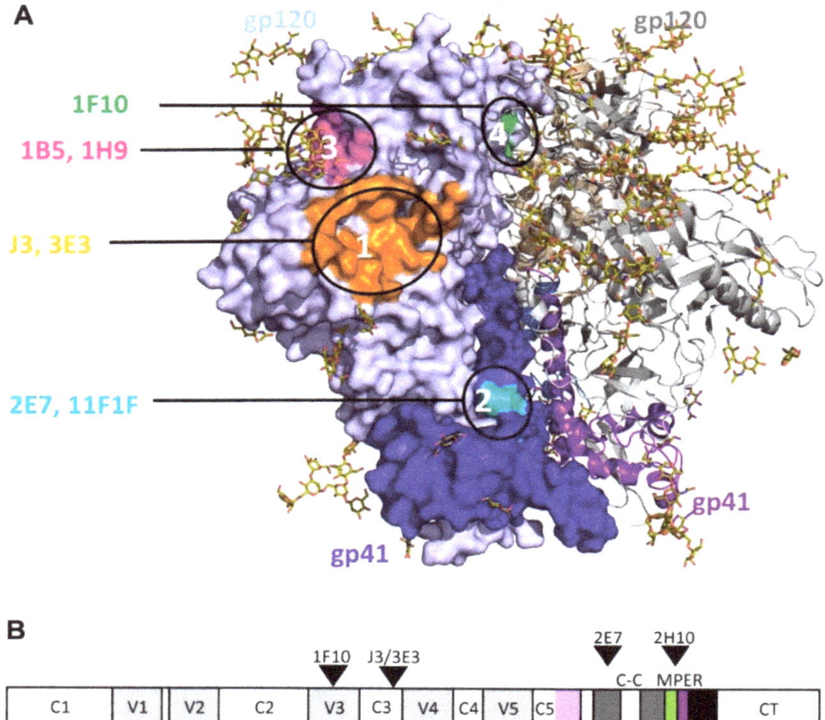


**Figure 6.** Mapping of the VHH epitopes onto the HIV-1 BG505 trimer (PDB ID 4TVP). (**A**) One monomer of the trimer is shown as a space filling model, carbohydrates are shown as sticks. The positions of the different VHH epitopes determined by crystallography and/or pepscan mapping are show in different colors. The distances between the epitopes were estimated and used to determine the linker length between VHH recognizing CD4bs (J3 and 3E3 and the epitopes recognized by the other VHH in the construction of a set of bi-specific VHH. The gp41 residues contacted by 2E7 are shown in Cyan, 1F10 in green. The 2E7 epitope located on gp41 HR1 is hidden in the trimer of the native Env gp140 conformation and not accessible for 2E7 binding. (**B**) Schematic of the domain organization of Env and location of the epitopes.

**Table 1.** Data collection and refinement statistics of 2E7GP41.

| | 2E7GP41 | |
|---|---|---|
| **Data Collection** | **Native** | **Anisotropic Scaling **** |
| Wavelength (Å) | 0.97239 | |
| Space group | $P22_12_1$ | |
| Cell dimensions *a*, *b*, *c* (Å) | 37.95, 121.26, 132.21 | |
| Resolution (Å) * | 44.68–2.95 (3.03–2.95) | 44.68–2.95 (3.03–2.95) |
| Unique reflections * | 13440 (993) | 12538 (327) |
| $R_{merge}$ (%) | 8.4 (76.4) | 7.9 (34.2) |
| $I/\sigma I$ * | 13.8 (2.3) | 14.6 (3.9) |
| Completeness (%) * | 99.0 (100.00) | 96.6 (34.5) |
| Redundancy * | 4.7 (5.0) | 4.3 (1.3) |

**Table 1.** *Cont.*

| | 2E7GP41 | |
|---|---|---|
| **Data Collection** | **Native** | **Anisotropic Scaling** ** |
| Wilson *B* factor (Å$^2$) | 63.85 | 58.0 |
| **Refinement** | | |
| Resolution | | 44.68–2.96 (3.07–2.96) |
| $R_{work}/R_{free}$ (%) * | | 19.42 (30.07)/24.91 (34.4) |
| **No. atoms** | | |
| Protein | | 3087 |
| Water | | 0 |
| ***B* factors (Å$^2$)** | | |
| Protein | | 55.4 |
| Water | | 0 |
| **r.m.s. deviations** | | |
| Bond lengths (Å) | | 0.006 |
| Bond angles (°) | | 1.26 |
| **Ramachandran** | | |
| Favored (%) | | 99.0 |
| Outliers (%) | | 0.0 |
| Clashscore *** | | 3.78 (100th percentile) |
| Molprobity score *** | | 1.47 (100th percentile) |

* Values in parentheses refer to highest resolution shell; ** The data were truncated according to the Diffraction Anisotropy Server of UCLA to 2.95 Å, 3.1 Å and 2.95 Å along *a*, *b*, *c*, respectively. M. Strong, M.R. Sawaya, S. Wang, M. Phillips, D. Cascio, D. Eisenberg, Proc. Natl. Acad. Sci. USA. 103, 8060-8-65, 2006. *** Percentiles are indicated for resolution range 2.962 Å ± 0.25 Å based on analysis with Molprobity.

#### 2.1.3. Epitope Group III Consisting of 1B5 and 1H9 and Family Members Recognize Part of the Co-Receptor Binding Site

Figure 3A shows that VHH 1B5 and 1H9 show a similar cross-competition pattern, moreover they compete with each other, indicating that they target a similar epitope. They compete with 17b, but not with b12 or sCD4 (Figure 3B). 1B5 and 1H9 were subjected to pepscan analysis, but neither bound to any of the peptides of the arrays of overlapping linear and cyclic 15-mer peptides covering gp160 proteins derived from various subtypes (data not shown). Escape mutant studies indicated that residues P417 and R419 [34] are involved in the interaction of 1B5 with gp120. We this suggest that group 3 VHH target the coreceptor site largely based on the competition assays (Figure 3B). The proposed locations of the 1B5 and 1H9 epitopes are shown in Figure 6, in which the residues P417 and R419 are shown in magenta.

#### 2.1.4. Epitope Group IV Consisting of 1F10 Binds to the Crown of the V3 Loop

1F10 neutralizes 18 out of the 26 viruses tested (69%). Based on the competition experiments it is clear that 1F10 competes with 17b, but not with sCD4 or b12. It competes marginally with all the other VHHs except for the CD4bs targeting VHH J3 and 3E3 (Figure 3 and Table S1). Furthermore, 1F10 does not compete with CD4bs antibody HJ16 either [8], supporting that this VHH binds a non-CD4bs epitope. However, 1F10 does compete with HGN194, a neutralizing Ab which binds to the crown of the V3 loop. Pepscan analysis with 1F10 was performed on subtype A, C and CRF BC viruses (strains UG037, ZM96 and CN54). Clear binding peaks were observed for peptides derived from the V3 region and were largely consistent between the three subtypes. (data not shown). The consensus sequence for the 1F10 epitope is IRIGPGQT (307–314 according to the HXB2 numbering) which overlaps with the HGN194 epitope RRSVRIGPGQTF (304–315). Single amino acid full replacement analyses were performed using a cyclic peptide (CRSVRIGPGQTFYAC) and two linear peptides (KRIRIGPGQTFY and KSINIGPGRAFA), each containing the sequence region recognized by 1F10. The replacement analysis identified the core epitope of 1F10 as IxIGPGxT (Figure 4B). The epitope of 1F10 is depicted in Figure 6A, where it is highlighted in green.

### 2.2. Construction of Bispecific anti-HIV VHH

Of the four CD4bs (group 1) VHH, J3 and 3E3 have the broadest and most potent HIV neutralization ability (Table S1). Thus, these two VHH were chosen for the construction of bi-specific molecules. VHH targeting distinct epitopes were paired with either 3E3 or J3 to maximize breadth. From group 2, VHH 2E7 and 11F1F were chosen and 1H9 is representing group 3 as it neutralizes CAP45.2.00.G3, a strain not neutralized by J3, and because its median IC50 is lower than that of 1B5 (5 μg/mL versus 13 μg/mL). The sole representative of group 4, VHH 1F10 (targeting the V3 loop), was also chosen as candidate for bi-specific VHH as it is broad and neutralizes at least two of the viruses (e.g., Du172.17 and CAP45.2.00.G3) J3 does not [16,17]. An overview of the neutralization $IC_{50}$ values for these VHH is given in supplementary Table S1.

VHH targeting the CD4bs were joined with VHH targeting the HR1 of gp41, the co-receptor binding site or the V3 loop (Figure 6) by a flexible 15 or 25 amino acid glycine-serine (GGGGS) repeat linker to form bi-specific VHH molecules. As the 15 and 25 linkers did not show a significant difference in their functionality (data not shown), the 15 linkers were chosen for further investigation. The bi-specific VHH with the CD4bs VHH on the N-terminal side seem to behave superior to those with the CD4bs VHH on the C-terminal side (data not shown). The bi-specific VHH bind to all tested Env proteins if either constituent VHH is able to bind (Figure 3C).

In general, the ability of the bi-specific VHH to compete with human bnAbs is equal to the combined competing abilities of both monomeric components. However, for bi-specific constructs containing 1F10, the ability to compete with 17b is reduced (J3-1F10) or turned into enhanced binding (3E3-1F10). This may be due to conformational changes induced by J3 and 3E3, as both individual VHH enhance 17b binding. Reduced competition with 17b is also seen with 3E3-1H9.

### 2.3. Broad and Potent HIV Neutralization by bi-Specific VHH Targeting a Combination of gp120 and gp41 Epitopes

Preliminary neutralization experiments indicated that the bi-specific constructs containing 1H9 did not yield a large improvement of breadth or potency and thus it was not characterized further. The remaining bi-specific VHH were tested for the ability to neutralize viruses resistant to either J3 or 3E3 to test whether the breadth of the bivalent construct was higher than that of J3 and 3E3 alone. Du172.17, a subtype C tier 3 virus, is not neutralized by J3, but is neutralized by all three J3-containing bi-specific VHH with IC50 values equal or lower than those of 1F10, 2E7 or 11F1F alone (Figure 7A). The same holds true for TV1.2, this subtype C tier 2 virus, is resistant to 3E3, but is neutralized by all three 3E3-containing bi-specific VHH or their mixes (Figure 7B). However, for some combinations the mixed monomers appear to be more potent than the linked constructs, therefore an additional six viruses from different subtypes and tiers were tested (Figure 7C). These viruses were in some cases susceptible to neutralization by the mix of both VHH constituting the bi-specific VHH, however, neutralization at least as potent as that of the most potent component VHH was seen for both linked and unlinked VHH mixtures in all cases (data not shown). Dependent on the HIV strain, some bi-specific VHH were substantially more potent than the equimolar mix of the constituent VHH. The bi-specific VHH containing 1F10 did not show greatly enhanced potencies compared to the equimolar mixes of the constituent VHH for any of the strains tested, whereas the bi-specific VHH with 2E7 and 11F1F have greatly enhanced potencies up to a 1400-fold, especially on the two C-clade viruses. For the subtype A and B viruses, approximately equivalent potency was seen for all bi-specific VHH compared to their respective unlinked equimolar mixtures, only J3-11F1F showed an increase in potency of a factor 6.3 against MS208.A1 and up to a 5-fold potency increase for Bal.26. In contrast, large increases in potency for the bi-specific VHH relative to the unlinked equimolar mixes were observed against the two subtype C viruses. A particularly large potency increase was seen for 3E3/J3-11F1F and J3-2E7 against the tier 2 ZM214M.PL15 virus and 3E3/J3-11F1F against the tier 1 96ZM651.02 virus. We conclude that the increase in breadth and potency was most efficient against clade C with a combination of anti-CD4bs VHH and anti-gp41 VHH.

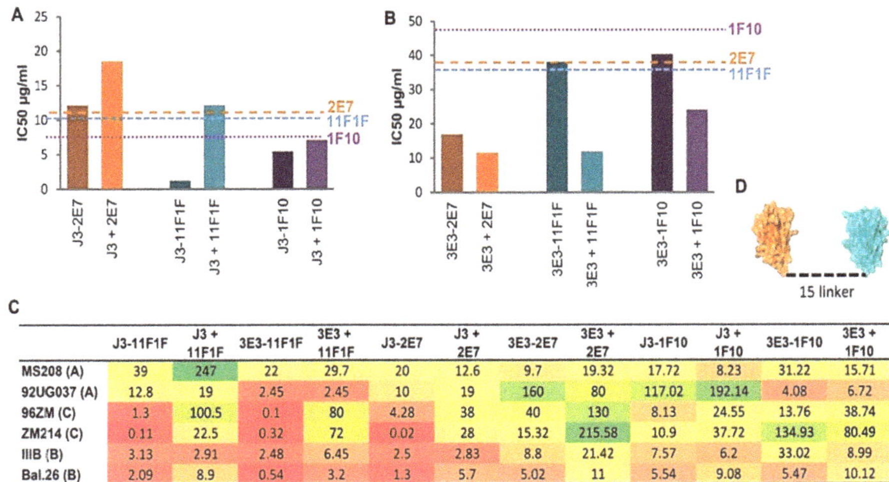


| | J3-11F1F | J3 + 11F1F | 3E3-11F1F | 3E3 + 11F1F | J3-2E7 | J3 + 2E7 | 3E3-2E7 | 3E3 + 2E7 | J3-1F10 | J3 + 1F10 | 3E3-1F10 | 3E3 + 1F10 |
|---|---|---|---|---|---|---|---|---|---|---|---|---|
| MS208 (A) | 39 | 247 | 22 | 29.7 | 20 | 12.6 | 9.7 | 19.32 | 17.72 | 8.23 | 31.22 | 15.71 |
| 92UG037 (A) | 12.8 | 19 | 2.45 | 2.45 | 10 | 19 | 160 | 80 | 117.02 | 192.14 | 4.08 | 6.72 |
| 96ZM (C) | 1.3 | 100.5 | 0.1 | 80 | 4.28 | 38 | 40 | 130 | 8.13 | 24.55 | 13.76 | 38.74 |
| ZM214 (C) | 0.11 | 22.5 | 0.32 | 72 | 0.02 | 28 | 15.32 | 215.58 | 10.9 | 37.72 | 134.93 | 80.49 |
| IIIB (B) | 3.13 | 2.91 | 2.48 | 6.45 | 2.5 | 2.83 | 8.8 | 21.42 | 7.57 | 6.2 | 33.02 | 8.99 |
| Bal.26 (B) | 2.09 | 8.9 | 0.54 | 3.2 | 1.3 | 5.7 | 5.02 | 11 | 5.54 | 9.08 | 5.47 | 10.12 |

**Figure 7.** Broad and Potent HIV Neutralization by bi-specific VHH. (**A**,**B**) IC50 values in µg/ml for the indicated bi-specific VHH in comparison with the unlinked component VHH against viruses ((**A**) Du172.17 or (**B**) TV1.2) that were resistant to one of the VHH. Dotted lines represent the IC50 (µg/mL) for each component VHH and are color-coded in line with the legend. (**C**) IC50 nM for the indicated bi-specific VHH and the unlinked component VHH against the viruses indicated. IC50 values were generated from duplicate titrations of VHH onto TZM-bl cells as described in the materials and methods. Schematic representations of J3-2E7 bi-specific VHH (**D**).

## 3. Discussion

To broaden neutralization capacities, but also to reduce the chance of the emergence of escape mutants, targeting two independent epitopes on Env is likely to be beneficial [35]. Recently, bi-specific antibodies, targeting amongst others, the CD4bs [36,37] and MPER [38,39] showed broad and potent neutralization. Furthermore, fusions of CD4 and mAb 17b revealed synergetic effects [36,40] as well as VHH fusions revealed important synergistic effects [22,40,41]. To determine optimal combinations for the construction of bi-specific VHH we first determined and characterized a number of new VHH, which could be combined with the most potent neutralizing VHH J3 or 3E3 [16,21]. 13 new VHH were classified into four groups recognizing non-overlapping epitopes. Only for 1F10 was some competition seen with the epitopes of other VHH (1B5/1H9 and 2E7/11F1F). This competition is most likely due to steric hindrance or because of conformational changes occurring after binding of one of the VHH, preventing the binding of the other.

To determine the epitopes of the VHH binding to the three epitope groups other than the CD4bs in more detail, we subjected the VHH of these groups to pepscan analysis. 2E7, as a representative of epitope group 2, binds to gp41. It binds to the linear sequence (I/V)ERYL(R/K)DQQLLG(L/I)W at position 583–596 according to the HXB2 numbering. Binding to the predicted gp41 epitope was confirmed by the crystal structure of 2E7 in complex with a peptide of this epitope. The epitope is part of the C-terminal part of the HR1 coiled coil that stabilizes the trimer interface in the native gp140 conformation [29,42,43]. In this native like conformation, the epitope is inaccessible to 2E7 due to steric clashes. Instead we show that 2E7 binds with high nanomolar affinity to the fusion intermediate conformation of gp41 that bridges the viral and cellular membranes during the initial step of membrane fusion [33] and is occluded by HR2 in the gp41 post fusion conformation [44]. Its mode of action is thus similar to HR1 mAbs D5, HK20 and 8066 whose epitope is located at the N-terminal part of HR1 [31,32,45]. Although these mAbs neutralize HIV-1, their breadth and potency is largely increased when they are used as smaller Fabs or single chain antibodies [31,32,45], indicating that the site is difficult to access for a conventional antibody during the fusion reaction. It therefore remains to be tested whether 2E7 has the same breadth and potency as a complete Fc containing antibody.

1B5 and 1H9, (epitope group 3) compete with 17b, an Ab that targets the co-receptor site. This in combination with preliminary data of co-crystallization experiments [manuscript in preparation] suggests an epitope overlapping with the co-receptor binding site.

High resolution pepscan analysis revealed that 1F10 (epitope group 4) binds to the sequence IRIGPGQT (HXB2 position 307–314) in the crown of the V3 loop, an epitope also targeted by the human mAbs 447-52D and HGN194. The latter is able to prevent the mucosal transmission of a subtype C SHIV when passively transferred [46]. The VHH described here thus target the following four distinct epitopes: the CD4bs, cluster1/HR1 of gp41, the co-receptor binding site or the V3 loop. All of the VHH epitopes are depicted in Figure 6A.

Bi-specific antibodies are currently investigated for a number of purposes and in general the bi-specific antibodies are performing quite well in vitro as is demonstrated by Asokan et al. [36]. In particular the construct VRC07-PG9-16 performed very well by neutralizing up 97% of the viruses tested with a median potency of 0.055 μg/mL. However, the situation in vivo is more complicated as there is a reasonably high chance that the bi-specific antibody initiates an immune response. Results of clinical trials of Ablynx show that VHH have a low risk for triggering immune responses in humans, because of their physical-chemical properties, the folding of CDR1 and 2 and their small size.

We constructed bi-specific VHH, containing either J3 or 3E3 (epitope group 1) in combination with a VHH that binds to an independent epitope. The constructs containing 1H9, did not show enhanced neutralization compared to the mix of the constituent VHH in preliminary experiments, so therefore they were not studied further. This observation agrees well with the fact that other bi-specific VHH targeting the CD4bs and the co-receptor binding site simultaneously did not show enhanced potency [20]. The increases in potency of the bi-specific VHH containing 1F10 were not spectacular for the viruses that were neutralized by either constituent VHH. This suggests that 1F10 hardly binds simultaneously with J3 or 3E3. The bi-specific VHH with the CD4bs VHH on the N-terminal side neutralize superior to those with the CD4bs VHH on the C-terminal side. The largest improvements in potencies, up to 1400-fold, are obtained with the constructs containing the gp41 targeting 2E7 or 11F1F as counterparts. The improvements are highest towards the viruses 96ZM651.02 and ZM214, both C-clade viruses, whereas towards the viruses from other clades, the improvements are less than 10-fold. Unexpectedly, J3-11F1F as well as 3E3-11F1F showed more than 200-fold increased potencies towards ZM214, whereas monomeric 11F1F was unable to neutralize this virus at a concentration of less than 1 μM. A plausible explanation for this may be that the binding of J3 or 3E3 causes conformational changes that allow 11F1F to bind or enhance 11F1F binding, confirming an epitope present on the intermediate conformation of HR1.

## 4. Methods

### *4.1. Materials and Methods*

#### 4.1.1. Proteins

Monoclonal antibodies b12 (EVA3065, by Dr D. P. Burton [47], 17b (ARP3071 by Dr J. Robinson) [48], 2F5 (EVA 3063, by Dr H. Katinger), 4E10 (ARP3239, by Polymun), and L120 ARP359, by Becton Dickinson) the recombinant proteins gp120IIIB (EVA607, by ImmunoDiagnostics), gp140UG37 (ARP698, by Polymun), gp140CN54 (ARP699 by Polymun), gp41 (ARP680) and human soluble CD4 (EVA609, by ImmunoDiagnostics) were obtained through the Centralized Facility for AIDS Reagents (CFAR), the National Institute for Biological Standards and Controls (NIBSC).

4.1.2. Viruses

Replication competent virus stocks were prepared from HIV-1 molecular clones by transfection of 293T cells. HIV-1 envelope pseudotyped viruses were produced in 293T cells by co-transfection with the pSG3Δenv plasmid. The subtype B and C HIV-1 Reference Panels of Env Clones [49,50] were obtained through the NIH AIDS Research and Reference Reagent Program, Division of AIDS, NIAID, NIH, USA. The 96ZM651.02 gp160 clone was kindly provided by Dr D. Montefiori (Duke University Medical Center, Durham, NC) through the Comprehensive Antibody Vaccine Immune Monitoring Consortium (CA2 VIMC) as part of the Collaboration for AIDS Vaccine Discovery (CAVD). All additional pseudoviruses were produced at the VIMC laboratory.

4.1.3. Cells

TZM-bl cells [49,51,52] were obtained through the NIH AIDS Research and Reference Reagent Program from J. C. Kappes, X. Wu, and Tranzyme, Inc., and cultured in Dulbecco's modified Eagle medium (Invitrogen) containing 10% (*v*/*v*) fetal calf serum (FCS).

4.1.4. VHH

A12, C8 and D7 were described by [15], and the 3D structure of D7 (highly homologous to A12) has been determined by Hinz et al. [53] A12 has been analyzed with EM tomography in complex with trimeric spikes [54]. 2E7, 1F10 and 1B5 are described in reference [17]. Selection of 1H9, 2B4F, 11F1F, 11F1B were not described earlier in detail (manuscript in preparation), J3 and 3E3 have previously been described [16,17,21]. VHH were purified as described previously [16,17].

### *4.2. Cross-Competition Assay*

To be able to detect only one of the two VHH that are present during this assay, part of them had to be biotinylated. NHS-LC-LC-biotin (Thermo scientific, Cat. No: 21343), was incubated 10:1 with VHH, 1 h, RT. Unbound biotin was removed by dialysis. Biotinylated VHH (bVHH) were titrated.

MaxiSorp plates were coated with 100 ng/well gp140UG37 or gp140CN54 (for J3 250 ng gp140CN54) or 250 ng for gp120IIIB 250. After blocking, competing (non-biotinylated) VHH was added for 1 h. As control, binding of all competing VHH was detected separately. Subsequently 10 μL b-VHH was added to the competing VHH in concentration determined previously. B-VHH was detected by horseradish peroxidase (HRP) conjugated Streptavidin and visualized by o-Phenylenediamine, supplemented with 0.03% $H_2O_2$. Reaction was stopped using 1 M $H_2SO_4$ and signals measured at 490 nm. These values were then converted to percentages, in which competition with itself was regarded as maximal competition, i.e., 0%, and the competition against an irrelevant VHH as the unhindered binding, i.e., 100%.

### *4.3. Binding to Various Env Proteins*

MaxiSorp plates were coated o/n with 100 ng gp140UG37, gp120IIIB or gp41. After blocking VHH (mono or bivalent) were added and detected by mouse anti Myc (9E10) and peroxidase conjugated donkey anti mouse (DAMPO). Visualization occurred as stated above.

### *4.4. Competition Assay with mAbs and sCD4*

MaxiSorp plates were coated o/n with 100 ng gp140UG37 (for b12, 17b, sCD4 and partially 2F5 competition) or gp41 (4E10 and partially 2F5 competition). After blocking, VHH (mono or bivalent) were added and allowed to bind for 1 h, RT. 10 μL of competitor was added to the wells in a final concentration of 0.4 μg/mL b12, 1 μg/mL 17b, 0.07 μg/mL 4E10, 0.07 μg/mL 2F5 (gp140UG37) or 0.6 μg/mL 2F5 (gp41). Competitors mAbs b12, 17b, 2F5 and 4E10 were detected with peroxidase conjugated goat anti human (Jackson Immunoresearch), sCD4 by L120 and peroxidase conjugated donkey anti mouse. Visualization occurred as stated above.

For the titrated competition assays, MaxiSorp plates were coated overnight at 4 °C with 200 ng per well gp140UG37 (for 3E3, 1H9 and 1F10 competition) or gp41 (2E7 and 11F1F competition). Following blocking, the binding of HRP conjugated streptavidin to biotinylated VHH was detected using TMB-ELISA substrate (Pierce).

### 4.5. Construction of bi-Specific VHH

Essentially the procedure described by Hultberg 2011 has been followed, in short N- and C-terminal VHH fragments were amplified from their expression vectors, by PCR using DreamTaq green (Fermentas). Primers were used that annealed just outside of the VHH coding region and would add a linker and restriction sites to allow cloning of the N and C terminal VHH respectively. The resulting PCR products were purified using the NucleoSpin® Extract II kit (Machery-Nagel, Düren, Germany), restricted with the appropriate enzymes and cloned into the VHH expression vector. Bacteria were transformed with the constructs by heatshock and subsequently clones were picked for sequence analysis. Bi-specific VHH were expressed in bacteria and purified by metal affinity chromatography in the same way as the monovalent VHH, which has been described before [16,17].

### 4.6. HIV Neutralization Assay

The HIV-1 neutralizing activities of the VHH were assessed in the TZM-bl cell based assay, as described previously [15–17]. No virus inactivation was observed with a negative control VHH. Cells were lysed with Bright-Glo luciferase reagent (Promega, Madison, WI, USA) and the luminescence measured. Fifty % inhibitory concentration (IC50) titers were calculated using the XLFit4 software (ID Business Solutions, Guildford, UK).

### 4.7. Epitope Mapping

The binding of VHH to arrays of peptides was assessed in a pepscan-based ELISA. Each well in the card contained covalently linked peptides that were incubated overnight at 4 °C with VHH, at a concentration of approximately 1 µg/mL, in PBS supplemented with 1% BSA and 0.1% Tween 80. After washing, the plates were incubated with a mouse anti-Histidine followed by HRP linked Rabbit anti-mouse (Southern Biotech, Birmingham, AL, USA) for 1 h at 25 °C. After further washing, peroxidase activity was assessed with an ABTS based substrate. The color development was quantified after 60 min using a charge-coupled device camera and an image-processing system.

### 4.8. Crystal Structure of VHH 2E7 in Complex with a gp41 Peptide

Purified VHH 2E7 was incubated with a 1.5 M excess of the gp41 peptide 582-AVERYLKD QQLLGIW-596 and crystallized at a concentration of 2.5 mg/mL. Crystals were obtained by the vapor diffusion method in sitting drops, with equal volumes of protein and reservoir solution (in 0.1M Hepes pH 7.0, 30% PEG 6000). The crystal was cryo-cooled at 100 K in absence of surrounding liquid [55]. A complete dataset was collected at the ESRF (Grenoble, France) beamline ID29. Data were processed and scaled with MOSFLM [56], and SCALA [57]. The crystals belong to space group $P2_12_12_1$ with unit cell dimensions of a = 37.95 Å, b = 121.26 Å, c = 132.21 Å and three copies of the 2E7-gp41 peptide complex. The structure was solved by molecular replacement using PHASER [58] and the VHH structure from Protein Data Bank (PDB) ID 3EZJ as a search model. An initial model was built with ARP/wARP [59] and completed by several cycles of manual model building with Coot [60] and refinement with Refmac [61] using data to 2.9 Å resolution. The final model contains 2E7 residues 1–120 and gp41 residues 582 to 596. The R and $R_{free}$ of the refined model are 19.5 and 24.6, respectively with 99% of the residues in the allowed regions of a Ramachandran plot. Molecular graphics figures were generated with PyMOL (W. Delano; http://www.pymol.org). Coordinates and structure factures have been deposited in the Protein Data Bank with accessions code 5HM1 (2E7).

### 4.9. SPR Analysis of 2E7 Binding to $gp41_{int}$

The fusion intermediate conformation of gp41, gp41int was purified as described before [30]. Surface plasmon resonance (SPR) analysis was performed with a Biacore 3000 (GE Healthcare). As a flow buffer 10 mM HEPES, 150 mM NaCl, pH 7.4 with 0.005% Tween-20 was used. $Gp41_{int}$ was immobilized to ~1200 response units using 50 µg/mL protein in flow buffer on an activated CM-5 sensor chip (GE Healthcare: BR-1000-50) according to the manufacturer's instructions. Specific binding to the target protein was corrected for nonspecific binding to the deactivated control channel. The flow rate was 50 µL/min. Regeneration of the sensor chip was achieved with 10mM HCl followed by 4 M $MgCl_2$ for 60 s at 60 µL/min. Data were analyzed with the BIA evaluation software version 4.1 and globally fit to t a 1:1 Langmuir model double referenced by subtraction of the blank surface and a blank injection.

## 5. Conclusions

Using various techniques we selected VHH against 4 different epitopes on HIV 1 gp140. This enabled us to rationally design bi-specific VHH. We show that increased potencies and extended breadths can be achieved by bi-specific VHH targeting two independent epitopes most likely on the same trimer, thereby producing important synergistic effects. Due to the lower risk of development of escape mutants and an improved efficacy, bi-specific anti-HIV-1 VHH may have great advantages over monovalent VHH in preventing HIV-1 transmission, either in a topical microbicide or when expressed from a gene therapy vector. Moreover, the lower production costs and higher stability of bi-specific VHH, make them superior to conventional antibodies. Because of its breadth and potency, VHH J3 was tested against SHIVs in a macaque challenge study quite successfully (manuscript in preparation).

At present the mono- and bispecific VHH are tested on their capability to recognize and destroy immune cells infected by HIV-1. Furthermore, an AAV based vector expressing VHH may provide prevention against infection and additional effects might be achieved by adding human Fc fragments to bi-specific VHH to suppress viremia, as was shown recently for the human antibody 3BNC117 [62].

**Supplementary Materials:** The following are available online at http://www.mdpi.com/2073-4468/8/2/38/s1, Table S1: Summary of neutralization data.

**Author Contributions:** N.M.S. Experimental setup and execution, data analysis, drafting of manuscript; M.H. Experimental setup and execution, data analysis; L.R. Experimental setup, data analysis, drafting of manuscript; L.E.M. Experimental setup and execution, data analysis, drafting of manuscript; J.W.B. Experimental setup and execution, data analysis; C.C. Experimental setup and execution, data analysis; M.H. Experimental setup and execution, data analysis; R.A.W. Study design and intellectual input, critical revising of the manuscript; W.W. data analysis, drafting of manuscript, critical revising of the manuscript; T.V. Study design and intellectual input, critical revising of the manuscript.

**Funding:** This research received no external funding.

**Acknowledgments:** This research received funding of the European Commission sixth Frame Work Programme as part of the European Vaccines and Microbicides Enterprise (EUROPRISE), the European Commission seventh Frame Work Programme as part of the Combined Highly Active Anti-retroviral Microbicides project (CHAARM), the Bill and Melinda Gates Foundation as part of the Collaboration for AIDS Vaccine Discovery (CAVD grant 38637 (to R.A. Weiss), grant 38619 (to M.S. Seaman)), the UK Medical Research Council (to R. Weiss), the Institute Universitaire de France (IUF) (to W. Weissenhorn) and a post doctoral fellowship from Sidaction (to M. Hock and W. Weissenhorn). Further, Winfried Weissenhorn acknowledges the platforms of the Grenoble INSTRUCT-ERIC center (ISBG; UMS 3518 CNRS-CEA-UJF-EMBL) supported by the French Infrastructure for Integrated Structural Biology Initiative FRISBI (ANR-10-INSB-05-02) and GRAL (ANR-10-LABX-49-01) within the Grenoble Partnership for Structural Biology (PSB), the ESRF-EMBL Joint Structural Biology Group for access and support at the ESRF beam lines and J. Marquez (EMBL) from the crystallization platform. The authors would like to thank Mohamed ElKhattabi for llama immunizations, Dennis Burton for the kind gift of b12 and Davide Corti and Antonio Lanzavecchia for the kind gifts of HJ16, HGN194 and HGP120 and Mike Seamen for the determination of neutralization breadth for individual VHH.

**Conflicts of Interest:** The authors declare that there are no competing financial interest, other than that N.M.S., L.R., L.E.M., R.A.W. and T.V. are inventors on a patent that is filed on the use of several VHH mentioned in this paper. In the period we carried out this research 4 persons were related to companies, Nika M Strokappe, Lucy Rutten and myself worked at UU or later at QVQ Holding b.v. and Jaap W. Back at Pepscan. All others worked either in the teams of Robin Weiss or Winfried Weissenhorn. Later on Jaap went to Eurofins, Miriam to Immunocore and Matthias to Novartis. However none of these companies were involved in this work. For all clarity, QVQ become member of the Gates project from 2011 onwards. Most of the work described in this paper was done in the period 2011–2015.

## References


1. Granelli-Piperno, A.; Pritsker, A.; Pack, M.; Shimeliovich, I.; Arrighi, J.F.; Park, C.G.; Trumpfheller, C.; Piguet, V.; Moran, T.M.; Steinman, R.M. Dendritic cell-specific intercellular adhesion molecule 3-grabbing nonintegrin/CD209 is abundant on macrophages in the normal human lymph node and is not required for dendritic cell stimulation of the mixed leukocyte reaction. *J. Immunol.* **2005**, *175*, 4265–4273. [CrossRef] [PubMed]
2. Abdool Karim, Q.; Abdool Karim, S.S.; Frohlich, J.A.; Grobler, A.C.; Baxter, C.; Mansoor, L.E.; Kharsany, A.B.; Sibeko, S.; Mlisana, K.P.; Omar, Z.; et al. CAPRISA 004 Trial Group Effectiveness and safety of tenofovir gel, an antiretroviral microbicide, for the prevention of HIV infection in women. *Science* **2010**, *329*, 1168–1174. [CrossRef] [PubMed]
3. Klein, F.; Gaebler, C.; Mouquet, H.; Sather, D.N.; Lehmann, C.; Scheid, J.F.; Kraft, Z.; Liu, Y.; Pietzsch, J.; Hurley, A.; et al. Broad neutralization by a combination of antibodies recognizing the CD4 binding site and a new conformational epitope on the HIV-1 envelope protein. *J. Exp. Med.* **2012**, *209*, 1469–1479. [CrossRef] [PubMed]
4. Klein, F.; Halper-Stromberg, A.; Horwitz, J.A.; Gruell, H.; Scheid, J.F.; Bournazos, S.; Mouquet, H.; Spatz, L.A.; Diskin, R.; Abadir, A.; et al. HIV therapy by a combination of broadly neutralizing antibodies in humanized mice. *Nature* **2012**, *492*, 118–122. [CrossRef] [PubMed]
5. Scheid, J.F.; Mouquet, H.; Feldhahn, N.; Seaman, M.S.; Velinzon, K.; Pietzsch, J.; Ott, R.G.; Anthony, R.M.; Zebroski, H.; Hurley, A.; et al. Broad diversity of neutralizing antibodies isolated from memory B cells in HIV-infected individuals. *Nature* **2009**, *458*, 636–640. [CrossRef] [PubMed]
6. Walker, L.M.; Phogat, S.K.; Chan-Hui, P.Y.; Wagner, D.; Phung, P.; Goss, J.L.; Wrin, T.; Simek, M.D.; Fling, S.; Mitcham, J.L.; et al. Broad and potent neutralizing antibodies from an African donor reveal a new HIV-1 vaccine target. *Science* **2009**, *326*, 285–289. [CrossRef] [PubMed]
7. Scheid, J.F.; Mouquet, H.; Ueberheide, B.; Diskin, R.; Klein, F.; Oliveira, T.Y.; Pietzsch, J.; Fenyo, D.; Abadir, A.; Velinzon, K.; et al. Sequence and structural convergence of broad and potent HIV antibodies that mimic CD4 binding. *Science* **2011**, *333*, 1633–1637. [CrossRef] [PubMed]
8. Corti, D.; Langedijk, J.P.; Hinz, A.; Seaman, M.S.; Vanzetta, F.; Fernandez-Rodriguez, B.M.; Silacci, C.; Pinna, D.; Jarrossay, D.; Balla-Jhagjhoorsingh, S.; et al. Analysis of memory B cell responses and isolation of novel monoclonal antibodies with neutralizing breadth from HIV-1-infected individuals. *PLoS ONE* **2010**, *5*, e8805. [CrossRef] [PubMed]
9. Wu, X.; Yang, Z.Y.; Li, Y.; Hogerkorp, C.M.; Schief, W.R.; Seaman, M.S.; Zhou, T.; Schmidt, S.D.; Wu, L.; Xu, L.; et al. Rational Design of Envelope Identifies Broadly Neutralizing Human Monoclonal Antibodies to HIV-1. *Science* **2010**, *329*, 856–861. [CrossRef] [PubMed]
10. Walker, L.M.; Huber, M.; Doores, K.J.; Falkowska, E.; Pejchal, R.; Julien, J.P.; Wang, S.K.; Ramos, A.; Chan-Hui, P.Y.; Moyle, M.; et al. Broad neutralization coverage of HIV by multiple highly potent antibodies. *Nature* **2011**, *477*, 466–470. [CrossRef] [PubMed]
11. Huang, J.; Ofek, G.; Laub, L.; Louder, M.K.; Doria-Rose, N.A.; Longo, N.S.; Imamichi, H.; Bailer, R.T.; Chakrabarti, B.; Sharma, S.K.; et al. Broad and potent neutralization of HIV-1 by a gp41-specific human antibody. *Nature* **2012**, *491*, 406–412. [CrossRef] [PubMed]
12. Brinckmann, S.; da Costa, K.; van Gils, M.J.; Hallengard, D.; Klein, K.; Madeira, L.; Mainetti, L.; Palma, P.; Raue, K.; Reinhart, D.; et al. Rational design of HIV vaccines and microbicides: Report of the EUROPRISE network annual conference 2010. *J. Transl. Med.* **2011**, *9*, 40. [CrossRef] [PubMed]
13. van de Laar, T.; Visser, C.; Holster, M.; Lopez, C.G.; Kreuning, D.; Sierkstra, L.; Lindner, N.; Verrips, T. Increased heterologous protein production by Saccharomyces cerevisiae growing on ethanol as sole carbon source. *Biotechnol. Bioeng.* **2007**, *96*, 483–494. [CrossRef] [PubMed]
14. Gorlani, A.; Brouwers, J.; McConville, C.; van der Bijl, P.; Malcolm, K.; Augustijns, P.; Quigley, A.F.; Weiss, R.; De Haard, H.; Verrips, T. Llama antibody fragments have good potential for application as HIV type 1 topical microbicides. *AIDS Res. Hum. Retroviruses* **2012**, *28*, 198–205. [CrossRef] [PubMed]

15. Forsman, A.; Beirnaert, E.; Aasa-Chapman, M.M.; Hoorelbeke, B.; Hijazi, K.; Koh, W.; Tack, V.; Szynol, A.; Kelly, C.; McKnight, A.; et al. Llama antibody fragments with cross-subtype human immunodeficiency virus type 1 (HIV-1)-neutralizing properties and high affinity for HIV-1 gp120. *J. Virol.* **2008**, *82*, 12069–12081. [CrossRef] [PubMed]
16. McCoy, L.E.; Quigley, A.F.; Strokappe, N.M.; Bulmer-Thomas, B.; Seaman, M.S.; Mortier, D.; Rutten, L.; Chander, N.; Edwards, C.J.; Ketteler, R.; et al. Potent and broad neutralization of HIV-1 by a llama antibody elicited by immunization. *J. Exp. Med.* **2012**, *209*, 1091–1103. [CrossRef] [PubMed]
17. Strokappe, N.; Szynol, A.; Aasa-Chapman, M.; Gorlani, A.; Forsman Quigley, A.; Hulsik, D.L.; Chen, L.; Weiss, R.; de Haard, H.; Verrips, T. Llama antibody fragments recognizing various epitopes of the CD4bs neutralize a broad range of HIV-1 subtypes A, B and C. *PLoS ONE* **2012**, *7*, e33298. [CrossRef] [PubMed]
18. Doria-Rose, N.A.; Louder, M.K.; Yang, Z.; O'Dell, S.; Nason, M.; Schmidt, S.D.; McKee, K.; Seaman, M.S.; Bailer, R.T.; Mascola, J.R. HIV-1 neutralization coverage is improved by combining monoclonal antibodies that target independent epitopes. *J. Virol.* **2012**, *86*, 3393–3397. [CrossRef] [PubMed]
19. Strokappe, N.M. *HIV-1, How Llamas Help Us Fight the AIDS Pandemic*; Utrecht University: Utrecht, The Netherlands, 2013.
20. Matz, J.; Kessler, P.; Bouchet, J.; Combes, O.; Ramos, O.H.; Barin, F.; Baty, D.; Martin, L.; Benichou, S.; Chames, P. Straightforward selection of broadly neutralizing single-domain antibodies targeting the conserved CD4 and coreceptor binding sites of HIV-1 gp120. *J. Virol.* **2013**, *87*, 1137–1149. [CrossRef] [PubMed]
21. McCoy, L.E.; Rutten, L.; Frampton, D.; Anderson, I.; Granger, L.; Bashford-Rogers, R.; Dekkers, G.; Strokappe, N.M.; Seaman, M.S.; Koh, W.; et al. Molecular Evolution of Broadly Neutralizing Llama Antibodies to the CD4-Binding Site of HIV-1. *PLoS Pathog.* **2014**, *10*, e1004552. [CrossRef] [PubMed]
22. Hultberg, A.; Temperton, N.J.; Rosseels, V.; Koenders, M.; Gonzalez-Pajuelo, M.; Schepens, B.; Itati Ibanez, L.; Vanlandschoot, P.; Schillemans, J.; Saunders, M.; et al. Llama-derived single domain antibodies to build multivalent, superpotent and broadened neutralizing anti-viral molecules. *PLoS ONE* **2011**, *6*, e17665. [CrossRef] [PubMed]
23. Jahnichen, S.; Blanchetot, C.; Maussang, D.; Gonzalez-Pajuelo, M.; Chow, K.Y.; Bosch, L.; De Vrieze, S.; Serruys, B.; Ulrichts, H.; Vandevelde, W.; et al. CXCR4 nanobodies (VHH-based single variable domains) potently inhibit chemotaxis and HIV-1 replication and mobilize stem cells. *Proc. Natl. Acad. Sci. USA* **2010**, *107*, 20565–20570. [CrossRef] [PubMed]
24. Mouquet, H.; Warncke, M.; Scheid, J.F.; Seaman, M.S.; Nussenzweig, M.C. Enhanced HIV-1 neutralization by antibody heteroligation. *Proc. Natl. Acad. Sci. USA* **2012**, *109*, 875–880. [CrossRef] [PubMed]
25. Balazs, A.B.; West, A.P., Jr. Antibody gene transfer for HIV immunoprophylaxis. *Nat. Immunol.* **2013**, *14*, 1–5. [CrossRef] [PubMed]
26. Forsman Quigley, A.; Strokappe, N.; McCoy, L.; Rutten, L.; Tan, S.; Aasa-Chapman, M.; Seaman, M.; Szynol, A.; Liu, Y.Y.; de Haard, H.; et al. Broadly neutralising single-chain llama antibody fragments targeting novel gp120 and gp41 epitopes on the human immunodeficiency virus type 1 (HIV-1) envelope spike. manuscript in preparation.
27. McCoy, L.; Rutten, L.; Strokappe, N.; Verrips, T.; Webb, B.; Weiss, R. Broadly neutralizing VHH againstHIV-1. WO Patent WO2013036130A1, 14 March 2013.
28. Xu, J.Y.; Gorny, M.K.; Palker, T.; Karwowska, S.; Zolla-Pazner, S. Epitope mapping of two immunodominant domains of gp41, the transmembrane protein of human immunodeficiency virus type 1, using ten human monoclonal antibodies. *J. Virol.* **1991**, *65*, 4832–4838. [PubMed]
29. Pancera, M.; Zhou, T.; Druz, A.; Georgiev, I.S.; Soto, C.; Gorman, J.; Huang, J.; Acharya, P.; Chuang, G.Y.; Ofek, G.; et al. Structure and immune recognition of trimeric pre-fusion HIV-1 Env. *Nature* **2014**, *514*, 455–461. [CrossRef] [PubMed]
30. Lai, R.P.; Hock, M.; Radzimanowski, J.; Tonks, P.; Hulsik, D.L.; Effantin, G.; Seilly, D.J.; Dreja, H.; Kliche, A.; Wagner, R.; et al. A fusion intermediate gp41 immunogen elicits neutralizing antibodies to HIV-1. *J. Biol. Chem.* **2014**, *289*, 29912–29926. [CrossRef]
31. Luftig, M.A.; Mattu, M.; Di Giovine, P.; Geleziunas, R.; Hrin, R.; Barbato, G.; Bianchi, E.; Miller, M.D.; Pessi, A.; Carfi, A. Structural basis for HIV-1 neutralization by a gp41 fusion intermediate-directed antibody. *Nat. Struct. Mol. Biol.* **2006**, *13*, 740–747. [CrossRef]

32. Sabin, C.; Corti, D.; Buzon, V.; Seaman, M.S.; Lutje Hulsik, D.; Hinz, A.; Vanzetta, F.; Agatic, G.; Silacci, C.; Mainetti, L.; et al. Crystal structure and size-dependent neutralization properties of HK20, a human monoclonal antibody binding to the highly conserved heptad repeat 1 of gp41. *PLoS Pathog.* **2010**, *6*, e1001195. [CrossRef]
33. Weissenhorn, W.; Hinz, A.; Gaudin, Y. Virus membrane fusion. *FEBS Lett.* **2007**, *581*, 2150–2155. [CrossRef] [PubMed]
34. Grupping, K. Inhibiting the CD4-gp120 Interaction to Prevent HIV Infection: Insights from Mutational Resistance Analysis. Ph.D. Thesis, Universiteit Antwerpen, Antwerp, Belgium, 2013.
35. Pennings, P.S. Standing Genetic Variation and the Evolution of Drug Resistance in HIV. *PLoS Comput. Biol.* **2012**, *8*, e1002527. [CrossRef] [PubMed]
36. Asokan, M.; Rudicell, R.S.; Louder, M.; McKee, K.; O'Dell, S.; Stewart-Jones, G.; Wang, K.; Xu, L.; Chen, X.; Choe, M.; et al. Bispecific Antibodies Targeting Different Epitopes on the HIV-1 Envelope Exhibit Broad and Potent Neutralization. *J. Virol.* **2015**, *89*, 12501–12512. [CrossRef] [PubMed]
37. Liu, L.; Patel, B.; Ghanem, M.H.; Bundoc, V.; Zheng, Z.; Morgan, R.A.; Rosenberg, S.A.; Dey, B.; Berger, E.A. Novel CD4-Based Bispecific Chimeric Antigen Receptor Designed for Enhanced Anti-HIV Potency and Absence of HIV Entry Receptor Activity. *J. Virol.* **2015**, *89*, 6685–6694. [CrossRef] [PubMed]
38. Bournazos, S.; Gazumyan, A.; Seaman, M.S.; Nussenzweig, M.C.; Ravetch, J.V. Bispecific Anti-HIV-1 Antibodies with Enhanced Breadth and Potency. *Cell* **2016**, *165*, 1609–1620. [CrossRef] [PubMed]
39. Huang, Y.; Yu, J.; Lanzi, A.; Yao, X.; Andrews, C.D.; Tsai, L.; Gajjar, M.R.; Sun, M.; Seaman, M.S.; Padte, N.N.; et al. Engineered Bispecific Antibodies with Exquisite HIV-1-Neutralizing Activity. *Cell* **2016**, *165*, 1621–1631. [CrossRef] [PubMed]
40. Dey, B.; Del Castillo, C.S.; Berger, E.A. Neutralization of human immunodeficiency virus type 1 by sCD4-17b, a single-chain chimeric protein, based on sequential interaction of gp120 with CD4 and coreceptor. *J. Virol.* **2003**, *77*, 2859–2865. [CrossRef] [PubMed]
41. Lutje Hulsik, D.; Liu, Y.Y.; Strokappe, N.M.; Battella, S.; El Khattabi, M.; McCoy, L.E.; Sabin, C.; Hinz, A.; Hock, M.; Macheboeuf, P.; et al. A gp41 MPER-specific llama VHH requires a hydrophobic CDR3 for neutralization but not for antigen recognition. *PLoS Pathog.* **2013**, *9*, e1003202. [CrossRef] [PubMed]
42. Julien, J.P.; Cupo, A.; Sok, D.; Stanfield, R.L.; Lyumkis, D.; Deller, M.C.; Klasse, P.J.; Burton, D.R.; Sanders, R.W.; Moore, J.P.; et al. Crystal structure of a soluble cleaved HIV-1 envelope trimer. *Science* **2013**, *342*, 1477–1483. [CrossRef]
43. Lyumkis, D.; Julien, J.P.; de Val, N.; Cupo, A.; Potter, C.S.; Klasse, P.J.; Burton, D.R.; Sanders, R.W.; Moore, J.P.; Carragher, B.; et al. Cryo-EM structure of a fully glycosylated soluble cleaved HIV-1 envelope trimer. *Science* **2013**, *342*, 1484–1490. [CrossRef] [PubMed]
44. Weissenhorn, W.; Dessen, A.; Harrison, S.C.; Skehel, J.J.; Wiley, D.C. Atomic structure of the ectodomain from HIV-1 gp41. *Nature* **1997**, *387*, 426–430. [CrossRef] [PubMed]
45. Gustchina, E.; Li, M.; Louis, J.M.; Anderson, D.E.; Lloyd, J.; Frisch, C.; Bewley, C.A.; Gustchina, A.; Wlodawer, A.; Clore, G.M. Structural basis of HIV-1 neutralization by affinity matured Fabs directed against the internal trimeric coiled-coil of gp41. *PLoS Pathog.* **2010**, *6*, e1001182. [CrossRef] [PubMed]
46. Sholukh, A.M.; Watkins, J.D.; Vyas, H.K.; Gupta, S.; Lakhashe, S.K.; Thorat, S.; Zhou, M.; Hemashettar, G.; Bachler, B.C.; Forthal, D.N.; et al. Defense-in-depth by mucosally administered anti-HIV dimeric IgA2 and systemic IgG1 mAbs: Complete protection of rhesus monkeys from mucosal SHIV challenge. *Vaccine* **2015**, *33*, 2086–2095. [CrossRef] [PubMed]
47. Burton, D.R.; Barbas, C.F., 3rd; Persson, M.A.; Koenig, S.; Chanock, R.M.; Lerner, R.A. A large array of human monoclonal antibodies to type 1 human immunodeficiency virus from combinatorial libraries of asymptomatic seropositive individuals. *Proc. Natl. Acad. Sci. USA* **1991**, *88*, 10134–10137. [CrossRef] [PubMed]
48. Thali, M.; Moore, J.P.; Furman, C.; Charles, M.; Ho, D.D.; Robinson, J.; Sodroski, J. Characterization of conserved human immunodeficiency virus type 1 gp120 neutralization epitopes exposed upon gp120-CD4 binding. *J. Virol.* **1993**, *67*, 3978–3988. [PubMed]
49. Li, M.; Gao, F.; Mascola, J.R.; Stamatatos, L.; Polonis, V.R.; Koutsoukos, M.; Voss, G.; Goepfert, P.; Gilbert, P.; Greene, K.M.; et al. Human immunodeficiency virus type 1 env clones from acute and early subtype B infections for standardized assessments of vaccine-elicited neutralizing antibodies. *J. Virol.* **2005**, *79*, 10108–10125. [CrossRef] [PubMed]

50. Li, M.; Salazar-Gonzalez, J.F.; Derdeyn, C.A.; Morris, L.; Williamson, C.; Robinson, J.E.; Decker, J.M.; Li, Y.; Salazar, M.G.; Polonis, V.R.; et al. Genetic and neutralization properties of subtype C human immunodeficiency virus type 1 molecular env clones from acute and early heterosexually acquired infections in Southern Africa. *J. Virol.* **2006**, *80*, 11776–11790. [CrossRef] [PubMed]
51. Derdeyn, C.A.; Decker, J.M.; Sfakianos, J.N.; Wu, X.; O'Brien, W.A.; Ratner, L.; Kappes, J.C.; Shaw, G.M.; Hunter, E. Sensitivity of human immunodeficiency virus type 1 to the fusion inhibitor T-20 is modulated by coreceptor specificity defined by the V3 loop of gp120. *J. Virol.* **2000**, *74*, 8358–8367. [CrossRef] [PubMed]
52. Wei, X.; Decker, J.M.; Liu, H.; Zhang, Z.; Arani, R.B.; Kilby, J.M.; Saag, M.S.; Wu, X.; Shaw, G.M.; Kappes, J.C. Emergence of resistant human immunodeficiency virus type 1 in patients receiving fusion inhibitor (T-20) monotherapy. *Antimicrob. Agents Chemother.* **2002**, *46*, 1896–1905. [CrossRef] [PubMed]
53. Hinz, A.; Lutje Hulsik, D.; Forsman, A.; Koh, W.W.; Belrhali, H.; Gorlani, A.; de Haard, H.; Weiss, R.A.; Verrips, T.; Weissenhorn, W. Crystal Structure of the Neutralizing Llama V(HH) D7 and Its Mode of HIV-1 gp120 Interaction. *PLoS ONE* **2010**, *5*, e10482. [CrossRef] [PubMed]
54. Meyerson, J.R.; Tran, E.E.; Kuybeda, O.; Chen, W.; Dimitrov, D.S.; Gorlani, A.; Verrips, T.; Lifson, J.D.; Subramaniam, S. Molecular structures of trimeric HIV-1 Env in complex with small antibody derivatives. *Proc. Natl. Acad. Sci. USA* **2013**, *110*, 513–518. [CrossRef] [PubMed]
55. Pellegrini, E.; Piano, D.; Bowler, M.W. Direct cryocooling of naked crystals: Are cryoprotection agents always necessary? *Acta Crystallogr. D Biol. Crystallogr.* **2011**, *67*, 902–906. [CrossRef] [PubMed]
56. Battye, T.G.; Kontogiannis, L.; Johnson, O.; Powell, H.R.; Leslie, A.G. iMOSFLM: A new graphical interface for diffraction-image processing with MOSFLM. *Acta Crystallogr. D Biol. Crystallogr.* **2011**, *67*, 271–281. [CrossRef] [PubMed]
57. Evans, P. Scaling and assessment of data quality. *Acta Crystallogr. D Biol. Crystallogr.* **2006**, *62*, 72–82. [CrossRef] [PubMed]
58. McCoy, A.J. Solving structures of protein complexes by molecular replacement with Phaser. *Acta Crystallogr. D Biol. Crystallogr.* **2007**, *63*, 32–41. [CrossRef] [PubMed]
59. Perrakis, A.; Morris, R.; Lamzin, V.S. Automated protein model building combined with iterative structure refinement. *Nat. Struct. Biol.* **1999**, *6*, 458–463. [CrossRef] [PubMed]
60. Emsley, P.; Cowtan, K. Coot: Model-building tools for molecular graphics. *Acta Crystallogr. D Biol. Crystallogr.* **2004**, *60*, 2126–2132. [CrossRef]
61. Murshudov, G.N.; Vagin, A.A.; Dodson, E.J. Refinement of macromolecular structures by the maximum-likelihood method. *Acta Crystallogr. D Biol. Crystallogr.* **1997**, *53*, 240–255. [CrossRef]
62. Caskey, M.; Klein, F.; Lorenzi, J.C.; Seaman, M.S.; West, A.P., Jr.; Buckley, N.; Kremer, G.; Nogueira, L.; Braunschweig, M.; Scheid, J.F.; et al. Viraemia suppressed in HIV-1-infected humans by broadly neutralizing antibody 3BNC117. *Nature* **2015**, *522*, 487–491. [CrossRef]

Article

# VHH-Photosensitizer Conjugates for Targeted Photodynamic Therapy of Met-Overexpressing Tumor Cells

**Raimond Heukers [1], Vida Mashayekhi [2], Mercedes Ramirez-Escudero [3], Hans de Haard [4], Theo C. Verrips [1], Paul. M.P. van Bergen en Henegouwen [2,]* and Sabrina Oliveira [2,5]**

1 QVQ Holding BV, Yalelaan 1, 3584 CL Utrecht, The Netherlands; r.heukers@qvquality.com (R.H.); theo.verrips@outlook.com (T.C.V.)
2 Cell Biology Division, Department of Biology, Faculty of Science, Utrecht University, Padualaan 8, 3584 CH Utrecht, The Netherlands; v.mashayekhi@uu.nl (V.M.); s.oliveira@uu.nl (S.O.)
3 Crystal & Structural Chemistry, Bijvoet Center for Biomolecular Research, Faculty of Science, Utrecht University, Padualaan 8, 3584 CH Utrecht, The Netherlands; m.ramirezescudero@uu.nl
4 Argenx BVBA, Industriepark-Zwijnaarde 7, 9052 Gent, Belgium; hdehaard@argenx.com
5 Pharmaceutics Division, Department of Pharmaceutical Sciences, Faculty of Science, Utrecht University, Universiteitsweg 99, 3584 CG Utrecht, The Netherlands
* Correspondence: p.vanbergen@uu.nl

Received: 27 February 2019; Accepted: 28 March 2019; Published: 4 April 2019

**Abstract:** Photodynamic therapy (PDT) is an approach that kills (cancer) cells by the local production of toxic reactive oxygen species upon the local illumination of a photosensitizer (PS). The specificity of PDT has been further enhanced by the development of a new water-soluble PS and by the specific delivery of PS via conjugation to tumor-targeting antibodies. To improve tissue penetration and shorten photosensitivity, we have recently introduced nanobodies, also known as VHH (variable domains from the heavy chain of llama heavy chain antibodies), for targeted PDT of cancer cells overexpressing the epidermal growth factor receptor (EGFR). Overexpression and activation of another cancer-related receptor, the hepatocyte growth factor receptor (HGFR, c-Met or Met) is also involved in the progression and metastasis of a large variety of malignancies. In this study we evaluate whether anti-Met VHHs conjugated to PS can also serve as a biopharmaceutical for targeted PDT. VHHs targeting the SEMA (semaphorin-like) subdomain of Met were provided with a C-terminal tag that allowed both straightforward purification from yeast supernatant and directional conjugation to the PS IRDye700DX using maleimide chemistry. The generated anti-Met VHH-PS showed nanomolar binding affinity and, upon illumination, specifically killed MKN45 cells with nanomolar potency. This study shows that Met can also serve as a membrane target for targeted PDT.

**Keywords:** targeted photodynamic therapy; hepatocyte growth factor receptor; HGFR; c-Met; Met; nanobodies; VHH; photosensitizer

## 1. Introduction

Photodynamic therapy (PDT) is a type of cancer treatment in which tumor cells are killed by reactive oxygen species, such as singlet oxygen, formed by the local and light-induced activation of a photosensitizer (PS) [1]. By locally reacting with proteins, lipids, and nucleic acids, the reactive oxygen species generated can induce cell death, vascular damage, and an inflammatory response [2]. It is for this mode of action that PDT is used in clinic to treat malignancies. Unfortunately, the PSs generally used in the clinic are relatively hydrophobic, are systemically applied, and are non-targeted. These factors combined can result in off-target toxicity and long lasting photosensitivity [2,3]. Improved strategies involve the development of more water-soluble PS, delivery of PS with nanocarriers and photo-immunotherapy (PIT), among others [4–6]. PIT is one of the strategies of targeted PDT, in which the PS is delivered selectively to tumors via conjugation to tumor-targeting antibodies.

In order to increase the efficacy of targeted PDT and reduce the period of photosensitivity, we have recently introduced smaller PS conjugates with enhanced tumor penetration in combination with reduced blood circulation time [7,8]. These smaller conjugates were generated by conjugating a water-soluble PS to small antibody fragments originating from heavy chain antibodies found in animals of the *Camelidae* family (i.e., nanobodies, single domain antibodies (sdAbs) or VHHs (variable domain of the heavy chain from heavy chain-only antibodies)) [9]. Compared to commonly used conventional antibodies of the IgG class, VHHs are 10 times smaller and consist of only a single domain with in general only one, rarely two, disulfide bridges [10,11]. These features favor the selection and production process and make VHHs very stable [11]. Also, because the C-terminus of a VHH is located opposite to its epitope-binding loops (i.e., complementary determining regions), C-terminal conjugation to effector molecules generally does not affect the binding properties [12–16].

Previously, we have described the nanobody-PS or VHH-PS conjugates specifically targeting the epidermal growth factor receptor (EGFR), a receptor tyrosine kinase (RTK) that is found overexpressed on a large variety of cancers, such as head and neck, lung, or colon cancer [17,18]. In vitro, upon illumination, VHH-PS conjugates selectively killed EGFR-overexpressing tumor cells with nanomolar $IC_{50}$ values. In vivo, VHH-targeted PDT induced 80%–90% tumor necrosis, as measured 24 h after illumination [7,8]. Another receptor tyrosine kinase, which is frequently overexpressed or deregulated in a large number of carcinomas, sarcomas, hematopoietic malignancies, and other neoplasms is the hepatocyte growth factor receptor (HGFR, c-Met or Met) [19,20]. Met is a 195 kDa hetero-dimeric single-membrane spanning receptor tyrosine kinase that is activated by hepatocyte growth factor (HGF, also known as scatter factor) [21]. The extracellular part of Met consists of three subdomains: an N-terminal 7-bladed β-propeller-like SEMA (semaphorin-like) subdomain, a PSI (plexin, semaphorin, integrin-like) subdomain, and four IPT (immune-globulin-like, plexins, transcription factors) subdomains [22,23]. Blades 2–3 of the SEMA subdomain and IPT 3–4 interact with its natural ligand HGF [24]. In a previous study, multiple VHHs targeting the extracellular part of Met were developed (i.e., E9, F5, and G2) [25]. Of these VHHs, G2 was used to target cross-linked albumin nanoparticles to Met-expressing cells [25]. By using the anti-Met VHH G2, the targeted nanoparticles showed specific binding and uptake into Met-expressing cells.

In this study, we have characterized these Met-targeting VHHs further by assessing the subdomains they bind to. Moreover, we have used the best of these VHHs for the specific delivery of water-soluble PS to Met-expressing cells for targeted PDT. That VHH was extended with a C-terminal C-Direct tag, allowing affinity purification from yeast supernatant and directional conjugation of the PS to an unpaired cysteine using maleimide chemistry. Subsequently, binding of this conjugate to cells and their ability to kill Met-expressing cells by PDT were evaluated.

## 2. Methods

### 2.1. Molecular Modeling

The molecular structure of G2c with the C-Direct tag was modeled with I-TASSER [26] and visualized using the PyMOL Molecular Graphics System, Version 2.0 (Schrödinger, LLC, Cambridge, USA). Nb10 from PDB ID 4DKA:A showed 85% homology to G2 and was taken as a reference [27].

### 2.2. VHH Production and Purification

Selection and characterization of VHH clones targeting Met have been described previously [25]. VHH protein for epitope mapping was obtained from production in *E. coli* TG1. For this, the VHH genes were cloned into the pMEK222 vector for productions in *E. coli*, which provides the VHH with a C-terminal FLAG-His tag. VHHs were produced and purified from *E. coli* TG1 using immobilized metal-affinity chromatography (IMAC, Thermo Fisher Scientific, Waltham, MA, USA) as previously described [7,28]. For production in yeast, VHH genes were recloned in the pYQVQ11 vector for VHH production in yeast, which provides the VHH with a C-terminal C-Direct tag containing a free thiol (cysteine) and an EPEA (Glu, Pro, Glu, Ala) purification tag (C-tag, Thermo Fisher Scientific). To improve production yields and facilitate purification from supernatant, C-Direct-tagged VHH were produced in *S. cerevisiae* strain VWK18 as described previously [28–31]. VHHs were purified from the yeast supernatant using an Äkta Start (GE Healthcare, Chicago, IL, USA), a Capture-Select affinity chromatography column (Thermo Fisher Scientific) and size-exclusion chromatography (Thermo Fisher Scientific) according to the manufacturer's protocols. Purified VHH was filter sterilized and stored in PBS (phosphate buffered saline).

### 2.3. Cell Culture

Met-overexpressing human gastric cancer cell line MKN45 (ACC-409) was obtained from the Deutsche Sammlung von Mikroorganismen und Zellkulturen (DSMZ, Braunschweig, Germany). The human ovarian carcinoma cell line TOV112D (CRL-11731) was obtained from the American Tissue Culture Collection (ATCC, LGC Standards GmbH, Wesel, Germany). TOV112D cells that stably express Met (TOV + Met) were previously described [25]. These cell lines were cultured as previously described in either DMEM (Dulbecco's Modified Eagle Medium) (Thermo Fisher Scientific) for TOV112D cells or RPMI 1640 (Thermo Fisher Scientific) for MKN45 cells [25]. HepG2 cells were obtained from ATCC (LGC Standards GmbH) and were cultured in DMEM low-glucose (Thermo Fisher Scientific). All media was supplemented with streptomycin, penicillin, l-glutamine, and FCS (fetal calf serum) as described previously [25].

### 2.4. VHH Binding ELISA

Binding affinity and binding epitopes were determined by ELISA on the Met ectodomain. The human Met, llama Met, llama/human chimeric Met ectodomains, and control antibody c224G11 were described previously [32]. Maxisorp 96-wells plates (Thermo Fisher Scientific) were coated overnight with 100 μL of 2 μg/mL of ectodomain and subsequently blocked with 4% milk in PBS. Bound c224G11 was detected with donkey-anti-mouse$^{HRP}$ (Bio-Rad, Hercules, CA, USA), and bound VHH was detected using mouse-anti-FLAG (clone M2, Sigma-Aldrich, St. Louis, MO, USA) followed by donkey-anti-rabbit$^{HRP}$ (Bio-Rad) with TMB Ultra as substrate (Thermo Fisher Scientific). Antibody incubations were performed in 1% milk in PBS and for 1 h at room temperature. Absorbance was measured using a Multiskan Go plate reader (Thermo Fisher Scientific). Binding of PS-conjugated VHH was carried out in cells in binding medium (DMEM without phenol red, 50 mM HEPES pH 7.5, supplemented with 1% bovine serum albumin) for 2 h at 4 °C to prevent internalization. Bound VHH-PS was detected using the Odyssey near-infrared scanner (LI-COR Biosciences, Lincoln, NE, USA). Data was plotted and fitted for one-site-specific binding using Prism 7 software (GraphPad, La Jolla, CA, USA).

### 2.5. Surface Plasmon Resonance (SPR)

Met ectodomain variants (human, llama and chimera LS1 and LS5) were amine-coupled on a pre-activated G-COOH sensor chip (SensEye) with a Continuous Flow Microspotter (CFM, Wasatch Microfluidics Inc., Salt Lake City, UT, USA) at pH 4.5 (50 mM Sodium acetate buffer, 0.005% Tween 20). Beforehand, the sensor chip was equilibrated in 50 mM MES (4-Morpholineethanesulfonic acid) buffer pH 5.5, then activated with 400/100 mM EDC/NHS [1-ethyl-3-(3-dimethylaminopropyl) carbodiimide hydrochloride/N-Hydroxysuccinimide]. After ectodomain coupling, excess reactive esters were deactivated with 100 mM ethanolamine pH 8. The amount of immobilized protein ranged from 700 to 3000 response units (RU); 1 RU corresponds to approximately 1 pg of protein per mm$^2$. Surface plasmon resonance (SPR) experiments were performed at a constant temperature of 25 °C on a MX96 (IBIS Technologies). G2 was flowed over the sensor chip as analyte at concentrations ranging from 20 pM to 200 nM for 60 min in buffer containing 25 mM phosphate-buffered saline pH 8.0 and 0.005% Tween 20. After each injection, an 8 min dissociation time was set, followed by a regeneration step of 30 s with 50 mM acetic acid buffer, pH 4.5. Calibration of sensor signals and reference subtraction was evaluated with SprintX (IBIS Technologies, Enschede, The Netherlands), and further analyses were performed in Scrubber 2 (BioLogic Software, Seyssinet-Pariset, France). Association- and dissociation-rate constants ($k_a/k_d$) were determined by globally fitting the SPR data to a 1:1 Langmuir binding model. The dissociation constant ($K_D$) was calculated from the $k_a$ and $k_d$ parameters. In addition, the $K_D$ was determined from the steady state analyte binding levels (averaged between 40 and 60 min association time) plotted against concentration, and by fitting a one-site saturation model.

### 2.6. Conjugation to PS

VHHs were site-directionally conjugated to the photosensitizer IRDye700DX–maleimide (LI-COR Biosciences). First, the VHHs were incubated with an 2.75-fold molar excess of TCEP (tris(2-carboxyethyl) phosphine hydrochloride) (VWR International, Radnor, PA, USA) to reduce the C-terminal cysteine upon which the VHH were incubated with a four-fold molar excess of IRDye700DX-maleimide for 16 h at 4 °C. Free label was removed by size-exclusion chromatography using three consequent Zeba Desalting Columns (Thermo Fisher Scientific) according to the manufacturer's protocols. Degree of conjugation was determined using the Multiskan Go spectrophotometer (Thermo Fisher Scientific), and the amount of free dye was determined after size separation by SDS-PAGE (Bio-Rad) on the Odyssey scanner (Li-COR Biosciences). Afterwards, the SDS-PAGE gel was stained with Page Blue (Thermo Fisher Scientific) to show total protein. For the internalization assay, G2c was conjugated to HiLyte Fluor 647 (HL647)–maleimide (Eurogentec, Liege, Belgium) according to the protocol described above.

### 2.7. Internalization Assay

MKN45 cells were seeded in eight-well chamber slides (Lab-Tek, Nunc, Thermo Fisher Scientific) two days before the experiment. Cells were incubated with 1 μM G2-Alexa647 conjugate for 2 h at 37 °C. Unbound conjugate was removed by washing three times with PBS. Cells were fixed in 4% PFA for 10 min at room temperate (RT). PFA-induced autofluorescence was quenched using 100 mM glycine in PBS (10 min, RT). Cells were washed with PBS and then incubated with DAPI (4′,6-diamidino-2-phenylindole, 0.25 μg/mL, Thermo Scientific) for 10 min at RT. The slides were mounted using SlowFade (Invitrogen) and pictures were taken with a LSM700 confocal microscope using a 63× oil immersion objective (Carl Zeiss Microscopy, Jena, Germany). The images were analyzed with ImageJ.

### 2.8. Photodynamic Therapy

Photodynamic therapy was performed similarly to the process previously described [7,8]. Cells (24,000/well) were seeded in 96-well cell culture plates (Greiner) and allowed to adhere overnight. Cells were then pulsed with VHH-PS in medium without phenol red for 2 h at 37 °C, after which unbound VHH-PS was removed by washing the cells twice with 100 μL medium. Bound VHH-PS was then detected using the Odyssey scanner and an EVOS fluorescence microscope (Thermo Fischer Scientific). Cells were illuminated for 1 h using a 690 diode laser (Modulight, Tampere, Finland) with a 5 mW/cm$^2$ fluence rate for a total light dose of 18 J/cm$^2$, then were incubated overnight at 37 °C in the cell culture incubator. The next day, the cells were screened for viability. For this, cells were incubated with calcein (Thermo Fisher Scientific) and propidium iodide (Thermo Fisher Scientific) to stain live and dead cells, after which they were imaged on an EVOS fluorescence microscope. Cell viability was quantified using the alamarBlue reagent (Thermo Fisher Scientific), which was quantified using a FLUOstar Optima plate reader (BMG Labtech, Ortenberg, Germany). Data were plotted and fitted using Prism 7 software (GraphPad, La Jolla, CA, USA).

## 3. Results

Three previously selected VHHs recognizing the Met ectodomain were taken for further characterization, i.e., E9, G2 and F5 [25]. These VHHs were recloned in an *Escherichia coli* production vector, and binding to the human Met ectodomain was assessed with ELISA (Figure 1A–C). The two VHH clones E9 and G2 bound to the human Met ectodomain with apparent binding affinities ($K_D$) of 7 ± 8 nM and 4 ± 3 nM, respectively. The $K_D$ value of F5 could not be determined properly, but was estimated to be at least higher than 50 nM. To investigate which subdomain these VHHs bind to, ELISAs on chimeric human/llama Met ectodomains were performed. In these chimeric constructs (LS, llama SEMA), either the first 122 (LS1) or 473 (LS5) amino acids of human Met were replaced by those of llama Met (Figure 1B) [32]. Integrity of these constructs was confirmed by the binding of the anti-Met antibody c224G11 (ABT-700 or Telisotuzumab), which is directed against the first IPT region (Figure 1C, left) [32–34]. c224G11 showed a similar binding affinity to both human Met and the two chimeric mutants, whereas no binding of c224G11 to llama Met was observed. Because the VHHs were raised in llama, no cross-reactivity of the VHHs with llama Met or parts of llama Met was expected. The binding affinity and $B_{max}$ of all three VHHs was completely lost for LS5 and strongly reduced on LS1, as compared to human Met, suggesting that the binding epitopes for all three VHHs involve propeller blades 2–6 of the SEMA subdomain. Due to its high affinity for Met and the additional characterization performed previously [25], G2 was selected as the lead clone for Met-targeted PDT.

In order to facilitate directional conjugation of photosensitizers to VHHs without affecting their binding integrity, an additional cysteine was introduced to the C-terminus of G2 via incorporation into the C-terminal C-Direct tag, thereby creating G2c (Figure 2A). Molecular modeling of G2c was performed based on the published structure of a VHH with high homology (Nb10, PDB ID 4DKA:A, C-score 0.13) [27]. This model showed that the unpaired cysteine in the C-Direct tag is located opposite of the CDRs (complementarity determining regions) and close to its framework, which should allow functionalization of VHHs without affecting their binding characteristics (Figure 2B). For higher yield productions, G2c was produced in *Saccharomyces cerevisiae* and purified from supernatant using affinity chromatography and size-exclusion chromatography.

We examined the propensity of G2c to interact with the four variants of the Met ectodomain (human, llama, LS1 and LS5) using surface plasmon resonance (SPR). In agreement with the ELISA data (Figure 1), the SPR experiments confirm that G2c interacts with both human Met and the LS1 variant but not with llama Met or the LS5 variant (Figure 3A), suggesting the involvement of propeller blades 2–6 of the Met SEMA subdomain to be involved in the binding of G2c. The association constant was four times lower towards LS1 compared with human Met (Figure 3B,C). Kinetic parameters calculated at equilibrium conditions (Figure 3D) showed that the binding affinity of G2c for LS1 is lower as compared with its affinity to human Met ($K_D$ of 2.3 ± 0.1 nM for human Met and 4.7 ± 0.5 nM for LS1, $n$ = 3). In conclusion, addition of a cysteine residue to the C-terminal region of G2 did not affect binding affinity to human Met nor did it affect domain specificity.

Subsequently, the free thiol in G2c was conjugated to the water-soluble PS IRDye700DX using maleimide chemistry, resulting in a degree of conjugation of ~1, as determined spectrophotometrically. However, because ~40% of the signal in the solution was still free PS (Figure 4A), a degree of conjugation of ~0.6 would be a more realistic estimation. Because of the hydrophilic nature of IRDye700DX and the lack of toxicity observed for free PS in our previous study [7], we decided to continue with the conjugated G2c-PS. Conjugation of PS to G2c only mildly affected its apparent binding affinity for Met, as determined by ELISA (with apparent affinity values of 2.2 ± 0.2 nM for G2, 1.8 ± 0.1 nM for G2c, and 5.9 ± 0.3 nM for G2c-PS, Figure 4B). In addition, G2c-PS was also still able to bind to the Met-overexpressing MKN45 cells (Figure 5A). No binding of G2c was detected on Met-negative TOV112D cells. For the in vitro PDT experiments, cells were pulsed with a concentration range of G2c-PS for 2 h at 37 °C, a pulse duration that reflects the circulation of VHHs in blood and the time required for tumor uptake of fluorescently labeled VHHs [35]. Besides acting as a PS via the production of reactive oxygen species, IRDye700DX is also a fluorophore and can therefore be used for detecting binding of the conjugate to cells. During the pulse, G2c-PS was able to associate with Met-expressing MKN45 cells with half-max signals being obtained at concentrations comparable to the binding affinity of the VHH for the Met ectodomain (Figure 5B). Little association of G2c-PS was observed for the Met-negative TOV112D cells. Uptake of G2c-PS by MKN45 cells was initially suggested in wide-field microscopy images (Figure 5C), which was confirmed by confocal microscopy imaging (Figure 5D). Exposure of MKN45 cells to different concentrations of G2c-PS, combined with illumination, resulted in ~0% cell viability at nanomolar concentrations (Figure 6A). This was also confirmed by fluorescence microscopy, in which almost all cells that were pulsed with G2c-PS and subsequently illuminated showed an uptake of propidium iodide and an absence of calcein staining (Figure 6B). We subsequently assessed the ability of G2c-PS to induce cell death of two other Met-expressing cell lines. Although showing lower EC50 values compared with MKN45 cells, G2c-PS could reduce the viability of the previously described TOV + Met cell line and the liver hepatocellular carcinoma cell line HepG2 in a dose-dependent manner (Figure 6C).

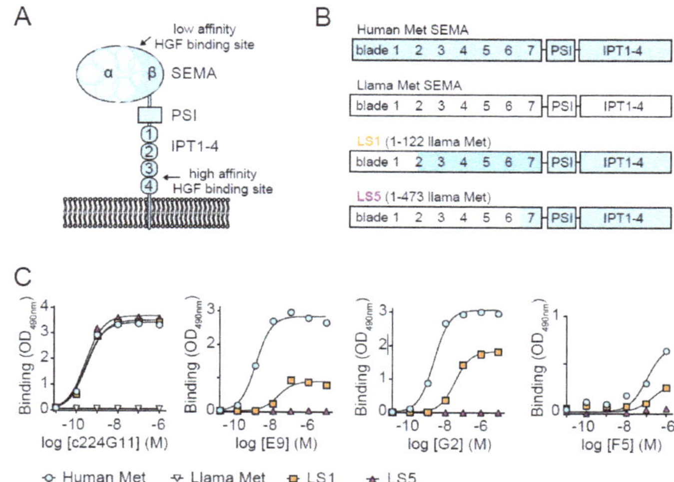


**Figure 1.** Selected VHHs (variable domains from the heavy chain of llama heavy chain antibodies) recognize SEMA (semaphorin-like) subdomain on human Met ectodomain. (**A**) Met ectodomain consisting of a large SEMA subdomain (a cleaved seven-bladed propeller alpha and beta subunit), a PSI (plexin, semaphorin, integrin-like) subdomain, and four IPT (immune-globulin-like, plexins, transcription factors) subdomains. HGF (hepatocyte growth factor) binds with low affinity to the SEMA subdomain and with high affinity to the IPT subdomains. (**B**) Schematic representation of human (blue) and llama (white) Met and the two chimeric Met variants LS1 and LS5 (llama SEMA) in which either the first 122 or 473 amino acids of the human Met were replaced with those from llama. (**C**) Representative figure of binding of either the conventional control antibody c224G11 or the VHHs E9, G2, or F5 to the four Met variants in ELISA ($n = 2$).

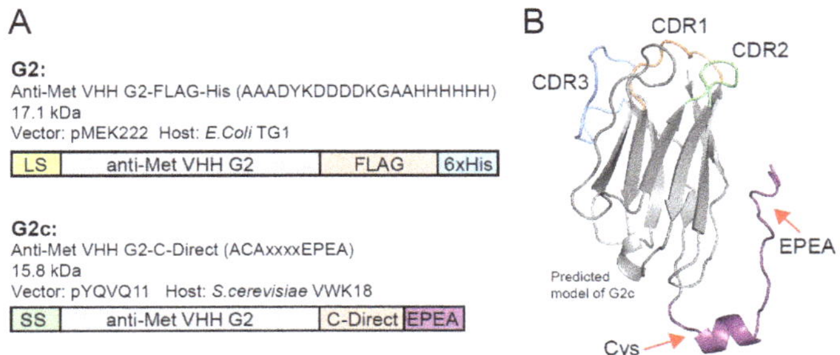


**Figure 2.** Production of G2 with a C-terminal tag containing a free thiol (G2c) and conjugation to PS. (**A**) Schematic representation of the anti-Met VHH G2, expressed as a FLAG-His tagged protein (top) and G2c carrying an additional cysteine in its C-terminal C-Direct tag. The molecular weights of the proteins are indicated, as well as the expression vector and production organism used. LS, pelB leader sequence. SS, Suc2 secretion sequence. (**B**) Predicted model of G2c in which the C-terminal tag (purple) containing the unpaired cysteine and the EPEA (Glu, Pro, Glu, Ala) affinity tag is indicated with red arrows. CDRs (complementarity determining regions) are located on the opposing end of the VHH. The model was based on the structure of Nb10 (85% amino acid sequence homology with QME-G2) as published by Park et al. (PDB ID 4DKA:A, C-score 0.13) [27].

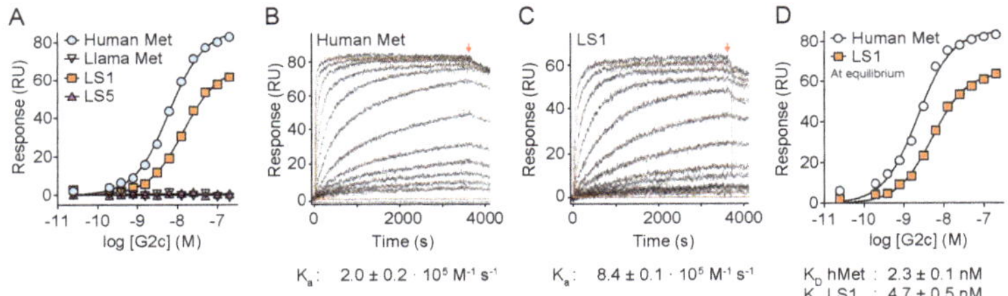


**Figure 3.** Binding analysis of G2c on Met using surface plasmon resonance (SPR). (**A**) Binding of G2c to wild-type human Met ectodomain and the LS1 llama/human chimera, but not the llama or the LS5 chimeric Met ectodomain proteins. (**B**,**C**) Representative SPR sensorgrams of the association phase and dissociation phase (8 min) (starting at arrow) of G2c binding to human Met (**B**) and LS1 chimera (**C**). Kinetic fitting is shown in orange. (**D**) Equilibrium-binding plot of G2c to human Met and LS1 chimeric variant. RU refers to response units. Kinetic parameters and equilibrium dissociation constants are the average ± SD of three independent measurements.

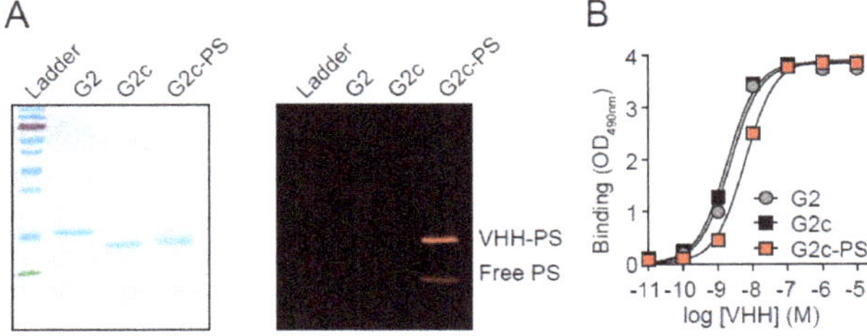


**Figure 4.** Conjugation of G2c to photosensitizer IRDye700DX. (**A left**) Reducing SDS-PAGE gel showing purified G2 from *Escherichia coli*, G2c from *Saccharomyces cerevisiae*, and G2c-PS. Size-separated proteins were stained with PageBlue stain. (**A right**) G2c-PS and free PS in the gel shown on the left, as detected with an Odyssey scanner before the PageBlue stain. (**B**) Binding of G2, G2c, and G2c-PS to Met ectodomain in ELISA (representative figure, mean ± SD).

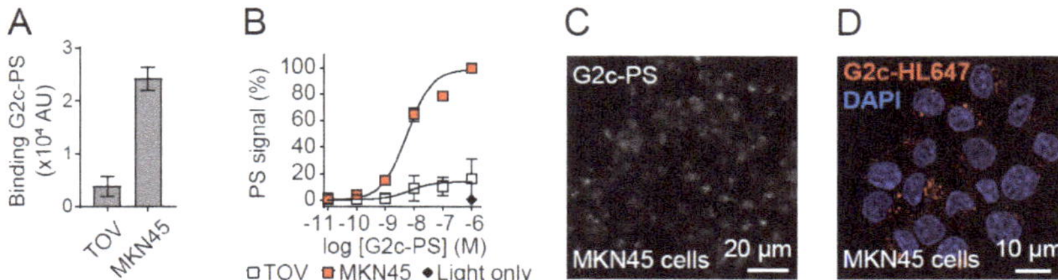


**Figure 5.** Met-targeting G2c-PS conjugates associate with Met-overexpressing MKN45 cells. (**A**) Binding of G2c-PS to MKN45 cells and not to TOV112D cells. Cells were incubated with 50 nM G2c-PS for 2 h on ice and PS fluorescence was detected with an Odyssey infrared scanner. Mean ± SD of $n = 2$ is shown. (**B**) Increased cell association of G2c-PS with MKN45 cells with increasing concentrations. MKN45 and TOV112D cells were pulsed with G2c-PS for 2 h at 37 °C. PS fluorescence was detected with an Odyssey infrared scanner. Mean ± SD of $n = 3$ is shown. (**C**) Association of G2c-PS in MKN45 cells. Cells were pulsed for 2 h with 1 µM of G2c-PS and imaged by wide-field fluorescent microscopy. (**D**) Uptake of G2c-HL647 in MKN45 cells. Cells were pulsed for 2 h with 1 µM of G2c-HL647 (red) and imaged by confocal laser scanning microscopy.

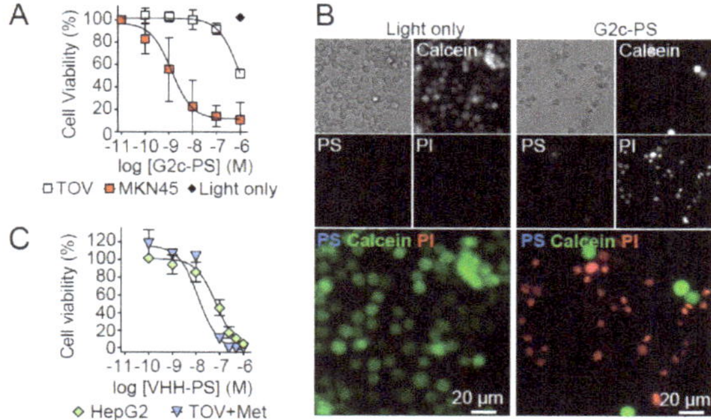


**Figure 6.** Met-targeting G2c-PS induce cell death of Met-expressing cells upon activation by illumination. (**A**) Selective killing of Met-overexpressing MKN45 tumor cells and not of Met-negative TOV112D cells by G2c-targeted PDT. Cells were pulsed for 2 h at 37 °C and subsequently illuminated to activate the PS. Cell viability was assessed one day later. Mean ± SD of $n = 3$ is shown. (**B**) Phase contrast and fluorescence microscopy (EVXOS) of PDT-treated MKN45 cells showing G2c-PS (blue in merge, visible as magenta (red + blue)), live cells (calcein, green), and dead cells (propidium iodide, red). (**C**) Induction of cell death of Met-expressing TOV + Met and HepG2 cells by G2c-PS upon a 2 h pulse followed by PS activation via illumination. Representative plot of $n = 2$.

## 4. Discussion

PDT is a valuable method for inducing local cell death of malignant cells by the local activation of PS. Unfortunately, clinically employed PS are administered systemically, are relatively hydrophobic, and passively accumulate in the tumor as a result of their hydrophobicity in combination with the enhanced permeability and retention (EPR) effect [36,37]. This results in a sub-optimal tumor uptake of PS, off-target toxicity, and long photosensitivity [38]. To enhance the specific tumor uptake of photosensitizers, PS-delivering nanocarriers have been developed and water-soluble PSs have been conjugated to tumor-specific peptides and antibodies (the latter is also known as photo-immunotherapy, or PIT) [39–41]. Because the extent of damage caused by PDT is correlated to the amount of PS delivered, ideal targets for these approaches are highly expressed on tumor cells and preferably in a homogeneous fashion throughout the solid tumor mass. Examples of tumor targets are the typical tumor-related receptors EGFR [42,43] and HER2 [44], although the interleukin-2 receptor [43] and carcinoembryonic antigen have also been used as valuable tumor biomarkers [45–47]. In some types of cancers (e.g., non-small cell lung cancer, or colorectal cancer), treatment with anti-EGFR therapy has resulted in the intratumoral clonal selection of therapy-resistant cells [48]. This can be the result of activating mutations of signaling proteins from the EGFR signaling pathways, such as KRAS. Alternatively, other receptor tyrosine kinases (RTK) can become activated and, in most of these cases, upregulation of the Met receptor tyrosine kinase is regarded as one of the main mechanisms for this type of resistance [49–52]. (The Met RTK is also considered a good tumor target, and therapeutic anti-Met antibodies have been developed [32,34,53].) In order to expand possible therapeutic strategies for certain cancers, we here evaluated whether Met could also serve as a target for targeted PDT.

Targeted PDT using antibodies as targeting agents (i.e., PIT) is currently being clinically evaluated for the treatment of recurrent head and neck cancer (ClinicalTrials.gov: NCT02422979). These results are eagerly awaited, as these are the first tests in patients and thus could significantly advance the field of targeted PDT. Nevertheless, we consider further improvements a necessity, and in the current study we have employed nanobodies or VHHs as targeting agents. Conventional antibodies of the IgG-type are large, dimeric molecules, typically designed for extended blood circulation. These characteristics may lead to slow tumor penetration/accumulation, poor tumor to normal tissue ratios, and long photosensitivity [54]. Reduction of the duration of photosensitivity might be achieved by the rapid tumor accumulation and subsequent rapid clearance of PS from the body. Antibody fragments smaller than IgG like Fabs, scFvs, or VHH/nanobodies have these properties without compromising their affinity for their tumor biomarkers [55]. As we have described before, VHHs in particular are characterized by rapid tumor accumulation and rapid blood clearance, combined with the ability to bind targets with high affinity [16,35]. Another argument for using VHHs over conventional antibody fragments might be the short distance between the conjugated PS and the paratope on the target (~4 nm). Because the reactive oxygen species generated by PS are short lived (<40 ns) and may travel only a short distance (<20 nm), the delivery of the PS close to the cell membrane or to vital cell organelles might favor its potency [56,57]. For that same reason, it would be of interest to assess the potency of Met-targeting PS conjugates recognizing epitopes closer to the membrane than the SEMA subdomain, such as the PSI or IPT subdomains [23]. This would require new PSI- or IPT-specific VHHs.

By using chimeric constructs consisting of different human and Llama Met fragments, we were able to show that all three Met-targeting VHHs used in this study bind to the SEMA subdomain of the Met ectodomain. The best SEMA binder (i.e., G2) was provided with a C-terminal tag containing an unpaired cysteine to allow conjugation to the PS using maleimide chemistry. The addition of this thiol-containing C-Direct tag did not affect the binding characteristics of the VHH, which is in line with previous studies of C-terminal-labeled VHHs via an unpaired cysteine [14–16]. Multiple methods have been described to site-directional functionalize antibody fragments [15,58]. In our experience, incorporation of an unpaired cysteine in the C-terminal tag allowed the straightforward production, purification, and conjugation to commercially available tracers with the use of a single tag [15,16,59].

The binding affinity of G2c for Met was determined on purified ectodomains using two different technologies: ELISA and SPR. In these experiments, the apparent binding affinity of G2c in ELISA was comparable (low nanomolar range) to the $K_D$ values determined under equilibrium by SPR. The SPR analysis does, however, allow the determination of association and dissociation rate constants, which provided additional information on the binding kinetics of G2c on wild-type and chimeric Met proteins. Moreover, similar affinities for G2 were found for binding to Met-expressing cells [25], indicating that cellular components do not influence VHH binding.

Application of the anti-Met conjugate G2c-PS in in vitro PDT resulted in the efficient and specific killing of the Met-overexpressing MKN45 cells and the Met-expressing TOV + Met and HepG2 cells, while the Met negative cell line TOV112D remained undisturbed. The observation that the treated cells could be stained with propidium iodide suggests that cells could have died through necrosis. Further studies would be needed to determine the exact mechanism of cell death. The potency of G2c-PS in killing MKN45 cells was in the nanomolar range. In the two other cell lines, the Met-expressing

HepG2 tumor cells and the TOV112D cells stably expressing Met (TOV + Met), the observed potencies were lower. These differences in potencies might be related to the relative Met expression levels of these cells. MKN45 is a highly Met-overexpressing cell line due to genomic amplification of the Met gene [60]. It would be interesting to assess the potencies and efficacies of Met-targeted PDT on a wider range of Met-expressing tumor cells types. As a consequence of the high expression in MKN45 cells, the receptor is constitutively auto-active and internalized, allowing uptake and intracellular accumulation of the Met-targeted PS [61]. This is in line with what we have observed for G2c-PS and G2c-HL647. As the subcellular localization of PS can influence the mechanism of cell death and the overall potency of the PDT, as suggested in our previous study with different formats of EGFR-targeted VHHs [62], it would be interesting to determine the contribution of PS uptake in the observed effects. The potency and efficacy of the VHH-PS conjugates could be further enhanced by employing a mixture of these Met-targeting conjugates with the previously developed EGFR-targeting ones. This could also affect a larger population of tumor cells, including cells that upregulate Met expression as a resistance mechanism against anti-EGFR therapy, and could be an advantage in tumors with heterogeneously expressed markers such as EGFR and Met. [24,51].

In conclusion, this study has demonstrated that targeting Met using site-directionally conjugated VHH-PS has the ability to specifically kill Met-overexpressing tumor cells. Follow up studies should evaluate the potency of this approach in more relevant models (in vivo). Taken together, Met-targeted PDT might serve as an alternative or complementary approach for combating cancer.

**Author Contributions:** Conceptualization, investigation and analysis: R.H., V.M., M.R.-E. and S.O. Resources: H.d.H. and T.C.V. Supervision: P.M.P.v.B.e.H. and S.O. Writing: R.H., V.M., T.C.V., P.M.P.v.B.e.H. and S.O.

**Funding:** This project received funding from the European Research Council (ERC) under the European Union's Horizon 2020 research and innovation program (grant agreement No. 677582).

**Conflicts of Interest:** R.H. and T.C.V. are affiliated with QVQ Holding BV, a company offering VHH services and VHH-based imaging molecules. P.M.P.v.B.e.H. owns shares of this company. H.d.H. is affiliated with argenx BVBA, a company developing therapeutic antibodies. V.M., M.R.-E., and S.O. declare no conflict of interest. The funders had no role in the design of the study; in the collection, analyses, or interpretation of data; in the writing of the manuscript, or in the decision to publish the results.

## References

1. Van Straten, D.; Mashayekhi, V.; de Bruijn, H.S.; Oliveira, S.; Robinson, D.J. Oncologic photodynamic therapy: Basic principles, current clinical status and future directions. *Cancers (Basel)* **2017**, *9*, 19. [CrossRef] [PubMed]
2. Agostinis, P.; Berg, K.; Cengel, K.A.; Foster, T.H.; Girotti, A.W.; Gollnick, S.O.; Hahn, S.M.; Hamblin, M.R.; Juzeniene, A.; Kessel, D.; et al. Photodynamic therapy of cancer: An update. *CA Cancer J. Clin.* **2011**, *61*, 250–281. [CrossRef]
3. Kwiatkowski, S.; Knap, B.; Przystupski, D.; Saczko, J.; Kedzierska, E.; Knap-Czop, K.; Kotlinska, J.; Michel, O.; Kotowski, K.; Kulbacka, J. Photodynamic therapy—Mechanisms, photosensitizers and combinations. *Biomed. Pharmacother.* **2018**, *106*, 1098–1107. [CrossRef] [PubMed]
4. Gomer, C.J. Preclinical examination of first and second generation photosensitizers used in photodynamic therapy. *Photochem. Photobiol.* **1991**, *54*, 1093–1107. [CrossRef]
5. Githaka, N.; Konnai, S.; Skilton, R.; Kariuki, E.; Kanduma, E.; Murata, S.; Ohashi, K. Genotypic variations in field isolates of Theileria species infecting giraffes (Giraffa camelopardalis tippelskirchi and Giraffa camelopardalis reticulata) in Kenya. *Parasitol. Int.* **2013**, *62*, 448–453. [CrossRef] [PubMed]
6. Bechet, D.; Couleaud, P.; Frochot, C.; Viriot, M.L.; Guillemin, F.; Barberi-Heyob, M. Nanoparticles as vehicles for delivery of photodynamic therapy agents. *Trends Biotechnol.* **2008**, *26*, 612–621. [CrossRef]
7. Heukers, R.; van Bergen en Henegouwen, P.M.P.; Oliveira, S. Nanobody-targeted photodynamic therapy for oncology. *Photodiagnosis Photodyn. Ther.* **2015**, *12*, 339. [CrossRef]
8. Van Driel, P.B.A.A.; Boonstra, M.C.; Slooter, M.D.; Heukers, R.; Stammes, M.A.; Snoeks, T.J.A.; De Bruijn, H.S.; Van Diest, P.J.; Vahrmeijer, A.L.; Van Bergen En Henegouwen, P.M.P.; et al. EGFR targeted nanobody-photosensitizer conjugates for photodynamic therapy in a pre-clinical model of head and neck cancer. *J. Control. Release* **2016**, *229*, 93–105. [CrossRef]

9. Hamers-Casterman, C.; Atarhouch, T.; Muyldermans, S.; Robinson, G.; Hammers, C.; Songa, E.B.; Bendahman, N.; Hammers, R. Naturally occurring antibodies devoid of light chains. *Nature* **1993**, *363*, 446–448. [CrossRef]
10. Muyldermans, S. Nanobodies: Natural Single-Domain Antibodies. *Annu. Rev. Biochem.* **2013**, *82*, 775–797. [CrossRef]
11. Harmsen, M.M.; De Haard, H.J. Properties, production, and applications of camelid single-domain antibody fragments. *Appl. Microbiol. Biotechnol.* **2007**, *77*, 13–22. [CrossRef]
12. De Groeve, K.; Deschacht, N.; De Koninck, C.; Caveliers, V.; Lahoutte, T.; Devoogdt, N.; Muyldermans, S.; De Baetselier, P.; Raes, G. Nanobodies as tools for in vivo imaging of specific immune cell types. *J. Nucl. Med.* **2010**, *51*, 782–789. [CrossRef] [PubMed]
13. Massa, S.; Vikani, N.; Betti, C.; Ballet, S.; Vanderhaegen, S.; Steyaert, J.; Descamps, B.; Vanhove, C.; Bunschoten, A.; van Leeuwen, F.W.B.; et al. Sortase A-mediated site-specific labeling of camelid single-domain antibody-fragments: A versatile strategy for multiple molecular imaging modalities. *Contrast Media Mol. Imaging* **2016**, *11*, 328–339. [CrossRef] [PubMed]
14. Massa, S.; Xavier, C.; De Vos, J.; Caveliers, V.; Lahoutte, T.; Muyldermans, S.; Devoogdt, N. Site-specific labeling of cysteine-tagged camelid single-domain antibody-fragments for use in molecular imaging. *Bioconjug. Chem.* **2014**, *25*, 979–988. [CrossRef] [PubMed]
15. Van Lith, S.A.M.; Van Duijnhoven, S.M.J.; Navis, A.C.; Leenders, W.P.J.; Dolk, E.; Wennink, J.W.H.; Van Nostrum, C.F.; Van Hest, J.C.M. Legomedicine—A Versatile Chemo-Enzymatic Approach for the Preparation of Targeted Dual-Labeled Llama Antibody-Nanoparticle Conjugates. *Bioconjug. Chem.* **2017**, *28*, 539–548. [CrossRef] [PubMed]
16. Kijanka, M.; Warnders, F.J.; El Khattabi, M.; Lub-De Hooge, M.; Van Dam, G.M.; Ntziachristos, V.; De Vries, L.; Oliveira, S.; Van Bergen En Henegouwen, P.M.P. Rapid optical imaging of human breast tumour xenografts using anti-HER2 VHHs site-directly conjugated to IRDye 800CW for image-guided surgery. *Eur. J. Nucl. Med. Mol. Imaging* **2013**, *40*, 1718–1729. [CrossRef] [PubMed]
17. Köstler, W.J.; Yarden, Y. The epidermal growth factor receptor family. In *Handbook of Cell Signaling*, 2nd ed.; Elsevier: Amsterdam, The Netherlands, 2010; Volume 2, pp. 435–441, ISBN 9780123741455.
18. Bublil, E.M.; Yarden, Y. The EGF receptor family: Spearheading a merger of signaling and therapeutics. *Curr. Opin. Cell Biol.* **2007**, *19*, 124–134. [CrossRef]
19. Peruzzi, B.; Bottaro, D.P. Targeting the c-Met signaling pathway in cancer. *Clin. Cancer Res.* **2006**, *12*, 3657–3660. [CrossRef]
20. Birchmeier, C.; Birchmeier, W.; Gherardi, E.; Vande Woude, G.F. Met, metastasis, motility and more. *Nat. Rev. Mol. Cell Biol.* **2003**, *4*, 915–925. [CrossRef]
21. Cooper, C.S.; Park, M.; Blair, D.G.; Tainsky, M.A.; Huebner, K.; Croce, C.M.; Vande Woude, G.F. Molecular cloning of a new transforming gene from a chemically transformed human cell line. *Nature* **1984**, *311*, 29–33. [CrossRef]
22. Stamos, J.; Lazarus, R.A.; Yao, X.; Kirchhofer, D.; Wiesmann, C. Crystal structure of the HGF β-chain in complex with the Sema domain of the Met receptor. *EMBO J.* **2004**, *23*, 2325–2335. [CrossRef] [PubMed]
23. Gherardi, E.; Youles, M.E.; Miguel, R.N.; Blundell, T.L.; Iamele, L.; Gough, J.; Bandyopadhyay, A.; Hartmann, G.; Butler, P.J.G. Functional map and domain structure of MET, the product of the c-met protooncogene and receptor for hepatocyte growth factor/scatter factor. *Proc. Natl. Acad. Sci. USA* **2003**, *100*, 12039–12044. [CrossRef] [PubMed]
24. Bradley, C.A.; Salto-Tellez, M.; Laurent-Puig, P.; Bardelli, A.; Rolfo, C.; Tabernero, J.; Khawaja, H.A.; Lawler, M.; Johnston, P.G.; Van Schaeybroeck, S. Targeting c-MET in gastrointestinal tumours: Rationale, opportunities and challenges. *Nat. Rev. Clin. Oncol.* **2017**, *14*, 562–576. [CrossRef] [PubMed]
25. Heukers, R.; Altintas, I.; Raghoenath, S.; De Zan, E.; Pepermans, R.; Roovers, R.C.; Haselberg, R.; Hennink, W.E.; Schiffelers, R.M.; Kok, R.J.; et al. Targeting hepatocyte growth factor receptor (Met) positive tumor cells using internalizing nanobody-decorated albumin nanoparticles. *Biomaterials* **2014**, *35*, 601–610. [CrossRef] [PubMed]
26. Roy, A.; Kucukural, A.; Zhang, Y. I-TASSER: A unified platform for automated protein structure and function prediction. *Nat. Protoc.* **2010**, *5*, 725–738. [CrossRef] [PubMed]

27. Park, Y.J.; Budiarto, T.; Wu, M.; Pardon, E.; Steyaert, J.; Hol, W.G.J. The structure of the C-terminal domain of the largest editosome interaction protein and its role in promoting RNA binding by RNA-editing ligase L2. *Nucleic Acids Res.* **2012**, *40*, 6966–6977. [CrossRef] [PubMed]
28. Gorlani, A.; Lutje Hulsik, D.; Adams, H.; Vriend, G.; Hermans, P.; Verrips, T. Antibody engineering reveals the important role of J segments in the production efficiency of llama single-domain antibodies in Saccharomyces cerevisiae. *Protein Eng. Des. Sel.* **2012**, *25*, 39–46. [CrossRef] [PubMed]
29. Thomassen, Y.E.; Meijer, W.; Sierkstra, L.; Verrips, C.T. Large-scale production of VHH antibody fragments by Saccharomyces cerevisiae. *Enzyme Microb. Technol.* **2002**, *30*, 273–278. [CrossRef]
30. Van De Laar, T.; Visser, C.; Holster, M.; López, C.G.; Kreuning, D.; Sierkstra, L.; Lindner, N.; Verrips, T. Increased heterologous protein production by Saccharomyces cerevisiae growing on ethanol as sole carbon source. *Biotechnol. Bioeng.* **2007**, *96*, 483–494. [CrossRef]
31. Gorlani, A.; De Haard, H.; Verrips, T. Expression of VHHs in saccharomyces cerevisiae. *Methods Mol. Biol.* **2012**, *911*, 277–286. [CrossRef]
32. Basilico, C.; Hultberg, A.; Blanchetot, C.; De Jonge, N.; Festjens, E.; Hanssens, V.; Osepa, S.I.; De Boeck, G.; Mira, A.; Cazzanti, M.; et al. Four individually druggable MET hotspots mediate HGF-driven tumor progression. *J. Clin. Investig.* **2014**, *124*, 3172–3186. [CrossRef] [PubMed]
33. Gonzalez, A.; Broussas, M.; Beau-Larvor, C.; Haeuw, J.F.; Boute, N.; Robert, A.; Champion, T.; Beck, A.; Bailly, C.; Corvaïa, N.; et al. A novel antagonist anti-cMet antibody with antitumor activities targeting both ligand-dependent and ligand-independent c-Met receptors. *Int. J. Cancer* **2016**, *139*, 1851–1863. [CrossRef] [PubMed]
34. Prat, M.; Oltolina, F.; Basilico, C. Monoclonal Antibodies against the MET/HGF Receptor and Its Ligand: Multitask Tools with Applications from Basic Research to Therapy. *Biomedicines* **2014**, *2*, 359–383. [CrossRef] [PubMed]
35. Oliveira, S.; van Dongen, G.A.M.S.; Stigter-van Walsum, M.; Roovers, R.C.; Stam, J.C.; Mali, W.; van Diest, P.J.; van Bergen en Henegouwen, P.M.P. Rapid visualization of human tumor xenografts through optical imaging with a near-infrared fluorescent anti-epidermal growth factor receptor nanobody. *Mol. Imaging* **2012**, *11*, 33–46. [CrossRef] [PubMed]
36. Fang, J.; Nakamura, H.; Maeda, H. The EPR effect: Unique features of tumor blood vessels for drug delivery, factors involved, and limitations and augmentation of the effect. *Adv. Drug Deliv. Rev.* **2011**, *63*, 136–151. [CrossRef] [PubMed]
37. Iyer, A.K.; Khaled, G.; Fang, J.; Maeda, H. Exploiting the enhanced permeability and retention effect for tumor targeting. *Drug Discov. Today* **2006**, *11*, 812–818. [CrossRef]
38. Allison, R.R.; Sibata, C.H. Oncologic photodynamic therapy photosensitizers: A clinical review. *Photodiagnosis Photodyn. Ther.* **2010**, *7*, 61–75. [CrossRef] [PubMed]
39. Shirasu, N.; Nam, S.O.; Kuroki, M. Tumor-targeted photodynamic therapy. *Anticancer Res.* **2013**, *33*, 2823–2832.
40. Bugaj, A.M. Targeted photodynamic therapy—A promising strategy of tumor treatment. *Photochem. Photobiol. Sci.* **2011**, *10*, 1097–1109. [CrossRef]
41. Lucky, S.S.; Soo, K.C.; Zhang, Y. Nanoparticles in photodynamic therapy. *Chem. Rev.* **2015**, *115*, 1990–2042. [CrossRef]
42. Savellano, M.D.; Hasan, T. Photochemical targeting of epidermal growth factor receptor: A mechanistic study. *Clin. Cancer Res.* **2005**, *11*, 1658–1668. [CrossRef] [PubMed]
43. Nakajima, T.; Sano, K.; Choyke, P.L.; Kobayashi, H. Improving the efficacy of photoimmunotherapy (PIT) using a cocktail of antibody conjugates in a multiple antigen tumor model. *Theranostics* **2013**, *3*, 357–365. [CrossRef] [PubMed]
44. Savellano, M.D.; Pogue, B.W.; Hoopes, P.J.; Vitetta, E.S.; Paulsen, K.D. Multiepitope HER2 targeting enhances photoimmunotherapy of HER2-overexpressing cancer cells with pyropheophorbide-a immunoconjugates. *Cancer Res.* **2005**, *65*, 6371–6379. [CrossRef] [PubMed]
45. Shirasu, N.; Yamada, H.; Shibaguchi, H.; Kuroki, M.; Kuroki, M. Potent and specific antitumor effect of CEA-targeted photoimmunotherapy. *Int. J. Cancer* **2014**, *135*, 2697–2710. [CrossRef] [PubMed]

46. Maawy, A.A.; Hiroshima, Y.; Zhang, Y.; Heim, R.; Makings, L.; Garcia-Guzman, M.; Luiken, G.A.; Kobayashi, H.; Hoffman, R.M.; Bouvet, M. Near infra-red photoimmunotherapy with anti-CEA-IR700 results in extensive tumor lysis and a significant decrease in tumor burden in orthotopic mouse models of pancreatic cancer. *PLoS ONE* **2015**, *10*, e0121989. [CrossRef] [PubMed]
47. Carcenac, M.; Larroque, C.; Langlois, R.; Van Lier, J.E.; Artus, J.C.; Pèlegrin, A. Preparation, phototoxicity and biodistribution studies of anti-carcinoembryonic antigen monoclonal antibody-phthalocyanine conjugates. *Photochem. Photobiol.* **1999**, *70*, 930–936. [CrossRef]
48. Brand, T.M.; Iida, M.; Wheeler, D.L. Molecular mechanisms of resistance to the EGFR monoclonal antibody cetuximab. *Cancer Biol. Ther.* **2011**, *11*, 777–792. [CrossRef]
49. Misale, S.; Di Nicolantonio, F.; Sartore-Bianchi, A.; Siena, S.; Bardelli, A. Resistance to Anti-EGFR therapy in colorectal cancer: From heterogeneity to convergent evolution. *Cancer Discov.* **2014**, *4*, 1269–1280. [CrossRef]
50. Bardelli, A.; Corso, S.; Bertotti, A.; Hobor, S.; Valtorta, E.; Siravegna, G.; Sartore-Bianchi, A.; Scala, E.; Cassingena, A.; Zecchin, D.; et al. Amplification of the MET receptor drives resistance to anti-EGFR therapies in colorectal cancer. *Cancer Discov.* **2013**, *3*, 658–673. [CrossRef]
51. Wang, S.; Song, Y.; Yan, F.; Liu, D. Mechanisms of resistance to third-generation EGFR tyrosine kinase inhibitors. *Front. Med.* **2016**, *10*, 383–388. [CrossRef]
52. Matsubara, D.; Ishikawa, S.; Sachiko, O.; Aburatani, H.; Fukayama, M.; Niki, T. Co-activation of epidermal growth factor receptor and c-MET defines a distinct subset of lung adenocarcinomas. *Am. J. Pathol.* **2010**, *177*, 2191–2204. [CrossRef] [PubMed]
53. Hultberg, A.; Morello, V.; Huyghe, L.; De Jonge, N.; Blanchetot, C.; Hanssens, V.; De Boeck, G.; Silence, K.; Festjens, E.; Heukers, R.; et al. Depleting MET-expressing tumor cells by ADCC provides a therapeutic advantage over inhibiting HGF/MET signaling. *Cancer Res.* **2015**, *75*, 3373–3383. [CrossRef] [PubMed]
54. Van Dongen, G.A.M.S.; Visser, G.W.M.; Vrouenraets, M.B. Photosensitizer-antibody conjugates for detection and therapy of cancer. *Adv. Drug Deliv. Rev.* **2004**, *56*, 31–52. [CrossRef]
55. Wu, A.M.; Senter, P.D. Arming antibodies: Prospects and challenges for immunoconjugates. *Nat. Biotechnol.* **2005**, *23*, 1137–1146. [CrossRef] [PubMed]
56. Redmond, R.W.; Kochevar, I.E. Spatially Resolved Cellular Responses to Singlet Oxygen. *Photochem. Photobiol.* **2006**, *82*, 1178. [CrossRef] [PubMed]
57. Louryan, S. La formation de la tête des vertébrés: Faits et hypothèses. *Rev. Med. Brux.* **2005**, *26*, 98–102. [CrossRef] [PubMed]
58. Tsuchikama, K.; An, Z. Antibody-drug conjugates: Recent advances in conjugation and linker chemistries. *Protein Cell* **2018**, *9*, 33–46. [CrossRef] [PubMed]
59. van Brussel, A.S.A.; Adams, A.; Oliveira, S.; Dorresteijn, B.; El Khattabi, M.; Vermeulen, J.F.; van der Wall, E.; Mali, W.P.T.M.; Derksen, P.W.B.; van Diest, P.J.; et al. Hypoxia-Targeting Fluorescent Nanobodies for Optical Molecular Imaging of Pre-Invasive Breast Cancer. *Mol. Imaging Biol.* **2016**, *18*, 535–544. [CrossRef]
60. Rege-Cambrin, G.; Scaravaglio, P.; Carozzi, F.; Giordano, S.; Ponzetto, C.; Comoglio, P.M.; Saglio, G. Karyotypic analysis of gastric carcinoma cell lines carrying an amplified c-met oncogene. *Cancer Genet. Cytogenet.* **1992**, *64*, 170–173. [CrossRef]
61. Smolen, G.A.; Sordella, R.; Muir, B.; Mohapatra, G.; Barmettler, A.; Archibald, H.; Kim, W.J.; Okimoto, R.A.; Bell, D.W.; Sgroi, D.C.; et al. Amplification of MET may identify a subset of cancers with extreme sensitivity to the selective tyrosine kinase inhibitor PHA-665752. *Proc. Natl. Acad. Sci. USA* **2006**, *103*, 2316–2321. [CrossRef]
62. Heukers, R.; van Bergen en Henegouwen, P.M.P.; Oliveira, S. Nanobody-photosensitizer conjugates for targeted photodynamic therapy. *Nanomed. Nanotechnol. Biol. Med.* **2014**, *10*, 1441–1451. [CrossRef] [PubMed]

*Article*

# A Collection of Single-Domain Antibodies that Crowd Ricin Toxin's Active Site

**Siva Krishna Angalakurthi [1,†], David J. Vance [2,†], Yinghui Rong [2], Chi My Thi Nguyen [3], Michael J. Rudolph [3], David Volkin [1], C. Russell Middaugh [1], David D. Weis [4] and Nicholas J. Mantis [2,*]**

1 Department of Pharmaceutical Chemistry and Macromolecule and Vaccine Stabilization Center, University of Kansas, Lawrence, KS 660451, USA; s214a774@ku.edu (S.K.A.); volkin@ku.edu (D.V.); middaugh@ku.edu (C.R.M.)
2 Division of Infectious Diseases, Wadsworth Center, New York State Department of Health, Albany, NY 12208, USA; david.vance@health.ny.gov (D.J.V.); yinghui.rong@health.ny.gov (Y.R.)
3 New York Structural Biology Center (NYSBC), New York, NY 10027, USA; ntchimy@gmail.com (C.M.T.N.); mrudolph@nysbc.org (M.J.R.)
4 Department of Chemistry and Ralph Adams Institute for Bioanalytical Chemistry, University of Kansas, Lawrence, KS 660451, USA; dweis@ku.edu
* Correspondence: nicholas.mantis@health.ny.gov; Tel.: +1-518-473-7487
† These authors contributed equally to this manuscript.

Received: 26 November 2018; Accepted: 11 December 2018; Published: 17 December 2018

**Abstract:** In this report, we used hydrogen exchange-mass spectrometry (HX-MS) to identify the epitopes recognized by 21 single-domain camelid antibodies ($V_HHs$) directed against the ribosome-inactivating subunit (RTA) of ricin toxin, a biothreat agent of concern to military and public health authorities. The $V_HHs$, which derive from 11 different B-cell lineages, were binned together based on competition ELISAs with IB2, a monoclonal antibody that defines a toxin-neutralizing hotspot ("cluster 3") located in close proximity to RTA's active site. HX-MS analysis revealed that the 21 $V_HHs$ recognized four distinct epitope subclusters (3.1–3.4). Sixteen of the 21 $V_HHs$ grouped within subcluster 3.1 and engage RTA α-helices C and G. Three $V_HHs$ grouped within subcluster 3.2, encompassing α-helices C and G, plus α-helix B. The single $V_HH$ in subcluster 3.3 engaged RTA α-helices B and G, while the epitope of the sole $V_HH$ defining subcluster 3.4 encompassed α-helices C and E, and β-strand h. Modeling these epitopes on the surface of RTA predicts that the 20 $V_HHs$ within subclusters 3.1–3.3 physically occlude RTA's active site cleft, while the single antibody in subcluster 3.4 associates on the active site's upper rim.

**Keywords:** toxin; antibody; camelid; vaccine; biodefense; hydrogen exchange-mass spectrometry

## 1. Introduction

Ricin is a member of the ribosome-inactivating protein (RIP) family of toxins and classified as a biothreat agent due to its high potential to induce morbidity and mortality after inhalation [1–3]. The toxin is a ~65 kDa heterodimeric glycoprotein from the castor bean plant (*Ricinus communis*) consisting of a binding subunit (RTB) and an enzymatic subunit (RTA). RTB is a galactose/N-acetyl galactosamine (Gal/GalNAc)-specific lectin that promotes toxin attachment and entry into mammalian cells [4]. RTA is an RNA N-glycosidase (EC 3.2.2.22) that depurinates a conserved adenosine within the sarcin-ricin loop (SRL) of 28S rRNA, thereby stalling ribosome translocation [5,6]. At the structural level, RTA is a globular protein with a total of 10 β-strands (A–J) and seven α-helices (A–G). RTA folds into three distinct domains: domain 1 (residues 1–117) is dominated by a six-stranded β-sheet, domain 2 (residues 118–210), by five α-helices, and domain 3 (residues 211–267), which interfaces with RTB through hydrophobic interactions and a single disulfide bond [7,8]. RTA's active site constitutes a shallow pocket formed at the interface of the three domains [8,9]. Active site residues include Tyr80, Tyr123, Glu177, Arg180, and Trp211 (Figure 1A) [10].

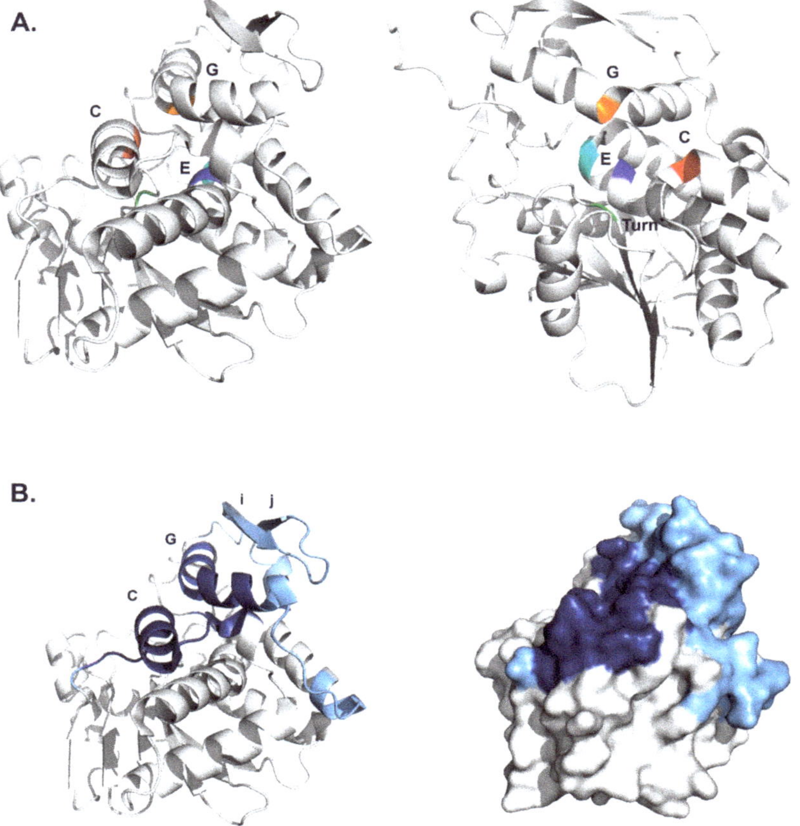


**Figure 1.** Structure of enzymatic subunit (RTA) with active site residues and IB2's epitope highlighted. (**A**) The residues that constitute the RTA's active site are in α-helices C, G and E, and a loop between β-strands e and f. The following residues are colored: Tyr80 (green); Tyr123 (red); Glu 177 (blue); Arg 180 (cyan) Trp211 (orange). (**B**) IB2's epitope on RiVax adapted from previous publication [11]. The color shading corresponds to strong (deep blue) and intermediate (light blue) protection. No significant interaction is colored gray.

Inhalation of ricin results in severe lung inflammation characterized by an influx of neutrophils, alveolar edema, and hemorrhage, presumably initiated by the intoxication of alveolar macrophages and lung epithelial cells [1,12,13]. Non-human primates (NHPs) exposed to ricin by aerosol succumb to the effects of the toxin within 24–52 h [12,14]. At the present time, medical intervention following ricin exposure is strictly supportive [15]. However, vaccination strategies have shown great promise in affording complete or near complete protection against ricin intoxicosis in mice and NHPs [16]. For example, intramuscular administration of RiVax, a non-toxic thermostabilized recombinant RTA-based subunit vaccine adjuvanted with aluminum salts, to Rhesus macaques was sufficient to confer immunity to a lethal dose (LD) ricin challenge delivered by aerosol [14]. In vivo neutralization of ricin toxin following vaccination is associated with onset of anti-RTA IgG antibodies in serum and bronchoalveolar lavage (BAL) fluid [13,14,17].

Monoclonal (mAb) and polyclonal (pAb) antibody responses in mice, rabbits, and NHPs elicited by RiVax vaccination are directed against four spatially distinct immunodominant regions on RTA, which we refer to as epitope clusters 1–4 [11,18–23]. A combination of competition ELISAs, X-ray crystallography, and hydrogen exchange-mass spectrometry (HX-MS) has revealed key secondary elements associated with each cluster. Cluster 1 encompasses RTA's β-strand h (residues 113–117) and α-helix B (94–107), a protruding immunodominant secondary structure element previously known to be a target of potent toxin-neutralizing antibodies [24]. Cluster 2 consists of two subclusters: one involving α-helix A (14–24) and α-helices F–G (184–207) and the other encompassing β-strands d-e (62–69) and parts of α-helices D–E (154–164). Cluster 3 involves α-helices C (121–135) and G (207–217) near RTA's active site, while Cluster 4 is proposed to form a diagonal sash from the front to back of RTA spanning β-strands b, c, and d (35–59). Our long-term goal is to generate a comprehensive molecular B-cell epitope map of each of these clusters and define the specific antibody-contact points on RTA that render the toxin inactive. Such information will be invaluable in efforts to deconvolute the complex human antibody response profile to ricin toxin and RiVax [25].

While much has been learned about clusters 1 and 2 over recent years, comparatively little is known about cluster 3, as it is defined by only a single mAb called IB2 [11]. IB2 was first identified as a toxin-neutralizing mouse mAb that, in competition ELISAs, proved to be distinct from other mAbs in our collection at the time [18,26]. IB2 can passively protect mice against a $5 \times LD_{50}$ ricin challenge by injection, indicating it has neutralizing activity in vivo, and must, by definition, interact with an important element on ricin toxin. As noted above, we recently demonstrated by HX-MS analysis that IB2 recognizes an epitope involving RTA's α-helix C (residues 121–135) and α-helix G (residues 207–217), which is not only in close proximity to RTA's active site but includes two active site residues, Tyr123 and Trp211 (Figure 1B). However, efforts to interrogate cluster 3 in more detail have been hindered by the absence other cluster 3-specific mAbs. Indeed, recent screens of B-cell hybridomas derived from RiVax and ricin toxoid immunized mice failed to identify additional cluster 3 antibodies [27].

Whereas isolation of additional IB2-like mouse mAbs has not been fruitful, we did recently identify 21 unique heavy chain-only single-domain camelid antibodies ($V_HHs$) that are competed by IB2 for binding to ricin toxin (D. Vance, C. Shoemaker, N. Mantis, manuscript in preparation) [23]. We wished to characterize these $V_HHs$ in detail with respect to their binding affinities, epitopes, and capacities to neutralize ricin. In this report, we localized by HX-MS the epitopes of all 21 of these $V_HHs$. We found that the 21 $V_HHs$ fall within one of four distinct but overlapping subclusters (3.1–3.4) that share at least one secondary element contacted by IB2. Only two of the 21 $V_HHs$, V6D4 and V1D3, have appreciable toxin-neutralizing activity (TNA), which we speculate is due to their epitope specificity along with strong binding affinity to toxin. This work furthers our overall goal of constructing a complete B-cell epitope map of ricin toxin.

## 2. Materials and Methods

### 2.1. *RiVax and $V_HH$ Production*

RiVax was expressed and purified from *E. coli*, as described [28]. Please note that RiVax differs from native RTA at two positions, which render the enzyme inactive: there is an Ala at position 80 substituted for Tyr, and a Met at position 76 in place of Val [29]. RiVax also lacks high mannose residues normally found on RTA, due to the fact that RiVax is expressed in *E. coli*. In addition, the RiVax used here has the addition of an Ala at the N-terminus, which we denoted as residue 0 for simplicity. $V_HHs$ were expressed in *E. coli* as either thioredoxin- and E-tagged constructs or tag-free variants [22].

### 2.2. Competition ELISA

NUNC microtiter plates (Fisher Scientific, Hampton, NH) were coated with competitor mAbs (1 μg/mL in Phosphate Buffered Saline (PBS)) overnight at 4 °C and then blocked for 2 h with 2% goat serum (Gibco, Gaithersburg, MD, USA) in 0.1% PBST. Ricin (1 μg/mL) (Vector Labs, Burlingame, CA, USA) was then captured by the mAbs and probed with $V_HH$ analytes at 330 nM. Bound $V_HHs$ were detected with an anti-E-tag-HRP secondary antibody (Bethyl Labs, Montgomery, TX, USA) and developed with SureBlue 3,3′,5,5′-tetramethylbenzidine (TMB) substrate (SeraCare, Milford, MA, USA). After quenching with 1 M phosphoric acid (Sigma Aldrich, Carlsbad, CA, USA), absorbance was read at 450 nm on a VersaMax microplate reader (Molecular Devices, Sunnyvale, CA, USA). % inhibition was calculated by comparing absorbance of captured $V_HHs$ on each mAb-ricin complex with that of the absorbance of each $V_HH$ captured onto SylH3-ricin, where SylH3 is an anti-RTB mAb that does not interfere with the binding of any $V_HHs$ to RTA's cluster 3.

### 2.3. Vero Cell Cytotoxicity Assay

Vero cells were detached from culture dishes with trypsin (Gibco), seeded into white 96-well cell culture treated plates (Fisher Scientific) (100 uL per well, $5 \times 10^4$ cells/mL) and allowed to adhere overnight. The cells were then treated with Dulbecco's Modified Eagle Medium (DMEM) alone, ricin alone (10 ng/mL), or a mixture of ricin with $V_HHs$ at five-fold dilutions. After 2 h at 37 °C, the culture medium was changed, and the cells were incubated at 37 °C for ~48 h. Viability was assessed using CellTiter-GLO (Promega, Madison, WI, USA). All treatments were performed in triplicate and repeated at least three times.

### 2.4. Affinity Determinations

$V_HH$ association and dissociation rates were determined by SPR using a ProteOn XPR36 system (Bio-Rad Inc., Hercules, CA, USA). Ricin was immobilized on a general layer compact (GLC) chip (Bio-Rad Inc.) equilibrated in PBS-0.005% Tween running buffer at a flow rate of 30 μL/min. Following EDAC [N-ethyl-N=-(3-dimethylaminopropyl) carbodiimide hydrochloride] (200 mM)–sulfo-NHS (N-hydroxysulfosuccinimide) (50 mM) activation (3 min), ricin was diluted in 10 mM sodium acetate (pH 5.0) at either 4 μg/mL or 2 μg/mL and coupled for 2 min. A third vertical channel received only acetate buffer and served as a reference channel. The surfaces were deactivated using 1 M ethanolamine for 5 min. A ProteOn array system multichannel module (MCM) was rotated to the horizontal orientation for affinity determination experiments. Each $V_HH$ was serially diluted in running buffer and then injected at 50 μL/min for 180 s, followed by 1 to 3 h of dissociation. After each experiment, the chip was regenerated with 10 mM glycine (pH 1.5) at 100 μL/min for 18 s, until the response unit (RU) values had returned to baseline. All kinetic experiments were performed at 25 °C. Kinetic constants for the antibody/ricin interactions were obtained with ProteOn Manager software 3.1.0 (Bio-Rad Inc.) using the Langmuir fit model.

### 2.5. HX-MS

HX-MS experiments for epitope mapping were conducted essentially as described previously [11]. Briefly, a H/DX PAL™ robotic system (LEAP Technologies, Morrisville, NC, USA) was used for sample preparation, mixing and injection. For the free RiVax, 4 μL of 20 μM RiVax stock solution was incubated with 36 μL of deuterated buffer (10 mM sodium phosphate, 150 mM sodium chloride, pD 7.4). For the bound states, the stock solution had a final concentration of 20 μM RiVax and 40 μM $V_HH$ resulting in 1:2 molar ratio of RiVax:$V_HH$. Four μL of the stock was incubated with 36 μL of deuterated buffer. Samples were incubated at 25 °C for five HX times between 13 s and 24 h and subsequently quenched using 200 mM phosphate-4 M guanidine hydrochloride solution (pH 2.5) held at 0 °C. The quenched samples were then injected onto an immobilized pepsin column where proteolysis occurs overlapping peptides from RiVax. The peptides were desalted using a C18 trap and separated using a segmented gradient with water/acetonitrile/0.1% formic acid on a C18 column (Zorbax 300SB-C18 2.1 × 50 mm, 1.8 μm particle diameter, Agilent, Santa Clara, CA, USA). The entire liquid chromatography system (immobilized pepsin column, C18 trap and a C18 RP-UHPLC column) was kept in a refrigerated cabinet that is maintained at 0 °C to minimize back exchange. Nevertheless, the first two residues in a peptide generally undergo rapid back exchange [30]. RiVax peptides were analyzed by an QTOF mass analyzer (model 6530, Agilent Technologies, Santa Clara, CA, USA) for their increase in mass i.e., for deuterium uptake. All HX-MS measurements were based on triplicate independent HX reactions of each labeling time.

### *2.6. Data Analysis*

The HX-MS data processing was carried out using HDExaminer (version 2.3, Sierra Analytics, Modesto, CA, USA). A total of 138 peptides (Table S1) that cover the entire sequence of RiVax were analyzed. For each peptide, the magnitude of protection from each HX time was averaged and normalized to its peptide length to obtain a ΔHX value, $\Delta HX = HX_{bound} - HX_{bound}$, as described previously [11]. The propagated standard error in delta HX was estimated as described in [31]. The magnitudes of delta HX values of overlapping peptides that span the entire RiVax are then classified using K-means clustering into three categories and were colored as follows: strong protection, intermediate protection, no significant protection. For visualization, the HX-MS results were mapped onto the crystal structure of RiVax (PDB: 3SRP) [32] using PyMoL (The PyMOL Molecular Graphics System; Schrodinger LLC, San Diego, CA, USA). For better visualization purpose, only overlapping peptides that fall in strong and intermediate protection category are colored.

## 3. Results

### *3.1. Identification and Characterization of Cluster 3 $V_HHs$*

Using a variety of screening strategies that are described in detail in separate manuscripts (D. Vance, J. Tremblay, C. Shoemaker, N. Mantis, manuscript in preparation) [23], we identified from different phage-displayed alpaca single chain libraries a total of 21 $V_HHs$ whose binding to ricin toxin was partially or completely inhibited by IB2 in a capture ELISA (Figure 2). The competitive ELISA was designed such that IB2 was immobilized on microtiter plates and then allowed to capture ricin in solution. The plates were washed to remove unbound ricin and then probed with query $V_HHs$, as described in the figure legend and Materials and Methods. The DNA sequences and mAb competition profiles of ten of the $V_HHs$ were reported in a recent study, although only two (JNM-D1 and V1B11) HX-MS epitopes were described [23].

To further differentiate among the 21 $V_HHs$, they were subjected to a more comprehensive competition array using a panel of nine additional RTA-specific mAbs representing cluster 1 (PB10, WECB2), cluster 1–2 (SWB1), cluster 2 (PH12, TB12, PA1, SyH7), and cluster IV (JD4, GD12) [11]. The competition ELISA revealed a wide range of profiles (Figure 2), indicating the 21 $V_HHs$, as a whole, represent a diversity of epitopes on RTA. Indeed, the predicted CDR3 amino acid sequences of the 21 $V_HHs$ suggest they represent at least 11 different B-cell lineages: five unique $V_HHs$ and 16 others that fell into one of six sequence families (Table 1; Figure S1).

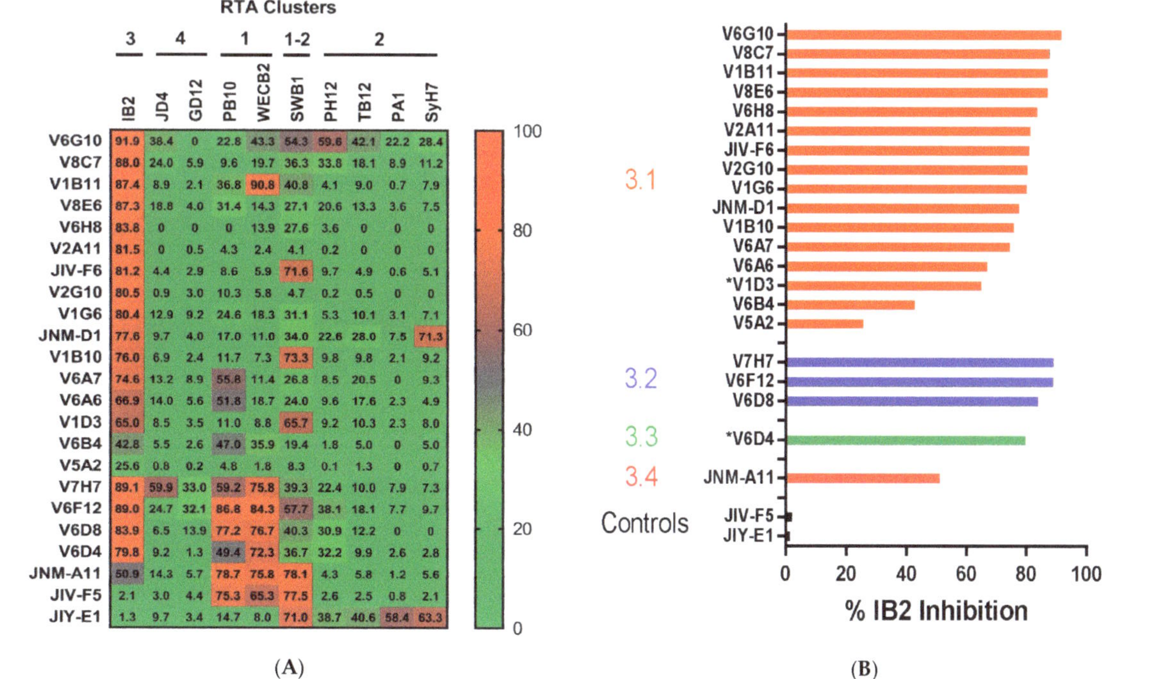


**Figure 2.** $V_H$H binning by competition ELISA. (**A**) Ricin was captured on microtiter plates by anti-RTA mAbs (indicated along the top panel) or an anti-RTB mAb (SylH3) as a control. The plates were then probed with individual $V_H$Hs as indicated on the left most column and detected with a secondary E-tag antibody. Binding inhibition was calculated as $100 - (100 \times (A_{mAb\text{-}Ricin}/A_{SylH3\text{-}Ricin}))$ where interference by SylH3 was assumed to be negligible. The colored scale bar on far right indicates % inhibition. $V_H$Hs JIV-F5 and JIY-E1 were used as controls, since they are known to bind epitopes on RTA outside of IB2's footprint. (**B**) The IB2 values from panel A are re-plotted for clarity to compare relative IB2 inhibition values and color coded based on subcluster designations described later in the manuscript. The two $V_H$Hs with toxin-neutralizing activity are denoted with an *.

**Table 1.** $V_HH$ Families based on CDR3 similarity.

| Family | Members |
|---|---|
| V1D3 * | JIV-F6, V1B10 |
| V2A11 | V6H8 |
| V2G10 | V1G6 |
| V6D8 | V6F12 |
| V6A6 | V6A7, V6G10, V8C7, V8E6 |
| V6D4 * | V6B4 |

*, indicates $V_HH$s with toxin-neutralizing activity; The following $V_HH$s were not assigned to a family: JNM-A11, JNM-D1, V1B11, V5A2, V7H7.

The binding kinetics of each $V_HH$ for ricin holotoxin was determined by surface plasmon resonance (SPR). Twelve of the 21 $V_HH$s had dissociation constants ($K_d$) of greater than 1 nM, while the remaining nine had dissociation constants ranging from 0.2–1 nM (Table 2; Figure S2). The $V_HH$s were also tested for ricin TNA in a Vero cell assay. Only two $V_HH$s, V6D4 ($IC_{50}$, 200 nM) and V1D3 ($IC_{50}$, 80 nM), had demonstrable TNA (Figure S3). Neutralizing activity was not solely a function of binding affinity, as several $V_HH$s with $K_D$s comparable to V6D4 and V1D3 lacked detectable neutralizing activity. For that reason, we sought to localize the epitopes on RTA recognized by each of the 21 $V_HH$s with the expectation that such information would offer insight into the basis of toxin-neutralizing activity.

**Table 2.** Cluster 3 $V_HH$ TNA and Binding Affinities.

| $V_HH$ | Subcluster | $IC_{50}$ (nM) | $K_D$ [a] (nM) | $k_{on}$ [b] | $k_{off}$ [c] |
|---|---|---|---|---|---|
| V1D3 | 3.1 | 80 | 0.460 | $3.15 \times 10^5$ | $1.45 \times 10^{-4}$ |
| V8C7 | | - | 0.597 | $1.58 \times 10^5$ | $9.40 \times 10^{-5}$ |
| V6B4 | | - | 0.652 | $1.70 \times 10^5$ | $1.11 \times 10^{-4}$ |
| V8E6 | | - | 0.830 | $1.26 \times 10^5$ | $1.04 \times 10^{-4}$ |
| V1B10 | | - | 0.917 | $8.29 \times 10^4$ | $7.60 \times 10^{-5}$ |
| V6A6 | | - | 0.996 | $5.06 \times 10^5$ | $5.04 \times 10^{-4}$ |
| V6H8 | | - | 1.150 | $6.63 \times 10^4$ | $7.66 \times 10^{-5}$ |
| V2G10 | | - | 1.160 | $8.48 \times 10^4$ | $9.84 \times 10^{-5}$ |
| JNM-D1 | | - | 1.190 | $1.80 \times 10^5$ | $2.15 \times 10^{-4}$ |
| V6G10 | | - | 1.270 | $1.77 \times 10^5$ | $2.24 \times 10^{-4}$ |
| V5A2 | | - | 1.460 | $2.15 \times 10^5$ | $3.14 \times 10^{-4}$ |
| V6A7 | | - | 1.760 | $7.70 \times 10^4$ | $1.36 \times 10^{-4}$ |
| V2A11 | | - | 1.820 | $2.97 \times 10^4$ | $5.41 \times 10^{-5}$ |
| JIV-F6 | | - | 1.860 | $1.94 \times 10^5$ | $3.61 \times 10^{-4}$ |
| V1G6 | | - | 5.340 | $3.05 \times 10^4$ | $1.63 \times 10^{-4}$ |
| V1B11 | | - | 8.840 | $2.76 \times 10^4$ | $2.44 \times 10^{-4}$ |
| V7H7 | 3.2 | - | 0.507 | $1.65 \times 10^5$ | $8.36 \times 10^{-5}$ |
| V6D8 | | - | 1.130 | $2.14 \times 10^5$ | $2.41 \times 10^{-4}$ |
| V6F12 | | - | 1.210 | $1.80 \times 10^5$ | $2.17 \times 10^{-4}$ |
| V6D4 | 3.3 | 200 | 0.222 | $1.44 \times 10^5$ | $3.21 \times 10^{-5}$ |
| JNM-A11 | 3.4 | - | 0.212 | $4.20 \times 10^5$ | $8.91 \times 10^{-5}$ |

[a], determined by SPR with Langmuir fit model; [b], 1/Ms; [c], 1/s.

### 3.2. $V_HH$ Epitope Mapping by HX-MS

We have previously used HX-MS to localize more than two dozen $V_HH$ and mAb epitopes on RTA or on RiVax, an attenuated recombinant RTA subunit vaccine antigen with point mutations at positions V76 and Y80 [11,23,27,31,33]. We used RiVax in place of RTA because it is non-toxic to humans and therefore poses no hazard to research staff. RiVax also assumes a tertiary structure essentially identical to RTA [32]. Therefore, HX-MS was performed with RiVax in the presence of two-fold molar excess of each of the cluster 3 $V_H$Hs at five exchange times between 13 s and 24 h. Epitope assignment was based on reduced (slower) HX exchange for peptides in the presence of a $V_H$H, as compared to RiVax alone. For example, in the presence of V6B4, the HX rate in the peptide corresponding to RiVax residues 57–60 was unaltered, whereas there was much slower exchange observed for the peptide corresponding to residues 206–218 (Figure 3). While reduced hydrogen exchange is generally attributed to direct antibody-protein interaction, we cannot necessarily exclude possible allosteric effects that may occur upon antibody engagement, especially when reduced exchange is observed at a distance not consistent with being part of a core epitope [11].

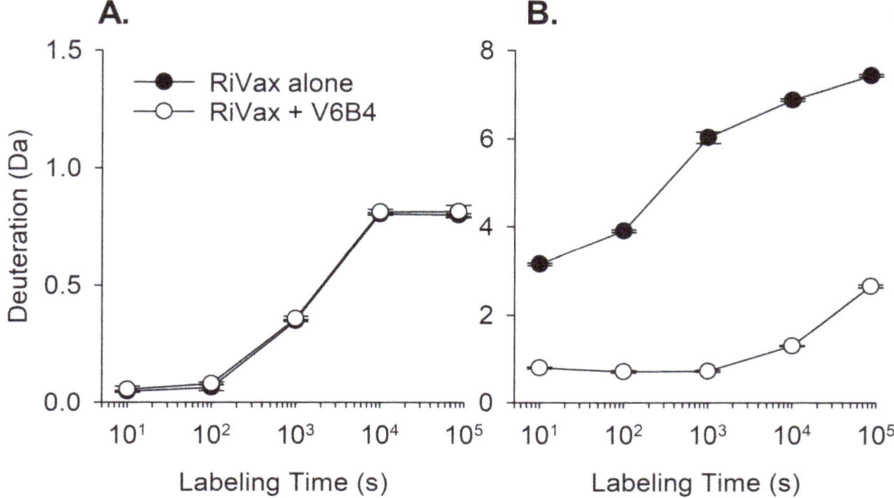


**Figure 3.** Hydrogen exchange (HX) kinetics of two representative RiVax peptides in presence of V6B4. Hydrogen deuterium exchange kinetics of two representative RiVax peptides in presence of V6B4. (**A**) Peptide 14 (56–59), where the HX rate was not affected by association with V6B4. (**B**) Peptide 94 (205–217) where the rate of HX was substantially slowed by V6B4. Significance limit for HX differences was defined as described in the Experimental section.

### 3.3. *Identification of Epitope Subclusters*

The results of epitope mapping studies revealed that the Cluster 3 $V_H$Hs grouped within four subclusters, referred to as 3.1–3.4 (Table 3; Table S2;). Subcluster 3.1 involves contact with RiVax α-helices C and G, a profile very similar to mAb IB2. Subcluster 3.2 encompasses α-helices B, C and G, while subcluster 3.3 covers α-helices B and G, but not α-helix C. Finally, subcluster 3.4 encompasses α-helices C and E, but not G. Each of these subclusters is now described in more detail.

**Table 3.** Localization of epitopes on RiVax recognized by representative Cluster 3 $V_HHs$.

| | | Strong and Intermediate Protected Elements in RiVax [a] | | |
|---|---|---|---|---|
| $V_HH$ | Subcluster | Peptides | Residues | Structure(s) |
| V6B4 | 3.1 | 48–51 | 119–133 | α-helix C |
| | | 91–102 | 205–217 | α-helix G |
| | | 132–134 | 249–255 | C-terminus |
| V1D3 * | | 54,55 | 127–135 | α-helix C |
| | | 91 | 205–210 | α-helix G |
| | | 112–116 | 226–243 | β-strands i, j |
| | | 132–134 | 249–255 | C-terminus |
| V6D8 | 3.2 | 35–39 | 92–107 | α-helix B |
| | | 49–54 | 123–135 | α-helix C |
| | | 102 | 211–217 | α-helix G |
| | | 129–134 | 247–255 | C-terminus |
| V6F12 | | 35–40 | 92–107 | α-helix B |
| | | 47,49 | 118–126 | α-helix C |
| | | 94,97,100,102–103 | 205–217 | α-helix G |
| | | 132–134 | 249–255 | C-terminus |
| V7H7 | | 35–39 | 92–107 | α-helix B |
| | | 49 | 123–126 | α-helix C |
| | | 94–95,97–98,100,102 | 205–217 | α-helix G |
| | | 129–131 | 249–255 | C-terminus |
| V6D4 * | 3.3 | 35–37,39 | 92–107 | α-helix B |
| | | 102 | 211–217 | α-helix G |
| | | 132–134 | 249–255 | C-terminus |
| JNM-A11 | 3.4 | 45,46 | 108–122 | β-strand h |
| | | 49–51 | 124–133 | α-helix C |
| | | 70,71 | 162–168 | α-helix E |

[a], Peptides on RiVax are indicated in supplementary Table S1. *, indicates $V_HHs$ with toxin-neutralizing activity. Underline indicates intermediate protection determined by HX-MS.

Subcluster 3.1: Sixteen of the 21 $V_HHs$ shared an HX-MS profile involving contact with α-helices C and G, which we refer to as subcluster 3.1 (Table 3; Table S2; Figures S4 and S5). While the HX-MS profiles of the $V_HHs$ within 3.1 were qualitatively similar, there were quantitative differences that may be significant in terms of neutralizing activity. For example, V1D3, one of the two $V_HHs$ with toxin-neutralizing activity, had a binding pattern virtually identical to IB2 in that it strongly protected α-helix C (peptides 54–55, residues 127–135) and the C-terminus region (peptides 132–134, residues 249–255) (Figures 4 and 5). Moreover, V1D3 demonstrated intermediate protection of α-helix G (peptide 91, residues 205–210), as well as strands i and j (peptides 112–116, residues 226–243). In contrast, V6B4, an antibody without toxin-neutralizing activity, strongly protected RiVax residues 119 to 133 (peptides 48–51), corresponding to α-helix C, and residues 205–217 (peptides 91 to 102), corresponding to α-helix G (Figures 4 and 5). However, V6B4 differed from V1D3 in three respects. First, V6B4 had stronger protection of α-helix G than C, as compared to V1D3. Second, V1D3 interacted with β-strands i and j, while V6B4 did not, possibility indicating that V1D3 overall contact interface with RiVax is larger than V6B4's. Finally, the patterns of protection in α-helix C are distinct. In case of V6B4, the entirety of α-helix C is strongly protected, while in the case of V1D3 it is only the C-terminal end that is strongly protected (Figures 4 and 5). V1D3 also caused intermediate protection in the N-terminal end of helix G, while V6B4 protected all of helix G. It is unclear if these differences in α-helix C and α-helix G protection explain V1D3's TNA.

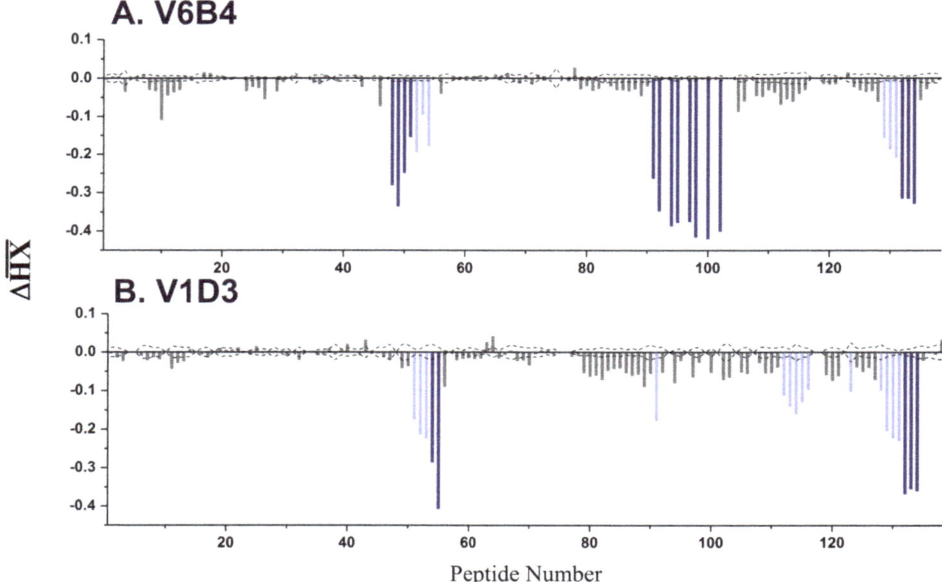


**Figure 4.** HX-MS analysis of RiVax bound to two $V_HH$s in subcluster 3.1. The $\Delta\overline{HX}$ values for each RiVax peptide are shown for $V_HH$s (**A**) V6B4 and (**B**) V1D3. The $\Delta\overline{HX}$ values are clustered using k-means clustering into three categories: strong (deep blue), intermediate (light blue) or no significant protection (gray). The dotted lines represent "3σ" confidence intervals for statistically significant changes in hydrogen exchange.

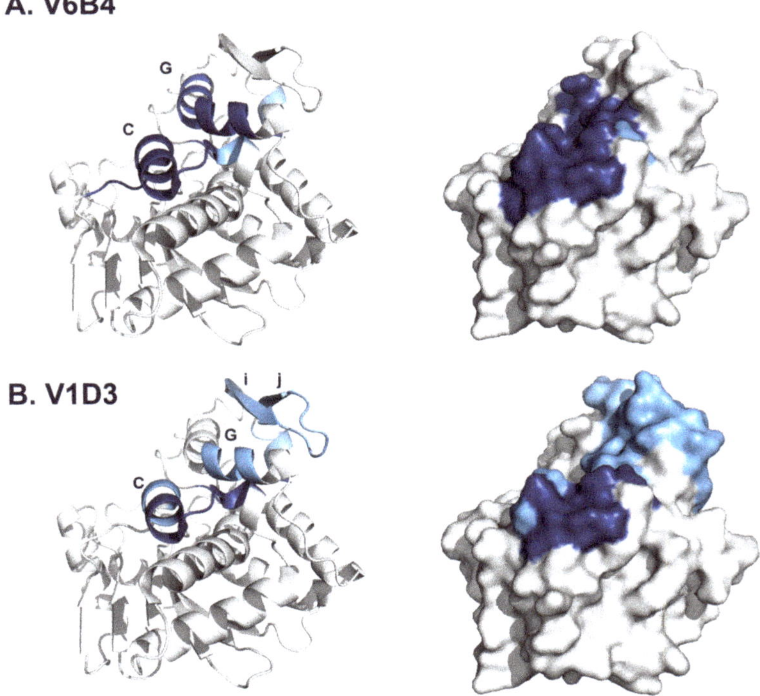


**Figure 5.** Epitope localization for two subcluster 3.1 $V_HHs$. HX protection categories shown in Figure 4 were mapped onto the crystal structure of RiVax for (**A**) V6B4 and (**B**) V1D3. The most relevant secondary structure elements, α-helices C and G and β-strands i and j, are labeled. The color shading corresponds to strong (deep blue), intermediate (light blue) or no significant protection (gray), as represented in Figure 4.

The competition ELISA with the panel of RTA-specific mAbs revealed additional degrees of difference among the 16 $V_HHs$ in subcluster 3.1 (Figure 2). Not only was there a clear gradation of competition with IB2 (range 25–90%), but there were marked disparities with other mAbs. For example, V1B11 is a potent inhibitor of WECB2, V1D3 stood out because of competition with SWB1, while JNM-D1 competes strongly with SyH7. Because the footprints of all 10 anti-RTA mAbs have been defined, we can infer from the various inhibition profiles how different $V_HHs$ engage RTA. Thus, looking directly at the RTA active site, with RTB oriented on the bottom, we predict that V1B11 likely approaches RTA from the top down, V1D3 likely from top left, and JNM-D1 likely from bottom left.

Subcluster 3.2: Three $V_HHs$, V6D8, V6F12 and V7H7, were grouped within subcluster 3.2 based on a common HX-MS profile encompassing α-helices B, C and G (Figures 6 and 7). For example, V6D8 strongly protected α-helices B (residues 92–107; peptides 35–39), C (residues 123–135; peptides 49–54), G (residues 211–217; peptides 102) and a short region near the C-terminus of α-helix G (residues 247–255, peptides 129–134). V6F12 shared a binding profile with V6D8, which was not surprising since the two $V_HHs$ are likely from the same B-cell lineage (Table 1; Figure S1). Although the protection profiles of V6D8, V6F12, and V7H7 were qualitatively similar, and all three $V_HHs$ were competed by IB2 to a similar degree, the magnitudes of protection in the three secondary structural features were distinct. V6D8 and V6F12 interacted primarily with α-helices B and C, and secondarily with α-helix G. V7H7, by contrast, primarily protected several overlapping peptides in α-helix G, and secondarily protected α-helices B and C (except for one peptide in α-helix C). Finally, HX-MS indicated that V6D8, V6F12, and V7H7 each contact α-helix B, which has been postulated as being a neutralizing hotspot on RTA [22]. However, none of the $V_H$Hs within this subcluster had any detectable TNA, possibly because their binding affinities do not achieve a minimum threshold required to inactivate ricin. Other previously described $V_H$Hs that engage α-helix B and have potent toxin-neutralizing activities each have binding affinities of less than 200 pM, including JIV-F5 (19 pM), JIY-E5 (191 pM), and JPF-A9 (102 pM) [21,22,33]. This contrasts with V7H7, the strongest binder in subcluster 3.2, which has a binding affinity of ~500 pM.

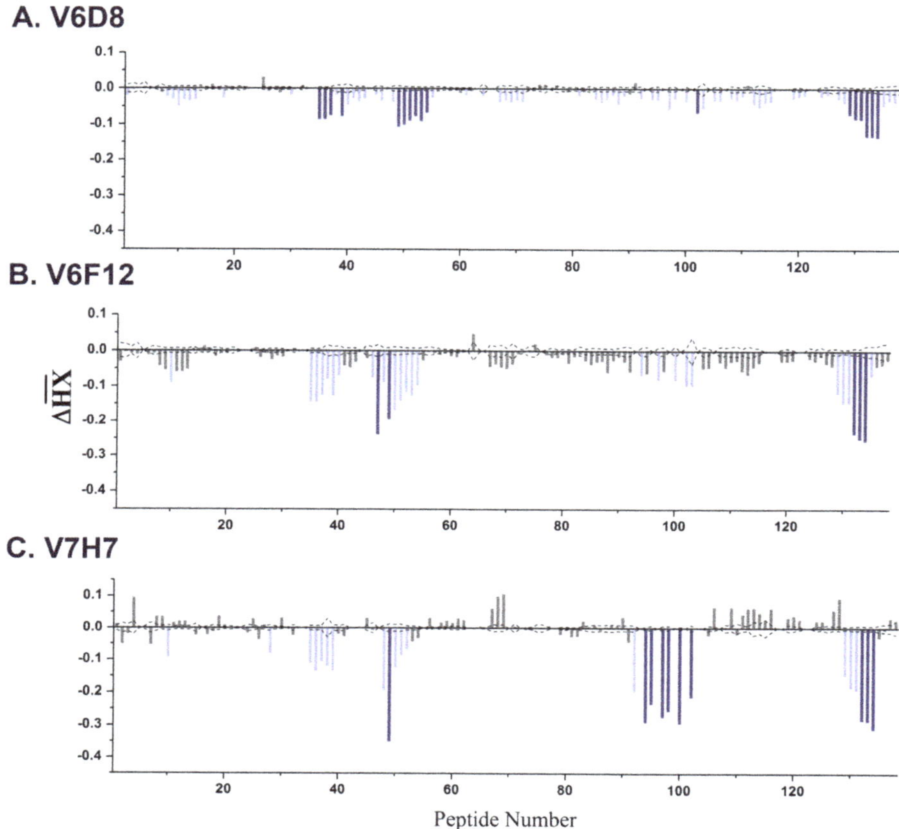


**Figure 6.** HX-MS analysis of RiVax bound to two $V_H$Hs in subcluster 3.2. The $\Delta\overline{HX}$ values for each RiVax peptide are shown for $V_H$Hs (**A**) V6D8 (**B**) V6F12 and (**C**) V7H7. The $\Delta\overline{HX}$ values are clustered using k-means clustering into three categories: strong (deep blue), intermediate (light blue) or no significant protection (gray). The dotted lines represent "3σ" confidence intervals for statistically significant changes in hydrogen exchange.

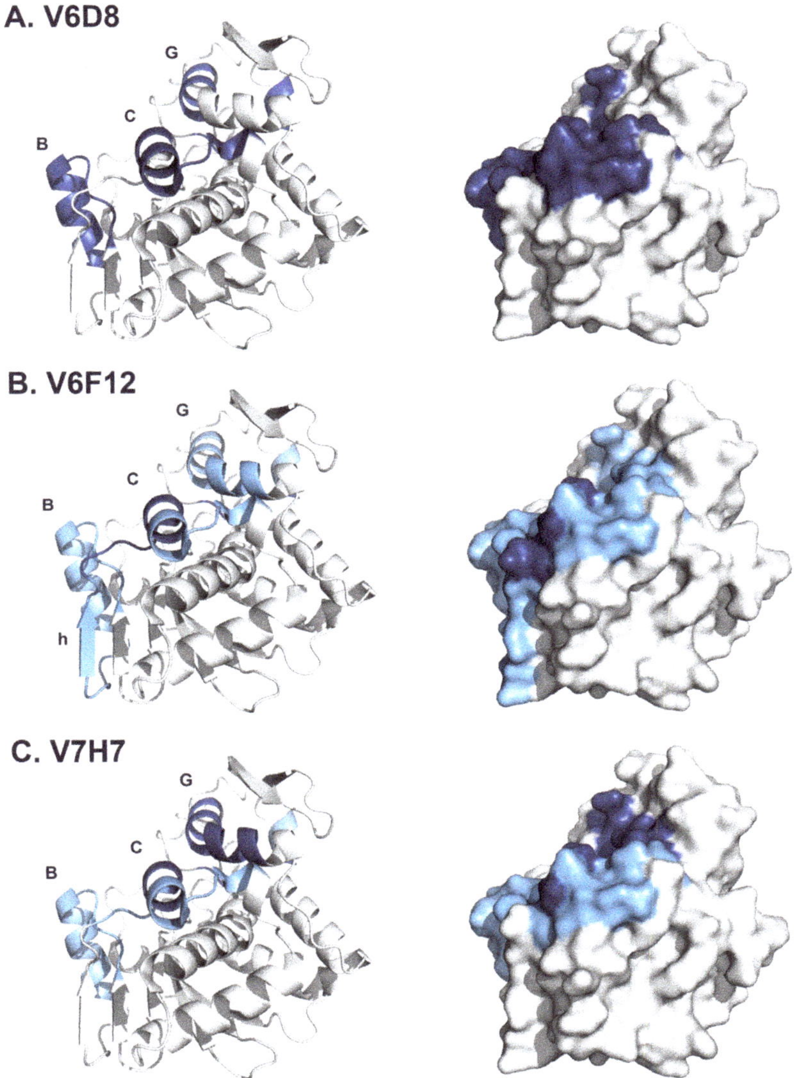


**Figure 7.** Epitope locations of V6D8, V6F12 and V7H7 on RiVax. HX protection categories shown in Figure 6 were mapped onto the structure of RiVax for (**A**) V6D8 (**B**) V6F12 and (**C**) V7H7. Secondary structure elements including α-helices B, C and G and β-strand h are labeled. Intermediate protection by V6D8 is spread over much of RiVax's surface and the magnitudes of protection are low. Therefore, only strongly protected elements are mapped onto the crystal structure of RiVax. The color shading corresponds to strong (deep blue) or no significant protection (gray).

Subcluster 3.3: The third subcluster is populated by V6D4, which had weak toxin-neutralizing activity ($IC_{50}$ ~200 nM) in the Vero cell cytotoxicity assay. HX-MS analysis demonstrated strong protection of α-helix G (peptide 102, residues 211–217) and intermediate protection of α-helix B (peptides 35–39, residues 92–107). V6D4 also protected a short region in the C-terminus of RiVax, but not α-helix C itself (Figure 8). Whether V6D4's neutralizing activity is a result of contact with α-helix B is unclear, though its high affinity for ricin ($K_d$ = 222 pM) may put it above any relevant affinity threshold.

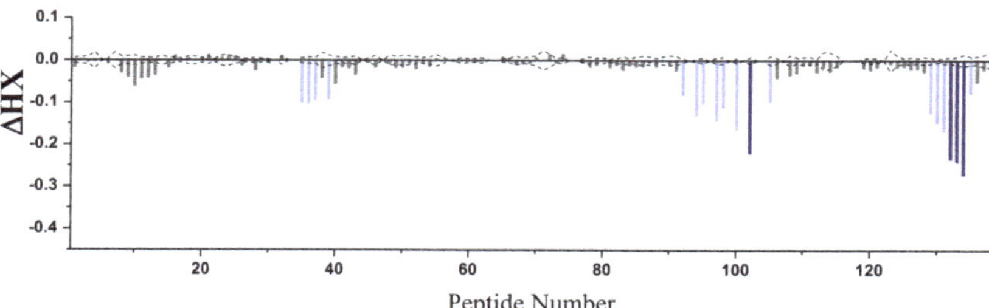


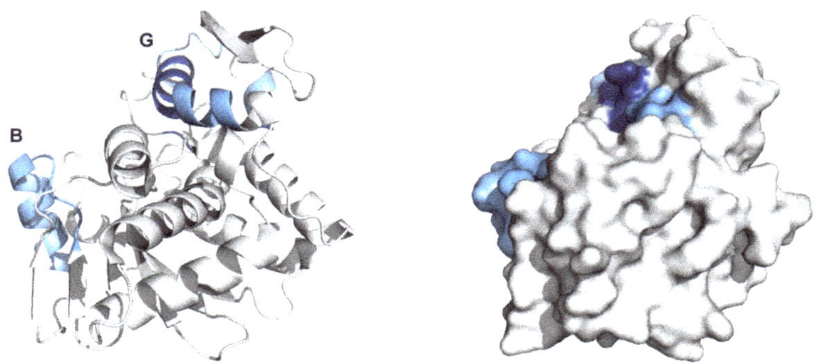


**Figure 8.** Epitope mapping of V6D4 from subcluster 3.3. (**A**) Relative levels of protection of RiVax peptides by V6D4 as defined by HX-MS. The color shading corresponds to strong (deep blue), intermediate (light blue) or no significant protection (gray), as represented in Figure 4. (**B**) The HX protection categories, as shown in panel A, were mapped onto the crystal structure of RiVax. Secondary structures α-helices B and G are labeled.

Subcluster 3.4: The fourth subcluster is also populated by a single antibody, JNM-A11. JNM-A11 showed strong protection of residues in RTA's α-helix C (peptides 49–51; residues 124–133) and intermediate protection of the N-terminal region of α-helix E (peptides 70 and 71; residues 162–168) and β-strand h (peptides 45 and 46; residues 108–122) (Figure 9). JNM-A11 did not protect α-helix G, which differentiates it from the 20 other $V_H$Hs in cluster 3. JNM-A11's competition profile against a panel of RTA-specific mAbs is consistent with results obtained by HX-MS. Namely, JNM-A11 competed with both Cluster 1 (PB10, WECB2) and cluster 1–2 (SWB1) mAbs (Figure 2A). Finally, JNM-A11 did not neutralize ricin, despite a strong binding affinity ($K_d$ = 212 pM). Since JNM-A11 appears to target α-helix C almost exclusively, we infer that contact with α-helix C alone is not sufficient to affect ricin function.

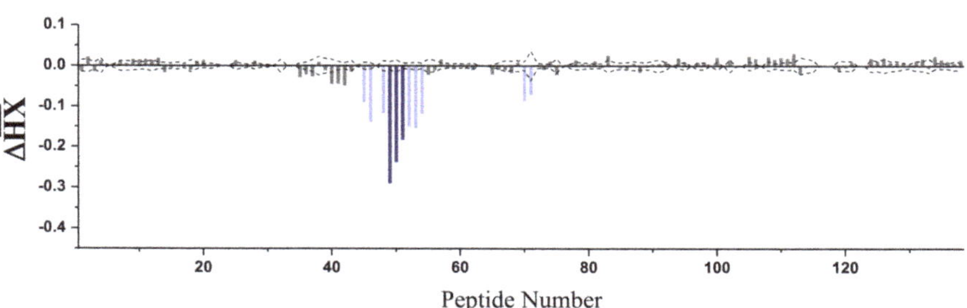


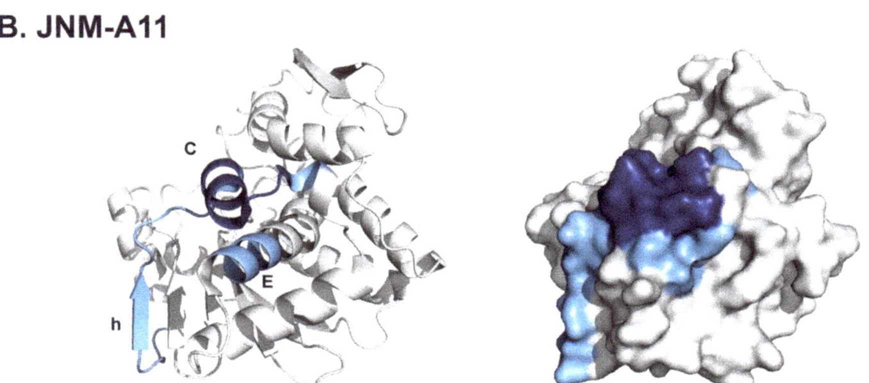


**Figure 9.** Epitope mapping of JNM-A11 from cluster 3.4. (**A**) Relative levels of protection of RiVax peptides by JNM-A11, as defined by HX-MS. The color shading corresponds to strong (deep blue), intermediate (light blue) or no significant protection (gray), as represented in Figure 4. (**B**) The HX protection categories, as shown in panel a, are mapped onto the crystal structure of RiVax. Secondary structures α-helices C and E and β-strand h are labeled.

## 4. Discussion

As part of our long-standing effort to generate a comprehensive B-cell epitope map of ricin toxin, we have characterized 21 unique $V_H$Hs that share the common property of being within the shadow of IB2 based on competition ELISAs. IB2 is a toxin-neutralizing mAb that engages with α-helices C and G on RiVax and defines so-called epitope cluster 3 [11,18]. Cluster 3 is of interest because it encompasses the residues on RTA involved in ribosome inactivation [10,34,35]. The 21 $V_H$Hs originated from different phage-displayed libraries, each generated from alpacas immunized with ricin toxin antigens, including RiVax (D. Vance, C. Shoemaker, N. Mantis, manuscript in preparation) [23]. As a result, of epitope mapping studies by HX-MS, the 21 $V_H$Hs were further grouped into four distinct subclusters (3.1–3.4) based on their interactions with RiVax α-helix C and α-helix G, as well as other local secondary structures including α-helix B, α-helix E, and β-strand h (Figure 10). The fact that all 21 $V_H$Hs engage RiVax via α-helix C and/or α-helix G explains the observed competition with IB2 by ELISA (Figure 2). However, we are unable to explain exactly why V1D3 and V6D4 are the only $V_H$Hs within cluster 3 that have toxin-neutralizing activity, since other $V_H$Hs have similar footprints on ricin and nearly identical binding affinities as V1D3 and V6D4 but are devoid of neutralizing activity. We can only speculate that neutralizing activity is due to specific residue contacts or combinations of contact that are not apparent by HX-MS epitope mapping methodologies (see below).

A

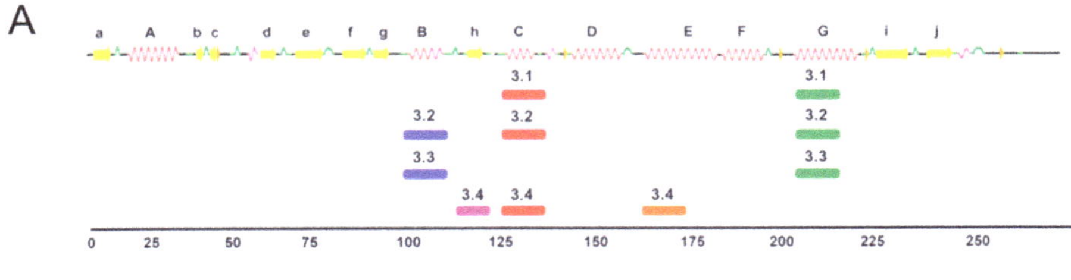


B

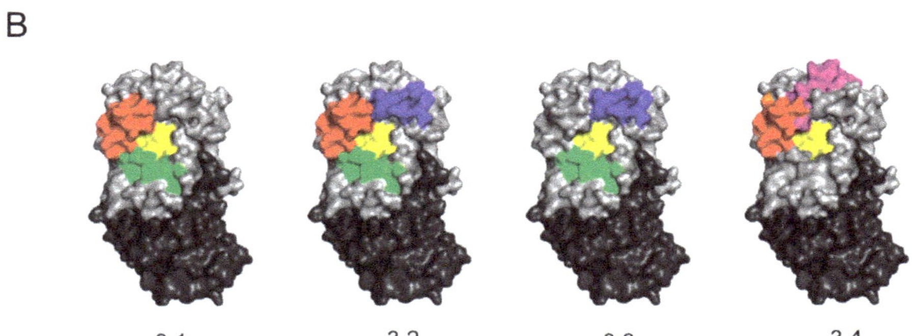


**Figure 10.** Visual representation subcluster 3 binding sites on ricin toxin. (**A**) Linear depiction of RTA with arrows denoting β-strand secondary structure and coils indicating α-helices, as per Protein Data Bank (PDB) format. Below, the colored bars denote epitope coverage for each of the $V_H$H subclusters 3.1–3.4. The colors correspond to secondary structures highlighted in panel B. Horizontal line below refers to RTA amino acid residue number. (**B**) Surface representations of ricin (PDB 2AAI) using PyMol showing the regions of protection for each of the four subclusters (3.1–3.4). Colors are as follows: RTA, light gray; RTB, dark gray; active site, yellow; α-helix B, blue; α-helix C, red; α-helix G, green; α-helix E, orange; β-strand h, purple.

RTA's active site consists of a large solvent-exposed cleft on one face of the molecule [10,34–37]. Active site residues include Tyr80, Tyr123, Arg180, and Glu177, which are involved in stacking the purine ring of target adenosine moiety (Tyr80, Tyr123) and transition state stabilization (Glu177, Arg180). Viewing the active site pocket head on, α-helices C (121–135) and G (207–217) would be located at 10 o'clock and 7 o'clock, respectively (Figure 10). Thus, antibodies in subclusters 3.1, 3.2 and 3.3 would be expected to physically occlude (straddle) or even occupy the active site pocket, whereas the single antibody (JNM-A11) in subcluster 3.4 is probably associated with upper rim (11 o'clock) of the active site (Figure 10). To examine these possibilities, efforts are ongoing to solve the X-ray crystal structures of all 21 of these $V_H$Hs in complex with RTA.

The current study also highlights both the advantages and shortcomings of HX-MS for use in B-cell epitope mapping. On the upside, the HX-MS pipeline proved to be robust and relatively high throughput due to the fact that we had already established a RiVax peptide map and baseline HX kinetics [11]. HX-MS was able to assess RiVax-$V_H$H binding in solution and parse cluster 3 epitopes into four subclusters that we are currently compared to interaction sites observed by X-ray crystallography. On the other hand, HX-MS provides only peptide level resolution in terms of defining actual antibody contacts on the target antigen and cannot reveal subtle interactions that may ultimately be of consequence to toxin-neutralizing activity. As a case in point, we recently described two $V_H$Hs (JPF-A9 and V8A7) with essentially identical HX-MS profiles but that differ in both binding affinity and toxin-neutralizing activity as a result of a single residue difference in CDR2 [33]. Coupling HX-MS with high density competition ELISAs and/or site-directed mutagenesis can significantly improve epitope definition [23,38–40]. The magnitude of HX protection will depend on the affinity and kinetics of binding. Lower affinity generally leads to weaker protection against HX, thereby making it more difficult to resolve the epitope from allosteric effects. However, in practice we have found that introduction of point mutations in $V_H$Hs that led to ~10-fold differences in binding affinity (e.g., 0.4 to 4 nM) did not notably alter their HX profiles [31]. Since each epitope mapping data set is treated independently, our analysis still finds the most strongly protected regions.

At this point in time, more than 30 alpaca B-cell epitopes and more than a dozen murine B-cell epitope on RTA have been reported [18,19,21–23,41–44]. The availability of this dense epitope map and a collection well characterized antibodies has already proven to have utility in terms of pre-clinical evaluation of RiVax and other candidate RTA-based vaccine antigens. In one instance the mAbs were used as tools in competition ELISAs to demonstrate epitope use within humans and non-human primates vaccinated with RiVax [14]. More recently, the mAbs were used to evaluate the integrity of key neutralizing epitopes on RiVax during long-term storage [25]. The 21 $V_H$Hs described here focused around RTA's active site now add to that growing list of critical reagents.

**Supplementary Materials:** The following are available online at http://www.mdpi.com/2073-4468/7/4/45/s1, Table S1: RiVax peptic peptides, Table S2: HX-MS analysis of $V_H$Hs in subcluster 3.1, Figure S1: Alignment of cluster 3 $V_H$H families, Figure S2: Representative sensorgrams of Cluster 3 $V_H$Hs, Figure S3: Representative toxin-neutralizing activities of cluster 3 $V_H$Hs, Figure S4: HX-MS analysis of RiVax bound to $V_H$Hs in subcluster 3.1, Figure S5: Epitope localization of subcluster 3.1 $V_H$Hs on the surface of RiVax.

**Author Contributions:** Study conceptualization, N.J.M., C.R.M. and D.D.W.; methodology, D.J.V., C.M.T.N., M.J.R., D.D.W. and N.J.M.; formal analysis, S.K.A., D.J.V., D.V., D.D.W. and N.J.M.; investigation, S.K.A., D.J.V., Y.R.; writing—original draft preparation, S.K.A., D.J.V.; writing—review and editing, D.J.V., D.V., D.D.W., N.J.M.; supervision, D.V., C.R.M., D.D.W., N.J.M.; project administration, N.J.M.; funding acquisition, N.J.M.

**Funding:** This work was supported by Contract No. HHSN272201400021C to N.J.M. from the National Institutes of Allergy and Infectious Diseases, National Institutes of Health.

**Acknowledgments:** We thank Chuck Shoemaker and Jacqueline Tremblay (Tufts University) for generating alpaca $V_H$H libraries used in this study. We gratefully acknowledge Beth Cavosie for administrative support. We thank the Wadsworth Center's Biochemistry and Immunology Core facility for assistance with SPR. DDW gratefully acknowledges an equipment loan from Agilent Technologies.

**Conflicts of Interest:** The authors declare no conflict of interest. The content is solely the responsibility of the authors and does not necessarily represent the official views of the National Institutes of Health. The funders had no role in study design, data collection and analysis, decision to publish, or preparation of the manuscript.

## References

1. Gal, Y.; Mazor, O.; Falach, R.; Sapoznikov, A.; Kronman, C.; Sabo, T. Treatments for pulmonary ricin intoxication: Current aspects and future prospects. *Toxins (Basel)* **2017**, *9*, 311. [CrossRef] [PubMed]
2. Griffiths, G.D. Understanding ricin from a defensive viewpoint. *Toxins (Basel)* **2011**, *3*, 1373–1392. [CrossRef] [PubMed]
3. Reisler, R.B.; Smith, L.A. The need for continued development of ricin countermeasures. *Adv. Prev. Med.* **2012**, *2012*, 149737. [CrossRef] [PubMed]
4. Rutenber, E.; Ready, M.; Robertus, J.D. Structure and evolution of ricin b chain. *Nature* **1987**, *326*, 624–626. [CrossRef]
5. Endo, Y.; Mitsui, K.; Motizuki, M.; Tsurugi, K. The mechanism of action of ricin and related toxic lectins on eukaryotic ribosomes. The site and the characteristics of the modification in 28 s ribosomal rna caused by the toxins. *J. Biol. Chem.* **1987**, *262*, 5908–5912. [PubMed]
6. Endo, Y.; Tsurugi, K. Rna n-glycosidase activity of ricin a-chain. Mechanism of action of the toxic lectin ricin on eukaryotic ribosomes. *J. Biol. Chem.* **1987**, *262*, 8128–8130. [PubMed]
7. Montfort, W.; Villafranca, J.E.; Monzingo, A.F.; Ernst, S.R.; Katzin, B.; Rutenber, E.; Xuong, N.H.; Hamlin, R.; Robertus, J.D. The three-dimensional structure of ricin at 2.8 a. *J. Biol. Chem.* **1987**, *262*, 5398–5403. [PubMed]

8. Rutenber, E.; Katzin, B.J.; Ernst, S.; Collins, E.J.; Mlsna, D.; Ready, M.P.; Robertus, J.D. Crystallographic refinement of ricin to 2.5 a. *Proteins* **1991**, *10*, 240–250. [CrossRef] [PubMed]
9. Katzin, B.J.; Collins, E.J.; Robertus, J.D. Structure of ricin a-chain at 2.5 A. *Proteins* **1991**, *10*, 251–259. [CrossRef]
10. Monzingo, A.F.; Robertus, J.D. X-ray analysis of substrate analogs in the ricin a-chain active site. *J. Mol. Biol.* **1992**, *227*, 1136–1145. [CrossRef]
11. Toth, R.T.I.; Angalakurthi, S.K.; Van Slyke, G.; Vance, D.J.; Hickey, J.M.; Joshi, S.B.; Middaugh, C.R.; Volkin, D.B.; Weis, D.D.; Mantis, N.J. High-definition mapping of four spatially distinct neutralizing epitope clusters on rivax, a candidate ricin toxin subunit vaccine. *Clin. Vaccine Immunol.* **2017**, *24*. [CrossRef] [PubMed]
12. Pincus, S.H.; Bhaskaran, M.; Brey, R.N., 3rd; Didier, P.J.; Doyle-Meyers, L.A.; Roy, C.J. Clinical and pathological findings associated with aerosol exposure of macaques to ricin toxin. *Toxins (Basel)* **2015**, *7*, 2121–2133. [CrossRef]
13. Smallshaw, J.E.; Richardson, J.A.; Vitetta, E.S. Rivax, a recombinant ricin subunit vaccine, protects mice against ricin delivered by gavage or aerosol. *Vaccine* **2007**, *25*, 7459–7469. [CrossRef] [PubMed]
14. Roy, C.J.; Brey, R.N.; Mantis, N.J.; Mapes, K.; Pop, I.V.; Pop, L.M.; Ruback, S.; Killeen, S.Z.; Doyle-Meyers, L.; Vinet-Oliphant, H.S.; et al. Thermostable ricin vaccine protects rhesus macaques against aerosolized ricin: Epitope-specific neutralizing antibodies correlate with protection. *Proc. Natl. Acad. Sci. USA* **2015**, *112*, 3782–3787. [CrossRef] [PubMed]
15. Audi, J.; Belson, M.; Patel, M.; Schier, J.; Osterloh, J. Ricin poisoning: A comprehensive review. *JAMA* **2005**, *294*, 2342–2351. [CrossRef] [PubMed]
16. Vance, D.J.; Mantis, N.J. Progress and challenges associated with the development of ricin toxin subunit vaccines. *Expert Rev. Vaccines* **2016**, *15*, 1213–1222. [CrossRef]
17. Smallshaw, J.E.; Richardson, J.A.; Pincus, S.; Schindler, J.; Vitetta, E.S. Preclinical toxicity and efficacy testing of rivax, a recombinant protein vaccine against ricin. *Vaccine* **2005**, *23*, 4775–4784. [CrossRef] [PubMed]
18. O'Hara, J.M.; Kasten-Jolly, J.C.; Reynolds, C.E.; Mantis, N.J. Localization of non-linear neutralizing b cell epitopes on ricin toxin's enzymatic subunit (rta). *Immunol. Lett.* **2014**, *158*, 7–13. [CrossRef]
19. O'Hara, J.M.; Neal, L.M.; McCarthy, E.A.; Kasten-Jolly, J.A.; Brey, R.N., 3rd; Mantis, N.J. Folding domains within the ricin toxin a subunit as targets of protective antibodies. *Vaccine* **2010**, *28*, 7035–7046. [CrossRef]
20. Rudolph, M.J.; Vance, D.J.; Cassidy, M.S.; Rong, Y.; Mantis, N.J. Structural analysis of single domain antibodies bound to a second neutralizing hot spot on ricin toxin's enzymatic subunit. *J. Biol. Chem.* **2017**, *292*, 872–883. [CrossRef]
21. Rudolph, M.J.; Vance, D.J.; Cassidy, M.S.; Rong, Y.; Shoemaker, C.B.; Mantis, N.J. Structural analysis of nested neutralizing and non-neutralizing b cell epitopes on ricin toxin's enzymatic subunit. *Proteins* **2016**, *84*, 1162–1172. [CrossRef]
22. Rudolph, M.J.; Vance, D.J.; Cheung, J.; Franklin, M.C.; Burshteyn, F.; Cassidy, M.S.; Gary, E.N.; Herrera, C.; Shoemaker, C.B.; Mantis, N.J. Crystal structures of ricin toxin's enzymatic subunit (rta) in complex with neutralizing and non-neutralizing single-chain antibodies. *J. Mol. Biol.* **2014**, *426*, 3057–3068. [CrossRef]
23. Vance, D.J.; Tremblay, J.M.; Rong, Y.; Angalakurthi, S.K.; Volkin, D.B.; Middaugh, C.R.; Weis, D.D.; Shoemaker, C.B.; Mantis, N.J. High-resolution epitope positioning of a large collection of neutralizing and nonneutralizing single-domain antibodies on the enzymatic and binding subunits of ricin toxin. *Clin. Vaccine Immunol.* **2017**, *24*. [CrossRef]
24. Lemley, P.V.; Amanatides, P.; Wright, D.C. Identification and characterization of a monoclonal antibody that neutralizes ricin toxicity in vitro and in vivo. *Hybridoma* **1994**, *13*, 417–421. [CrossRef] [PubMed]
25. Westfall, J.; Yates, J.L.; Van Slyke, G.; Ehrbar, D.; Measey, T.; Straube, R.; Donini, O.; Mantis, N.J. Thermal stability and epitope integrity of a lyophilized ricin toxin subunit vaccine. *Vaccine* **2018**, *36*, 5967–5976. [CrossRef]
26. O'Hara, J.M.; Mantis, N.J. Neutralizing monoclonal antibodies against ricin's enzymatic subunit interfere with protein disulfide isomerase-mediated reduction of ricin holotoxin in vitro. *J. Immunol. Methods* **2013**, *395*, 71–78. [CrossRef] [PubMed]
27. Van Slyke, G.; Angalakurthi, S.K.; Toth, R.T.; Vance, D.J.; Rong, Y.; Ehrbar, D.; Shi, Y.; Middaugh, C.R.; Volkin, D.B.; Weis, D.D.; et al. Fine-specificity epitope analysis identifies contact points on ricin toxin recognized by protective monoclonal antibodies. *ImmunoHorizons* **2018**, *2*, 262–273. [CrossRef]

28. Thomas, J.C.; O'Hara, J.M.; Hu, L.; Gao, F.P.; Joshi, S.B.; Volkin, D.B.; Brey, R.N.; Fang, J.; Karanicolas, J.; Mantis, N.J.; et al. Effect of single-point mutations on the stability and immunogenicity of a recombinant ricin a chain subunit vaccine antigen. *Hum. Vaccin. Immunother.* **2013**, *9*, 744–752. [CrossRef] [PubMed]
29. Smallshaw, J.E.; Firan, A.; Fulmer, J.R.; Ruback, S.L.; Ghetie, V.; Vitetta, E.S. A novel recombinant vaccine which protects mice against ricin intoxication. *Vaccine* **2002**, *20*, 3422–3427. [CrossRef]
30. Bai, Y.; Milne, J.S.; Mayne, L.; Englander, S.W. Primary structure effects on peptide group hydrogen exchange. *Proteins* **1993**, *17*, 75–86. [CrossRef]
31. Bazzoli, A.; Vance, D.J.; Rudolph, M.J.; Rong, Y.; Angalakurthi, S.K.; Toth, R.T.t.; Middaugh, C.R.; Volkin, D.B.; Weis, D.D.; Karanicolas, J.; et al. Using homology modeling to interrogate binding affinity in neutralization of ricin toxin by a family of single domain antibodies. *Proteins* **2017**, *85*, 1994–2008. [CrossRef] [PubMed]
32. Legler, P.M.; Brey, R.N.; Smallshaw, J.E.; Vitetta, E.S.; Millard, C.B. Structure of rivax: A recombinant ricin vaccine. *Acta Crystallogr. D Biol. Crystallogr.* **2011**, *67*, 826–830. [CrossRef] [PubMed]
33. Rudolph, M.J.; Vance, D.J.; Kelow, S.; Angalakurthi, S.K.; Nguyen, S.; Davis, S.A.; Rong, Y.; Middaugh, C.R.; Weis, D.D.; Dunbrack, R., Jr.; et al. Contribution of an unusual cdr2 element of a single domain antibody in ricin toxin binding affinity and neutralizing activity. *Protein Eng. Des. Sel.* **2018**, *31*, 277–287. [CrossRef] [PubMed]
34. Frankel, A.; Welsh, P.; Richardson, J.; Robertus, J.D. Role of arginine 180 and glutamic acid 177 of ricin toxin a chain in enzymatic inactivation of ribosomes. *Mol. Cell. Biol.* **1990**, *10*, 6257–6263. [CrossRef] [PubMed]
35. Ready, M.P.; Kim, Y.; Robertus, J.D. Site-directed mutagenesis of ricin a-chain and implications for the mechanism of action. *Proteins* **1991**, *10*, 270–278. [CrossRef] [PubMed]
36. Wahome, P.G.; Robertus, J.D.; Mantis, N.J. Small-molecule inhibitors of ricin and shiga toxins. *Curr. Top. Microbiol. Immunol.* **2012**, *357*, 179–207. [PubMed]
37. Weston, S.A.; Tucker, A.D.; Thatcher, D.R.; Derbyshire, D.J.; Pauptit, R.A. X-ray structure of recombinant ricin a-chain at 1.8 a resolution. *J. Mol. Biol.* **1994**, *244*, 410–422. [CrossRef]
38. Fernandez, E.; Kose, N.; Edeling, M.A.; Adhikari, J.; Sapparapu, G.; Lazarte, S.M.; Nelson, C.A.; Govero, J.; Gross, M.L.; Fremont, D.H.; et al. Mouse and human monoclonal antibodies protect against infection by multiple genotypes of japanese encephalitis virus. *MBio* **2018**, *9*. [CrossRef]
39. Gribenko, A.V.; Parris, K.; Mosyak, L.; Li, S.; Handke, L.; Hawkins, J.C.; Severina, E.; Matsuka, Y.V.; Anderson, A.S. High resolution mapping of bactericidal monoclonal antibody binding epitopes on staphylococcus aureus antigen mntc. *PLoS Pathog.* **2016**, *12*, e1005908. [CrossRef]
40. Lim, X.X.; Chandramohan, A.; Lim, X.E.; Crowe, J.E., Jr.; Lok, S.M.; Anand, G.S. Epitope and paratope mapping reveals temperature-dependent alterations in the dengue-antibody interface. *Structure* **2017**, *25*, 1391–1402 e1393. [CrossRef]
41. Maddaloni, M.; Cooke, C.; Wilkinson, R.; Stout, A.V.; Eng, L.; Pincus, S.H. Immunological characteristics associated with the protective efficacy of antibodies to ricin. *J. Immunol.* **2004**, *172*, 6221–6228. [CrossRef] [PubMed]
42. Noy-Porat, T.; Rosenfeld, R.; Ariel, N.; Epstein, E.; Alcalay, R.; Zvi, A.; Kronman, C.; Ordentlich, A.; Mazor, O. Isolation of anti-ricin protective antibodies exhibiting high affinity from immunized non-human primates. *Toxins (Basel)* **2016**, *8*, 64. [CrossRef] [PubMed]
43. Vance, D.J.; Mantis, N.J. Resolution of two overlapping neutralizing b cell epitopes within a solvent exposed, immunodominant alpha-helix in ricin toxin's enzymatic subunit. *Toxicon* **2012**, *60*, 874–877. [CrossRef] [PubMed]
44. Vance, D.J.; Tremblay, J.M.; Mantis, N.J.; Shoemaker, C.B. Stepwise engineering of heterodimeric single domain camelid vhh antibodies that passively protect mice from ricin toxin. *J. Biol. Chem.* **2013**, *288*, 36538–36547. [CrossRef] [PubMed]

Article

# Genetic Fusion of an Anti-BclA Single-Domain Antibody with Beta Galactosidase

**George P. Anderson [1], Lisa C. Shriver-Lake [1], Scott A. Walper [1], Lauryn Ashford [2], Dan Zabetakis [1], Jinny L. Liu [1], Joyce C. Breger [1], P. Audrey Brozozog Lee [3] and Ellen R. Goldman [1,*]**

1 Naval Research Laboratory, Center for Biomolecular Science and Engineering, Washington, DC 20375, USA; george.anderson@nrl.navy.mil (G.P.A.); lisa.shriverlake@nrl.navy.mil (L.C.S.-L.); scott.walper@nrl.navy.mil (S.A.W.); daniel.zabetakis@nrl.navy.mil (D.Z.); jinny.liu@nrl.navy.mil (J.L.L.); joyce.breger@nrl.navy.mil (J.C.B.)

2 The Washington Center for Internships and Academic Seminars, 1333 16th Street N.W., Washington, DC 20036, USA; laurynashford96@gmail.com

3 Nova Research Inc., Alexandria, VA 22308, USA; plee142@su.edu

* Correspondence: ellen.goldman@nrl.navy.mil; Tel.: +1-202-404-6052

Received: 12 September 2018; Accepted: 27 September 2018; Published: 29 September 2018

**Abstract:** The Bacillus collagen-like protein of anthracis (BclA), found in *Bacillus anthracis* spores, is an attractive target for immunoassays. Previously, using phage display we had selected llama-derived single-domain antibodies that bound to *B. anthracis* spore proteins including BclA. Single-domain antibodies (sdAbs), the recombinantly expressed heavy domains from the unique heavy-chain-only antibodies found in camelids, provide stable and well-expressed binding elements with excellent affinity. In addition, sdAbs offer the important advantage that they can be tailored for specific applications through protein engineering. A fusion of a BclA targeting sdAb with the enzyme Beta galactosidase (β-gal) would enable highly sensitive immunoassays with no need for a secondary reagent. First, we evaluated five anti-BclA sdAbs, including four that had been previously identified but not characterized. Each was tested to determine its binding affinity, melting temperature, producibility, and ability to function as both capture and reporter in sandwich assays for BclA. The sdAb with the best combination of properties was constructed as a fusion with β-gal and shown to enable sensitive detection. This fusion has the potential to be incorporated into highly sensitive assays for the detection of anthrax spores.

**Keywords:** *Bacillus anthracis*; immunoassay; single-domain antibody; genetic fusion; Beta galactosidase

## 1. Introduction

*Bacillus anthracis,* the causative agent of anthrax, is a biothreat of grave concern [1,2]. Capable of lethality in both animals and humans, *B. anthracis* has been investigated since the early 1930s for use as a potential bioweapon by several countries around the world. The letter-based attacks of 2001 in the United States is an example of the impact this bacterium has when exploited as a bioweapon. *B. anthracis* spores are easily produced and once aerosolized and disseminated can remain dormant and viable for extended periods. Additionally, cleanup of contaminated areas requires harsh chemical agents and repeated treatments to ensure complete inactivation of the bacterial spores. Much research is centered on developing decontamination methods that are both effective and gentle [3,4], as well as biosensors and reagents for the rapid detection of spores [5–8]. The Bacillus collagen-like protein of anthracis (BclA), a spore protein, is a good target for antibody development; BclA is an immunodominant glycoprotein and the major component of the hair-like projections that cover the exosporium of *B. anthracis* spores [9–12].

By nature, antibodies can target and bind to specific antigens. Heavy-chain-only antibodies are found in camelids (camels, llamas, and alpacas) and sharks and lack the light chains that pair with the heavy chains in conventional antibodies [13,14]. Binding takes place through a single unpaired variable heavy domain, which in camelids is known as a VHH. Recombinantly produced VHH are termed single-domain antibodies (sdAbs), or nanobodies [15]. At ~15 kDa sdAbs are about a tenth the size of conventional antibodies; however sdAbs are highly effective in targeting and binding to antigens, while also possessing robust thermal stability and good production characteristics [16–18]. Another advantage of sdAbs is that they can readily be engineered and produced as fusions with other protein domains to introduce additional functionalities [19–24]. A popular type of fusion is the pairing of a sdAb with the enzyme alkaline phosphatase (AP) [19,20,25,26]. These fusions have two advantages. First, when using a sdAb-AP fusion, there is no need for a secondary antibody, eliminating a step from immunoassays. Secondly, AP is a dimer, so it yields a dimeric binding element with improved apparent affinity due to avidity. The enzyme Beta galactosidase (β-gal) can also be used with secondary reagents in immunoassays [27,28]. Additionally, β-gal is a tetramer with a molecular weight of 464 kDa, so fusions with this enzyme would also benefit from avidity. Previously, it had been reported that the enzyme β-gal is able to function with a scFv (linked variable heavy and variable light chain from a conventional antibody) inserted at the N-terminus of the enzyme [29]. Unlike fusions with AP, the β-gal fusions need to be produced in the cytoplasm. This is because periplasmicly directed β-gal fusions cause lethality which, depending on the fusion, can be due to jamming the translocation pore or misfolding of the β-gal in the periplasm [30,31].

Previously, we described the isolation of sdAbs that recognize several *B. anthracis* spore proteins including BclA [6]. In that work, we constructed a library of phage displayed sdAbs derived from llamas that had been immunized with recombinant spore proteins. Numerous sdAbs that appeared to bind BclA were identified and they fell into several families based on sequence similarity. However, only three of the BclA binding sdAbs were produced and characterized. In the current work, we re-visited the previously isolated BclA binding sdAbs, and characterized four additional clones that had been identified by phage display but neither produced nor tested. These new sdAbs, along with one previously characterized clone, were assessed for their binding kinetics and ability to be integrated into an immunoassay for the detection of BclA. Each was produced in both the periplasm and cytoplasm, and their binding ability and melting temperatures measured. Clone A5, which had been previously characterized, offered the best combination of properties, and thus was further developed as a genetic fusion with β-gal. This sdAb-β-gal fusion was incorporated into an enzyme linked immunosorbent assay (ELISA) for the detection of BclA. We showed that genetic fusions of sdAbs with β-gal provide a route to generate detection reagents.

## 2. Materials and Methods

### 2.1. Reagents

The BclA binding sdAbs (A4, A5, C5, D4, E6) had been isolated previously [6]. Sequences are shown in Figure 1.

A C-terminal fragment of BclA (iBclA) was the kind gift of Dr. Michael Weiner (AxioMx, Inc., Branford, CT, USA). The iBclA includes both a His tag as well as a biotinylation tag and was produced and purified as previously described [21]. The iBclA protein is more soluble than non-truncated recombinant BclA and was used for all BclA assays described in this work.

Unless otherwise specified, chemical reagents were from Sigma Aldrich (St. Louis, MO, USA), Thermo Fisher Scientific (Waltham, MA, USA), or VWR International (Radnor, PA, USA). Restriction endonucleases and ligation reagents were from New England Biolabs (Ipswich, MA, USA).

DNA amplification was accomplished with the Roche Expand High Fidelity DNA polymerase kit (Sigma Aldrich, St. Louis, MO, USA). Specific kits and assays are defined where applicable.

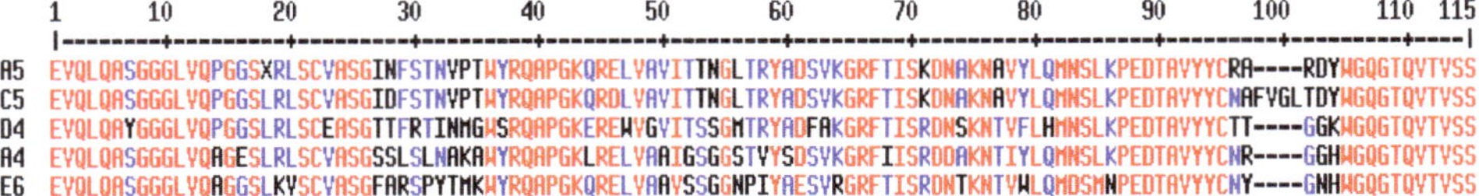

**Figure 1.** Deduced protein sequences of the five single-domain antibodies (sdAbs) that were evaluated. Sequences have been aligned using Multalin [32] to help identify similarities and differences in the protein sequence of the sdAbs. Red denotes high consensus and blue low consensus. Sequences are given in single letter amino acid code.

### 2.2. Periplasmic and Cytoplasmic Production of sdAbs

The coding sequences for the sdAbs were each mobilized from the pecan21 phage display vector into pET22b as NcoI-NotI fragments as described previously [33], or into pET28b using an analogous protocol. The sdAb expression plasmids were transformed into Tuner (DE3) for protein production. Freshly transformed colonies were used to start overnight cultures in 50 mL terrific broth (TB) with appropriate antibiotics (for pET22b: ampicillin 100 μg/mL; for pET28b kanamycin 30 μg/mL) at 25 °C. The next day the overnight cultures were poured into 450 mL of TB with appropriate antibiotics and grown for 2 h at 25 °C prior to induction with isopropyl-β-D-1 thiogalactoside (IPTG, 0.5 mM) and a further 2 h growth.

Purification of sdAbs expressed from pET22b, the periplasmic expression vector, were carried out through an osmotic shock protocol as described previously [34]. For cytoplasmic expression of the sdAbs, purification was similar to the protocol described previously [21] with some minor modifications. Cell pellets were suspended in phosphate-buffered saline (PBS) containing 0.05% Tween 20 prior to sonication, and the immobilized metal affinity chromatography (IMAC) resin was eluted with 0.25 M imidazole prior to purification by fast protein liquid chromatography (FPLC) on a ENrich SEC 70 10 × 300 mm column (Bio-Rad, Hercules, CA, USA).

### 2.3. Surface Plasmon Resonance

Affinity and kinetics measurements were performed using the ProteOn XPR36 (Bio-Rad, Hercules, CA, USA). Lanes of a general layer compact (GLC) chip were individually coated with BclA and measurements were essentially as described previously [6]. Data analysis was performed with ProteOn Manager 2.1 software, corrected by subtraction of the zero-antibody concentration column as well as interspot correction. The standard error of the fits was less than 10%. Binding constants were determined using the Langmuir model built into the analysis software.

### 2.4. Determining Melting Temperature by Fluorescent Dye Melt Assay

The fluorescent dye melt assay was performed as described previously [35]. Each sdAb was first diluted to a concentration of 500 μg/mL in a final volume of 20 μL PBS. Next a 1:1000 dilution of Sypro Orange dye was added to each sample. Samples were measured in triplicate using a Step One Real-Time polymerase chain reaction (PCR) machine (Applied Biosystems, Foster City, CA, USA). The heating program was run in continuous mode from 25 °C–99 °C at a heating rate of 1% (~2 °C per minute), and data was recorded using the carboxy-X-rhodamine (ROX) filter. The melting point was determined to be the peak of the first derivative of the fluorescence intensity.

### 2.5. Producing Fusion of sdAbs with β-gal

The *E. coli* β-gal was first synthesized and cloned as described previously [36]. Briefly, the β-gal gene from *E. coli* K12 was synthesized by Genscript (Piscataway, NJ, USA) to include a 5′ NotI site and a 3′ XhoI site. The NotI-XhoI fragment was purified and cloned into the pET28b expression vector (Millipore Sigma, Burlington, MA, USA). Next, the sdAb A5 was cloned into the pET28 vector containing β-gal as a NcoI-NotI fragment. Sequencing (Eurofins Genomics, Louisville, KY, USA) confirmed that the constructs were correct.

For protein production, plasmid DNA was transformed to the *E. coli* expression strain BL21(DE3). Colonies were inoculated into 3 mL TB with kanamycin (30 µg/mL) and grown overnight at 25 °C. The next day, 1 mL of the overnight culture was added to 450 mL TB with kanamycin and grown for ~7 h at 30 °C prior to induction (IPTG, 0.5 mM) and then grown overnight at 25 °C. The next morning cells were pelleted and processed as described above for the cytoplasmically grown sdAb.

### 2.6. ELISA

Each sdAb was biotinylated using a 10-fold excess of EZ-Link NHS-LC-LC-Biotin (Thermo Fisher Scientific, Waltham, MA, USA) for 30 min and then the excess biotin was removed using Zeba spin columns. The sdAb concentration was determined by absorbance at 280 nm. ELISAs were performed similar to those described previously [34,37] to determine optimum sdAb pairs. Briefly, wells of 96-well plates (Nunc Maxisorb from Thermo Fisher Scientific, Waltham, MA, USA) were coated by incubating 100 µL capture sdAb (2 µg/mL in PBS) overnight at 4 °C. Wells were washed with PBS containing 0.05% Tween 20 (PBST) and blocked for an hour at room temperature with 4% powdered milk in PBS. After blocking, BclA was added to the wells. For the checkerboard assay 5 µg/mL was used in each well, whereas for dose response, dilution series of BclA were employed. All conditions were measured at least in duplicate, and wells containing just PBS were included as a no-antigen control. After washing with PBST, biotinylated sdAb reporter was added at 2 µg/mL. Wells were washed and then incubated for an hour with streptavidin-horseradish peroxidase (HRP) (100 µL, 1 µg/mL). After a final washing with PBST, signal was developed by the addition of SureBlue (SeraCare, Gaithersburg, MD, USA). Finally, 100 µL 1 N HCl was added to stop color development and the plate read at 450 nm with Tecan Plate Reader (Morrisville, NC, USA).

A dose-response sandwich ELISA was performed with the optimum fusion pair. The plates were coated overnight with 10 µg/mL A5 and blocked for 2 h the next day. Starting with 10 µg/mL β-gal, 1:3 serial dilutions were performed. For the tracer steps, either biotin-labeled A5 (1 µg/mL) followed by streptavidin-HRP and SureBlue or A5-β-gal (1 µg/mL) followed with MuGal (50 µg/mL) to generate a fluorescent product was used. The substrate for HRP is colorimetric and the absorption was read at 450 nm after the color development had been stopped by addition of 1 N HCl. The β-gal assay used a florescent substrate, the fluorescent mixture was excited at 365 nm and the emission collected at 445 nm.

## 3. Results

### 3.1. Evaluating BclA Binding sdAbs

The BclA binding sdAbs (A4, A5, C5, D4, E6) had been isolated previously [6]; however only A5 from the group had been moved to an expression vector and characterized. In prior work it was determined that A5 could be expressed and produced both in the periplasm as well as in the cytoplasm [6,21,38]. Here, in addition to A5, we chose to revisit anti-BclA sdAbs that had been identified through rounds of phage display panning [6], but were not characterized. Each was cloned into pET22b for periplasmic expression and pET28b for cytoplasmic expression. Table 1 shows representative yields, as well as binding kinetics and melting temperatures for both expression protein variants.

**Table 1.** Single-domain antibody (sdAb) properties: yields, binding kinetics, affinity, and melting temperature (Tm).

| SdAb | Production | Yield (mg/L) | $k_a$ (1/Ms) | $k_d$ (1/s) | $K_D$ (M) | Tm (°C) |
|---|---|---|---|---|---|---|
| A4 | periplasmic | 15 | $9.3 \times 10^5$ | $1.8 \times 10^{-4}$ | $2.0 \times 10^{-10}$ | 57 |
| | cytoplasmic | 0.3 | $1.6 \times 10^5$ | $5.8 \times 10^{-5}$ | $3.7 \times 10^{-10}$ | 58 |
| A5 | periplasmic | 7 | $3.7 \times 10^5$ | $3.7 \times 10^{-5}$ | $1.0 \times 10^{-10}$ | 67 |
| | cytoplasmic | 6 | $2.4 \times 10^5$ | $2.8 \times 10^{-5}$ | $1.1 \times 10^{-10}$ | 56 |
| C5 | periplasmic | 6 | $2.5 \times 10^5$ | $2.7 \times 10^{-4}$ | $1.1 \times 10^{-9}$ | 58 |
| | cytoplasmic | 4 | $2.8 \times 10^5$ | $7.5 \times 10^{-4}$ | $2.7 \times 10^{-9}$ | 45 |
| D4 | periplasmic | 11 | $1.5 \times 10^6$ | $1.9 \times 10^{-4}$ | $1.3 \times 10^{-10}$ | 67 |
| | cytoplasmic | 3 | $8.6 \times 10^4$ | $2.1 \times 10^{-4}$ | $2.5 \times 10^{-9}$ | 50 |
| E6 | periplasmic | 20 | $3.4 \times 10^5$ | $1.2 \times 10^{-4}$ | $3.6 \times 10^{-10}$ | 60 |
| | cytoplasmic | 5 | $1.9 \times 10^3$ | $5.7 \times 10^{-5}$ | $2.9 \times 10^{-8}$ | 59 |

Each of the sdAbs was tested by surface plasmon resonance (SPR) to determine its binding affinity. The on and off rate constants, as well as the calculated dissociation constant ($K_D$) from a typical measurement are shown in Table 1. The representative SPR data from which these numbers are derived is shown in Appendix A. Measurements were made at least twice, and generally agreed within a factor of 2 (see Appendix B for a table of average values and the average deviations). The periplasmic version of these sdAbs all had excellent affinity, with low nM or sub nM $K_D$ values. Some of the sdAbs, in particular E6, were measured to have much poorer affinity when produced in the cytoplasm.

As an indicator of thermal stability, melting temperatures were measured using a fluorescent-based dye melt assay [39,40]. Results from both periplasmically and cytoplasmically produced sdAbs are reported in Table 1. Three out of the five sdAbs (A5, C5, and D4) showed lower melting temperatures when produced in the cytoplasm.

Finally, we evaluated how the five sdAbs perform as both capture and tracer reagents for the detection of BclA in a sandwich format. First, a checkerboard-type assay with a single concentration of BclA was performed (Figure 2). For this experiment, only periplasmically produced sdAbs were examined. Clone C5, with the poorest affinity, also was the worst performer both as a capture and tracer. Overall A4 and A5 performed the best in this assay. An assay using combinations of A4 and A5 capture tracer pairs was performed to determine dose response curves for BclA detection, and to choose the best capture tracer pair (Figure 3). As was seen in the checkerboard assay, the sdAb A5 performed very well as both a capture and tracer, while A4 performed well as a capture or tracer in combination with A5.

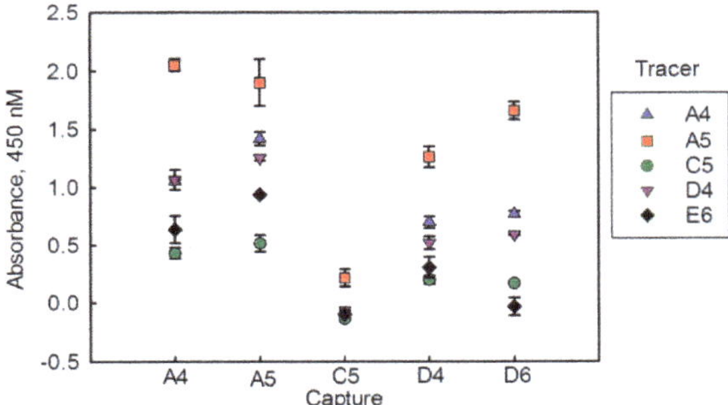


**Figure 2.** Checkerboard-enzyme linked immunosorbent assay (ELISA) to determine the optimal capture and tracer pairs. Each of the five sdAb captures are shown on the *X* axis and tracers indicated by the symbol (key to the right of the graph). Each tracer sdAb was biotinylated, while the capture sdAbs were adsorbed on the ELISA plate. Measurements were performed in duplicate. The average background (0.41) was subtracted from the average values; error bars represent the average deviation.

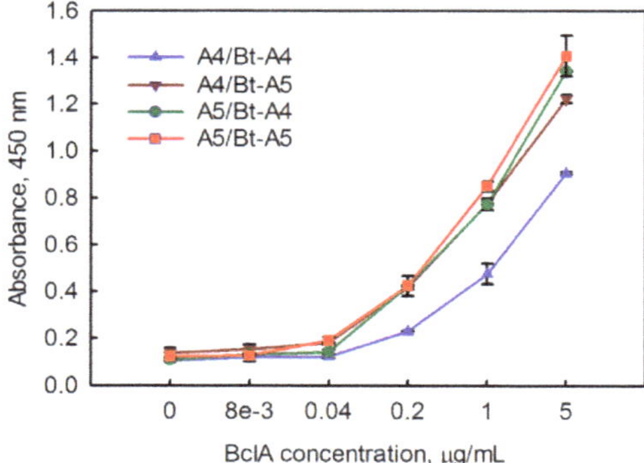


**Figure 3.** Dose response curves using combinations of A4 and A5 capture/tracer pairs. In each case the tracer is biotinylated (Bt). Measurements were performed in duplicate; the error bars represent the average deviation.

From the characterization of the sdAbs, the originally characterized clone, A5, exhibited the best combination of producibility in both the periplasm and cytoplasm, affinity, melting temperature, and function as a reporter in sandwich assays for BclA. Therefore, we chose A5 to produce a fusion with β-gal.

### 3.2. Construction and Evaluation of β-gal Fusion

Before constructing a sdAb fusion, the β-gal gene from *E. coli* was cloned into pET28 for cytoplasmic expression, and the enzymatic function of the construct verified [36]. Once enzymatic functionality had been confirmed, the A5 anti-BclA sdAb was cloned before the N-terminus of β-gal with no linker sequence other than the "AAA" encoded by the NotI restriction site separating the two functional units of the genetic fusion. The fusion construct produced well, with a protein yield of ~25 mg/L.

The function of the β-gal component of the fusion was demonstrated as previously described [36]. Assessing function of the A5 component was achieved through a sandwich ELISA. The BclA was captured through A5 adsorbed on wells of a 96-well plate. Results are show in Figure 4 in comparison to a standard ELISA using a secondary reagent. Both methods enabled the detection of BclA down to the lowest levels tested (0.14 µg/mL).

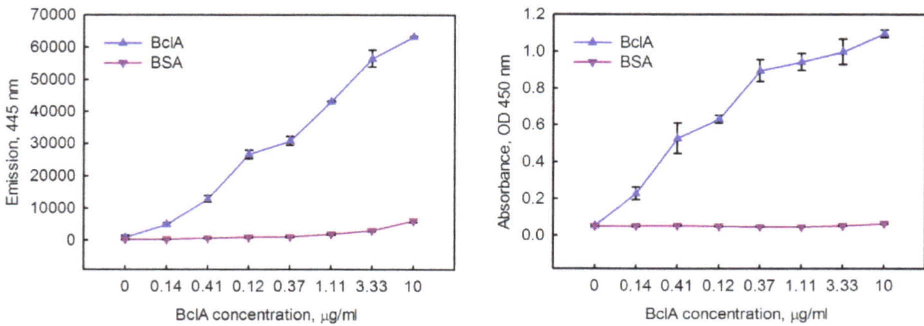


**Figure 4.** Dose response curves using A5 capture with A5-β-gal tracer (**left**) and A5 capture with Bt-A5 tracer (**right**). Measurements were performed in triplicate; the error bars represent the standard deviation.

## 4. Discussion

Immunoassays based on the detection of spore surface antigen have the potential to be a valuable asset for the detection of *B. anthracis*, the causative agent of anthrax, providing a rapid and facile method to detect spores. Previously, we had isolated several families of sdAbs that recognized BclA [6], a spore protein of *B. anthracis* that has been a target for immunoassays for sensitive spore detection [7] though only a small subset of the isolated sdAbs were initially characterized. One goal of this current work was the construction of a genetic fusion between β-gal and sdAbs. Therefore, it was necessary to examine the ability of each sdAb to be produced and to function when expressed in the cytoplasm. We examined additional anti-BclA sdAbs spanning four additional sequence families, and characterized each in terms of producibility, affinity, melting temperature, and ability to function in immunoassays for BclA.

As with variable domains from conventional antibodies, VHH contain a pair of cysteines that form an intra-chain disulfide bond. It has been demonstrated that deletion of this conserved disulfide bond, either through mutagenesis or cytoplasmic expression can lead to loss of protein stability, reflected in a lower melting temperature [41–44]. We observed a lower melting temperature in cytoplasmically produced sdAbs for three out of the five clones examined, suggesting the disulfide fails to form in the reducing environment of the cytoplasm. For clones A4 and E6 the melting temperature measured for cytoplasmically and periplasmically produced sdAb was essentially identical. This may indicate that disulfide bond is either formed or not formed in these two clones regardless of environment. Alternatively, it is possible that the canonical disulfide bond does not contribute significantly to the stability of these particular sdAb clones. Loss of the disulfide bond can also, in some instances, lead to a decrease in affinity for antigen [42]. Among the five BclA binding sdAbs examined, two (D4 and E6) showed a decrease in affinity for BclA when the sdAb was purified from the cytoplasm as determined by SPR. Oddly, E6 was one of the sdAbs for which a decreased melting temperature in the cytoplasmically produced protein was not observed. Perhaps the disulfide bond of E6 contributes more to its affinity than its stability. A definitive assessment of the disulfide bond status in the purified sdAbs could be determined through classical titration methods or a more sensitive mass spectrometry assay to quantify the state of the disulfide bonds in these sdAbs.

For the majority of the sdAbs examined in this study, we observed sdAb production was compromised when produced in the cytoplasm. In particular, clone A4 decreased from 15 mg/L production in the periplasm to less than 1 mg/mL for cytoplasmic expression. Cytoplasmic expression of A4 was attempted three separate times with low protein expression consistent between the biological replicates. Several researchers reporting high cytoplasmic expression of sdAbs [45] use a specialized *E. coli* strain that promotes disulfide bond formation in the cytoplasm [46,47]. Future work could involve testing expression of sdAb-β-gal genetic fusions in these strains.

After characterizations of their producibility, binding kinetics, and melting temperatures, sdAbs were evaluated for their ability to be employed as either a capture or tracer recognition molecule in an ELISA-type assay. From these studies, A5 functioned well as either capture or tracer reagent; clone A5 also showed good production in the cytoplasm with cytoplasmically produced protein showing the same high affinity as protein produced in the periplasm. This sdAb was then used to form the β-gal fusion. Specific detection of BclA with A5-β-gal fusion as tracer demonstrated the feasibility of using an enzyme-sdAb fusion that can be employed to reduce the number of immunoassay steps and possibly increase sensitivity.

The availability of β-gal fusions with sdAbs offers an alternative to fusions with AP that could be of great benefit to sensitive detection techniques such as the single-molecule array method [28,48–51]. Currently these assays require use of secondary reagents; often the sdAb is biotinylated and used in conjunction with streptavidin-β-gal [28,51]. Biotinylation of sdAbs is straightforward; however occasionally it can lead to labeling within the binding loops (complementarity determining regions, CDRs) that participate in antigen binding. In addition, the biotinylated reagent can vary from batch to batch depending on the extent of biotin labeling which can provide inconsistent results.

## 5. Conclusions

When examining the combination of properties (cytoplasmic expression, affinity, melting temperature, and function in immunoassays for the detection of BclA), clone A5 was the best choice for construction of a β-gal fusion. Previously, researchers demonstrated that β-gal was tolerant to N-terminal fusions and could be produced genetically fused to a scFv for use in immunoassays [29]. We have expanded upon this prior work, demonstrating a functional genetic fusion between β-gal and an sdAb. In general, we have found that sdAb fusion proteins often are more soluble and better produced than fusions with scFv [22]. The fusion of A5 and β-gal expressed well and functioned for the detection of antigen in immunoassays, demonstrating the ability of sdAbs to function in cytoplasmically produced fusions with the enzyme β-gal. Using a sdAb fusion with β-gal eliminates the need to biotinylate the sdAb or to use a secondary antibody, which simplifies the optimization of the assay and reduces assay time. This work demonstrates the feasibility of fusing sdAbs with β-gal for use in detection assays, highlighting the ability to engineer sdAbs. The fusion of A5 with β-gal has the potential to provide sensitive detection of *B. anthracis* spores when integrated into an assay, such as the single-molecule array method, that uses β-gal to generate signal [28,51].

**Author Contributions:** Conceptualization: G.P.A., J.L.L., and E.R.G.; Investigation: G.P.A., L.C.S.-L., S.A.W., L.A., D.Z., J.C.B., P.A.B.L. and E.R.G.; Methodology: G.P.A. and L.C.S.-L.; Project administration: E.R.G.; Resources: S.A.W.; Supervision: G.P.A. and E.R.G.; Visualization: L.C.S.-L., D.Z. and E.R.G.; Writing—original draft: G.P.A., L.A. and E.R.G.; Writing—review & editing: G.P.A., L.C.S.-L., S.A.W., J.C.B., J.L.L. and E.R.G.

**Funding:** This work was supported by Naval Research Laboratory base funds.

**Acknowledgments:** L.A. was a participant in an Office of Naval Research sponsored summer internship program for students attending historically black colleges and universities/minority institutions.

**Conflicts of Interest:** The authors declare no conflict of interest. The funding sponsors had no role in the design of the study; in the collection, analyses, or interpretation of data; in the writing of the manuscript, and in the decision to publish the results.

## Appendix A

Representative surface plasmon resonance (SPR) data from periplasmically and cytoplasmically produced sdAbs. Concentrations, in nM are noted to the right of each plot.

**Representative SPR data, periplasmically expressed sdAbs**

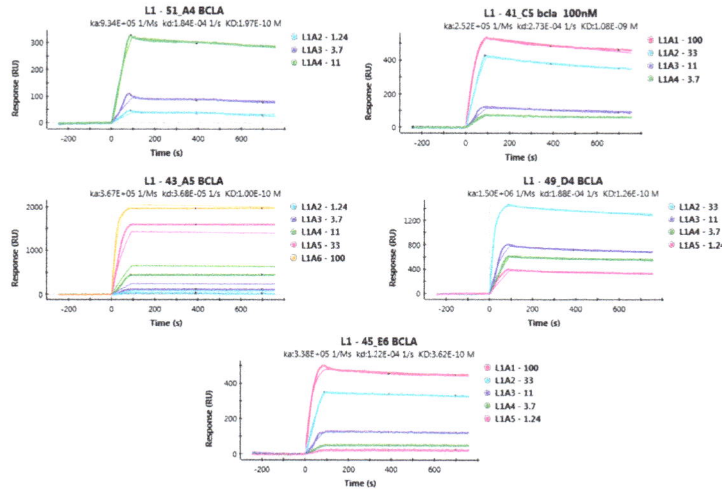

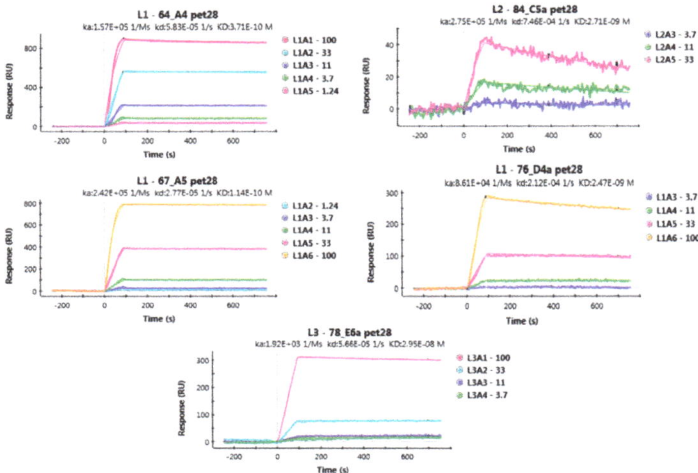


## Appendix B

Kinetic constants reported as average value/average deviation, based on two measurements.

| SdAb | Production | $k_a$ (1/Ms) | $k_d$ (1/s) | $K_D$ (M) |
|---|---|---|---|---|
| A4 | periplasmic | $9.1 \times 10^5/2.2 \times 10^4$ | $1.4 \times 10^{-4}/4.1 \times 10^{-5}$ | $1.6 \times 10^{-10}/4.1 \times 10^{-11}$ |
| | cytoplasmic | $1.6 \times 10^5/1.5 \times 10^3$ | $6.6 \times 10^{-5}/9.5 \times 10^{-6}$ | $4.3 \times 10^{-10}/5.6 \times 10^{-11}$ |
| A5 | periplasmic | $3.6 \times 10^5/9.5 \times 10^3$ | $3.7 \times 10^{-5}/7.0 \times 10^{-7}$ | $1.0 \times 10^{-10}/5.0 \times 10^{-12}$ |
| | cytoplasmic | $2.2 \times 10^5/2.2 \times 10^4$ | $2.4 \times 10^{-5}/3.7 \times 10^{-6}$ | $1.1 \times 10^{-10}/6.0 \times 10^{-12}$ |
| C5 | periplasmic | $2.0 \times 10^5/5.3 \times 10^4$ | $2.2 \times 10^{-4}/5.3 \times 10^{-5}$ | $1.1 \times 10^{-9}/2.5 \times 10^{-11}$ |
| | cytoplasmic | $3.0 \times 10^5/2.8 \times 10^4$ | $6.9 \times 10^{-4}/5.4 \times 10^{-5}$ | $2.4 \times 10^{-9}/3.0 \times 10^{-10}$ |
| D4 | periplasmic | $1.5 \times 10^6/3.0 \times 10^4$ | $2.0 \times 10^{-4}/1.3 \times 10^{-5}$ | $1.4 \times 10^{-10}/1.1 \times 10^{-11}$ |
| | cytoplasmic | $9.5 \times 10^4/9.0 \times 10^3$ | $2.3 \times 10^{-4}/2.1 \times 10^{-5}$ | $2.5 \times 10^{-9}/1.0 \times 10^{-11}$ |
| E6 | periplasmic | $3.0 \times 10^5/3.2 \times 10^4$ | $1.2 \times 10^{-4}/4.0 \times 10^{-6}$ | $4.0 \times 10^{-10}/3.3 \times 10^{-11}$ |
| | cytoplasmic | $9.8 \times 10^2/9.4 \times 10^2$ | $4.7 \times 10^{-5}/9.0 \times 10^{-6}$ | $6.0 \times 10^{-7}/5.7 \times 10^{-7}$ |

## References


1. Spencer, R.C. Bacillus anthracis. *J. Clin. Pathol.* **2003**, *56*, 182–187. [CrossRef] [PubMed]
2. Inglesby, T.V.; O'Toole, T.; Henderson, D.A.; Bartlett, J.G.; Ascher, M.S.; Eitzen, E.; Friedlander, A.M.; Gerberding, J.; Hauer, J.; Hughes, J.; et al. Anthrax as a biological weapon, 2002: Updated recommendations for management. *JAMA* **2002**, *287*, 2236–2252. [CrossRef] [PubMed]
3. Buhr, T.L.; Young, A.A.; Barnette, H.K.; Minter, Z.A.; Kennihan, N.L.; Johnson, C.A.; Bohmke, M.D.; DePaola, M.; Cora-Laó, M.; Page, M.A. Test methods and response surface models for hot, humid air decontamination of materials contaminated with dirty spores of bacillus anthracis Δsterne and bacillus thuringiensis al hakam. *J. Appl. Microbiol.* **2015**, *119*, 1263–1277. [CrossRef] [PubMed]
4. Buhr, T.L.; Young, A.A.; Bensman, M.; Minter, Z.A.; Kennihan, N.L.; Johnson, C.A.; Bohmke, M.D.; Borgers-Klonkowski, E.; Osborn, E.B.; Avila, S.D.; et al. Hot, humid air decontamination of a c-130 aircraft contaminated with spores of two acrystalliferous bacillus thuringiensis strains, surrogates for bacillus anthracis. *J. Appl. Microbiol.* **2016**, *120*, 1074–1084. [CrossRef] [PubMed]
5. Walper, S.A.; Anderson, G.P.; Brozozog Lee, P.A.; Glaven, R.H.; Liu, J.L.; Bernstein, R.D.; Zabetakis, D.; Johnson, L.; Czarnecki, J.M.; Goldman, E.R. Rugged single domain antibody detection elements for bacillus anthracis spores and vegetative cells. *PLoS ONE* **2012**, *7*, e32801. [CrossRef] [PubMed]

6. Walper, S.A.; Lee, P.A.B.; Anderson, G.P.; Goldman, E.R. Selection and characterization of single domain antibodies specific for bacillus anthracis spore proteins. *Antibodies* **2013**, *2*, 152–167. [CrossRef]
7. Morel, N.; Volland, H.; Dano, J.; Lamourette, P.; Sylvestre, P.; Mock, M.; Creminon, C. Fast and sensitive detection of bacillus anthracis spores by immunoassay. *Appl. Environ. Microbiol.* **2012**, *78*, 6491–6498. [CrossRef] [PubMed]
8. Kuehn, A.; Kovác, P.; Saksena, R.; Bannert, N.; Klee, S.R.; Ranisch, H.; Grunow, R. Development of antibodies against anthrose tetrasaccharide for specific detection of bacillus anthracis spores. *Clin. Vaccine Immunol. CVI* **2009**, *16*, 1728–1737. [CrossRef] [PubMed]
9. Redmond, C.; Baillie, L.W.; Hibbs, S.; Moir, A.J.; Moir, A. Identification of proteins in the exosporium of bacillus anthracis. *Microbiology* **2004**, *150*, 355–363. [CrossRef] [PubMed]
10. Steichen, C.; Chen, P.; Kearney, J.F.; Turnbough, C.L., Jr. Identification of the immunodominant protein and other proteins of the bacillus anthracis exosporium. *J. Bacteriol.* **2003**, *185*, 1903–1910. [CrossRef] [PubMed]
11. Sylvestre, P.; Couture-Tosi, E.; Mock, M. A collagen-like surface glycoprotein is a structural component of the bacillus anthracis exosporium. *Mol. Microbiol.* **2002**, *45*, 169–178. [CrossRef] [PubMed]
12. Todd, S.J.; Moir, A.J.; Johnson, M.J.; Moir, A. Genes of bacillus cereus and bacillus anthracis encoding proteins of the exosporium. *J. Bacteriol.* **2003**, *185*, 3373–3378. [CrossRef] [PubMed]
13. Hamers-Casterman, C.; Atarhouch, T.; Muyldermans, S.; Robinson, G.; Hamers, C.; Songa, E.B.; Bendahman, N.; Hamers, R. Naturally occurring antibodies devoid of light chains. *Nature* **1993**, *363*, 446–448. [CrossRef] [PubMed]
14. Greenberg, A.S.; Avila, D.; Hughes, M.; Hughes, A.; McKinney, E.C.; Flajnik, M.F. A new antigen receptor gene family that undergods rearrangement and extensive somatice diversificatino in sharks. *Nature* **1995**, *374*, 168–173. [CrossRef] [PubMed]
15. Ghahroudi, M.A.; Desmyter, A.; Wyns, L.; Hamers, R.; Muyldermans, S. Selection and identification of single domain antibody fragments from camel heavy-chain antibodies. *FEBS Lett.* **1997**, *414*, 521–526. [CrossRef]
16. Muyldermans, S. Nanobodies: Natural single-domain antibodies. *Annu. Rev. Biochem.* **2013**, *82*, 775–797. [CrossRef] [PubMed]
17. De Marco, A. Biotechnological applications of recombinant single-domain antibody fragments. *Microb. Cell Fact.* **2011**, *10*, 44. [CrossRef] [PubMed]
18. Wesolowski, J.; Alzogaray, V.; Reyelt, J.; Unger, M.; Juarez, K.; Urrutia, M.; Cauerhff, A.; Danquah, W.; Rissiek, B.; Scheuplein, F.; et al. Single domain antibodies: Promising experimental and therapeutic tools in infection and immunity. *Med. Microbiol. Immunol.* **2009**, *198*, 157–174. [CrossRef] [PubMed]
19. Sherwood, L.J.; Osborn, L.E.; Carrion, R.; Patterson, J.L.; Hayhurst, A. Rapid assembly of sensitive antigen-capture assays for marburg virus, using in vitro selection of llama single-domain antibodies, at biosafety level 4. *J. Infect. Dis.* **2007**, *196*, S213–S219. [CrossRef] [PubMed]
20. Liu, J.L.; Zabetakis, D.; Brozozog Lee, P.A.; Goldman, E.R.; Anderson, G.P. Single domain antibody alkaline phosphatase fusion proteins for antigen detection—Analysis of affinity and thermal stability of single domain antibody. *J. Immunol. Methods* **2013**, *393*, 1–7. [CrossRef] [PubMed]
21. Walper, S.A.; Battle, S.R.; Lee, P.A.B.; Zabetakis, D.; Turner, K.D.; Buckley, P.E.; Calm, A.M.; Welsh, H.S.; Warner, C.R.; Zacharko, M.A.; et al. Thermostable single domain antibody–maltose binding protein fusion for *Bacillus anthracis* spore protein bcla detection. *Anal. Biochem.* **2014**, *447*, 64–73. [CrossRef] [PubMed]
22. Liu, J.L.; Zabetakis, D.; Walper, S.A.; Goldman, E.R.; Anderson, G.P. Bioconjugates of rhizavidin with single domain antibodies as bifunctional immunoreagents. *J. Immunol. Methods* **2014**, *411*, 37–42. [CrossRef] [PubMed]
23. Pleschberger, M.; Saerens, D.; Weigert, S.; Sleytr, U.B.; Muyldermans, S.; Sara, M.; Egelseer, E.M. An s-layer heavy chain camel antibody fusion protein for generation of a nanopatterned sensing layer to detect the prostate-specific antigen by surface plasmon resonance technology. *Bioconjugate Chem.* **2004**, *15*, 664–671. [CrossRef] [PubMed]
24. Sherwood, L.J.; Hayhurst, A. Hapten mediated display and pairing of recombinant antibodies accelerates assay assembly for biothreat countermeasures. *Sci. Rep.* **2012**, *2*, 807. [CrossRef] [PubMed]
25. Liu, X.; Xu, Y.; Wan, D.-B.; Xiong, Y.-H.; He, Z.-Y.; Wang, X.-X.; Gee, S.J.; Ryu, D.; Hammock, B.D. Development of a nanobody–alkaline phosphatase fusion protein and its application in a highly sensitive direct competitive fluorescence enzyme immunoassay for detection of ochratoxin a in cereal. *Anal. Chem.* **2015**, *87*, 1387–1394. [CrossRef] [PubMed]

26. Sherwood, L.J.; Hayhurst, A. Ebolavirus nucleoprotein C-termini potently attract single domain antibodies enabling monoclonal affinity reagent sandwich assay (MARSA) formulation. *PLoS ONE* **2013**, *8*, e61232. [CrossRef] [PubMed]
27. Liu, Z.; Gurlo, T.; von Grafenstein, H. Cell-elisa using β-galactosidase conjugated antibodies. *J. Immunol. Methods* **2000**, *234*, P153–P167. [CrossRef]
28. Gaylord, S.T.; Dinh, T.L.; Goldman, E.R.; Anderson, G.P.; Ngan, K.C.; Walt, D.R. Ultrasensitive detection of ricin toxin in multiple sample matrixes using single-domain antibodies. *Anal. Chem.* **2015**, *87*, 6570–6577. [CrossRef] [PubMed]
29. Alcalá, P.; Ferrer-Miralles, N.; Villaverde, A. Engineering of *Escherichia coli* β-galactosidase for solvent display of a functional scfv antibody fragment. *FEBS Lett.* **2003**, *533*, 115–118. [CrossRef]
30. Snyder, W.B.; Silhavy, T.J. Beta-galactosidase is inactivated by intermolecular disulfide bonds and is toxic when secreted to the periplasm of *Escherichia coli*. *J. Bacteriol.* **1995**, *177*, 953–963. [CrossRef] [PubMed]
31. Dwyer, R.S.; Malinverni, J.C.; Boyd, D.; Beckwith, J.; Silhavy, T.J. Folding lacz in the periplasm of *Escherichia coli*. *J. Bacteriol.* **2014**, *196*, 3343–3350. [CrossRef] [PubMed]
32. Corpet, F. Multiple sequence alignment with hierarchical-clustering. *Nucleic Acids Res.* **1988**, *16*, 10881–10890. [CrossRef] [PubMed]
33. Walper, S.A.; Liu, J.L.; Zabetakis, D.; Anderson, G.P.; Goldman, E.R. Development and evaluation of single domain antibodies for vaccinia and the L1 antigen. *PLoS ONE* **2014**, *9*, e106263. [CrossRef] [PubMed]
34. Shriver-Lake, L.C.; Zabetakis, D.; Goldman, E.R.; Anderson, G.P. Evaluation of anti-botulinum neurotoxin single domain antibodies with additional optimization for improved production and stability. *Toxicon* **2017**, *135*, 51–58. [CrossRef] [PubMed]
35. Liu, J.L.; Zabetakis, D.; Goldman, E.R.; Anderson, G.P. Selection and evaluation of single domain antibodies toward ms2 phage and coat protein. *Mol. Immunol.* **2013**, *53*, 118–125. [CrossRef] [PubMed]
36. Brown Iii, C.W.; Oh, E.; Hastman, D.A.; Walper, S.A.; Susumu, K.; Stewart, M.H.; Deschamps, J.R.; Medintz, I.L. Kinetic enhancement of the diffusion-limited enzyme beta-galactosidase when displayed with quantum dots. *RSC Adv.* **2015**, *5*, 93089–93094. [CrossRef]
37. Anderson, G.P.; Bernstein, R.D.; Swain, M.D.; Zabetakis, D.; Goldman, E.R. Binding kinetics of antiricin single domain antibodies and improved detection using a b chain specific binder. *Anal. Chem.* **2010**, *82*, 7202–7207. [CrossRef] [PubMed]
38. Goldman, E.R.; Brozozog-Lee, P.A.; Zabetakis, D.; Turner, K.B.; Walper, S.A.; Liu, J.L.; Anderson, G.P. Negative tail fusions can improve ruggedness of single domain antibodies. *Protein Expr. Purif.* **2014**, *95*, 226–232. [CrossRef] [PubMed]
39. Lavinder, J.J.; Hari, S.B.; Sullivan, B.J.; Magliery, T.J. High-throughput thermal scanning: A general, rapid dye-binding thermal shift screen for protein engineering. *J. Am. Chem. Soc.* **2009**, *131*, 3794–3795. [CrossRef] [PubMed]
40. McConnell, A.D.; Spasojevich, V.; Macomber, J.L.; Krapf, I.P.; Chen, A.; Sheffer, J.C.; Berkebile, A.; Horlick, R.A.; Neben, S.; King, D.J.; et al. An integrated approach to extreme thermostabilization and affinity maturation of an antibody. *Protein Eng. Des. Sel.* **2013**, *26*, 151–163. [CrossRef] [PubMed]
41. Hagihara, Y.; Mine, S.; Uegaki, K. Stabilization of an immunoglobulin fold domain by an engineered disulfide bond at the buried hydrophobic region. *J. Biol. Chem.* **2007**, *282*, 36489–36495. [CrossRef] [PubMed]
42. Saerens, D.; Conrath, K.; Govaert, J.; Muyldermans, S. Disulfide bond introduction for general stabilization of immunoglobulin heavy-chain variable domains. *J. Mol. Biol.* **2008**, *377*, 478–488. [CrossRef] [PubMed]
43. George, J.; Compton, J.R.; Leary, D.H.; Olson, M.A.; Legler, P.M. Structural and mutational analysis of a monomeric and dimeric form of a single domain antibody with implications for protein misfolding. *Proteins Struct. Funct. Bioinform.* **2014**, *82*, 3101–3116. [CrossRef] [PubMed]
44. Saerens, D.; Pellis, M.; Loris, R.; Pardon, E.; Dumoulin, M.; Matagne, A.; Wyns, L.; Muyldermans, S.; Conrath, K. Identification of a universal vhh framework to graft non-canonical antigen-binding loops of camel single-domain antibodies. *J. Mol. Biol.* **2005**, *352*, 597–607. [CrossRef] [PubMed]
45. Veggiani, G.; de Marco, A. Improved quantitative and qualitative production of single-domain intrabodies mediated by the co-expression of erv1p sulfhydryl oxidase. *Protein Expr. Purif.* **2011**, *79*, 111–114. [CrossRef] [PubMed]

46. Zarschler, K.; Witecy, S.; Kapplusch, F.; Foerster, C.; Stephan, H. High-yield production of functional soluble single-domain antibodies in the cytoplasm of *Escherichia coli*. *Microb. Cell Fact.* **2013**, *12*, 97. [CrossRef] [PubMed]
47. Djender, S.; Schneider, A.; Beugnet, A.; Crepin, R.; Desrumeaux, K.E.; Romani, C.; Moutel, S.; Perez, F.; de Marco, A. Bacterial cytoplasm as an effective cell compartment for producing functional vhh-based affinity reagents and camelidae igg-like recombinant antibodies. *Microb. Cell Fact.* **2014**, *13*, 140. [CrossRef] [PubMed]
48. Li, Z.; Hayman, R.B.; Walt, D.R. Detection of single-molecule DNA hybridization using enzymatic amplification in an array of femtoliter-sized reaction vessels. *J. Am. Chem. Soc.* **2008**, *130*, 12622–12623. [CrossRef] [PubMed]
49. Walt, D.R. Optical methods for single molecule detection and analysis. *Anal. Chem.* **2013**, *85*, 1258–1263. [CrossRef] [PubMed]
50. Walt, D.R. Protein measurements in microwells. *Lab Chip* **2014**, *14*, 3195–3200. [CrossRef] [PubMed]
51. Dinh, T.L.; Ngan, K.C.; Shoemaker, C.B.; Walt, D.R. Rapid and ultrasensitive detection of botulinum neurotoxin serotype A1 in human serum and urine using single-molecule array method. *Forensic Toxicol.* **2017**, *35*, 179–184. [CrossRef]

Article

# Selection and Characterization of a Nanobody Biosensor of GTP-Bound RHO Activities

**Laura Keller [1,2,†], Nicolas Bery [3,†], Claudine Tardy [1], Laetitia Ligat [1], Gilles Favre [1,2], Terence H. Rabbitts [3] and Aurélien Olichon [1,2,*]**

1 Centre de Recherche en Cancérologie de Toulouse (CRCT), Inserm, Université Paul Sabatier-Toulouse III, CNRS, 31037 Toulouse, France; keller.laura@iuct-oncopole.fr (L.K.); claudine.tardy@inserm.fr (C.T.); laetitia.ligat@inserm.fr (L.L.); favre.gilles@iuct-oncopole.fr (G.F.)

2 Institut Claudius Regaud (ICR), Institut Universitaire du Cancer de Toulouse-Oncopole (IUCT-O), Laboratoire de Biologie Médicale Oncologique (LBMO), 31059 Toulouse, France

3 Weatherall Institute of Molecular Medicine, MRC Molecular Haematology Unit, University of Oxford, Oxford OX3 9DS, UK; nicolas.bery@ndcls.ox.ac.uk (N.B.); terence.rabbitts@imm.ox.ac.uk (T.H.R.)

* Correspondence: aurelien.olichon@inserm.fr

† These authors contributed equally to this work

Received: 29 November 2018; Accepted: 20 December 2018; Published: 9 January 2019

**Abstract:** RHO (Ras HOmologous) GTPases are molecular switches that activate, in their state bound to Guanosine triphosphate (GTP), key signaling pathways, which involve actin cytoskeleton dynamics. Previously, we selected the nanobody RH12, from a synthetic phage display library, which binds the GTP-bound active conformation of RHOA (Ras Homologous family member A). However, when expressed as an intracellular antibody, its blocking effect on RHO signaling led to a loss of actin fibers, which in turn affected cell shape and cell survival. Here, in order to engineer an intracellular biosensor of RHOA-GTP activation, we screened the same phage nanobody library and identified another RHO-GTP selective intracellular nanobody, but with no apparent toxicity. The recombinant RH57 nanobody displays high affinity towards GTP-bound RHOA/B/C subgroup of small GTPases in vitro. Intracellular expression of the RH57 allowed selective co-precipitation with the GTP-bound state of the endogenous RHOA subfamily. When expressed as a fluorescent fusion protein, the chromobody GFP-RH57 was localized to the inner plasma membrane upon stimulation of the activation of endogenous RHO. Finally, the RH57 nanobody was used to establish a BRET-based biosensor (Bioluminescence Resonance Energy Transfer) of RHO activation. The dynamic range of the BRET signal could potentially offer new opportunities to develop cell-based screening of RHOA subfamily activation modulators.

**Keywords:** nanobody; phage display; intrabody; intracellular antibody; GTPase RHO; BRET; RAS

## 1. Introduction

The RAS HOmologous RHO GTPases are small G proteins that act as molecular switches. These GTPases cycle between two conformational states depending on their binding to GDP (Guanosine diphosphate) or GTP (Guanosine triphosphate). A plethora of guanine nucleotide exchange factors (GEF) can stimulate the GTP loading of the small GTPases in a spatio-temporal manner, leading to local activation of effector proteins. A number of GTPase-activating proteins (GAP) can then catalyze the nucleotide hydrolysis to switch off the signaling [1]. Like for RAS, a subtle conformational change involving the two switch loops in the GTP-bound conformation of RHO enables the binding of the signaling effectors [2]. In contrast to the GTPase RAS-GTP, which can be massively induced [3], co-precipitation of the RHO-GTP pool by recombinant effector binding domains (RHO binding domain, RBD) showed that only a small fraction of the total RHO GTPases cellular pool is stimulated [4].

The RHOA-like subfamily includes RHOA/B/C members, which have been extensively studied for their involvement in actin cytoskeleton dynamics regulation, in cell proliferation and migration, or in development [5]. According to the physiological context, they can drive several other key signaling pathways such as YAP (Yes associated protein) for RHOA [6,7] or AKT for RHOB (RAS Homologous family member B) [8,9]. Their potential involvement in cancer cell migration and metastasis [5,10,11], or in targeted therapy resistance [9,12], suggests that the development of small molecule inhibitors targeting these GTPases would be valuable. For example, the small molecule RHOSIN inhibits the interaction of RHO with several GEFs (LARG, DBL, LBC, p115-RHOGEF, PDZ-RHOGEF), but this inhibitor presents limited efficiency in most cellular contexts [13].

The difficulty in identifying potent cellular inhibitors of RHO is partly the result of the lack of quantitative tools to precisely monitor their cellular activation. To date, all the molecular tools available to study RHO activity states are based on the use of effector RBDs either in pull down or capture ELISA (Enzyme-linked Immunosorbent assay) assays on cellular lysates, as intramolecular FRET-based sensors (Fluorescent probes based on Förster Resonance Energy Transfer) [14,15], or as tripartite split-GFP protein–protein interaction reporters in cells [16]. However, the poor stability of the RBD, as well as its low affinity towards the GTP-bound RHOs, has hampered the dynamic range of these assays [17,18]. In addition, the use of such effector domains in cells could be per se a potential competitor of the endogenous effectors, thus these tools require optimization of expression level in stable cell lines. Therefore, artificial affinity binding domains with higher stability and selectivity offer an attractive alternative to develop biosensors of RHO activation.

Nanobodies or stable single domain antibodies have emerged as useful molecular reagents to sense or track antigens in the reducing intracellular environment when used as intracellular antibodies [19,20]. Their high solubility in the reducing cytosol retains their conformational specificity and high affinity required for the selective recognition of antigens in living cells. We and others have already reported such alternative binding domains selective towards the GTP-bound conformation of a GTPase such as H-RAS [21], Dynamin [22], or RHO/RAC subfamilies [23]. In particular, we identified a synthetic nanobody (designated RH12) with high affinity towards the RHOA-like subfamily and RAC1 proteins, and with high specificity to their GTP-bound form in vitro and in cells. When expressed as an intracellular antibody, RH12 nanobody induced a dramatic effect on cell shape that was associated with actin polymerization defects [23]. We assumed that the RH12 nanobody acts as a macrodrug by blocking GTP-bound RHO and RAC signaling and inhibiting the RHO-RAC/effector interactions. Therefore, such a blocking intracellular antibody would not be suitable to build a biosensor.

Here, we report the characterization of another high affinity nanobody (referred to hereafter as RH57) that is specific to the GTP-bound fraction of RHOA and RAC1 subfamilies in vitro, but with no apparent competition with RHOA-like effectors when used as an intracellular antibody. We expressed it as a chromobody and observed its re-localization to the plasma membrane upon RHO stimulation. We functionalized the nanobody in order to produce a versatile dynamic BRET biosensor of RHO activation in cells. This clone not only provides an additional example of the diversity and functionality of the synthetic nanobody phage library from which it has been generated, but also illustrates the ability of intracellular antibodies to study the activity of challenging proteins.

## 2. Material and Methods

### 2.1. Plasmids

p-IB-6His-Myc and pIB-GFP intracellular antibodies and chromobodies expression vectors, in which any humanized synthetic single domain antibody (hs2dAb) from the NaLi-H1 library can be inserted by NcoI and NotI cloning sites, were described previously [23]. The 2x-Strep-tag® (similar to Twin-Strep-tag®from IBA-Lifesciences®) followed by HA tag (2SHA)-RHO plasmid expression was described previously [23]. pHEN-hs2dAb 6xhis-myc-6xhis was constructed as following: hs2dAb-6xhis-myc was digested from pIB-GFP and inserted into a modified pHEN6-VHH-6xhis that was previously described [24], thus creating a periplasmic expression vector pHEN6-hs2dAb-6xhis-myc-6xhis. The tag downstream of the NotI site was replaced by a synthetic Flag-Ctag DNA fragment encoding the following translated sequence: A A A G G G S G G D Y K D D D D K G Y Q D Y E P E A *, thus producing pHEN-hs2dAb-Flag-Ctag. The hs2dAb-Flag-Ctag was then sub-cloned into the previously described pAOT7 vector [23] using NcoI and EcoRI sites to produce the pAOT7-hs2dAb-Flag-Ctag, which allowed the production in *E.coli* cytoplasm.

### 2.2. Cell Lines, Transfection Method, and Reagents

HeLa cells (cervix adenocarcinoma; ATCC® CRM-CCL-2™, ATCC, Manassas USA) were grown in Dulbecco's Modified Eagle Medium (DMEM) (Lonza®, Basel, Switzerland) supplemented with 10% FBS (Foetal Bovine Serum) (Sigma Aldrich®, St. Louis, MO, USA). Transient transfections of DNA plasmids were performed using the Jet Prime method, as indicated by the supplier (PolyPlus Transfection®, Illkirch, France).

HEK293T human embryonic kidney cells (ATCC® CRL-3216) were grown in DMEM medium (Life Technologies®, Carlsbad, CA, USA) supplemented with 10% FBS (Sigma Aldrich®) and 1% Penicillin/Streptomycin (Life Technologies®). HEK293T cells were transfected with Lipofectamine 2000 (Thermo-Fisher®, Waltham, MA, USA, see BRET2 section). All cells were grown at 37 °C in a humidified incubator with 5% $CO_2$.

Western blots were probed with a mouse monoclonal 26C4 anti-RHOA (1/500, O/N, 4 °C, Santa Cruz Biotechnology®, Dallas, TX, USA), goat polyclonal anti-myc tag HRP conjugated (1/3000, 1 h, room temperature (RT), Novus Biologicals®, Centennial, CO, USA), and mouse monoclonal anti-RAC1 (1/1000, O/N, 4°C, Millipore). Detection was performed using peroxidase-conjugated secondary antibodies and a chemiluminescence detection kit (Biorad®, Hercules, CA, USA). F-Actin was stained with Alexa568-conjugated phalloidin (Molecular Probes, Eugene, USA).

### 2.3. Subtractive Phage Display Panning for Isolating RHO-GTP Specific hs2dAb

The NaLi-H1 library of humanized synthetic single domain antibody [23] was used for this study. A subtractive panning protocol was designed to isolate hs2dAb selective for RHOA-GTP conformation. The chitin binding domain (CBD) from chitinase A1 (New England Biolabs®, Ipswich, USA) or 2x-Strep-tag® (IBA-Lifesciences®, Göttingen, Germany) fusion of RHOA GTPase active mutant (RHOA L63) were expressed transiently for 24 h in HEK293 cells and captured freshly after cell lysis on magnetic beads before incubation with the library phages. Chitin magnetic beads (New England Biolabs®) or Strep-Tactin®-coated beads (IBA-Lifesciences®) were used alternatively for the capture of antigens for the four rounds of phage display. Phages were previously adsorbed on empty chitin or Strep-Tactin®-coated magnetic beads to remove nonspecific binders. From the second round of panning, depletion steps on GDP-loaded wild type RHOA or N19 inactive mutant and on RHOB L63, RHOC L63, and RAC1 L61 active mutants were included (Figure 1B). The adequate amount of antigen-coated beads was incubated for 2 h with the phage library ($10^{13}$ phages diluted in 1 mL of PBS + 0.1% Tween 20 + 2% non-fat milk). Phages and antigens-bound Strep-Tactin®-coated beads or Chitin beads were recovered on a magnet. Beads were washed with PBS–Tween 0.1% 10 times (round 1), 15 times (round 2), or overnight (rounds 3 and 4), and in the presence of an excess of untagged RHOA and RHOC L63 to further deplete in binders with a high dissociation rate. Bound phages were eluted using triethylamine (Sigma Aldrich®) and *E. coli* (TG1 strain) were infected with the eluted phages. For rounds 2, 3, and 4, only $10^{12}$ phages were used as input.

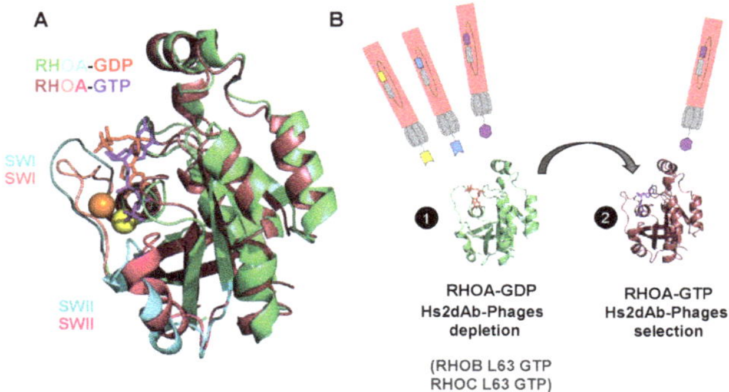


**Figure 1.** Antibody phage display selection of GTP-bound RHO conformational nanobodies. (**A**) View of the structure of RHOA-GTP V14 mutant (shown in wine-red and pink, Protein Data Base (PDB): 1a2b) superimposed with the structure of RHOA-GDP (shown in green and cyan, PDB: 1ftn). RHOA G14V mutant in the active state is bound to the GTP (purple-blue nucleotide) and $Mg^{2+}$ (shown as a yellow sphere). RHOA displays the inactive conformation bound to GDP (red nucleotide) and $Mg^{2+}$ (shown as an orange sphere). The structural alignment in this view shows a subtle closure of the switch I and switch II loops (SWI and SWII loops currently in cyan in RHOA-GDP and in pink in RHOA-GTP) around the phosphate gamma and the $Mg^{2+}$ (orange to yellow). (**B**) Scheme of the subtractive phage display enrichment of hs2dAb to GTP-bound RHOA (wine-red) by depletion with the inactive GDP-bound state (green) and with the GTP-bound state of RHOB and RHOC (see Methods). The phages presenting hs2dAb are shown in pink/grey. The RHOA structures were produced using PyMOL software 2.1 (Schrödinger, Mannheim, Germany).

### 2.4. Hs2dAb Purification

Briefly, hs2dAb-6 × His-Myc-6 × His was produced in the periplasm of XL1BLUE *E. coli* grown in TB/ampicillin (100 µg/mL) medium supplemented with 1% glucose in the start culture and 0.5% glucose during induction with 1 mM IPTG (Isopropyl β-D-1-thiogalactopyranoside). After overexpression for 16 h at 28 °C, the cells were harvested, suspended in 15 mL ice-cold TES buffer (200 mM Tris (Sigma Aldrich®) pH 8.5, 0.5 mM EDTA, 500 mM Sucrose), and stored at −80°C. Thirty milliliters of a one-quarter dilution of TES buffer was briefly added to the re-suspended pellets prior to vortex and kept at 4 °C for 30 min. After centrifugation (30 min, 13000× *g*, 4 °C), the periplasmic extract containing hs2dAb was purified by affinity chromatography. The protein extract was incubated for 2 h in the presence of complete His-Tag purification beads (ROCHE®, Basel, Switzerland) previously equilibrated with an equilibration buffer (50 mM Tris pH 8.0, 0.125 mM EDTA, 125 mM sucrose, 100 mM NaCl, 10 mM imidazole pH 7.0). The beads were washed with 30 mL of washing buffer (10 mM Tris pH 8.0, 150 mM NaCl, 10 mM imidazole pH 7.0). Hs2dAb were then eluted with elution buffer (500 mM imidazole pH 7.0, 25 mM Tris pH 6.8, 300 mM NaCl) and dialysed against PBS containing 20% glycerol for 16 h at 4 °C, and purity was assessed by SDS-PAGE followed by InstantBlue™ (Expedeon, Cambridgeshire, UK) Coomassie staining (Figure 2B).

Cytosolic expression of hs2dAb-Flag-Ctag was performed in BL21(DE3) *E.coli* cells from the pAOT7 vector. Transformed bacteria cells were used to grow 3 mL TB/kanamycin (35 µg/mL) cultures overnight at 37 °C prior to dilution of the pre-culture in baffled flasks containing 1 L of the same media. Cells were allowed to grow at 37 °C until OD600 reached 0.5 to 0.7. Cells were then induced with IPTG at a final concentration of 100 µM and grown for an additional 16 h at 18 °C. Cells were harvested by centrifugation at 4000× *g* for 20 min. The pellets were re-suspended in lysis buffer (PBS pH 7.4, 1X lysozyme and DNase I, proteases inhibitors) and lysed by sonication on ice prior to centrifugation (30 min, 15,000× *g*, 4 °C). A CaptureSelect C-tag affinity matrix (Thermo-fisher®) was equilibrated in PBS pH 7.4. After loading clarified lysate and washing in PBS pH 7.4, 300 mM NaCl, 20 mM $MgCl_2$,

C-tagged antibodies were eluted in 20 mM Tris pH 7.4 in the presence of 2 M $MgCl_2$. Dialysis was performed overnight against PBS containing 20% glycerol, and purity was assessed by SDS-PAGE followed by InstantBlue™ (Expedeon, Cambridgeshire, UK) Coomassie staining (Figure 2B).

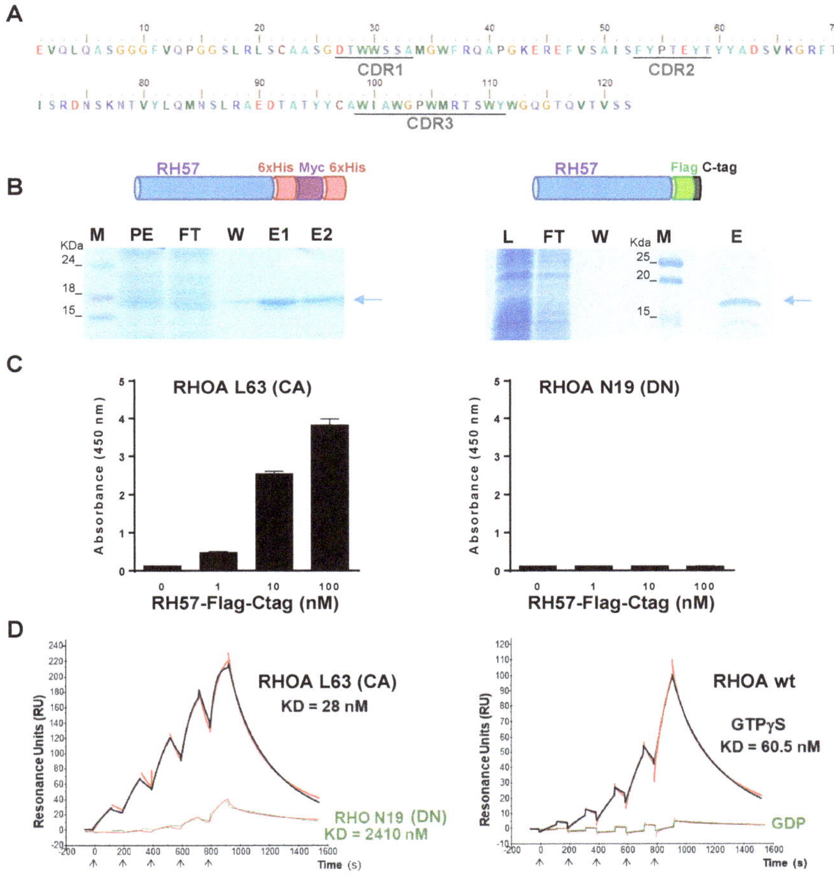


**Figure 2.** In vitro characterization of recombinant RH57 hs2dAb. (**A**) Amino acid sequence of the RH57 hs2dAb. Complementary-determining region (CDR) are underlined. (**B**) Representative Coomassie staining of SDS-PAGE showing the purification of RH57 fused to its carboxy terminal end with 6xHis-Myc-6xHis tag expressed in the periplasm (left) or RH57 fused to its carboxy terminal end with Flag tag and C-tag (right) used in the study. PE = periplasmic extract, L = clarified lysate, FT = flow through, W = wash, E = elution, and M = molecular weight marker. **(C)** ELISA detection of recombinant RHOA L63 constitutively active (CA) mutant versus RHO N19 dominant negative (DN) mutant. RHO proteins were captured on a Strep-Tactin coated plate and incubated with increasing concentrations of recombinant RH57-Flag-Ctag. Anti-Flag antibody was used to reveal the bound fraction of recombinant RH57 hs2dAb. **(D)** Single cycle kinetics (SCK) surface plasmon resonance (SPR) measurements on RH57-6xhis-myc-6xhis captured on an Ni-NTA (Nickel affinity capture of 6xHis) chip. Red lines correspond to the raw data measurements, and the fit curves are displayed in black or green according to the condition. Analytes were recombinant RHOA L63 CA mutant (black line, left SCK); RHOA N19 DN mutant (green line, left SCK); or RHOA WT loaded with either GTPγS (black line, right SCK) or GDP (green line, right SCK), injected at increasing concentrations (arrows, see Methods).

### 2.5. RHO GTPases Purification

RHO proteins deleted for the carboxy terminal four amino acids referred to as CAAX motif were expressed with an amino terminal tag that consisted of 2x-Strep-tag followed by HA tag (2SHA). 2SHA-RHO were expressed in BL21(DE3) *E.coli* cells from a pET vector. Transformed bacteria cells were used to grow 3 mL LB/carbenicillin (100 µg/mL) cultures overnight at 37 °C prior to inoculation in baffled flasks containing 1 L of the same media. Cells were allowed to grow at 37 °C until OD600 reached 0.5 to 0.7. Cells were then induced with IPTG at a final concentration of 100 µM and grown for an additional 20 h at 25 °C. Cells were harvested by centrifugation at 4000× *g* for 20 min. The pellets were re-suspended in lysis buffer (50 mM Tris HCl pH 8.0, 150 mM NaCl, 5 mM $MgCl_2$, 0.1% triton, 1 mM DTT, 1X lysozyme and DNase I, proteases inhibitors) and lysed by sonication on ice prior to centrifugation (15 min, 10,000× *g*, 4 °C). Strep-Tactin SuperFlow Plus (IBA-Lifesciences ®) matrix was equilibrated in buffer A (50 mM Tris HCl pH 8.0, 150 mM NaCl, 5 mM $MgCl_2$) and was incubated with supernatant for 2 h at 4 °C. Then, the supernatant and matrix were loaded on a simple column in order to maximize the capture of 2SHA-RHO proteins. The matrix was washed by 15 mL of washing buffer (300 mM NaCl, 50 mM Tris pH 8.0, 5 mM $MgCl_2$, 0.1% Tween20). RHO proteins were then eluted in buffer A containing 10 mM Biotin (Sigma Aldrich®). Dialysis was performed overnight against buffer A containing 15% glycerol.

### 2.6. ELISA Assays

Wells of Strep-Tactin-coated plates (IBA-Lifesciences®) were coated with 100 nM of recombinant 2S-HA-tagged RHOA protein (200 µL in TBS by well) during 2 h at RT and then blocked with 5% milk in TBS–Tween 0.05% (blocking buffer) for 1 h at RT. Several dilutions of hs2dAb in blocking buffer were applied to the ELISA plates in duplicates for 1 h at RT. Next, we added 0.5 µg/mL anti-Flag HRP antibody (F1804, Sigma-Aldrich®) in blocking buffer for 1 h at RT and the reaction was visualized by the addition of 100 µL chromogenic substrate (Thermoscientific®, 1-step ultraTMB) for 3 min. The reaction was stopped with 50 µL $H_2SO_4$ 1N and absorbance at 450 nm was measured using a FLUOstar OPTIMA microplate reader (BMG LABTECH, Ortenberg, Germany). Plates were washed three times with washing buffer (TBS–Tween20 0.05%) after each step. All steps were performed under agitation (400 rpm).

### 2.7. Loading Recombinant Proteins with GTPγS/GDP

We added 10 mM EDTA and 1 mM GDP or 100 µM GTPγS to recombinant proteins 2S-HA-RHOA wt. We incubated the reaction for at least 30 min at 30 °C for recombinant proteins. The reaction was stopped by adding 60 mM $MgCl_2$, and the mix was vortexed and put on ice.

### 2.8. Immunofluorescence Staining

Cells transfected with intracellular antibodies fused to a monomeric GFP (GFP carrying mutation Cys48Val) were fixed in 3.7% paraformaldehyde and permeabilized with PBS–Triton 0.1%, blocked with PBS–BSA 8%. Actin fibers were stained with Alexa 568-Phalloidin (1/40, 1 h, RT, Invitrogen®). All coverslips were mounted in Mowiol. Data acquisition was carried out on a Zeiss axiovert inverted microscope and analysed using Image J software (Fiji open-source platform).

### 2.9. Endogenous RHO Proteins Intracellular Antibodies Co-Immunoprecipitation Assays

The hs2dabs RH12, RH57, or NR with a 6xhis and carboxy Myc tags were transiently expressed in HeLa cells. After 24 h, cells were harvested and lysed in buffer (50 mM Tris pH 7.4, 500 mM NaCl, 10 mM $MgCl_2$, 1% TritonX100, supplemented with proteases and phosphatases inhibitors). To load the GTPases (Figure 3A), 10 mM EDTA and 1 mM GDP or 100 µM GTPγS were then added to cell lysates, and incubated for at least 30 min at 37 °C. The reaction was stopped by adding 60 mM $MgCl_2$, and the mix was vortexed and put on ice. To pull down the intracellular antibody from the cell lysate, the loaded lysate or the fresh lysate (in Figure 3A,B, respectively) were incubated on Complete His-Tag $Ni^{2+}$-NTA Purification Resin (ROCHE®) that had previously been equilibrated in the lysis buffer. After one hour at 4 °C, the beads were washed three times in buffer (50 mM Tris-HCl pH 7.4, 150 mM NaCl, 10 mM $MgCl_2$, 0.1% Tween20) and denatured in 1X Laemmli reducing sample buffer, boiled for 5 min, and separated on 12.5% SDS-PAGE for Western blot analysis.

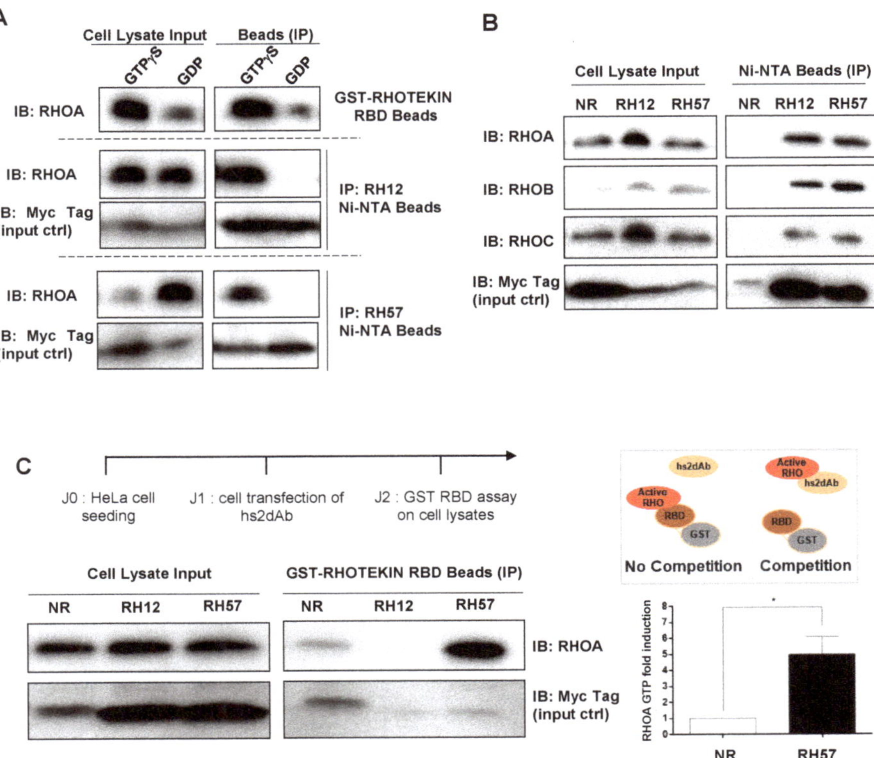


**Figure 3.** Intracellular antibody characterization of RH57 hs2dAb. (**A**) The hs2dabs RH12 and RH57 were expressed in HeLa cells with a 6xhis and carboxy Myc tag and the endogenous RHO GTPases were loaded with either GTPγS or GDP in the cell lysate. The intracellular antibodies were immuno-precipitated with Capture Select His tag affinity beads and an anti-RHOA antibody was used to monitor which pool of RHO (GTP or GDP) was bound by the intracellular antibodies. The recovered intracellular antibodies were monitored by immuno-blotting with anti-Myc tag. The GST-RHOTEKIN RHO binding domain (RBD), which allows to pull down GTP-bound RHO proteins on glutathione beads, was used as positive control. (**B**) Intracellular antibody immuno-precipitation assays on endogenous RHO proteins. Assays were performed as in panel A, but on non-loaded endogenous RHO proteins. The non-related NR hs2dAb was used as a negative control. **(C)** GST-RHOTEKIN RBD interference assay intracellular antibodies. The hs2dabs NR, RH12, and RH57 were expressed in HeLa cells with a 6xHis and carboxy Myc-tag. Intracellular expression of RH12 impedes the GST-RBD capture of RHOA-GTP. RH57 induced an increase of RHOA-GTP. NR binds to the beads. Representative blot of three independent experiments. Right: scheme of the binding competition events. Quantification of RHOA-GTP fold increase in RH57 conditions compared with the NR condition reveals potential stabilization of RHOA-GTP or GAP inhibition activity by the RH57. Error bars are mean ± SD of the biological repeats. * $p < 0.05$. IP = immuno-precipitation; IB = immuno-blot.

### 2.10. GST-RBD Assay

The hs2dabs RH12, RH57, or NR with a 6xhis and carboxy Myc tags were transiently expressed in HeLa cells. After 24 h, cells were harvested and lysed on ice in buffer (50 mM Tris-HCl pH 7.4, 500 mM NaCl, 10 mM $MgCl_2$, 1% TritonX100, supplemented with proteases and phosphatases inhibitors). To pull down the GTP-bound RHOA/B/C with the reference method, the lysates were incubated with GST-RBD beads (Figure 3A,C). After 45 min at 4 °C, the beads were washed three times in buffer (50 mM Tris-HCl pH 7.4, 150 mM NaCl, 10 mM $MgCl_2$, 0.1% Tween20) and denatured in 1X Laemmli reducing sample buffer, boiled for 5 min, and separated on 12.5% SDS-PAGE for Western blot analysis. Quantification of the RHOA immunoblot signal was done with ImageLab™ version 6.0 (Biorad®, Hercules, CA, USA) software for each condition to determine the ratio of RHOA-GTP/total RHOA. The results for the RH57 condition were normalized to NR. Statistical analysis (*t*-test) was calculated using GraphPad Prism software version 6.0 (California corporation, USA) * for $p < 0.05$ (Figure 3C).

### 2.11. Affinity Measurement

Hs2dAb binding studies based on SPR (surface plasmon resonance) technology were performed on a BIAcore T200 optical biosensor instrument (GE Healthcare®, Uppsala, Sweden). Capture of recombinant hs2dAb-6xHis was performed on a nitrilotriacetic acid (NTA) sensor chip in HBS-P + buffer (10 mM HEPES pH 7.4, 150 mM NaCl, and 0.05% surfactant P20) (GE Healthcare®). The four flow cells (Fc) of the sensor chip were used: one (Fc 1) to monitor nonspecific binding and to provide background corrections for analyses and the other three flow cells (Fc 2, 3, and 4) containing immobilized hs2dAb-6xHis for measurement.

For immobilization strategies, flow cells were loaded with nickel solution (Sigma Aldrich®) (10 µL/min for 60 s) in order to saturate the NTA surface with $Ni^{2+}$ and an extra wash using running buffer containing 3 mM EDTA after the nickel injection. His-tagged hs2dAb in running buffer was injected in flow cells at a flow-rate of 10 µL/min. The total amount of immobilized hs2dAb-6xHis was 250–300 resonance units (RUs; 1 RU corresponds to approximately 1 pg/mm$^2$ of protein on the sensor chip). A single-cycle kinetics (SCK) analysis to determine association, dissociation, and affinity constants (ka, kd, and $K_D$, respectively) was carried out. The SCK method prevents potential inaccuracy due to sensor chip regeneration between cycles, which is necessary in the conventional multiple cycle kinetics (MCK). The SCK binding parameters are evaluated for each injection according to the tools and fit models of the BIAevaluation software, giving similar values to MCK. As hs2dAb were smaller proteins than their respective antigens, hs2dAb were captured on the sensor chip, then the recombinant antigens were used as analyte and were injected sequentially with two-fold increased concentrations in a single cycle without regeneration of the censorship between injections. Concentrations of recombinant 2S-HA-RHOA L63 or N19 mutants ranged between 3.125 nM and 50 nM, and concentrations of 2S-HA-RHOA WT loaded with GDP or GTPγS ranged between 12.5 nM and 200 nM. Binding parameters were obtained by fitting the overlaid sensorgrams with 1:1. Langmuir binding model of the BIAevaluation software version 1.0 (GE Healthcare®, Uppsala, Sweden).

### 2.12. BRET

#### 2.12.1. Molecular Cloning

RHOA WT, RHOA L63, RHOB L63, RHOC L63, and RHOA N19 were cloned between NotI/XbaI of the pEF-RLuc8-(GGGGS)1-MCS described elsewhere [25]. The RH57 nanobody was inserted between NcoI/XhoI into pEF-GFP$^2$-(GGGGS)1-MCS [25].

#### 2.12.2. BRET2 Titration Curves and Stimulation Assays

For all BRET experiments (titration curves and stimulation assays), 650,000 HEK293T were seeded in each well of six-well plates. After 24 h at 37 °C, cells were transfected with a total of 1.6 g of DNA mix, containing the donor (RLuc8 plasmid) + acceptor (GFP$^2$ plasmid) using Lipofectamine

2000 transfection reagent (Thermo-Fisher). Cells were detached 24 h later, washed with PBS, and seeded in a white 96-well plate (clear bottom, PerkinElmer) in OptiMEM without phenol red medium complemented with 4% FBS or no FBS for the stimulation assays. Cells were incubated for an additional 20–24 h at 37 °C before the BRET assay readings. Cells were stimulated for 12 min with 20% serum (20 L of 100% serum added in 80 L of BRET medium without FBS) before the BRET readings.

#### 2.12.3. BRET2 Measurements

The BRET2 signal was determined immediately after addition of coelenterazine 400a substrate (10 M final) to cells (Cayman Chemicals), using a CLARIOstar instrument (BMG Labtech) including the luminescence module (filter settings: 515 nm–30 nm and 410 nm–80 nm). Total GFP$^2$ fluorescence was detected with excitation and emission peaks set at 405 nm and 515 nm respectively. Total RLuc8 luminescence was measured with the filter set up at 410 nm–80 nm.

The BRET signal (or BRET ratio) corresponds to the light emitted by the GFP$^2$ acceptor constructs (515 nm–30 nm) upon addition of coelenterazine 400a divided by the light emitted by the RLuc8 donor constructs (410 nm–80 nm). The background signal is subtracted from that BRET ratio using the donor-only negative control, where only the RLuc8 plasmid is transfected into the cells. Total GFP$^2$ and RLuc8 signals were used to control the protein expression level from each plasmid.

## 3. Results

### 3.1. Phage Display Selection of a New High Affinity GTP-Bound RHO Conformational hs2dAb

We previously selected, by phage display, the RH12 (hs2dAb from the synthetic nanobody library NaLi-H1 [23]. The RH12 pan-RHO-GTP blocking intracellular antibody was the only specific clone after five rounds of phage display selection and enrichment cycles. Albeit, RHOA conformational change according to its bound nucleotide involves only a small interface that comprises the switch domains (SWI and SWII, Figure 1A), we hypothesize that several other epitopes can be targeted by affinity reagents as small as nanobodies for an alternative selective recognition of the small GTPase active state. In order to obtain other RHO-GTP conformational nanobodies with different properties than those of the RH12, we slightly modified our subtractive phage display selection protocol that included depletion steps on an excess of inactive GDP-bound RHOA before capture on the active RHOA bound to non-hydrolysable GTP$\gamma$S (Figure 1B). First, we reduced the number of washing steps in the first two rounds of panning to initially select more diversity of binding domains, then we increased the washing time in the third and fourth rounds up to overnight to keep high affinity nanobodies. We also increased the excess of competitors RHOA-GDP or the active mutant of the closest RHO (RHOB L63, RHOC L63, and RAC1 L61). After phage and fragment ELISA preliminary screening of monoclonal hs2dAb on both conformations of RHOA, sequencing revealed that one of the positive clones selective to RHOA-GTP was different from the RH12. This new RHOA-GTP nanobody is referred to as RH57 (Figure 2A).

We proceeded with the production and characterization of the binding properties of the RH57 clone in vitro. The nanobody was expressed in *E.coli* periplasm with a carboxy terminal 6xhis-Myc-6xhis fusion or in *E.coli* cytosol with a carboxy terminal Flag-Ctag fusion (see Methods), both fusions including a purification tag and a second epitope tag for further detection (Figure 2B). We focused on the selectivity of the RH57 towards RHOA-GTP by producing and purifying mutants of RHOA that mimic the active and inactive conformations. RHOA L63 is a constitutively active mutant of RHOA (RHOA-CA), which is locked in the GTP-bound conformation, whereas RHOA N19 is considered as a dominant negative one (RHOA-DN). These mutants were expressed with an amino-terminal tag that consists of 2-Strep-Tag® and HA tags. Strep-Tag® was used to capture the GTPase mutants on Strep-Tactin® coated ELISA wells in order to address the RH57 capacity to detect RHOA L63 or N19 mutants. Figure 2C shows a selective detection of RHOA L63 protein in a dose response manner down to 1 nM RH57-Flag-Ctag. This result was consistent with the differential screening on loaded RHOA that led to the identification of the nanobody RH57. As the ELISA sensitivity suggested a relatively high affinity of the RH57 towards RHOA, we evaluated its binding kinetics properties by surface plasmon resonance (SPR). His tagged purified RH57 nanobody was captured on an Ni-NTA chip to perform single cycle kinetics measurements with the two RHOA mutants as analytes or RHOA WT loaded with either GTPγS or GDP. The $K_D$ values for RHOA L63 were two-fold lower than for the GTP-bound RHOA WT at 28 nM and 60.5 nM (Figure 2D), respectively, indicating a relatively high affinity of the RH57, but one log lower than the one previously reported for the RH12 [23]. Barely any resonance unit was obtained while RHOA was loaded with GDP, and a low signal was measured with the inactive RHOA N19, allowing an assessment of a $K_D$ value above 2 μM, which is a drop of two orders of magnitude compared with the L63 active conformation (Figure 2D). These data demonstrated the high affinity and selectivity of the RH57 nanobody towards the GTP-bound conformation of recombinant RHOA GTPase in vitro, regardless of its production in the oxidative periplasm or in the reducing cytosol of *E.coli*.

### 3.2. Characteriation of the RH57 Intracellular Antibody Binding Properties

Next, we characterized the biochemical properties of the RH57 nanobody when transiently expressed in HeLa cells as an intracellular antibody. For this new molecular construct, we added a carboxy-terminal 6xhis tag for its capture on an IMAC (Nickel ion metal affinity chromatography) matrix followed by a Myc-Tag for detection. We first investigated, by co-precipitation assays, whether the conformational selectivity towards RHO-GTP was preserved while the RH57 was expressed in the intracellular reducing environment. As RHO GTPases are activated at a relatively low basal level in cells cultivated in serum containing medium, we artificially increased the fraction of endogenous proteins in active conformation by loading the cellular extract with GTPγS or GDP as a negative control. We used the GST-RHOTEKIN RBD pulldown, which is the standard biochemical assay to trap the GTP-bound RHO-GTPases, as a positive control. As expected, the RH57 intracellular antibody co-precipitated RHOA only in the GTPγS loaded condition, as did RH12 (Figure 3A). The clear conformational selectivity towards cellular RHOA-GTP led us to perform co-precipitation of the endogenous RHOA subfamily members. The RH57 intracellular antibody captured similar amounts of the active conformation of RHOA/B/C proteins to the RH12 (Figure 3B). To further characterize the binding properties of RH57 intracellular antibody on RHOA protein, we investigated competition with GST-RHOTEKIN RBD/RHOA interaction. Unlike RH12, which totally depleted the RHOA-GTP pull down by the GST-RHOTEKIN RBD, RH57 did not interfere with GST-RHOTEKIN RBD/RHOA interaction (Figure 3C), suggesting that this intracellular antibody did not block endogenous RHO activity inside the cells. Moreover, the RHOA-GTP fraction trapped by the GST-RHOTEKIN RBD in the RH57 expressing cells lysate appeared significantly increased compared with the negative control NR (Figure 3C).

### 3.3. Evaluating the RH57 intracellular antibody as a biosensor of RHOA-GTP.

After the initial biochemical validation of the RH57 intracellular antibody, we investigated its behaviour when fused to GFP at its carboxy terminal end. Fluorescent nanobodies, also called chromobodies, have been mainly developed to trace the endogenous localization of the recognized antigen [26,27]. However, the intracellular localization of a chromobody depends on the binding kinetics of the antibody fragment, the expression level of the intracellular antibody as well as its target, and mostly if the target displays a specific localization where it is concentrated enough to re-localize the chromobody. One of the representative cells shown in Figure 4A exemplified the cellular re-localization to the nucleus of the chromatibody S12, which binds Histone H2A/H2B [27]. This condition contrasted with the whole cell staining obtained with the RH57 chromobody or with the non-related negative NR control. Nevertheless, we observed at low confluency that the signals at the cell edge displayed irregular intensities in 60% of the cells expressing the RH57-GFP at a moderate level. Thin areas of higher signal intensity with length of 2 to 5 μm could be observed at the cell periphery (Figure 4A and Figure S1). Such areas were barely detectable in cells expressing the NR-GFP control. The stronger intensity at the cell periphery may reflect membrane dynamics areas where RHO GTPases activity occurs to regulated local actin dynamics in the formation of lamellipodia or membrane ruffles [28]. Moreover, we compared the shape of cells expressing the RH57-GFP chromobody with the one expressing the RH12-GFP. As we have previously reported [23], the RH12 intracellular antibody expression led to a typical phenotype of RHO/RAC inhibition, with impeded acto-myosin dynamics, actin fiber loss, and cell cytoplasm shrinkage, associated with cell death. The cellular shape of cells expressing the RH57-GFP did not induce such dramatic perturbation of the cell even at a high expression level (Figure 4A), indicating that this domain could be a building block to generate a biosensor.

To complete our characterization of the chromobody RH57, we analyzed the signal on fixed HeLa cells transiently transfected, starved for 24 h, and then stimulated with 20% serum for 30 min (Figure 4B). We focused on the fluorescence staining at the edge of the cell and clearly found areas of higher intensity upon serum-mediated RHO stimulation. Lateral cut plot profiles on more than 20 transfected cells showed very heterogeneous pattern of GFP chromobody fluorescence according to the thickness and width of the cells, as well as the expression level in transient transfection. As shown in representative plot profiles (Figure 4B and Figure S2), a peak at one edge of the cell was observed in the profile plot of the stimulated cells expressing the RH57-GFP, whereas the profile of the lateral cut in starved, but not serum stimulated cells displayed a bell-like shape similar to the one found in the NR-GFP control conditions. The peak of RH57-GFP at the cell border co-localized with the cortical actin fibers peak (Figure 4B). This accumulation could correspond to RHO activities, as RHOA-like activity or GEF activities are known to occur close to plasma membrane zones of protrusion or retractions linked to acto-myosin dynamic networks [28,29]. These results suggest that the RH57 intracellular antibody could be used to establish a biosensor of the GTP-bound state of the RHOA-like subfamily.

Finally, we used the RH57 nanobody to produce a BRET biosensor of RHOA activation (Figure 5). We previously developed a BRET-based RAS biosensor toolbox [25] between full-length RAS GTPases and their effectors or binders (such as the anti-RAS intracellular domain antibody, iDAb RAS [20]). Using a similar approach, we have established BRET-based RHO biosensors using the nanobody RH57. The full-length RHO GTPases are fused on their amino terminal end with the donor molecule *Renilla* Luciferase variant 8 (RLuc8). The anti-RHO-GTP binder RH57 is tagged to its amino terminal end with the acceptor molecule $GFP^2$. Both donor and acceptor plasmids are transfected into HEK293T cells and if the RLuc8-RHO fusion protein does not interact with the $GFP^2$-hs2dAb, the excitation of the donor molecule (RHO) by the luciferase substrate (coelenterazine 400a) will not excite the acceptor fusion molecule (nanobody) (Figure 5A, left panel). However, if an interaction occurs between RLuc8-RHO and a $GFP^2$-hs2dAb fusion, bringing the RLuc8 and $GFP^2$ within 10 nm, an energy transfer occurs from the RLuc8-RHO donor to the $GFP^2$ acceptor and a BRET2 signal is achieved (Figure 5A, right panel). The BRET signal (or BRET ratio) is the ratio between the light emitted by the $GFP^2$ acceptor constructs (at 510 nm) upon addition of coelenterazine 400a, divided by the light emitted by the RLuc8 donor constructs (at 395 nm) [30] (see Methods). We performed BRET donor saturation assays to determine whether we could detect an interaction between the RHO and the RH57. If the two proteins interact, the BRET signal will increase and reach saturation, whereas if they do not interact, the BRET signal will increase linearly [31]. Here, RH57 interacts specifically with RHOA, RHOB, and RHOC L63 active mutants, but not with the RHOA N19 dominant negative mutant (Figure 5B), confirming our in vitro data.

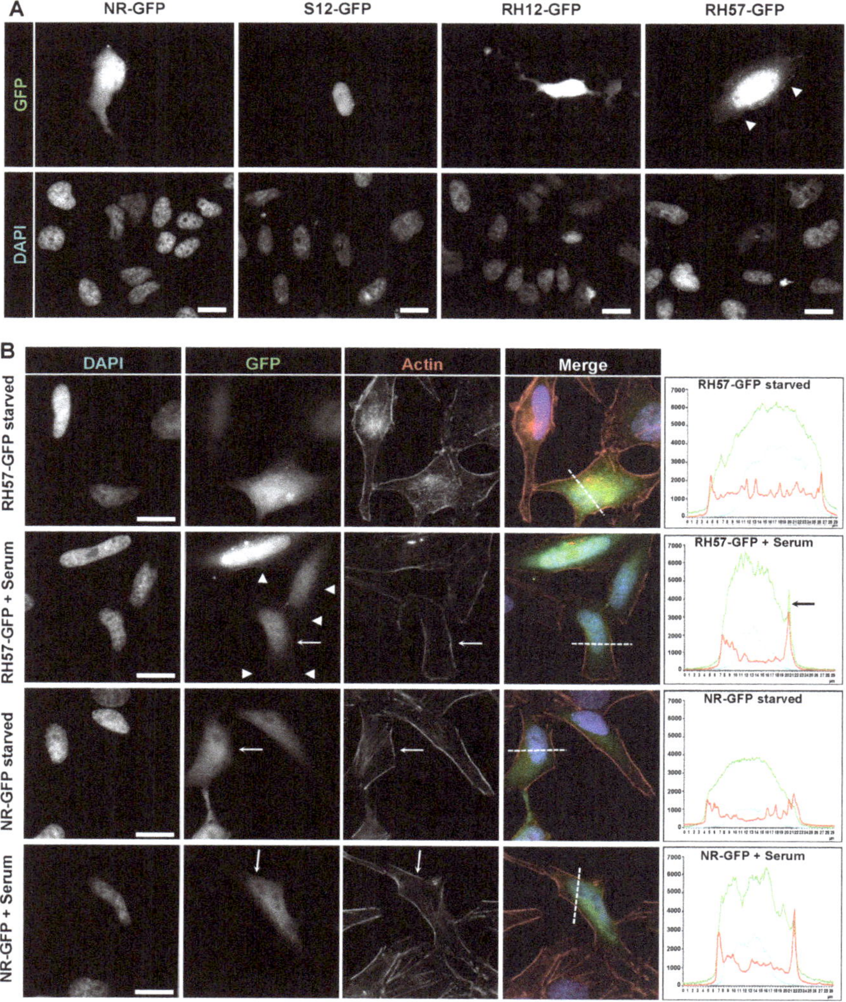


**Figure 4.** Chromobodies expression in HeLa cells. (**A**) Nanobodies NR, S12 (anti-Histone2A/B), RH12, and RH57 hs2dAb were fused to GFP and expressed in HeLa cells as chromobodies. While RH12 hs2dAb significantly alters cellular shape, potentially by interfering with RHO/RAC pathways, RH57 hs2dAb does not induce such a phenotype on HeLa cells. Arrowheads indicate areas of the plasma membrane with localized increased concentration of the chromobody (enlarged in Figure S1). The scale bars shown in the DAPI panels represent 20 μm. (**B**) HeLa cells transfected with RH57-GFP or NR-GFP and serum starved for 24 h were stimulated with 20% serum to activate RHO GTPases. Arrowheads indicate areas of the plasma membrane with a localized increased concentration of chromobody RH57-GFP upon treatment (enlarged in Figure S2). Cut profile plots display the DAPI (blue), the GFP chromobody (green), and F-Actin (red) intensities corresponding to the dash line shown in the merge channel and arrows in GFP and actin staining channels. F-actin was stained with Alexa 568-Phalloidin. Co-localization of RH57-GFP pixels with F-actin is indicated by large arrows on the plot profile and the channels. Images are representative of transfected cells at moderate expression level. The scale bars shown in the DAPI panels represent 20 μm.

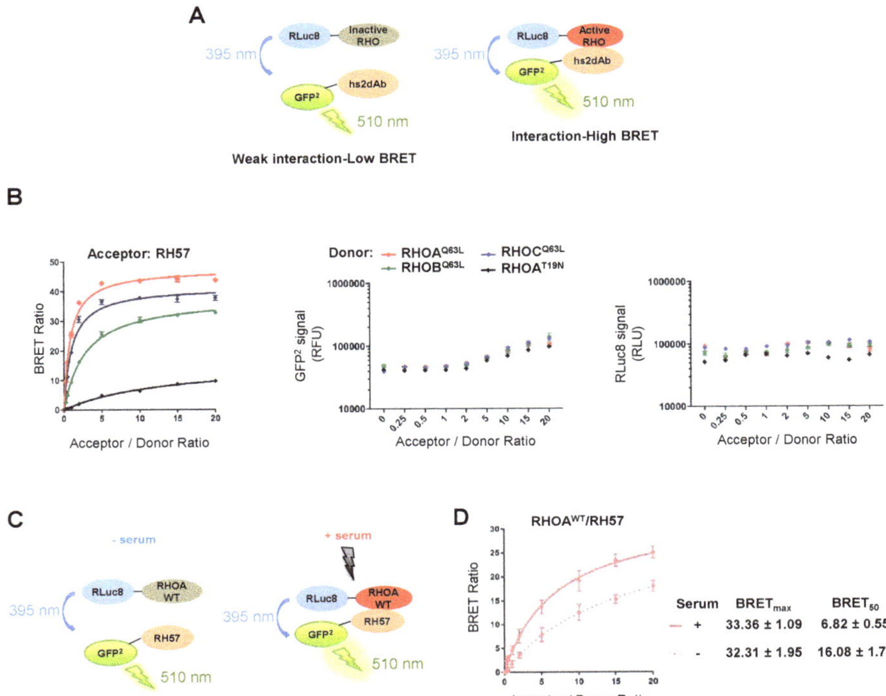


**Figure 5.** Set up of BRET-based RHO biosensors with anti-RHO hs2dAb to detect RHOA activation. (**A**) Schematic of the BRET2-based RHO biosensor system involving anti-RHO hs2dAb. The full-length RHO proteins are fused at their amino termini to the donor molecule RLuc8. The anti-RHO RH57, selected against the active RHO conformation (RHO-GTP), are fused at their amino termini to the acceptor molecule $GFP^2$. When the inactive RHO is expressed with the $GFP^2$-hs2dAb, only a weak BRET signal will be detected (left), while when the $GFP^2$-hs2dAb interact with the active RHO, they will induce a strong BRET signal (right). The BRET signal or BRET ratio is the ratio between the light emitted at 510 nm divided by the light emitted at 395 nm minus a background correction from the donor only construct. (**B**) BRET donor saturation assays between $RHOA^{Q63L}$, $RHOB^{Q63L}$, $RHOC^{Q63L}$, and $RHOA^{T19N}$ (donors) and the acceptors RH57. Acceptor/donor ratio is the ratio between the quantity of plasmid expressing the acceptor construct divided by the quantity of plasmid expressing the donor construct transfected into HEK293T cells. A ratio of 0 represents the donor only transfected cells condition. Control of the expression of the acceptor ($GFP^2$-RH57, $GFP^2$ signal) and donors (RLuc8-RHO, RLuc8 signal) from the BRET donor saturation assays are shown. The $GFP^2$ level at the acceptor/donor ratio of 0 shows the autofluorescence of the cells [25]. (**C**) Schematic of the BRET2-based RHO biosensor with RH57 to detect $RHOA^{WT}$ activation. Upon serum stimulation, $RHOA^{WT}$ will be activated and interact with RH57. (**D**) BRET donor saturation assay of $RHOA^{WT}$ with RH57. After 20% serum stimulation for 12 min, RH57 contacts $RHOA^{WT}$, as indicated by an increase of the $BRET_{max}$ and a decrease of the $BRET_{50}$ values. The experiments were performed twice (biological repeats) in quadruplicates (technical repeats). Error bars are mean ± SD of the biological repeats.

We have implemented the BRET-based biosensor using the RH57 nanobody to detect an activation of RHOA WT upon RHO stimulation with serum (Figure 5C). The serum treatment leads to RHOA activation and to its interaction with RH57 (Figure 5D). Indeed, it largely decreased the $BRET_{50}$ (an approximation of the relative affinity of the acceptor fusion for the donor fusion proteins, corresponding to the acceptor/donor ratio necessary to reach 50% of the $BRET_{max}$) and slightly increased the $BRET_{max}$ (an approximation for the total number of complex RHOA/RH57 and the distance between the donor and the acceptor within the dimer), which together are consistent with an increased affinity of the RH57 nanobody towards RHOA-GTP (Figure 5D). These results suggest that our $RHOA^{WT}$/RH57 biosensor could detect an activation of RHOA in living cells.

## 4. Discussion

Targeting the conformational active state of intracellular proteins requires high affinity and high specificity reagents such as intracellular antibodies. Antibody phage display selection has been useful to isolate very selective intracellular antibodies that discriminate between different cellular pools of a target protein according to its conformational state [32–34]. Small RHO GTPase cellular activities rely on such a conformational switch. Their spatio-temporal activation patterns regulate cytoskeleton dynamics and cell morphogenesis in various physiological contexts [35]. Therefore, intracellular antibodies appear as appropriate molecular reagents to develop biosensors of the GTP-bound state of RHO-GTPases, provided that their binding properties are not blocking the cellular function of the target. As a previous selection led to a blocking intracellular antibody that was unsuitable for the production of a sensor, in our current work, we screened the same synthetic nanobody phage display library to identify another intracellular domain antibody that was also selective of a GTP-bound RHOA-like subfamily, but with no apparent inhibitory effect. We demonstrated its high affinity and selectivity in immunoassays and validated its functionality as an intracellular antibody to produce a BRET biosensor of RHO activation.

RHO GTPase activation networks are tightly regulated to avoid sustained activation of morphogenetic signalling pathways [14,15]. Overall, the GTP-bound state is maintained as a minor proportion within the total pool of RHO proteins, which is mostly inactive as judged by biochemical precipitation assays [4,17]. Therefore, the ability of nanobodies to specifically recognize the RHO-GTP target relies on both a strong selectivity towards the GTP-bound conformation of the GTPase and appropriate on- and off-rates of interaction. As the structural superimposition of the two states suggests that the interface of the nanobody interaction should be close to the switch domains, the boundary between a blocking intracellular antibody and a sensing intracellular antibody will depend mainly on their affinity and the epitope overlap with binding domains of natural partners in the intracellular complexity. As the nanobody RH12 was unsuitable for biosensor purposes, we screened the phage display library and identified RH57, which does not compete with the RBD domain, indicating that its binding epitope is unlikely to involve the same interface as the RHO/RH12 interaction. However, we cannot clearly correlate the identification of the RH57 RHO-GTP sensor with any specific modified conditions we have implemented during the phage display selection. For instance, we introduced a depletion step with L63 CA mutants of RHOB and RHOC during the phage display, but we still selected RH12 and the RH57 clones, which are pan-RHO/RAC GTPases antibodies. This indicates that epitopes on the GTP-bound state may not favour the discrimination between RHOs, or that the depletion steps should have been implemented from the first round of selection.

Chromobodies are powerful tools to trace endogenous antigens within the cell in live cell imaging [19,27]. However, the intracellular localization of a chromobody depends on the binding kinetics of the antibody fragment to the recognized antigen and the expression level of the intracellular antibody as well as its target. Importantly, locally concentrated targets that display a specific localization will give a signal background ratio favourable to distinguish the bound and unbound chromobodies pools. The low expression level of the RHO-GTP state, as well as their transient activation mechanism, may explain the weak signal background ratio of potentially bound to unbound RH57-GFP chromobodies. Although we observed an increased signal at the border of fixed cells, acute sensing and tracking of RHO activities in live cells may require more advanced microscopy approaches dedicated to membrane dynamics imaging such as TIRF (total internal reflction fluorescence) [22,36]. Alternatively, engineering the hs2dAb such that the bound pool displays a shift in fluorescence intensity, such as coupling with solvent sensitive dyes [37], could improve the signal to background ratio. As the RH57 chromobody cannot be easily used as a sensor of endogenous RHO activity for the study of regulators or for the screening of compounds that would modulate RHO activation, we developed a BRET-based RHO biosensor using the RH57 nanobody. BRET-based biosensors have been successfully used to monitor the modulation of interactions with ligands [38–40]. Here, we show that the RH57 could be used to detect an activation of RHOA by serum stimulation of the cells. However, the determination of the binding interface of the RH57 on RHOA should be performed by resolving the crystal structure of the complex before using the BRET-based biosensor as a screening platform of ligand/small molecules modulating RHO activation. Indeed, even though the RH57 does not compete with the RBD, we cannot rule out that the RH57 interferes with GAP proteins or other regulators of RHO activity, as we noticed that this intracellular antibody led to an increase of the RHOA-GTP ratio in the GST-ROTHEKIN RBD assay.

In conclusion, we have identified a nanobody from a synthetic library suitable for the production of RHOA subfamily activation biosensors. This genetically encoded molecular tool will allow a broad range of applications to target RHO activity without interfering with RHO signaling. The RH57 domain could also be compared with some RBD used in BRET-based biosensors to monitor RHO activation with a fine spatio-temporal resolution. In the future, the BRET biosensor could allow the screening of small molecules that modulate RHOA/B/C activation. Small molecules could also be selected with this nanobody, adapting analogous methods to those used for intracellular antibody-derived (Abd) compounds targeting RAS [41].

## 5. Patents

L.K., N.B., G.F., and A.O. are co-inventors on the patent PTC/EP2016/052136, concerning the RHO GTP single domain antibodies discovery and their applications.

**Supplementary Materials:** The following are available online at http://www.mdpi.com/2073-4468/8/1/8/s1, Figure S1: Chromobodies RH57-GFP expression in HeLa cells, Figure S2: Chromobodies RH57-GFP expression in HeLa cells stimulated with 20% serum after 24h starvation.

**Author Contributions:** Conceptualization: L.K., N.B. C.T., T.H.R., and A.O. Methodology: L.K., N.B. T.H.R., and A.O. Investigation: L.K., N.B, C.T., L.L., and A.O. Data Curation: L.K., N.B, C.T., G.F., T.H.R., and A.O. Writing—Original Draft Preparation: L.K., N.B., T.H.R., and A.O. Writing—Review & Editing: L.K., N.B., C.T., G.F., T.H.R., and A.O. Supervision: G.F., T.H.R., and A.O. Funding Acquisition: G.F. and T.H.R.

**Funding:** This work was supported by the grant "Equipe labellisée par la Fondation pour la Recherche Médicale". The work of N.B. & T.H.R. was funded by a Wellcome Trust Investigator Award 099246/Z/12/Z and an MRC programme grant MR/J000612/1.

**Acknowledgments:** We acknowledge P. Lemaire and P. Chinestra for critical advice. We thank R. Gence for providing reagents.

**Conflicts of Interest:** The authors declare no conflict of interest.

## References

1. Wennerberg, K. The Ras superfamily at a glance. *J. Cell Sci.* **2005**, *118*, 843–846. [CrossRef] [PubMed]
2. Bishop, A.L.; Hall, A. Rho GTPases and their effector proteins. *Biochem. J.* **2000**, *348 Pt 2*, 241–255. [CrossRef]
3. Satoh, T.; Kaziro, Y. Ras in signal transduction. *Semin. Cancer Biol.* **1992**, *3*, 169–177. [PubMed]
4. Ren, X.D.; Schwartz, M.A. Determination of GTP loading on Rho. *Methods Enzymol.* **2000**, *325*, 264–272. [PubMed]
5. Sahai, E.; Marshall, C.J. RHO—GTPases and cancer. *Nat. Rev. Cancer* **2002**, *2*, 133–142. [CrossRef] [PubMed]
6. Kim, J.-G.; Islam, R.; Cho, J.Y.; Jeong, H.; Cap, K.-C.; Park, Y.; Hossain, A.J.; Park, J.-B. Regulation of RhoA GTPase and various transcription factors in the RhoA pathway. *J. Cell. Physiol.* **2018**, *233*, 6381–6392. [CrossRef]
7. Valon, L.; Etoc, F.; Remorino, A.; di Pietro, F.; Morin, X.; Dahan, M.; Coppey, M. Predictive Spatiotemporal Manipulation of Signaling Perturbations Using Optogenetics. *Biophys. J.* **2015**, *109*, 1785–1797. [CrossRef]
8. Adini, I.; Rabinovitz, I.; Sun, J.F.; Prendergast, G.C.; Benjamin, L.E. RhoB controls Akt trafficking and stage-specific survival of endothelial cells during vascular development. *Genes Dev.* **2003**, *17*, 2721–2732. [CrossRef]

9. Calvayrac, O.; Mazières, J.; Figarol, S.; Marty-Detraves, C.; Raymond-Letron, I.; Bousquet, E.; Farella, M.; Clermont-Taranchon, E.; Milia, J.; Rouquette, I.; et al. The RAS-related GTPase RHOB confers resistance to EGFR-tyrosine kinase inhibitors in non-small-cell lung cancer via an AKT-dependent mechanism. *EMBO Mol. Med.* **2017**, *9*, 238–250. [CrossRef]
10. Vega, F.M.; Ridley, A.J. Rho GTPases in cancer cell biology. *FEBS Lett.* **2008**, *582*, 2093–2101. [CrossRef]
11. Zandvakili, I.; Lin, Y.; Morris, J.C.; Zheng, Y. Rho GTPases: Anti- or pro-neoplastic targets? *Oncogene* **2017**, *36*, 3213–3222. [CrossRef]
12. Delmas, A.; Cherier, J.; Pohorecka, M.; Medale-Giamarchi, C.; Meyer, N.; Casanova, A.; Sordet, O.; Lamant, L.; Savina, A.; Pradines, A.; et al. The c-Jun/RHOB/AKT pathway confers resistance of BRAF-mutant melanoma cells to MAPK inhibitors. *Oncotarget* **2015**, *6*, 15250–15264. [CrossRef] [PubMed]
13. Shang, X.; Marchioni, F.; Sipes, N.; Evelyn, C.R.; Jerabek-Willemsen, M.; Duhr, S.; Seibel, W.; Wortman, M.; Zheng, Y. Rational design of small molecule inhibitors targeting RhoA subfamily Rho GTPases. *Chem. Biol.* **2012**, *19*, 699–710. [CrossRef] [PubMed]
14. Reinhard, N.R.; van Helden, S.F.; Anthony, E.C.; Yin, T.; Wu, Y.I.; Goedhart, J.; Gadella, T.W.J.; Hordijk, P.L. Spatiotemporal analysis of RhoA/B/C activation in primary human endothelial cells. *Sci. Rep.* **2016**, *6*, 25502. [CrossRef] [PubMed]
15. Pertz, O.; Hodgson, L.; Klemke, R.L.; Hahn, K.M. Spatiotemporal dynamics of RhoA activity in migrating cells. *Nature* **2006**, *440*, 1069–1072. [CrossRef] [PubMed]
16. Koraïchi, F.; Gence, R.; Bouchenot, C.; Grosjean, S.; Lajoie-Mazenc, I.; Favre, G.; Cabantous, S. High-content tripartite split-GFP cell-based assays to screen for modulators of small GTPase activation. *J. Cell Sci.* **2017**. [CrossRef] [PubMed]
17. Stofega, M.; DerMardirossian, C.; Bokoch, G.M. Affinity-Based Assay of Rho Guanosine Triphosphatase Activation. In *Transmembrane Signaling Protocols*; Humana Press: Totowa, NJ, USA, 2006; Volume 332, pp. 269–280. ISBN 978-1-59745-048-5.
18. Pellegrin, S.; Mellor, H. Rho GTPase activation assays. *Curr. Protoc. Cell Biol.* **2008**. [CrossRef] [PubMed]
19. Rothbauer, U.; Zolghadr, K.; Tillib, S.; Nowak, D.; Schermelleh, L.; Gahl, A.; Backmann, N.; Conrath, K.; Muyldermans, S.; Cardoso, M.C.; et al. Targeting and tracing antigens in live cells with fluorescent nanobodies. *Nat. Methods* **2006**, *3*, 887–889. [CrossRef]
20. Tanaka, T.; Lobato, M.N.; Rabbitts, T.H. Single domain intracellular antibodies: A minimal fragment for direct in vivo selection of antigen-specific intrabodies. *J. Mol. Biol.* **2003**, *331*, 1109–1120. [CrossRef]
21. Tanaka, T.; Rabbitts, T.H. Intrabodies based on intracellular capture frameworks that bind the RAS protein with high affinity and impair oncogenic transformation. *EMBO J.* **2003**, *22*, 1025–1035. [CrossRef]
22. Galli, V.; Sebastian, R.; Moutel, S.; Ecard, J.; Perez, F.; Roux, A. Uncoupling of dynamin polymerization and GTPase activity revealed by the conformation-specific nanobody dynab. *eLife* **2017**, *6*. [CrossRef] [PubMed]
23. Moutel, S.; Bery, N.; Bernard, V.; Keller, L.; Lemesre, E.; de Marco, A.; Ligat, L.; Rain, J.-C.; Favre, G.; Olichon, A.; et al. NaLi-H1: A universal synthetic library of humanized nanobodies providing highly functional antibodies and intrabodies. *eLife* **2016**, *5*. [CrossRef] [PubMed]
24. Olichon, A.; Surrey, T. Selection of genetically encoded fluorescent single domain antibodies engineered for efficient expression in Escherichia coli. *J. Biol. Chem.* **2007**, *282*, 36314–36320. [CrossRef] [PubMed]
25. Bery, N.; Cruz-Migoni, A.; Bataille, C.J.; Quevedo, C.E.; Tulmin, H.; Miller, A.; Russell, A.; Phillips, S.E.; Carr, S.B.; Rabbitts, T.H. BRET-based RAS biosensors that show a novel small molecule is an inhibitor of RAS-effector protein-protein interactions. *eLife* **2018**, *7*. [CrossRef]
26. Kaiser, P.D.; Maier, J.; Traenkle, B.; Emele, F.; Rothbauer, U. Recent progress in generating intracellular functional antibody fragments to target and trace cellular components in living cells. *Biochim. Biophys. Acta* **2014**. [CrossRef] [PubMed]
27. Jullien, D.; Vignard, J.; Fedor, Y.; Béry, N.; Olichon, A.; Crozatier, M.; Erard, M.; Cassard, H.; Ducommun, B.; Salles, B.; et al. Chromatibody, a novel non-invasive molecular tool to explore and manipulate chromatin in living cells. *J. Cell Sci.* **2016**, *129*, 2673–2683. [CrossRef] [PubMed]
28. Ridley, A.J. Rho GTPases and actin dynamics in membrane protrusions and vesicle trafficking. *Trends Cell Biol.* **2006**, *16*, 522–529. [CrossRef]
29. Machacek, M.; Hodgson, L.; Welch, C.; Elliott, H.; Pertz, O.; Nalbant, P.; Abell, A.; Johnson, G.L.; Hahn, K.M.; Danuser, G. Coordination of Rho GTPase activities during cell protrusion. *Nature* **2009**, *461*, 99–103. [CrossRef]

30. Pfleger, K.D.G.; Seeber, R.M.; Eidne, K.A. Bioluminescence resonance energy transfer (BRET) for the real-time detection of protein-protein interactions. *Nat. Protoc.* **2006**, *1*, 337–345. [CrossRef]
31. Mercier, J.-F.; Salahpour, A.; Angers, S.; Breit, A.; Bouvier, M. Quantitative assessment of beta 1- and beta 2-adrenergic receptor homo- and heterodimerization by bioluminescence resonance energy transfer. *J. Biol. Chem.* **2002**, *277*, 44925–44931. [CrossRef]
32. Haque, A.; Andersen, J.N.; Salmeen, A.; Barford, D.; Tonks, N.K. Conformation-sensing antibodies stabilize the oxidized form of PTP1B and inhibit its phosphatase activity. *Cell* **2011**, *147*, 185–198. [CrossRef] [PubMed]
33. Nizak, C.; Monier, S.; del Nery, E.; Moutel, S.; Goud, B.; Perez, F. Recombinant antibodies to the small GTPase Rab6 as conformation sensors. *Science* **2003**, *300*, 984–987. [CrossRef] [PubMed]
34. Dimitrov, A.; Quesnoit, M.; Moutel, S.; Cantaloube, I.; Poüs, C.; Perez, F. Detection of GTP-tubulin conformation in vivo reveals a role for GTP remnants in microtubule rescues. *Science* **2008**, *322*, 1353–1356. [CrossRef] [PubMed]
35. Fritz, R.D.; Pertz, O. The dynamics of spatio-temporal Rho GTPase signaling: Formation of signaling patterns. *F1000Research* **2016**, *5*, 749. [CrossRef] [PubMed]
36. Poulter, N.S.; Pitkeathly, W.T.E.; Smith, P.J.; Rappoport, J.Z. The Physical Basis of Total Internal Reflection Fluorescence (TIRF) Microscopy and Its Cellular Applications. In *Advanced Fluorescence Microscopy*; Verveer, P.J., Ed.; Springer: New York, NY, USA, 2015; Volume 1251, pp. 1–23. ISBN 978-1-4939-2079-2.
37. Nalbant, P.; Hodgson, L.; Kraynov, V.; Toutchkine, A.; Hahn, K.M. Activation of endogenous Cdc42 visualized in living cells. *Science* **2004**, *305*, 1615–1619. [CrossRef] [PubMed]
38. Namkung, Y.; Le Gouill, C.; Lukashova, V.; Kobayashi, H.; Hogue, M.; Khoury, E.; Song, M.; Bouvier, M.; Laporte, S.A. Monitoring G protein-coupled receptor and β-arrestin trafficking in live cells using enhanced bystander BRET. *Nat. Commun.* **2016**, *7*, 12178. [CrossRef] [PubMed]
39. Beautrait, A.; Paradis, J.S.; Zimmerman, B.; Giubilaro, J.; Nikolajev, L.; Armando, S.; Kobayashi, H.; Yamani, L.; Namkung, Y.; Heydenreich, F.M.; et al. A new inhibitor of the β-arrestin/AP2 endocytic complex reveals interplay between GPCR internalization and signalling. *Nat. Commun.* **2017**, *8*, 15054. [CrossRef] [PubMed]
40. Bellot, M.; Galandrin, S.; Boularan, C.; Matthies, H.J.; Despas, F.; Denis, C.; Javitch, J.; Mazères, S.; Sanni, S.J.; Pons, V.; et al. Dual agonist occupancy of AT1-R-α2C-AR heterodimers results in atypical Gs-PKA signaling. *Nat. Chem. Biol.* **2015**, *11*, 271–279. [CrossRef]
41. Quevedo, C.E.; Cruz-Migoni, A.; Bery, N.; Miller, A.; Tanaka, T.; Petch, D.; Bataille, C.J.R.; Lee, L.Y.W.; Fallon, P.S.; Tulmin, H.; et al. Small molecule inhibitors of RAS-effector protein interactions derived using an intracellular antibody fragment. *Nat. Commun.* **2018**, *9*. [CrossRef]

*Article*

# A Strategy to Optimize the Generation of Stable Chromobody Cell Lines for Visualization and Quantification of Endogenous Proteins in Living Cells

**Bettina-Maria Keller [1], Julia Maier [1], Melissa Weldle [1], Soeren Segan [2], Bjoern Traenkle [1] and Ulrich Rothbauer [1,2,*]**

[1] Pharmaceutical Biotechnology, Eberhard Karls University, 72076 Tuebingen, Germany; Bettina.Keller@nmi.de (B.-M.K.); J.Maier@med.uni-tuebingen.de (J.M.); melissa.weldle@student.uni-tuebingen.de (M.W.); Bjoern.Traenkle@nmi.de (B.T.)

[2] Natural and Medical Sciences Institute at the University of Tuebingen, Markwiesenstr. 55, 72770 Reutlingen, Germany; Soeren.Segan@nmi.de

* Correspondence: ulrich.rothbauer@uni-tuebingen.de; Tel.: +49-7121-51530-415; Fax: +49-7121-51530-816

Received: 4 December 2018; Accepted: 7 January 2019; Published: 10 January 2019

**Abstract:** Single-domain antibodies have emerged as highly versatile nanoprobes for advanced cellular imaging. For real-time visualization of endogenous antigens, fluorescently labelled nanobodies (chromobodies, CBs) are introduced as DNA-encoded expression constructs in living cells. Commonly, CB expression is driven from strong, constitutively active promoters. However, high expression levels are sometimes accompanied by misfolding and aggregation of those intracellular nanoprobes. Moreover, stable cell lines derived from random genomic insertion of CB-encoding transgenes bear the risk of disturbed cellular processes and inhomogeneous CB signal intensities due to gene positioning effects and epigenetic silencing. In this study we propose a strategy to generate optimized CB expressing cell lines. We demonstrate that expression as ubiquitin fusion increases the fraction of intracellularly functional CBs and identified the elongation factor 1α (EF1-α) promoter as highly suited for constitutive CB expression upon long-term cell line cultivation. Finally, we applied a CRISPR/Cas9-based gene editing approach for targeted insertion of CB expression constructs into the adeno-associated virus integration site 1 (AAVS1) safe harbour locus of human cells. Our results indicate that this combinatorial approach facilitates the generation of fully functional and stable CB cell lines for quantitative live-cell imaging of endogenous antigens.

**Keywords:** nanobodies; chromobodies; live-cell imaging; compound screening; cellular models

## 1. Introduction

Cell culture models provide substantial information on various cellular responses ranging from exposure to small chemical compounds to genetically mediated target depletion [1]. In combination with advanced microscopy techniques such as quantitative live-cell imaging they can be directly employed to monitor cellular events and phenotypic changes with spatial and temporal resolution. In vitro cell models are essential tools in basic biomedical research and are applied to identify novel cellular targets and compound candidates in pharmaceutical development [2,3]. For imaging-based analysis, expression of fluorescent fusion proteins (FP fusions) is the most commonly used approach to study localization and dynamic changes of proteins in living cells. Commercially available FP-based live-cell assays enable the investigation of processes such as cell proliferation, apoptosis or DNA damage. However, this approach is limited to the visualization of ectopically expressed FP fusions, which may considerably differ from their endogenous counterparts in terms of expression level, activity, localization and protein half-life [4–8]. Recently evolved gene editing methods such as the

CRISPR/Cas9 or the ZFN/TALEN technology can now be used to generate cell lines expressing fluorescently tagged proteins under their endogenous expression control. While these technologies offer novel opportunities for protein analysis the possibility of functional interference by the attached fluorescent moiety still remains.

Intracellular affinity reagents provide a straightforward approach to overcome the drawbacks of FP fusions as they exclusively visualize and trace the dynamics of endogenous target structures [9,10]. Due to their robustness and structural simplicity, fluorescently labelled nanoprobes derived from single-domain antibody fragments of camelids (chromobodies, CBs) can be selected to detect antigens in their native surroundings in living cells [9,11]. Since their first description in 2006, numerous CBs have been established and applied to visualize and monitor their target molecules in various cell models and whole organisms [12–17].

While transient expression of CBs is sufficient to visualize the dynamics and relocalization of endogenous proteins on single-cell level, larger screening campaigns require the development of stable cell lines with a homogenous and consistent CB expression [16–18]. Although various stable CB cell lines have been reported by us and others, the selection and characterization of cell lines compatible for quantitative live-cell imaging is still very cumbersome. Most importantly, quantitative image analysis often suffers from inconsistent CB expression levels, aggregation and a strong cell-to-cell variance of CB fluorescence intensities (Figure 1). In this context it has to be considered that in most available cell models CB expression is driven from the strong and constitutively active cytomegalovirus-(CMV) promoter. Although high CB levels are desired in general, elevated transgene expression is sometimes accompanied by misfolding and aggregation [19,20]. Additionally, following the conventional workflow of stable cell line generation CB transgenes are usually randomly inserted in the cellular genome (Figure 1). Notably, this not only bears the risk of unintended genomic manipulation which might affect cellular processes but can also lead to inconsistent CB signal intensities due to the integration of different copy numbers of the CB transgene, chromatin positioning effects [21] and epigenetic silencing of the promoter upon continuous sub-culturing [22].

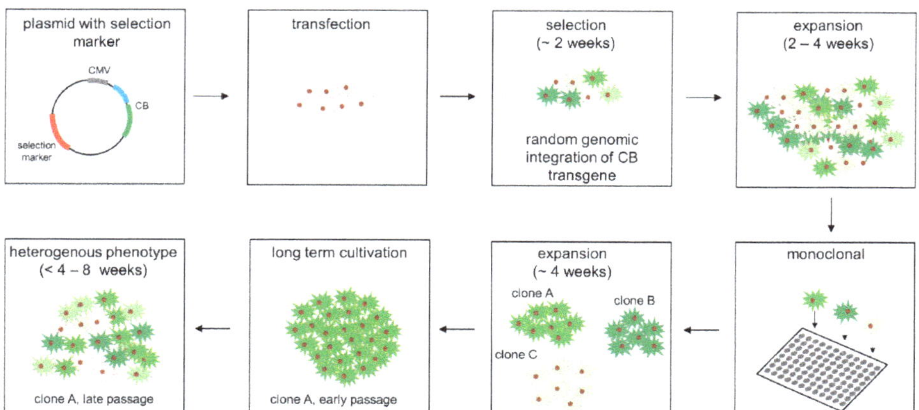


**Figure 1.** Typical workflow of stable chromobody cell line generation following random genomic integration of chromobody (CB) transgene.

Here, we explore a combination of previously established methods to improve the generation of stable CB cell lines. Based on comparative analyses we propose to (i) implement our recently developed turnover-accelerated CBs expressed as ubiquitin fusions in order to monitor changes in the antigen concentration and to avoid intracellular CB aggregation, (ii) select a promoter, which is less prone to epigenetic silencing and (iii) insert these optimized CB expression constructs into the adeno-associated virus integration site 1 (AAVS1) safe harbour locus of human cells using a targeted CRISPR/Cas9 gene editing approach.

## 2. Material and Methods

### 2.1. Expression Constructs

All primer sequences, synthesized DNA fragments and plasmids are listed in Table S1. All expression constructs used in this study are listed in Table S2.

The expression constructs encoding for Ub-M-LMN-CB and Ub-R-LMN-CB were generated by amplification the Lamin-NB from the Lamin-CB plasmid [23] using the following primer set: NB-ubi-for and NB-ubi-rev. The amplified DNA fragment was purified and ligated into PstI and BspEI restriction site of Ub-M-BC1-eGFP and Ub-R-BC1-eGFP (both [24]), respectively. To generate the Ub-R-BC1-CB containing the EF1-α promoter (referred to as EF1-α_Ub-R-BC1-eGFP), the EF1-α promoter was synthesized as gBlock® gene fragment (IDT, integrated DNA technologies) and Gibson Assembly [25] was performed after restriction digest of Ub-R-BC1-eGFP [24] using the restriction enzymes AseI und NheI. Fragments were assembled using Gibson-Assembly Master Mix (New England Biolabs, Ipswich, MA, USA) according to the manufacturer's protocol. The BC1-CB expression construct containing the human β-actin promoter (referred to as h-βact_Ub-R-BC1-eGFP) was generated by amplification of the promoter from pDRIVE-hβ-Actin plasmid ([26], kindly provided by Hiroyuki Konishi, Aichi Medical University School of Medicine, Japan) using the primer set β-actin-promoter-for and β-actin-promoter-rev and ligation into AseI und NheI digested Ub-R-BC1-eGFP [24]. To eliminate the PstI restriction site within the hβ-act promoter site-directed mutagenesis was performed utilizing the primer set β-actin-promoter-mutPstI-for and β-actin-promoter-mutPstI-rev using the Q5 Site-Directed Mutagenesis Kit (New England Biolabs) according to the manufacturer's protocol. For molecular cloning of the AAVS1-CB-donor plasmid (as described in Figure 4A) two DNA fragments were used, which were produced by gene synthesis. At first, AAVS1-CB-donor fragment 1 (gene synthesis, plasmid DNA, IDT) was digested using PciI and MfeI and directly ligated into a PciI and MfeI digested pEGFP-N1backbone (Clontech, Mountain View, CA, USA). Secondly, the resulting plasmid was digested with MfeI and XbaI and completed by AAVS1-CB-donor fragment 2 insertion (gBlock® gene fragment), which was amplified with the following primer set: AAVS1-CB-donor-fragment-2-for and AAVS1-CB-donor-fragment-2-rev. In a last step the complete cassette was sequence verified using primers Seq-AAVS1-CB-donor-1 - 4 for sequencing. To generate an AAVS1-CB-donor plasmid containing the EF1-α_Ub-R-BC1-CB (resulting plasmid AAVS1_Ub-R-BC1-CB), we used the AseI and XbaI restriction site to integrate the CB cassette including the promoter. For the AAVS1-CB-donor plasmid containing the EF1-α_Ub-R-ACT-CB (AAVS1_Ub-R-ACT-CB) the actin NB was amplified with the primer set NB-ubi-for and NB-ubi-rev and ligated into AAVS1_Ub-R-BC1-CB backbone using PstI and BspEI restriction site. All generated constructs in this study were sequence analysed after cloning.

### 2.2. Cell Culture, Transfection, Stable Cell Line Generation and Compound Treatment

HeLa Kyoto cells (Cellosaurus no. CVCL_1922) were obtained from S. Narumiya (Kyoto University, Japan), whereas DLD-1 (ATCC® Number CCL-221™) and HCT116 (ATCC® Number CCL-247™) were obtained from ATCC. All cell lines were tested negative for mycoplasma using the PCR mycoplasma kit Venor GeM Classic (Minerva Biolabs, Berlin, Germany) and the Taq DNA polymerase (Minerva Biolabs). Since this study does not include cell line-specific analysis, all cell lines were used without additional authentication. All cell lines were maintained according to standard protocols. Briefly, growth media containing DMEM (high glucose, pyruvate, ThermoFisher Scientific, Waltham, MA, USA) for HeLa Kyoto and HCT116 cells and RPMI 1640 (ThermoFisher Scientific) for DLD-1 cells supplemented with 10% ($v/v$) foetal bovine serum (FCS, ThermoFisher Scientific) and penicillin-streptomycin (ThermoFisher Scientific) were used for cultivation. Cells were routinely passaged using 0.05% trypsin-EDTA (ThermoFisher Scientific) and were cultivated at

37 °C in a humidified chamber with a 5% $CO_2$ atmosphere. Plasmid DNA was transfected with Lipofectamine 2000 (ThermoFisher Scientific) in HeLa cells, whereas DLD-1 and HCT116 were transfected with TransIT-X2® (Mirus Bio, Madison, WI, USA) according to the manufacturer's instructions. For site-directed integration of the CB into AAVS1 genomic locus, $5 \times 10^5$ cells were co-transfected with 2.5 µg of the respective donor plasmid and 2.5 µg plasmid encoding for Cas9 nuclease and gRNA specific for the AAVS1 locus. 24 h post transient transfection cells were subjected to a 48 h selection period using 0.6 µg/mL puromycin dihydrochloride (Sigma-Aldrich, St. Louis, MO, USA). Puromycin-resistant cells were allowed to grow for one week before single clones were isolated. Single clones were analysed regarding the CB expression level by fluorescence microscopy. To verify site-directed integration of the CB-donor plasmid at the AAVS1 locus, genomic DNA of the single clones and the respective parental cell line was isolated using QIAamp DNA mini Kit (QIAGEN, Venlo, the Netherlands). Next, the primer pair genPCR-AAVS1-int-for and genPCR-AAVS1-int-rev [27] was used for PCR-based genotyping. CB integration into the AAVS1 locus results in an amplicon of ~1400 bps (strategy outlined in Figure 4B). The resulting amplicon was purified and verified via sequence analysis. Compound treatment with FH535 (Sigma-Aldrich) and XAV939 (MedChemExpress) was performed for up to 24 h. For dose-response experiments cells were treated with 1 µM, 10 µM and 50 µM for FH535 and 1 µM, 5 µM and 10 µM for XAV939.

### 2.3. Fluorescence Imaging, Image Segmentation and Analysis

For fluorescence imaging 5000 cells per well were plated in a black µClear 96-well plate (Greiner Bio-One, Kremsmünster, Austria). For staining of the nuclei in PFA-fixed cells 0.02 µg/mL 4′,6-diamidino-2-phenylindole (DAPI, Sigma-Aldrich) was used while living cells were continuously incubated with 2 µg/mL Hoechst33258 (Sigma). Images were acquired with an ImageXpress micro XL system (Molecular Devices, San Jose, CA, USA) and analysed by MetaXpress software (64 bit, 6.2.3.733, Molecular Devices). Fluorescence images comprising a statistically relevant number of cells (>200) were acquired for each condition. For quantitative fluorescence analysis the mean fluorescence of the respective CB expression construct in mCherry or mCherry-CTNNB1 transfected cells was determined. Using the Custom Module Editor (version 2.5.13.3) of the MetaXpress software, we established an image segmentation algorithm that identifies areas of interest based on the parameters of size, shape and fluorescence intensity above local background. To segment the whole cell including the nucleus and the cytosolic compartment, the ectopically antigen mCherry-CTNNB1 or its respective control mCherry was used to generate the corresponding segmentation mask. The average fluorescence intensities were determined for each image followed by subtraction of background fluorescence. From these values the mean fluorescence and standard errors were calculated for three independent biological replicates and student's *t*-test was used for statistical analysis.

### 2.4. Western Blot

DLD-1_AAVS1_Ub-R-BC1-CB cells were seeded in a 10 $cm^2$ cell culture dish (Corning) with $3 \times 10^6$ cells per dish. After two days the cells were treated with DMSO or 10 µM FH535 for 24 h. For the lysis cells were harvested with a cell scraper and cold PBS and centrifuged at 500× *g* and 4 °C for 3 min. Per 50 µL pellet 100 µL lysis buffer (10 mM Tris/HCl, pH 7.5, 150 mM NaCl, 0.5 mM EDTA, 0.5% NP40, 1mM PMSF, 1× protease inhibitor cocktail (Serva, Heidelberg, Germany), 1× phosphatase inhibitor (PhosSTOP, Roche, Basel, Switzerland) 250 µg/µL DNase, 2.5 mM $MgCl_2$) was added. The samples were pipetted 30 times every 10 min for 30 min and centrifuged at 16,000× *g* for 10 min at 4 °C. The samples were boiled in 2× reducing SDS-sample buffer (60 mM Tris/HCl, pH 6.8, 2% (*w*/*v*) SDS, 5% (*v*/*v*) 2-mercaptoethanol, 10% (*v*/*v*) glycerol, 0.02% bromophenol blue) for 15 min at 95 °C. The SDS-PAGE and western blot were performed according to standard procedure. For western blotting the proteins were transferred on nitrocellulose membrane (Amersham, GE Healthcare, Chicago, IL, USA). The blots were scanned on a Typhoon-Trio laser scanner (GE Healthcare). The analysis was done with Image Quant TL Toolbox (GE Healthcare, version 7.0).

For immunoblot detection following antibodies were used: Total-CTNNB1 (mouse monoclonal, BD, #610154, 1:1000), Active-CTNNB1 (mouse monoclonal, 8E7, Millipore, #05-665, 1:500), Tubulin (mouse monoclonal, Sigma-Aldrich, #T9026, 1:2000). For the detection of the primary antibodies fluorescently labelled secondary antibodies (goat-anti-mouse, Alexa Fluor 647, ThermoFisher Scientific, 1:1000) was used.

## 3. Results

### 3.1. Expression of CBs as Ubiquitin Fusions Reduces Intracellular Aggregation

Recently, we observed that CBs are stabilized in the presence of their antigens and conceived a strategy to reduce the levels of non-bound CBs [24]. We utilized the N-end rule describing the N-terminal amino acid as one of the key determinants for the half-life of proteins [28,29]. Based on the ubiquitin fusion technique, we designed novel CB expression constructs comprising a N-terminal ubiquitin fusion, which is co-translationally cleaved by deubiquitinases. By this technique we were able to generate CBs displaying any desired amino acid at their N-termini and identified that arginine (Arg, R) mediated the fastest CB turnover when exposed at the N-terminus. Functionally, we showed that CBs containing a N-terminal Arg were highly antigen responsive and suitable to monitor and quantify dynamic changes of the concentration of endogenous antigens in real-time [24]. Besides the generation of proteins with a desired N-terminus, the ubiquitin fusion technique has also been reported to increase solubility and functionality of ectopically expressed proteins [30]. According to these findings we noticed that ubiquitin fusions of CBs are less prone to aggregation irrespective of the adjacent amino acid. Thus we asked, whether these potential benefits can be transferred to CBs such as the lamin-specific CB (LMN-CB), which has previously been shown to form aggregates upon high expression levels in HeLa cells [23,31].

To test whether the ubiquitin fusion approach is able to reduce the amount of aggregated LMN-CB we generated expression constructs encoding for Ub-M-LMN-CB and the turnover-accelerated version Ub-R-LMN-CB (Figure 2A). For microscopic analysis HeLa cells were transiently transfected either with the modified LMN-CB constructs or the original LMN-CB expression plasmid. 24 h post transfection HeLa cells were subjected to fluorescence imaging and the percentage of cells containing aggregates were determined for each LMN-CB variant (Figure 2B). In most of the cells expressing the original LMN-CB construct we observed the formation of fluorescent aggregates (~70% of all analysed cells). Upon transient expression of the LMN-CB as an ubiquitin fusion followed by a methionine (Ub-M-LMN-CB) the number of cells displaying aggregates was significantly reduced to ~27% and to only 8% in cells transfected with the turnover-accelerated version of the LMN-CB (Ub-R-LMN-CB) (Figure 2C). Notably, while the nuclear lamina was hardly detectable in the majority of cells expressing the original LMN-CB construct, the modified CB versions displayed the typical nuclear rim structure more prominently. These results confirmed our previous observations showing that the addition of the ubiquitin moiety increased the intracellular solubility of CBs while leaving the binding properties unaffected (Figure 2B).

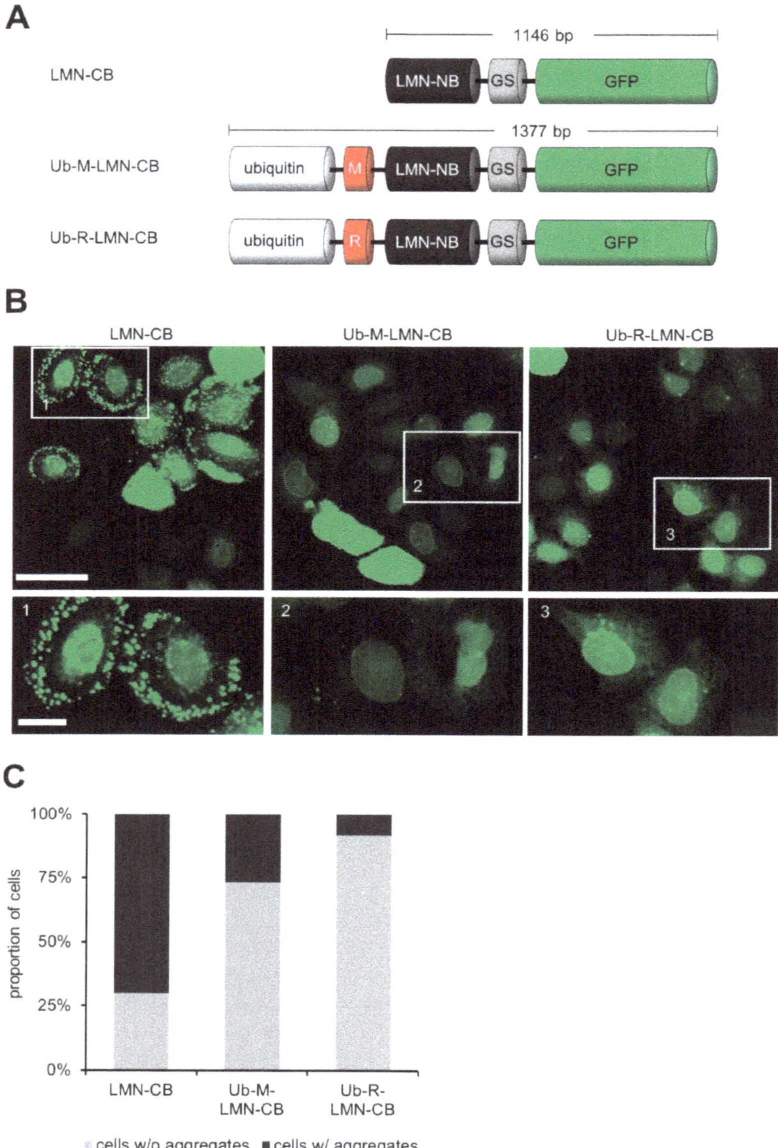


**Figure 2.** N-terminal fusion to ubiquitin reduces aggregation of the LMN-CB upon expression in HeLa cells. (**A**) Schematic illustration of different expression constructs: one encoding for an unmodified LMN-CB and two constructs encoding the CB as ubiquitin fusion displaying either methionine (Ub-M-LMN-CB) or arginine (Ub-R-LMN-CB) as the N-terminal amino acid. (**B**) Representative fluorescence images of HeLa cells transiently expressing either LMN-CB, Ub-M-LMN-CB or Ub-R-LMN-CB. Fluorescence images were acquired 24 h post transfection, scale bars images: 50 μm, insets: 10 μM. (**C**) Quantification of HeLa cells showing aggregates upon transient expression of the indicated LMN-CB variants. Number of analysed cells: LMN-CB: 198; Ub-M-LMN-CB: 142; Ub-R-LMN-CB: 155.

### *3.2. Comparison of CB Expression Driven by the CMV-, EF1-α or β-actin Promoter to Monitor Changes in Antigen Concentration*

As recently described, high expression levels of turnover-accelerated CBs are necessary to optically monitor rapid changes in antigen concentration within living cells via antigen-mediated CB stabilization (AMCBS) [24]. For larger screening campaigns using quantitative live-cell imaging the expression levels have to be consistently high over longer cultivation periods. One major critical factor causing reduction and thus heterogeneities in transgene expression lies within the epigenetic silencing of the promoter of the transgene, mainly caused by DNA methylation [32,33] and/or histone modification [34]. While the constitutive CMV promoter provides suitable expression levels, this viral promoter is reported to be highly sensitive to DNA methylation [35]. Indeed, after prolonged cultivation times we noticed heterogeneous CB fluorescence intensities in cell lines comprising a stable integration of the CB transgene controlled by the CMV promoter (Figure S1).

To optimize the stability of CB expression suitable for monitoring changes of endogenous antigen concentration we aimed for a strong promoter that is less prone to epigenetic silencing. Thus, we compared CMV-driven expression with the expression mediated by the human elongation factor 1α (EF1-α) promoter and the human β-actin (h-βact) promoter. While all three candidates were described to convey medium-to-high expression levels within different cell lines [26,36] both, the EF1-α and h-βact promoter were previously reported to be less prone to epigenetic silencing and maintain stable transgene expression levels over several passages [26,37].

For comparative analysis we replaced the CMV promoter in our previously reported expression construct encoding the turnover-accelerated β-catenin (CTNNB1)-specific BC1-CB (Ub-R-BC1-CB, [24]) by the EF1-α or h-βact promoter (Figure 3A).

We compared the expression levels and the performance with regard to antigen-mediated stabilization of the original CMV-driven and the newly generated EF1-α or h-βact-driven CB constructs by transfecting HeLa cells either in combination with mCherry as control or mCherry-CTNNB1 as the corresponding antigen. Quantitative fluorescence imaging revealed substantial differences in CB expression levels (Figure 3B,C). For the CMV-driven expression we observed the highest expression levels within HeLa cells with a mean fluorescence intensity (MFI) of ~700 in mCherry-transfected control cells and a MFI of ~5000 in the presence of mCherry-CTNNB1. An intermediate strength in CB expression was determined for the EF1-α-containing variant indicated by a MFI of ~130 in control cells and a MFI of ~1000 in the presence of the antigen. For the h-βact-driven expression we detected rather weak signals, which were close to background level. Interestingly, similar stabilization factors (8.5–9.7) were calculated for all constructs, indicating that AMCBS was not affected by the exchange of the promoter (Figure 3C). Considering that EF1-α promoter is less sensitive to DNA methylation [37] but provides similar CB expression levels compared to the original CMV promoter, we decided to implement the EF1-α promoter in our strategy to generate optimized stable CB cell lines.

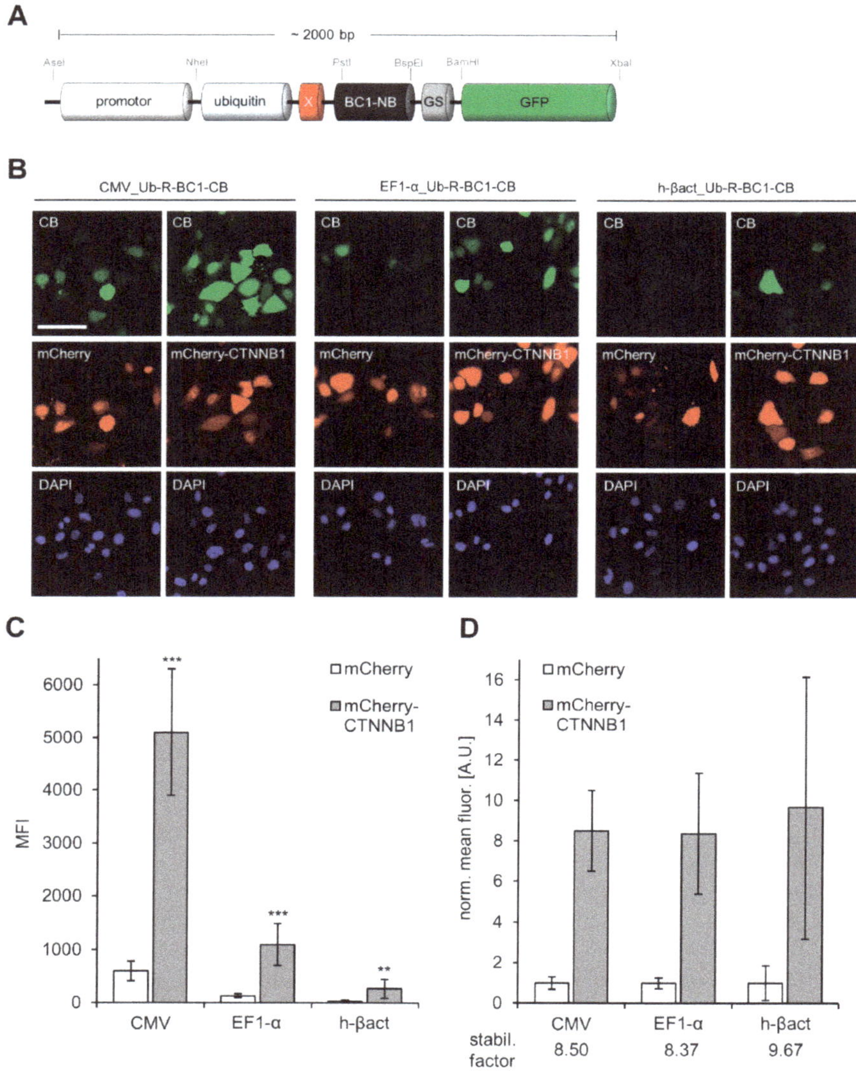


**Figure 3.** Quantitative image analysis of promoter-driven Ub-R-BC1-eGFP expression. (**A**) Schematic illustration of the CB expression construct for promoter testing. Promoter sequences can easily be exchanged using unique indicated restriction sites (AseI, NheI). (**B**) Representative images of HeLa cells transiently expressing Ub-R-BC1-CB driven by different promoters (CMV, EF1-α and h-βact) in combination with either mCherry (control) or mCherry-CTNNB1 (antigen). Nuclei were stained with DAPI, scale bar: 50 µm. (**C**) Bar chart represents mean CB fluorescence intensity (MFI) detected by quantitative fluorescence imaging in either control (mCherry) or antigen (mCherry-CTNNB1) expressing cells ($n$ = 3, >200 cells each). (**D**) For every promoter construct, MFI of the CB in antigen expressing cells was normalized to the respective CB-signal determined in cells co-expressing mCherry as control, leading to the indicated stabilization factors. Error bars: S.D. Statistical analysis was performed using student's $t$-test, *** $p < 0.001$, ** $p < 0.01$.

### 3.3. Design and Construction of AAVS1 Donor Vector for Site-Directed Stable Integration of Turnover-Accelerated CBs

Typically, the generation of stable CB cell models is based on the transfection of a cell line with a CB expression vector comprising a selection marker, which for example, confers resistance to antibiotics. Subsequently, cells are continuously cultivated in the presence of appropriate antibiotics to select clones that comprise a stable genomic integration of the CB transgene (Figure 1). Although this workflow was successfully applied to generate numerous stable CB cell lines, some pitfalls have to be considered. As the integration of the CB transgene occurs randomly, neither a prediction about the chromatin structure at the integration site can be made nor the number of CB transgene copies within the cellular host can be foreseen. Notably, the site of integration has a major effect on the expression levels of the transgene summarized as positioning effect [38]. Additionally, such stable cell lines have to be continuously cultured under constant selective pressure, which has been reported to affect host cell physiology, genetic stability and metabolism [39–41]. To address these shortcomings, we aimed to establish a new protocol that allows site-directed integration of turnover-accelerated CBs into the host cell DNA by applying the CRISPR/Cas9 gene editing technology.

Recently, the adeno-associated virus site 1 (AAVS1, position 19q13.42), located in the first intron of the protein phosphatase 1 regulatory subunit 12C (PPP1R12C), was described as genomic safe-harbour (GSH) integration site [27,42–46]. As transgene expression from this GSH integration site was previously reported to result in robust and persistent protein levels [47], we chose the AAVS1 locus to integrate our turnover-accelerated CBs. For targeted engineering we constructed a donor plasmid containing a turnover-accelerated CB expression cassette, which is driven by an EF1-$\alpha$ promoter and flanked by AAVS1-specific homology arms (HA-L/R) (Figure 4A). Additionally, a puromycin resistance gene containing a splice acceptor site (SA) linked to a self-cleaving peptide sequence (T2A) was added. Upon correct genomic integration, the expression of the puromycin N-acetyl-transferase will be driven by the endogenous PPP1R12C promoter (Figure 4A), which allows the selection of clones that underwent the desired CRISPR event. All fragments within the construct were sequence-optimized and the indicated restriction sites (Figure 4A) allow an easy exchange of the different components including promoter, nanobody binding moiety and fluorescent marker. In addition, we adapted a PCR-based genotyping strategy to verify clones comprising a correct transgene integration (illustrated in Figure 4B [27]).

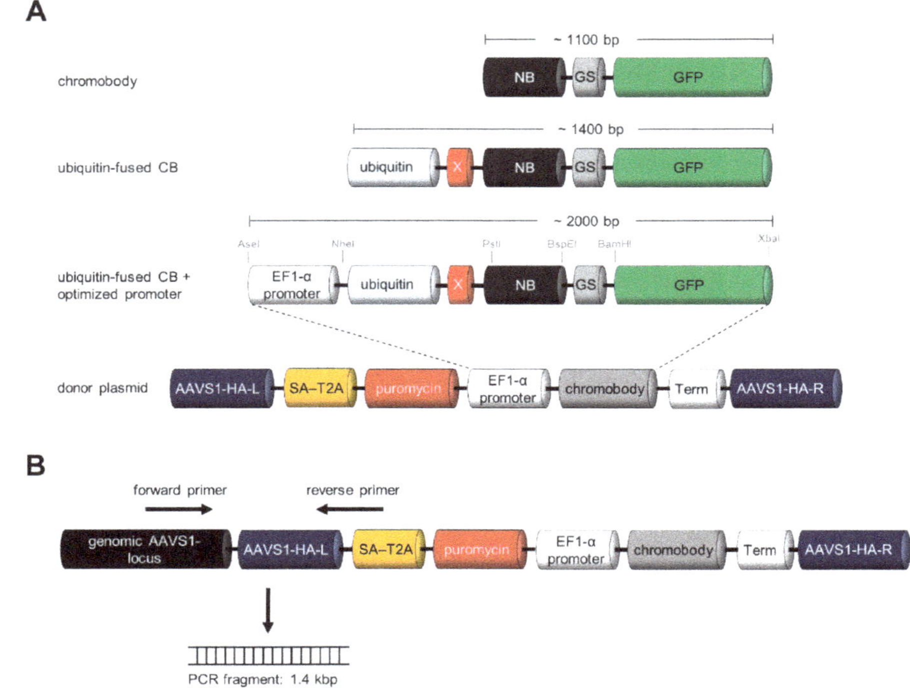


**Figure 4.** Strategy for stable integration of turnover-accelerated CBs into the human AAVS1 locus. (**A**) Schematic illustration of CB containing donor plasmid for stable integration into the human AAVS1 locus. (**B**) Strategy for PCR-based genotyping of host cell DNA to verify site-directed CB integration.

### *3.4. Site-Directed Integration of the Turnover-Accelerated Actin-Specific CB into the AAVS1 Locus*

We applied this strategy to generate HeLa cells stably expressing the actin-specific CB (ACT-CB). The ACT-CB has been previously shown to bind to F-actin without affecting the dynamic reorganization of the cytoskeleton [15], thus the detection of filamentous actin provides a suitable read-out to validate the functionality of the CB transgene upon CRISPR/Cas9-mediated integration. In a first step, we co-transfected HeLa cells with plasmids encoding for (i) Cas9 nuclease and the respective gRNA for the AAVS1 locus [27] and (ii) the donor plasmid containing the turnover-accelerated ACT-CB (Ub-R-ACT-CB, Figure 4A). After transfection, cells were cultured for 48 h in the presence of puromycin to enrich cells that underwent stable AAVS1 integration of the CB transgene. After clonal expansion, we identified two clones (referred to as clone B5 and clone B6) showing filamentous structures indicative for the actin cytoskeleton (Figure 5A). Both clones were similar in cell morphology and size and displayed a homogenous expression of the ACT-CB. However, we noticed that the fluorescence intensity of clone B5 was slightly higher compared to clone B6 (Figure 5B). PCR-based genotyping (as outlined in Figure 4B) revealed for clone B6 an amplicon at the expected size of ~1400 bps (Figure 5C), indicating that a correct CB transgene integration at the AAVS1 locus was only obtained for clone B6, which was further verified by sequence analysis of the PCR fragment. Next, we qualitatively compared the morphology of the novel CRISPR-engineered HeLa_AAVS1_Ub-R-ACT-CB cell line with a stable ACT-CB expressing HeLa cell line previously generated by random integration of a CMV-driven, non-turnover-accelerated ACT-CB (Figure 5D). Fluorescence live-cell imaging revealed a more homogenous CB expression and cell morphology for the CRISPR-modified cell clone, while we detected a quite heterogeneous cell morphology along with some CB aggregates in cells stably expressing the non-modified ACT-CB transgene (Figure 5D).

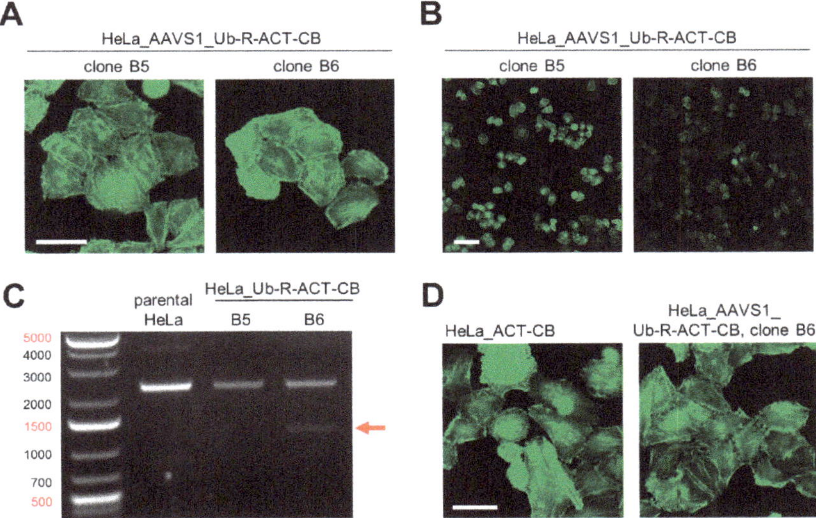


**Figure 5.** Generation of a HeLa cell line expressing Ub-R-ACT-CB stably integrated into the AAVS1 locus by using CRISPR/Cas9 technology. (**A**) Fluorescence images of two CRISPR-engineered HeLa cell clones expressing ACT-CB, scale bar: 50 µM. (**B**) Comparison of fluorescence intensity between HeLa_AAVS1_Ub-R-ACT-CB clone B5 and B6, scale bar: 100 µm. (**C**) PCR-based genotyping of HeLa_AAVS1_Ub-R-BC1-CB clones B5 and B6 in comparison to parental HeLa cells. Genomic DNA of the monoclonal cells was extracted and subjected to PCR using the genotyping strategy illustrated in Figure 4B. Shown are the resulting PCR products (indicated by arrow) on a 1% agarose gel stained with ethidiumbromid. (**D**) Representative fluorescence images of living HeLa cells stably expressing the respective ACT-CB. Left image illustrates HeLa_ACT-CB cells generated by random integration of the non-modified ACT-CB. The right image shows HeLa_AAVS1_Ub-R-ACT-CB cells generated by site-directed integration of the ubiquitin-modified ACT-CB using CRISPR/Cas9 technology, scale bar: 50 µm.

### *3.5. Site-Directed Integration of the Turnover-Accelerated BC1-CB into the AAVS1 Locus of Human CRC Cell Lines*

In a next step, we aimed to apply our approach to generate cell lines, which are suitable to monitor changes in endogenous protein concentration upon compound treatment by quantitative image analysis of the CB signal. In previous studies we demonstrated that our CTNNB1-specific CB (BC1-CB) traces changes in the levels of transcriptionally active, hypophosphorylated CTNNB1 upon induction of the β-catenin/WNT pathway [16,24]. These findings motivated us to generate more sophisticated CB cell models to monitor the effects of compounds on the reduction of endogenous CTNNB1 levels by following the BC1-CB signal. In over 90% of all colorectal cancer the β-catenin/WNT signalling pathway is mutated [48] resulting in an accumulation of transcriptionally active CTNNB1 involved in the initiation and progression of this cancer type. Consequently, cellular models which can be employed to screen for compound candidates affecting the endogenous levels of this particular CTNNB1 fraction would be advantageous for preclinical drug development. Thus, we chose two human colorectal carcinoma (CRC) cell lines DLD-1 and HCT116, which exhibit elevated level of active CTNNB1 to generate disease relevant CB-based screening cell lines [49].

For generation of the donor plasmid we replaced the actin binding moiety (ACT-NB) with the CTNNB1 binding nanobody BC1 using PstI and BspEI restriction site as outlined in Figure 4A and applied our CRISPR strategy as described above. Following puromycin selection we identified single clones of DLD-1 and HCT116 cells displaying a green fluorescence, which were isolated and expanded as monoclonal cell lines (Figure 6A). To confirm correct CB integration at the AAVS1 locus we performed PCR-based genotyping as illustrated in Figure 6B (strategy outlined in Figure 4B). By this we detected PCR products at ~1400 bps in DLD-1_AAVS1_Ub-R-BC1-CB and HCT116_AAVS1_Ub-R-BC1-CB cells, which were absent in the parental cells. Sequence analysis of the amplicons further verified that in both cell lines the CB transgene was successfully integrated into the AAVS1 locus. Next, we compared the phenotype of a monoclonal DLD-1 cell line generated by random integration of the non-modified BC1-CB with our newly generated DLD-1_AAVS1_Ub-R-BC1-CB cell line by fluorescence imaging (Figure 6C, Figure S2). For the DLD-1_AAVS1_Ub-R-BC1-CB cells we observed a more homogenous phenotype regarding cell area and CB fluorescence, while the conventional DLD-1_BC1-CB cells displayed a more heterogeneous morphology accompanied by miscellaneous CB signal intensities (Figure S2). Finally, we applied the novel generated DLD-1_AAVS1_Ub-R-BC1-CB cell line to monitor the effect of two compounds, FH535 and XAV939, which were previously reported to affect the level of endogenous CTNNB1 [50,51]. We performed time lapse imaging of DLD-1_AAVS1_Ub-R-BC1-CB for 24 h followed by quantitative analysis of CB fluorescence intensity in the nucleus upon treatment with different inhibitor concentrations (Figure 6D, Figure S3). By monitoring the BC1-CB fluorescence over time we observed a strong reduction of CTNNB1 to ~50% upon treatment with 10 μM and 50 μM of FH535 after 6 h. For XAV939 we detected milder effects as indicated by a reduction of the CB fluorescence to ~80% at the highest concentration of 10 μM after 12 h (Figure 6D). To verify whether the decreased CB fluorescence actually reflects a reduction of hypophosphorylated CTNNB1, we performed immunoblot analysis of DLD-1_AAVS1-Ub-R-BC1-CB cells treated either with DMSO as control or 10 μM FH535 for 24 h (Figure S4). While we observed only a minor effect of compound treatment for total CTNNB1 we observed a clear reduction of hypophosphorylated CTNNB1 in the soluble protein fraction upon treatment with FH535 (Figure S4).

Taken together, these data demonstrate that our CRISPR-based strategy to introduce optimized turnover-accelerated CB at a defined genomic locus results in the generation of highly versatile and disease relevant CB cell models. In combination with quantitative live-cell imaging these models can now be applied to monitor even subtle dose- and time-dependent compound effects on the level of endogenous proteins.

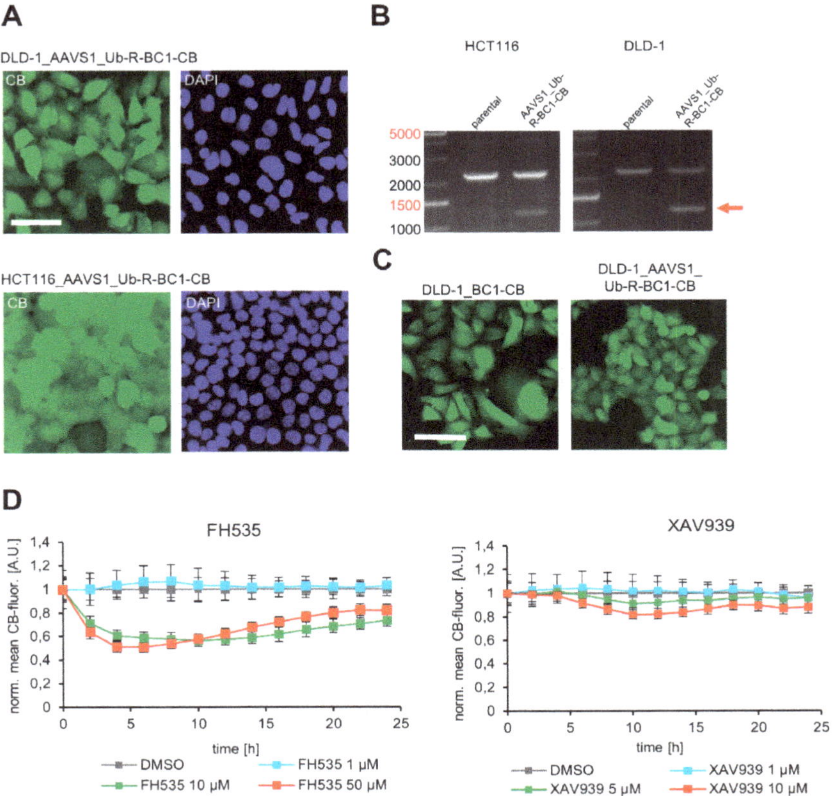


**Figure 6.** Generation of colorectal carcinoma cell lines with stable AAVS1 integration of the turnover-accelerated Ub-R-BC1-CB encoding transgene. (**A**) Representative fluorescence images of isolated DLD-1 (upper panel) and HCT116 (lower panel) clones stably expressing Ub-R-BC1-CB. Nuclei were stained with DAPI, scale bar: 50 μm. (**B**) Genotyping of HCT116_AAVS1_Ub-R-BC1-CB and DLD-1_AAVS1_Ub-R-BC1-CB. Genomic DNA of host cells was extracted and subjected to PCR using the primer strategy illustrated in Figure 4B. Corresponding PCR products are indicated by arrow. (**C**) Representative fluorescence images of DLD-1 cell lines stably expressing the indicated BC1-CB. Left image illustrates DLD-1_BC1-CB cells generated by random integration of the non-modified BC1-CB. The right image shows DLD-1_AAVS1_Ub-R-BC1-CB cells generated by CRISPR/Cas9-mediated site-directed integration of the ubiquitin fused BC1-CB, scale bar: 50 μm. (**D**) Quantification of nuclear CB fluorescence in DLD-1_AAVS1_Ub-R-BC1-CB cells upon compound treatment. DLD-1_AAVS1_Ub-R-BC1-CB were either treated with the indicated concentrations of FH535 or XAV939 or treated with 0.01% DMSO as control. Cells were continuously imaged every 2 h for up to 24 h. Fluorescence intensity of nuclear CB-signal was quantified and the fluorescence values were normalized to the DMSO control and plotted against time ($n$ = 2, >200 cells each). Error bars: S.D.

## 4. Discussion

Considering the importance of cellular imaging in biomedical research and preclinical drug development there is a continuous need for advanced labelling strategies to reliably visualize cellular components in a physiologically meaningful state [3,52]. During the last decade, fluorescently labelled nanobodies (chromobodies, CBs) emerged as versatile nanoprobes for cellular imaging of endogenous targets in living cells [9,10,12,53]. Recently, we demonstrated that quantitative analysis of

CB fluorescence can additionally be employed to monitor changes in the concentration of endogenous proteins due to a mechanism termed antigen-mediated CB stabilization (AMCBS) [24]. To monitor changes in antigen concentration with high precision a strong CB expression has to be combined with a fast CB turnover [24]. However, for some CBs we previously observed that high expression levels are accompanied with the formation of intracellular aggregates. Here, we showed that expression of CBs as ubiquitin fusions can not only be used to generate turnover-accelerated CBs but also can be applied to reduce the fraction of intracellularly aggregated CBs. Our results are consistent with previous findings reporting an increased solubility and functionality of recombinant proteins expressed as ubiquitin fusions in mammalian cells and bacteria [30,54]. While the underlying mechanism is not fully understood, a chaperone-similar function is suggested, where a partially unfolded nascent protein weakly interacts with the nearby, upstream located ubiquitin moiety and thereby transiently precludes unspecific intermolecular interactions [30].

To date most stable CB cell lines are still generated by transfection of an expression plasmid followed by antibiotic selection of cells comprising a stable genomic integration of the CB transgene. However, this method is very imprecise since neither the integration site nor the number of integrated copies of the transgene can be adjusted. Additionally, the most widely used CMV promoter is prone to epigenetic silencing [35]. Accordingly, we noticed heterogeneous and overall weaker CB fluorescence intensities for a multitude of stable CB cell lines upon long-term cultivation. With the EF1-$\alpha$ promoter, we identified a suitable alternative promoter providing strong CB expression and stabilization ratios in the presence of the antigen which is less prone to epigenetic silencing [37]. However, with the CMV, EF1-$\alpha$ and h-$\beta$act promoter we compared only three different promoter types. A more comprehensive analysis of further alternatives including a determination of the methylation status of the promoter regions upon long term cultivation might reveal promoter constructs which are even better suited for a more stable CB expression.

Besides epigenetic modulation, CB expression and thus CB fluorescence is affected by the number of integrated transgenes and chromatin positioning effects [21,38]. Here, the CRISPR/Cas9 gene editing technology has been demonstrated as a highly superior approach since it enables the integration of one (heterozygous) or two (homozygous) transgene copies at a predefined genomic site [27]. In the search of transcriptionally active insertion sites, so-called genomic safe harbour sites (GSH) have been described, which allow robust and stable transgene expression. Most importantly, transgene insertion at GSH does not have adverse effect on the host cell genome. In this context, the adeno-associated virus integration site (AAVS1) on human chromosome 19 was identified as a safe genomic location for integration and high yields of transgene expression [27,42–46]. Based on emerging knowledge regarding CRISPR-targeted genome editing using homology-dependent repair [27,55] we used the AAVS1 locus to insert CB-based nanoprobes in human cell models. To validate the versatility and flexibility of this approach, we designed ubiquitin-fused, turnover-accelerated CB expression constructs flanked by AAVS1-specific homology arms (HA-L/R) and introduced actin- (ACT-CB) and CTNNB1-specific CB (BC1-CB) in three human cell model systems.

Although this approach is experimentally straightforward, we faced some problems, which remain to be addressed. Firstly, while the puromycin resistance should only allow the selection of clones that underwent a correct transgene insertion at the AAVS1 locus, we obtained a stable HeLa cell clone with an unspecific integration of the ACT-CB transgene. It can be speculated that due to the increased genomic instability of HeLa cells the CB transgene was inserted randomly [56]. Notably, for HCT116 and DLD-1 cells our PCR-based genotyping approach revealed only clones with a correct CB insertion at the AAVS1 integration site. Secondly, in this study CB integration was exclusively analysed by PCR-based genotyping which provides no information regarding homo- or heterozygosity of the transgene at the AAVS1 locus or possible off-target integration. This could be analysed by junction PCR using primers binding outside the homology arms or Southern blot analysis [57].

Besides targeted genomic integration and expression of a CB transgene from a GSH to visualize an endogenous antigen, gene editing can also be used to directly add a fluorescent protein (FP) to the endogenous protein of interest (PoI) [58–60]. However such modifications are restricted either to the N- or the C-terminus of the PoI and from our experience it is not always possible to identify suitable gRNAs to target the intended genomic loci without affecting the integrity of the endogenous protein. Additionally, as repeatedly described, FP tagging can interfere with crucial protein parameters such as turnover, subcellular localization and participation in multi-protein complexes [6–8].

In summary, here we combined previously established methods and conceived a strategy to generate optimized CB expressing cell lines. We illustrate that the expression of CBs as ubiquitin fusion constructs substantially increases solubility and functionality of these intracellular nanoprobes. In addition, by implementing the EF1-$\alpha$ promoter for stable CB expression it can be assumed that unwanted epigenetic silencing during long-term cultivation will be reduced. Lastly, we established a protocol for site-directed integration of turnover-accelerated CBs into the AAVS1 locus by using the CRISPR/Cas9 gene editing technology. By applying this approach, we engineered the AAVS1 locus of three different cell lines (DLD-1, HCT116 and HeLa cells) to stably express turnover-accelerated CBs which are suitable not only to visualize the subcellular localization and dynamics of their respective antigens but also allows the quantification of changes of the endogenous protein levels. Notably, the generated CRISPR donor CB expression vector can be easily modified for integration of different CBs or FPs.

Considering the successful demonstration we are convinced that this approach is a substantial improvement over currently applied strategies to generate stable cell lines comprising intracellularly functional nanoprobes such as CBs. To our knowledge it is the first study describing a targeted insertion of an intrabody into a GSH loci for live-cell imaging. Although not tested yet, it is conceivable that this approach is easy transferable to other live-cell imaging probes such as fluorescently labelled single chain variable fragments (scFvs) [61]. As those nanoprobes have a high tendency to aggregate, fusion to ubiquitin might be particularly beneficial. However, we have to acknowledge that several aspects such as a comprehensive analysis of other promoters for CB expression, determination of epigenetic modification of the CB transgene upon long term cell cultivation and comparative analysis of additional GSH loci are still lacking. Thus we cannot not judge whether this approach is already optimal or can be further improved. Nevertheless, we expect that our strategy will facilitate the generation of more reliable CB cell models for biomedical research and preclinical compound screening campaigns in the future using advanced cellular imaging.

**Supplementary Materials:** The following are available online at http://www.mdpi.com/2073-4468/8/1/10/s1. Table S1: List of DNA oligonucleotides and synthesized gene fragments used in this study; Table S2: List of mammalian expression constructs used in this study; Figure S1: Stable HeLa_BC1-CB cell line displays heterogeneity in CB expression at high cell passages; Figure S2: Population-wide analysis of BC1-CB signal in DLD-1_BC1-CB cells and DLD-1_AAVS1_Ub-R-BC1-CB; Figure S3: Fluorescence images of DLD-1_AAVS1_Ub-R-BC1-CB cells upon compound treatment; Figure S4: FH535 reduces the amount of active CTNNB1 in DLD-1_AAVS1_Ub-R-BC1-CB cells.

**Author Contributions:** B.M.K., B.T. and U.R. conceived the study. B.M.K., J.M., M.W., S.S. and B.T. performed all experiments. B.M.K., B.T. and U.R. analysed the data. B.M.K., B.T. and U.R. wrote the manuscript.

**Funding:** The authors gratefully acknowledge the Ministry of Science, Research and Arts of Baden-Württemberg (V.1.4.-H3-1403-74) for financial support.

**Acknowledgments:** The authors thank Marion Jung (ChromoTek GmbH) for providing reagents and the original Lamin-CB expression construct. The authors gratefully acknowledge the Ministry of Science, Research and Arts of Baden-Württemberg (V.1.4.-H3-1403-74) for financial support.

**Conflicts of Interest:** U.R. is shareholder of the commercial company ChromoTek GmbH.

## References

1. Clemons, P.A. Complex phenotypic assays in high-throughput screening. *Curr. Opin. Chem. Boil.* **2004**, *8*, 334–338. [CrossRef] [PubMed]
2. Zock, J.M. Applications of high content screening in life science research. *Comb. Chem. High Throughput Screen* **2009**, *12*, 870–876. [CrossRef] [PubMed]

3. Bickle, M. The beautiful cell: High-content screening in drug discovery. *Anal. Bioanal. Chem.* **2010**, *398*, 219–226. [CrossRef] [PubMed]
4. Hosein, R.E.; Williams, S.A.; Haye, K.; Gavin, R.H. Expression of gfp-actin leads to failure of nuclear elongation and cytokinesis in tetrahymena thermophila. *J. Eukaryot. Microbiol.* **2003**, *50*, 403–408. [CrossRef] [PubMed]
5. Mendez, M.G.; Kojima, S.; Goldman, R.D. Vimentin induces changes in cell shape, motility and adhesion during the epithelial to mesenchymal transition. *FASEB J.* **2010**, *24*, 1838–1851. [CrossRef] [PubMed]
6. Stadler, C.; Rexhepaj, E.; Singan, V.R.; Murphy, R.F.; Pepperkok, R.; Uhlen, M.; Simpson, J.C.; Lundberg, E. Immunofluorescence and fluorescent-protein tagging show high correlation for protein localization in mammalian cells. *Nat. Methods* **2013**, *10*, 315–323. [CrossRef] [PubMed]
7. Snapp, E.L. Fluorescent proteins: A cell biologist's user guide. *Trends Cell Boil.* **2009**, *19*, 649–655. [CrossRef] [PubMed]
8. Virant, D.; Traenkle, B.; Maier, J.; Kaiser, P.D.; Bodenhofer, M.; Schmees, C.; Vojnovic, I.; Pisak-Lukats, B.; Endesfelder, U.; Rothbauer, U. A peptide tag-specific nanobody enables high-quality labelling for dstorm imaging. *Nat. Commun.* **2018**, *9*, 930. [CrossRef] [PubMed]
9. Kaiser, P.D.; Maier, J.; Traenkle, B.; Emele, F.; Rothbauer, U. Recent progress in generating intracellular functional antibody fragments to target and trace cellular components in living cells. *Biochim. Biophys. Acta* **2014**, *1844*, 1933–1942. [CrossRef] [PubMed]
10. Helma, J.; Cardoso, M.C.; Muyldermans, S.; Leonhardt, H. Nanobodies and recombinant binders in cell biology. *J. Cell Boil.* **2015**, *209*, 633–644. [CrossRef] [PubMed]
11. Moutel, S.; Bery, N.; Bernard, V.; Keller, L.; Lemesre, E.; de Marco, A.; Ligat, L.; Rain, J.C.; Favre, G.; Olichon, A.; et al. NaLi-H1: A universal synthetic library of humanized nanobodies providing highly functional antibodies and intrabodies. *eLife* **2016**, *5*, e16228. [CrossRef] [PubMed]
12. Rothbauer, U.; Zolghadr, K.; Tillib, S.; Nowak, D.; Schermelleh, L.; Gahl, A.; Backmann, N.; Conrath, K.; Muyldermans, S.; Cardoso, M.C.; et al. Targeting and tracing antigens in live cells with fluorescent nanobodies. *Nat. Methods* **2006**, *3*, 887–889. [CrossRef] [PubMed]
13. Irannejad, R.; Tomshine, J.C.; Tomshine, J.R.; Chevalier, M.; Mahoney, J.P.; Steyaert, J.; Rasmussen, S.G.; Sunahara, R.K.; El-Samad, H.; Huang, B.; et al. Conformational biosensors reveal GPCR signalling from endosomes. *Nature* **2013**, *495*, 534–538. [CrossRef] [PubMed]
14. Burgess, A.; Lorca, T.; Castro, A. Quantitative live imaging of endogenous DNA replication in mammalian cells. *PLoS ONE* **2012**, *7*, e45726. [CrossRef] [PubMed]
15. Panza, P.; Maier, J.; Schmees, C.; Rothbauer, U.; Sollner, C. Live imaging of endogenous protein dynamics in zebrafish using chromobodies. *Development* **2015**, *142*, 1879–1884. [CrossRef] [PubMed]
16. Traenkle, B.; Emele, F.; Anton, R.; Poetz, O.; Haeussler, R.S.; Maier, J.; Kaiser, P.D.; Scholz, A.M.; Nueske, S.; Buchfellner, A.; et al. Monitoring interactions and dynamics of endogenous beta-catenin with intracellular nanobodies in living cells. *Mol. Cell. Proteom.* **2015**, *14*, 707–723. [CrossRef] [PubMed]
17. Maier, J.; Traenkle, B.; Rothbauer, U. Real-time analysis of epithelial-mesenchymal transition using fluorescent single-domain antibodies. *Sci. Rep.* **2015**, *5*, 13402. [CrossRef] [PubMed]
18. Schorpp, K.; Rothenaigner, I.; Maier, J.; Traenkle, B.; Rothbauer, U.; Hadian, K. A multiplexed high-content screening approach using the chromobody technology to identify cell cycle modulators in living cells. *J. Biomol. Screen.* **2016**, *21*, 965–977. [CrossRef] [PubMed]
19. Halff, E.F.; Versteeg, M.; Brondijk, T.H.; Huizinga, E.G. When less becomes more: Optimization of protein expression in HEK293-EBNA1 cells using plasmid titration—A case study for NLRs. *Protein Expr. Purif.* **2014**, *99*, 27–34. [CrossRef] [PubMed]
20. Vavouri, T.; Semple, J.I.; Garcia-Verdugo, R.; Lehner, B. Intrinsic protein disorder and interaction promiscuity are widely associated with dosage sensitivity. *Cell* **2009**, *138*, 198–208. [CrossRef] [PubMed]
21. Akhtar, W.; de Jong, J.; Pindyurin, A.V.; Pagie, L.; Meuleman, W.; de Ridder, J.; Berns, A.; Wessels, L.F.; van Lohuizen, M.; van Steensel, B. Chromatin position effects assayed by thousands of reporters integrated in parallel. *Cell* **2013**, *154*, 914–927. [CrossRef] [PubMed]
22. Xia, W.; Bringmann, P.; McClary, J.; Jones, P.P.; Manzana, W.; Zhu, Y.; Wang, S.; Liu, Y.; Harvey, S.; Madlansacay, M.R.; et al. High levels of protein expression using different mammalian cmv promoters in several cell lines. *Protein Expr. Purif.* **2006**, *45*, 115–124. [CrossRef] [PubMed]

23. Zolghadr, K.; Gregor, J.; Leonhardt, H.; Rothbauer, U. Case study on live cell apoptosis-assay using lamin-chromobody cell-lines for high-content analysis. *Methods Mol. Boil.* **2012**, *911*, 569–575.
24. Keller, B.M.; Maier, J.; Secker, K.A.; Egetemaier, S.M.; Parfyonova, Y.; Rothbauer, U.; Traenkle, B. Chromobodies to quantify changes of endogenous protein concentration in living cells. *Mol. Cell. Proteom.* **2018**, *17*, 2518–2533. [CrossRef] [PubMed]
25. Gibson, D.G.; Young, L.; Chuang, R.Y.; Venter, J.C.; Hutchison, C.A., III; Smith, H.O. Enzymatic assembly of DNA molecules up to several hundred kilobases. *Nat. Methods* **2009**, *6*, 343–345. [CrossRef] [PubMed]
26. Damdindorj, L.; Karnan, S.; Ota, A.; Takahashi, M.; Konishi, Y.; Hossain, E.; Hosokawa, Y.; Konishi, H. Assessment of the long-term transcriptional activity of a 550-bp-long human β-actin promoter region. *Plasmid* **2012**, *68*, 195–200. [CrossRef] [PubMed]
27. Oceguera-Yanez, F.; Kim, S.I.; Matsumoto, T.; Tan, G.W.; Xiang, L.; Hatani, T.; Kondo, T.; Ikeya, M.; Yoshida, Y.; Inoue, H.; et al. Engineering the aavs1 locus for consistent and scalable transgene expression in human ipscs and their differentiated derivatives. *Methods* **2016**, *101*, 43–55. [CrossRef] [PubMed]
28. Varshavsky, A. Ubiquitin fusion technique and its descendants. *Methods Enzymol.* **2000**, *327*, 578–593. [PubMed]
29. Varshavsky, A. The n-end rule pathway and regulation by proteolysis. *Protein Sci.* **2011**, *20*, 1298–1345. [CrossRef] [PubMed]
30. Varshavsky, A. Ubiquitin fusion technique and related methods. *Methods Enzymol.* **2005**, *399*, 777–799. [PubMed]
31. Schmidthals, K.; Helma, J.; Zolghadr, K.; Rothbauer, U.; Leonhardt, H. Novel antibody derivatives for proteome and high-content analysis. *Anal. Bioanal. Chem.* **2010**, *397*, 3203–3208. [CrossRef] [PubMed]
32. Krishnan, M.; Park, J.M.; Cao, F.; Wang, D.; Paulmurugan, R.; Tseng, J.R.; Gonzalgo, M.L.; Gambhir, S.S.; Wu, J.C. Effects of epigenetic modulation on reporter gene expression: Implications for stem cell imaging. *FASEB J.* **2006**, *20*, 106–108. [CrossRef] [PubMed]
33. Yang, Y.; Mariati; Chusainow, J.; Yap, M.G. DNA methylation contributes to loss in productivity of monoclonal antibody-producing CHO cell lines. *J. Biotechnol.* **2010**, *147*, 180–185. [CrossRef] [PubMed]
34. Paredes, V.; Park, J.S.; Jeong, Y.; Yoon, J.; Baek, K. Unstable expression of recombinant antibody during long-term culture of cho cells is accompanied by histone H3 hypoacetylation. *Biotechnol. Lett.* **2013**, *35*, 987–993. [CrossRef] [PubMed]
35. Hsu, C.C.; Li, H.P.; Hung, Y.H.; Leu, Y.W.; Wu, W.H.; Wang, F.S.; Lee, K.D.; Chang, P.J.; Wu, C.S.; Lu, Y.J.; et al. Targeted methylation of cmv and e1a viral promoters. *Biochem. Biophys. Res. Commun.* **2010**, *402*, 228–234. [CrossRef] [PubMed]
36. Qin, J.Y.; Zhang, L.; Clift, K.L.; Hulur, I.; Xiang, A.P.; Ren, B.Z.; Lahn, B.T. Systematic comparison of constitutive promoters and the doxycycline-inducible promoter. *PLoS ONE* **2010**, *5*, e10611. [CrossRef] [PubMed]
37. Norrman, K.; Fischer, Y.; Bonnamy, B.; Wolfhagen Sand, F.; Ravassard, P.; Semb, H. Quantitative comparison of constitutive promoters in human ES cells. *PLoS ONE* **2010**, *5*, e12413. [CrossRef]
38. Wurm, F.M. Production of recombinant protein therapeutics in cultivated mammalian cells. *Nat. Biotechnol.* **2004**, *22*, 1393–1398. [CrossRef]
39. Rodolosse, A.; Barbat, A.; Chantret, I.; Lacasa, M.; Brot-Laroche, E.; Zweibaum, A.; Rousset, M. Selecting agent hygromycin B alters expression of glucose-regulated genes in transfected Caco-2 cells. *Am. J. Physiol.* **1998**, *274*, G931–G938. [CrossRef]
40. Valera, A.; Perales, J.C.; Hatzoglou, M.; Bosch, F. Expression of the neomycin-resistance (neo) gene induces alterations in gene expression and metabolism. *Hum. Gene Ther.* **1994**, *5*, 449–456. [CrossRef]
41. McDaniel, L.D.; Schultz, R.A. Elevation of sister chromatid exchange frequency in transformed human fibroblasts following exposure to widely used aminoglycosides. *Environ. Mol. Mutagen.* **1993**, *21*, 67–72. [CrossRef] [PubMed]
42. Kotin, R.M.; Linden, R.M.; Berns, K.I. Characterization of a preferred site on human chromosome 19q for integration of adeno-associated virus DNA by non-homologous recombination. *EMBO J.* **1992**, *11*, 5071–5078. [CrossRef] [PubMed]
43. Luo, Y.; Liu, C.; Cerbini, T.; San, H.; Lin, Y.; Chen, G.; Rao, M.S.; Zou, J. Stable enhanced green fluorescent protein expression after differentiation and transplantation of reporter human induced pluripotent stem cells generated by aavs1 transcription activator-like effector nucleases. *Stem Cells Transl. Med.* **2014**, *3*, 821–835. [CrossRef] [PubMed]

44. Sekine, K.; Takebe, T.; Taniguchi, H. Fluorescent labeling and visualization of human induced pluripotent stem cells with the use of transcription activator-like effector nucleases. *Transplant. Proc.* **2014**, *46*, 1205–1207. [CrossRef]
45. Zhang, P.W.; Haidet-Phillips, A.M.; Pham, J.T.; Lee, Y.; Huo, Y.; Tienari, P.J.; Maragakis, N.J.; Sattler, R.; Rothstein, J.D. Generation of gfap::Gfp astrocyte reporter lines from human adult fibroblast-derived i PS cells using zinc-finger nuclease technology. *Glia* **2016**, *64*, 63–75. [CrossRef]
46. Sadelain, M.; Papapetrou, E.P.; Bushman, F.D. Safe harbours for the integration of new DNA in the human genome. *Nat. Rev. Cancer* **2011**, *12*, 51–58. [CrossRef] [PubMed]
47. Smith, J.R.; Maguire, S.; Davis, L.A.; Alexander, M.; Yang, F.; Chandran, S.; ffrench-Constant, C.; Pedersen, R.A. Robust, persistent transgene expression in human embryonic stem cells is achieved with AAVS1-targeted integration. *Stem Cells* **2008**, *26*, 496–504. [CrossRef] [PubMed]
48. Cancer Genome Atlas, N. Comprehensive molecular characterization of human colon and rectal cancer. *Nature* **2012**, *487*, 330–337. [CrossRef] [PubMed]
49. Ahmed, D.; Eide, P.W.; Eilertsen, I.A.; Danielsen, S.A.; Eknaes, M.; Hektoen, M.; Lind, G.E.; Lothe, R.A. Epigenetic and genetic features of 24 colon cancer cell lines. *Oncogenesis* **2013**, *2*, e71. [CrossRef]
50. Handeli, S.; Simon, J.A. A small-molecule inhibitor of Tcf/β-catenin signaling down-regulates PPARγ and PPARδ activities. *Mol. Cancer Ther.* **2008**, *7*, 521–529. [CrossRef]
51. Yao, H.; Ashihara, E.; Maekawa, T. Targeting the Wnt/β-catenin signaling pathway in human cancers. *Expert Opin. Ther. Targets* **2011**, *15*, 873–887. [CrossRef] [PubMed]
52. Boutros, M.; Heigwer, F.; Laufer, C. Microscopy-based high-content screening. *Cell* **2015**, *163*, 1314–1325. [CrossRef] [PubMed]
53. Traenkle, B.; Rothbauer, U. Under the microscope: Single-domain antibodies for live-cell imaging and super-resolution microscopy. *Front. Immunol.* **2017**, *8*, 1030. [CrossRef] [PubMed]
54. Baker, R.T.; Smith, S.A.; Marano, R.; McKee, J.; Board, P.G. Protein expression using cotranslational fusion and cleavage of ubiquitin. Mutagenesis of the glutathione-binding site of human Pi class glutathione S-transferase. *J. Boil. Chem.* **1994**, *269*, 25381–25386.
55. Papapetrou, E.P.; Schambach, A. Gene insertion into genomic safe harbors for human gene therapy. *Mol. Ther.* **2016**, *24*, 678–684. [CrossRef] [PubMed]
56. Frattini, A.; Fabbri, M.; Valli, R.; De Paoli, E.; Montalbano, G.; Gribaldo, L.; Pasquali, F.; Maserati, E. High variability of genomic instability and gene expression profiling in different hela clones. *Sci. Rep.* **2015**, *5*, 15377. [CrossRef]
57. Koch, B.; Nijmeijer, B.; Kueblbeck, M.; Cai, Y.; Walther, N.; Ellenberg, J. Generation and validation of homozygous fluorescent knock-in cells using CRISPR-Cas9 genome editing. *Nat. Protoc.* **2018**, *13*, 1465–1487. [CrossRef]
58. Ratz, M.; Testa, I.; Hell, S.W.; Jakobs, S. CRISPR/Cas9-mediated endogenous protein tagging for resolft super-resolution microscopy of living human cells. *Sci. Rep.* **2015**, *5*, 9592. [CrossRef]
59. Zerjatke, T.; Gak, I.A.; Kirova, D.; Fuhrmann, M.; Daniel, K.; Gonciarz, M.; Muller, D.; Glauche, I.; Mansfeld, J. Quantitative cell cycle analysis based on an endogenous all-in-one reporter for cell tracking and classification. *Cell Rep.* **2017**, *19*, 1953–1966. [CrossRef]
60. Lackner, D.H.; Carre, A.; Guzzardo, P.M.; Banning, C.; Mangena, R.; Henley, T.; Oberndorfer, S.; Gapp, B.V.; Nijman, S.M.; Brummelkamp, T.R.; et al. A generic strategy for CRISPR-Cas9-mediated gene tagging. *Nat. Commun.* **2015**, *6*, 10237. [CrossRef]
61. Specht, E.A.; Braselmann, E.; Palmer, A.E. A critical and comparative review of fluorescent tools for live-cell imaging. *Annu. Rev. Physiol.* **2017**, *79*, 93–117. [CrossRef] [PubMed]

*Article*

# Properties of Fluorescent Far-Red Anti-TNF Nanobodies

**Ekaterina N. Gorshkova [1,*], Grigory A. Efimov [2], Ksenia D. Ermakova [1], Ekaterina A. Vasilenko [1], Diana V. Yuzhakova [3], Marina V. Shirmanova [3], Vladislav V. Mokhonov [1,4], Sergei V. Tillib [5,6], Sergei A. Nedospasov [1,4,5] and Irina V. Astrakhantseva [1,*]**

1 Center of Molecular Biology and Biomedicine, Institute of Biology and Biomedicine, Lobachevsky State University, Nizhniy Novgorod 603950, Russia; tworogowa.kseniya@yandex.ru (K.D.E.); kat802@rambler.ru (E.A.V.); vlad.mokhonov@gmail.com (V.V.M.); sergei.nedospasov@gmail.com (S.A.N.)
2 Laboratory of Transplantation Immunology, National Research Center for Hematology, Moscow 125167, Russia; gefimova@gmail.com
3 Institute of Biomedical Technologies, Nizhny Novgorod State Medical Academy, Nizhniy Novgorod 603005, Russia; yuzhakova-diana@mail.ru (D.V.Y.); shirmanovam@mail.ru (M.V.S.)
4 Engelhardt Institute of Molecular Biology, Russian Academy of Sciences, Moscow 119991, Russia
5 Lomonosov Moscow State University, Moscow 119991, Russia; sergei.tillib@gmail.com
6 Institute of Gene Biology, Russian Academy of Sciences, Moscow 119334, Russia
* Correspondence: e.n.gorshkova@gmail.com (E.N.G.); astrakhantsevairina@gmail.com (I.V.A.)

Received: 23 November 2018; Accepted: 13 December 2018; Published: 15 December 2018

**Abstract:** Upregulation of the expression of tumor necrosis factor (TNF-α, TNF) has a significant role in the development of autoimmune diseases. The fluorescent antibodies binding TNF may be used for personalized therapy of TNF-dependent diseases as a tool to predict the response to anti-TNF treatment. We generated recombinant fluorescent proteins consisting of the anti-TNF module based on the variable heavy chain (VHH) of camelid antibodies fused with the far-red fluorescent protein Katushka (Kat). Two types of anti-TNF VHH were developed: one (BTN-Kat) that was bound both human or mouse TNF, but did not neutralize their activity, and a second (ITN-Kat) that was binding and neutralizing human TNF. BTN-Kat does not interfere with TNF biological functions and can be used for whole-body imaging. ITN-Kat can be evaluated in humanized mice or in cells isolated from humanized mice. It is able to block human TNF (hTNF) activities both in vitro and in vivo and may be considered as a prototype of a theranostic agent for autoimmune diseases.

**Keywords:** TNF; fluorescent; nanobodies; sensor; anti-cytokine therapy; autoimmune disease

## 1. Introduction

Therapeutic neutralization of the inflammatory cytokines, in particular TNF, has revolutionized the treatment of autoimmune diseases including rheumatoid arthritis (RA), Crohn's disease, spondyloarthritis (SpA), and others. However, a significant number of the patients (~50%) with RA and SpA do not respond, showing only marginal improvement, or initially responding but then relapsing [1–4]. Molecular imaging may be employed to investigate the mechanisms of the disease pathogenesis [5]. Several attempts to monitor the disease activity and the localization of therapeutic antibodies in the inflamed joints using molecular imaging with radiolabeled monoclonal anti-TNF antibodies (certolizumab pegol, adalimumab, and infliximab) in rheumatic diseases were reported [6–10]. At the same time, real-time images could be captured in vivo using a fairly simple equipment and appropriate fluorescent proteins [11]. The use of far-red fluorescent proteins, such as Katushka [12], allows for in vivo imaging of fluorescence in the deep tissues. To be able to visualize TNF in living tissue, fluorescent proteins can be genetically fused to a binding moiety, so that the resultant protein meets three criteria: (1) stable folding, (2) small size, and (3) affinity that is sufficient for specific antigen binding [13].

Since 1989, when a novel type of antibody devoid of the light chains was identified in the sera of various members of the *Camelidae* family [14], a number of therapeutic proteins and tools based on the variable heavy chain (VHH) of camelid antibodies were developed and evaluated. VHHs are the smallest functional antigen-binding domains of these heavy-chain-only antibodies, which are also called "nanobodies". Their distinctive features are a small size, good stability and solubility, and high levels of expression in bacterial systems [15]. Also, it was shown that nanobodies are able to bind the epitopes inaccessible for conventional antibodies [16,17] and could display binding affinities in the lower nanomolar or even picomolar range [18]. All these features made nanobodies the ideal modules to be used in genetically encoded fluorescent proteins for successful molecular imaging. Several imaging techniques such as SPECT, PET, optical imaging, and ultrasound were successfully employed for nanobody-based imaging [19,20]. Nanobodies fused to fluorescent proteins were termed "chromobodies" and were used to visualize endogenous cellular structures in real-time studies of live cellular processes [21].

In this study two fluorescent sensors specific to TNF, both fused to far-red protein Katushka, were evaluated. One is based on VHH derived from Bactrian camel [22] and binds TNF without interfering with its functions. Therefore, it can be used to study the role of TNF in both normal and pathological conditions. The other protein is based on llama VHH, which binds and neutralizes human TNF bioactivity and can be used for the experimental therapy of the TNF-dependent autoimmune conditions in humanized mice with simultaneous visualization of the pathological processes in real time.

## 2. Materials and Methods

### 2.1. Design, Expression, and Purification of BTN-Kat and ITN-Kat Recombinant Proteins

DNA fragments encoding single-domain anti-TNF antibodies VHH41 [23] or ahTNF-VHH (GenBank: KU695528.1) [24,25] were cloned into the expression vector pET-28b (Merck Millipore, Darmstadt, Germany) between restriction sites BamHI and NcoI. To assemble the fluorescent-binding TNF nanobody (BTN-Kat) and the fluorescent-inhibiting TNF nanobody (ITN-Kat), the corresponding expression vector (which contained VHH41 or ahTNF-VHH, respectively) was digested with BamHI and NcoI and then ligated to PCR-amplified DNA fragment containing far-red fluorescent protein Katushka (TurboFP635, excitation/emission maxima at 588/635 nm) excised by XhoI- NcoI and DNA fragment encoded flexible glycine-serine linker. The C-terminal 6XHis tag sequence was in the same reading frame as the rest of the cDNA (Figure S1). Detailed information about expression and purification of the proteins is provided in Supplemental Material (Method S1, Table S1, Result S1).

### 2.2. Size Exclusion Chromatography

ITN-Kat (5 mg/mL) or BTN-Kat (5 mg/mL) was applied to a Superose 6 column (GE Healthcare, Amersham, UK) equilibrated in gel-filtration buffer (20 mM NaPi, pH 7.4, 150 mM NaCl). The flow rate was kept at 0.5 mL/min. Experimental and standard proteins were solubilized in gel-filtration buffer. The Superose 6 column was calibrated with standard proteins (Protein Standard Mix 15–600 kDa, Sigma-Aldrich, St. Louis, MO, USA). Elution volumes of ITN-Kat were found to be 12 mL, 4 mL, and 11 mL, and those of BTN-Kat were 11 mL, 3 mL, and 10.9 mL. To estimate their molecular masses, we used the plot of $\log_{10}$ molecular mass against elution volume. The same data were used to calculate the predicted molecular mass of the complex using the plot of molecular mass against elution volumes.

### 2.3. Mice

Human TNF knock-in mice (huTNFKI) described earlier [26,27] were bred in the SPF animal facility in the Institute of Biology and Biomedicine, Lobachevsky State University, in Nizhniy Novgorod on 12-h light/dark cycle at room temperature. All animal procedures were approved by the Scientific Council of the Institute of Biology and Biomedicine, Lobachevsky State University.

### 2.4. ELISA Measurement of the TNF Concentration in Murine Blood

Blood was taken from the buccal sinus of mice using sterile medical needles. To isolate the serum, blood was incubated at room temperature for 20–30 min and centrifuged at +4 °C and 14,000 rpm in clot activator tubes (Becton Dickinson, Franklin Lakes, NJ, USA). After that, the serum was transferred to fresh test tubes. TNF concentration was measured in sera by the Human-TNF ELISA Ready-SET-Go®(Fisher Scientific, Hampton, NH, USA).

### 2.5. Cytotoxic Assay

The inhibiting activity of BTN-Kat and ITN-Kat in TNF-mediated cytotoxicity was analyzed on WEHI-164 cell line [28]. The cells were plated at 5000 cells/well in 96-well culture plates. Recombinant hTNF was added at constant concentration (100 U/mL). The fluorescent antibodies were applied at serial dilutions 1 mM–2 pM. After 24 h of incubation, 3-(4,5-dimethylthiazol-2-yl)-2,5-diphenyltetrazolium bromide (MTT) (Sigma-Aldrich, St. Louis, MO, USA) was added at concentration 4 µg/mL. After 4 h of incubation, formazan crystals were solubilized in 10% $w/v$ SDS solution in DMSO, and OD was measured at 560 nm. The percentage of living cells was calculated and fitted to a nonlinear regression curve using Prism 5 (GraphPad Software, San Diego, CA, USA) software. One unit was defined as the amount of TNF inhibitor sufficient to mediate half-maximal protection from cytotoxicity in the presence of 100 U/mL human TNF.

### 2.6. LPS/D-Galactosamine-Induced Acute Hepatotoxicity Model

TNF-humanized mice were individually weighted and received intraperitoneal injection of 150, 300, or 450 pmol/g of BTN-Kat; 75, 150, or 300 pmol/g of ITN-Kat; or 150 pmol/g of infliximab or PBS buffer followed 30 min later by an otherwise lethal dose of lipopolysaccharide and D-Galactosamine (LPS/D-Gal) (Sigma-Aldrich, St. Louis, MO, USA) 400 ng/g and 800 µg/g, respectively. Mice were euthanized when moribund. Kaplan–Meier survival curves were plotted, and pairwise statistical comparison of BTN-Kat, ITN-Kat, and infliximab was performed.

### 2.7. Flow Cytometry Analysis of Bone Marrow-Derived Macrophages

Bone marrow-derived macrophages (BMDM) from TNF-humanized mice or from C57Bl/6 mice were cultured for 10 days in DMEM, L-glutamune, Pen/Strep, 20% ($v/v$) horse serum, and 30% ($v/v$) of L929 conditioned medium, and then detached with ice-cold PBS and counted. Activation of TNF production by the macrophages was performed using LPS (100 ng/mL, 4 h, 37 °C, 5%, $CO_2$). The cells mortality was obtained by measuring the percentage of dead macrophages by Trypan blue exclusion. Fcγ receptor was blocked by anti-Fcγ receptor antibodies (Biolegend, San Diego, CA, USA), and then cells were stained with anti-F4/80-FITC antibodies (Thermo Fischer, Waltham, MA, USA) and BTN-Kat or ITN-Kat by intracellular cytokine staining protocol using BD Fixation and Permeabilization Solution Kit with BD GolgiPlug™ (Becton Dickinson, Franklin Lakes, NJ, USA). The samples were analyzed on Cytoflex S flow cytometer (Beckman Coulter, Brea, CA, USA). The fluorescent signal from Katushka was detected with 585 nm laser excitation and 660/20 nm emission filter. Twenty thousand cells were evaluated per test; in a list mode, data were analyzed using CytExpert 2.0 software (Beckman Coulter, Brea, CA, USA).

### 2.8. Flow Cytometry Analysis of Murine Blood

Blood, collected from buccal sinus and treated with heparin (10 U/mL), was incubated with 1× RBC lysis buffer (Biolegend, San Diego, CA, USA) and then was resuspended in 100 µL of FACS buffer (PBS with 2% FBS), with Fixable Live/Dead Stain (Life Technologies, Carlsbad, CA, USA) and an appropriate combination of fluorescent antibodies specific to CD45, F4/80, CD3, CD45R (Thermo Fischer, Waltham, MA, USA), and FcR block (Biolegend, San Diego, CA, USA). After that, cells were incubated with BTN-Kat and ITN-Kat on ice without permeabilization, or stained by intracellular cytokine staining protocol using BD Fixation and Permeabilization Solution Kit with BD GolgiPlug™ (Becton Dickinson, Franklin Lakes, NJ, USA). Data were analyzed using CytExpert 2.0 (Beckman Coulter, Brea, CA, USA).

### 2.9. Fluorescence Whole-Body Imaging

150 pmol/g of the fluorescent sensors were injected intraperitonealy (i/p), and after 30 min animals were i/p injected with LPS (400 ng/g) and D-Gal (800 μg/g) or the same volume of PBS. As a control, mice were injected with PBS 30 min prior to the injection of LPS/D-Gal instead of the proteins or the PBS only. Fluorescence imaging was performed on the IVIS-Spectrum system (Caliper Life Sciences, Waltham, MA, USA) in the epi-luminescence mode; the fluorescence was excited at 605/30 nm and detected at 660/20 nm [29]. Before the procedure, the mice were shaven using a shaving machine and additionally depilated with the cream. In 1, 3, and 6 h after the injection of LPS/ D-Gal, the mice were euthanized by isoflurane, and whole-body fluorescence images of the animals were acquired. Quantitative analysis was performed in the Living Image Software 4.2 (PerkinElmer, Waltham, MA, USA) by calculation of the fluorescence intensity averaged over the abdominal region and normalizing to the values of autofluorescence measured from the mice injected with PBS alone. To analyze the uptake of the BTN-Kat or ITN-Kat by the liver, ex vivo fluorescence imaging was performed.

### 2.10. Statistical Analysis

Statistically significant differences of values were determined using STATISTICA 10 (StatSoft, Moscow, Russia) using the Mann–Whitney U-test. Differences were considered statistically significant at $p \leq 0.05$.

## 3. Results

### 3.1. Anti-TNF Antibodies Fused to Far-Red Katushka Protein Form Oligomers

Two genetically encoded fluorescent sensors based on two distinct TNF-binding modules and far-red protein Katushka (Figure 1A) were designed and successfully expressed in prokaryotic system (Result S1, Figure S2). Native electrophoresis of purified proteins showed that both proteins formed a single structure under native conditions (Figure S3). Profiles from size exclusion chromatography showed that fluorescent proteins had a tendency to form oligomers (Figure 1C). Apparent molecular weight values of 113,260 kDa and 36,413 kDa for ITN-Kat corresponded to trimeric and monomeric forms, respectively. For BTN-Kat, apparent MW values were 85,979 kDa and 118,904 kDa, which corresponded to the dimeric and trimeric proteins (Figure 1C). Previously the ability of the fluorescent protein Katushka to form dimers and tetramers were reported [12,30], thus, we believe that oligomerization is due to the presence of the Katushka module. However, in our experiments a trimeric form for both proteins was also observed. The mixture of the protein's oligomer forms was used for all subsequent experiments.

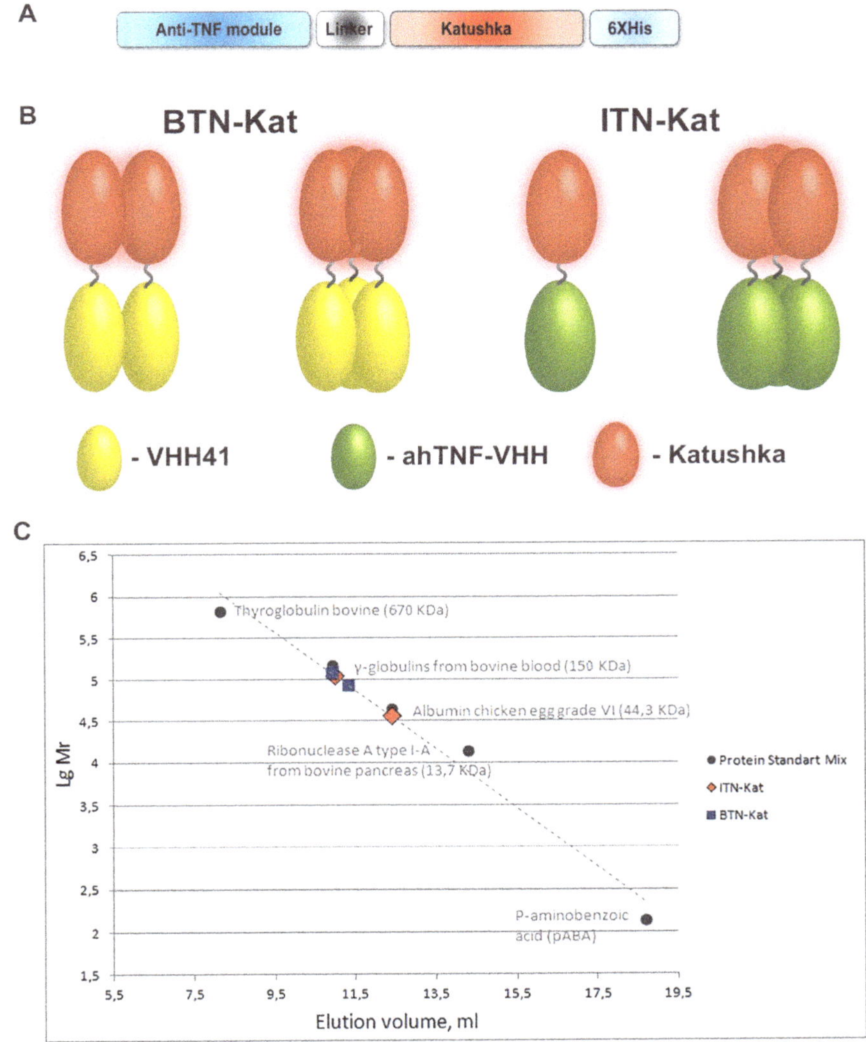


**Figure 1.** Fluorescent far-red anti-tumor necrosis factor (TNF) nanobodies spontaneously form oligomers. (**A**) Map of genetic constructs that encode fluorescent TNF sensors; (**B**) expected structure of BTN-Kat and ITN-Kat proteins in native conditions; (**C**) profile of size exclusion chromatography of ITN-Kat and BTN-Kat, imposed on the chromatogram of molecular weight markers (protein standard mix).

## *3.2. Fluorescent Nanobodies Interact with TNF In Vitro*

Inhibitory activities of BTN-Kat and ITN-Kat toward recombinant human TNF were evaluated in TNF-dependent colorimetric MTT cytotoxicity assay with WEHI 164 murine fibrosarcoma cell line, which is sensitive to human TNF. The analysis was prepared in 96-well plates. Each well contained 20,000 cells, the concentration of TNF was 100 U/mL in accordance with a predetermined TNF DL50 (1U). Cells were incubated with a mixture of TNF and inhibitor in a range of concentrations overnight. In contrast to BTN-Kat, ITN-Kat demonstrated a dose-dependent hTNF inhibitory activity (half maximal effective concentration ($EC_{50}$) = 9.73·$10^3$ pM). The clinically utilized chimeric monoclonal antibody infliximab was used as a positive control ($EC_{50}$ = 2.83·$10^3$ pM) (Figure 2).

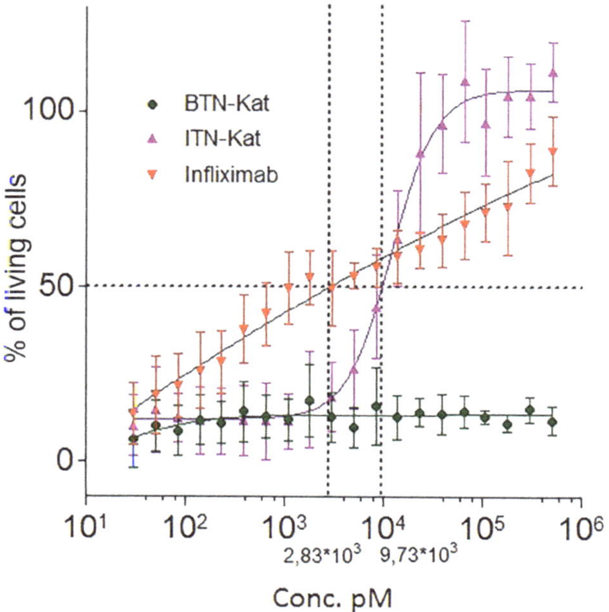


**Figure 2.** ITN-Kat, but not BTN-Kat, inhibited human TNF (hTNF) activity in vitro. TNF-neutralizing activities of BTN-Kat, ITN-Kat, and infliximab were evaluated by MTT cytotoxicity test using the WEHI 164 cell line. Percentage of living cells ±SD is plotted. Dashed lines indicate half maximal effective concentration ($EC_{50}$)

Additionally, in vitro TNF binding activity of BTN-Kat and ITN-Kat was examined using flow cytometry. Bone marrow-derived macrophages (BMDM) derived from TNF humanized mice were activated by bacterial LPS and then stained with BTN-Kat and ITN-Kat using an intracellular cytokine staining protocol that resulted in specific staining of the TNF-expressing cells. BTN-Kat demonstrated lower staining than ITN-Kat in BMDM cells from huTNFKI mice (Figure 3A); however, BTN-Kat was also able to interact with mouse TNF derived from the BMDM of C57Bl6 mice while ITN-Kat did not (Figure 3B). This observation suggested that imaging with BTN-Kat may be feasible in wild-type mice.

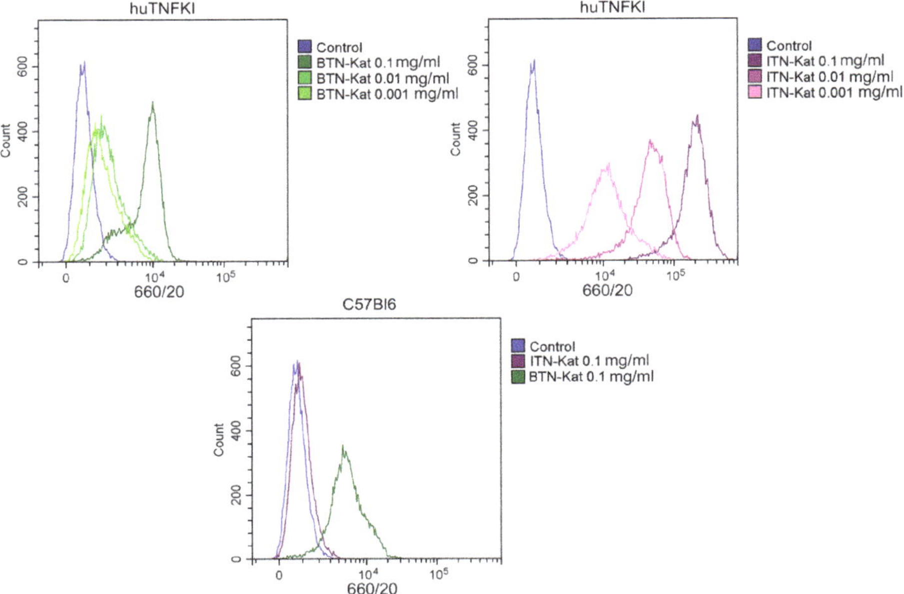


**Figure 3.** ITN-Kat is specific for human TNF while BTN-Kat interacts with both human and murine TNF. (**A**) BTN-Kat and ITN-Kat binding activity to bone marrow-derived macrophages (BMDM) from TNF-humanized (huTNFKI) mice. (**B**) BTN-Kat binding activity to BMDM from C57Bl6 mice (WT). Macrophages were activated by lipopolysaccharide (LPS) (100 ng/mL for 4 h) in the presence of brefeldin A and stained for TNF by BTN-Kat or ITN-Kat using intracellular cytokine staining protocol. Shown cells are gated on F4/80 expression.

### *3.3. ITN-Kat Showed TNF Neutralizing Activity In Vivo, while BTN-Kat Did not*

The ability of BTN-Kat and ITN-Kat to inhibit systemic TNF was evaluated in TNF-humanized mice in the experimental model of acute hepatotoxicity (Figure 4).

Mice injected with ITN-Kat at concentration 150 pmol/g survived (Figure 4B), as did mice treated with infliximab at the same dose. BTN-Kat was not able to protect mice from lethality at the same dose nor at higher doses (Figure 4A).

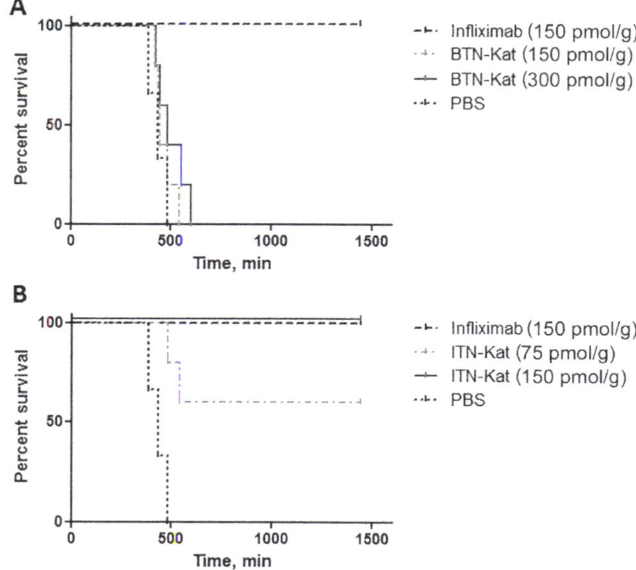


**Figure 4.** Protection from LPS/D-Galactoseamine (LPS/D-Gal) toxicity in vivo is provided by ITN-Kat, but not by BTN-Kat. TNF-humanized mice were injected either with ITN-Kat, BTN-Kat, infliximab, or PBS. Thirty minutes later, mice were injected with the otherwise lethal dose of LPS/D-Gal. (**A**) Survival curves of mice injected with 300 pmol/g and 150 pmol/g of BTN-Kat compared with the buffer or 150 pmol/g infliximab; (**B**) survival curves of mice injected with 150 pmol/g and 75 pmol/g of ITN-Kat as compared with the buffer or 150 pmol/g infliximab.

### *3.4. The In Vivo Fluorescence of Anti-TNF Nanobodies Correlates with TNF Levels in Mice*

TNF plays a critical role in liver injury induced by LPS/D-Gal. Thus, the model of acute hepatotoxicity is characterized by an inflammatory process in the abdominal cavity, and, as expected, the highest level of fluorescence of the sensors was detected in the abdominal area. The peak of the fluorescence intensity was observed at 1 h post-injection of the LPS/D-Gal. In this group fluorescence signal was significantly higher ($p < 0.05$) than in the control mice injected with PBS and LPS/D-Gal or PBS alone (Figure 5A–C). These results are consistent with the data obtained by ELISA, in which the highest level of TNF concentration in the blood was observed 1 h after the LPS/D-Gal injection (Figure 5D). However, at the 3-h time point after the LPS/D-Gal injection, the level of the BTN-Kat and ITN-Kat fluorescence did not differ from the mice injected with the sensor alone without LPS/D-Gal. At 6 h, the fluorescence intensity in mice injected with BTN-Kat fell to the baseline level of autofluorescence (Figure 5B), which suggested either complete excretion of BTN-Kat protein from the body or its degradation. However, at 6-h point, the fluorescent signals of control ITN-Kat-injected mice and ITN-Kat-injected mice challenged with LPS/D-Gal were significantly distinct from autofluorescence ($p < 0.05$) (Figure 5C). The results of ex vivo imaging confirmed the accumulation the fluorescent anti-TNF antibodies in the liver with the maximum of fluorescence signals at 1 h after LPS/D-Gal injection (Figure 5E).

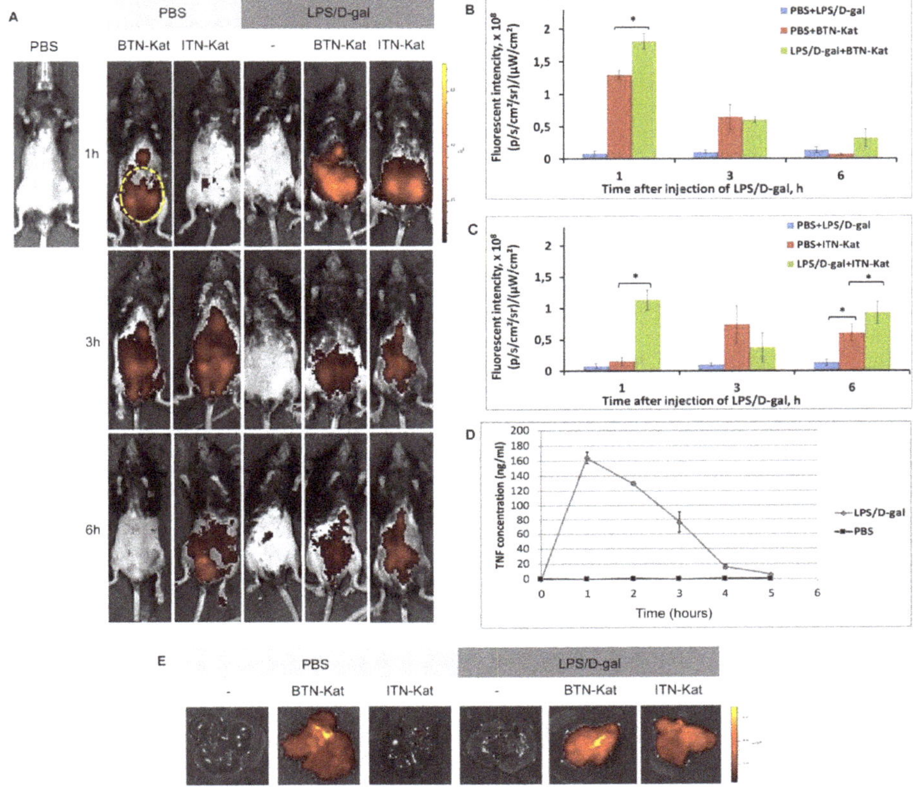


**Figure 5.** Fluorescent anti-TNF nanobodies accumulate in the liver of mice after the LPS/D-Gal injection. Visualization of TNF production in adult humanized TNF knock-in mice using BTN-Kat and ITN-Kat during LPS/D-Gal acute hepatotoxicity. (**A**) Fluorescence intensity images in mice during acute hepatotoxicity in comparison with the control; the abdominal region for quantitative analysis is displayed by the yellow circle; (**B**) fluorescence signal analysis of mice after injection of BTN-Kat followed by LPS/D-gal or PBS injection. Intensity of fluorescence ±SD is plotted; (**C**) fluorescence signal analysis of mice after injection of ITN-Kat followed by LPS/D-gal or PBS injection. Mean ± SD is plotted; (**D**) TNF level in mice serum after LPS/D-gal or PBS injection measured by ELISA; (**E**) fluorescence intensity ex vivo images of mice liver 1 h after LPS/D-gal injection.

### 3.5. LPS/D-Galactosamine-Induced Acute Hepatotoxicity Depends on TNF Expression by Monocytes

We then evaluated the expression levels of TNF by F4/80, CD3, and CD45R-positive cells (Figures 6 and S5) using flow cytometry protocols for both surface and intracellular staining during LPS/D-gal-induced acute hepatotoxicity. The results showed that the main source of TNF corresponded to F4/80-positive cells. At the same time, the maximal number of TNF-positive cells was observed at 1 h after the LPS/D-gal injection, in correlation with the TNF level in the blood (Figure 5D). Of note, we did not observe any significant changes in the TNF levels produced by other cell types (Figure S5).

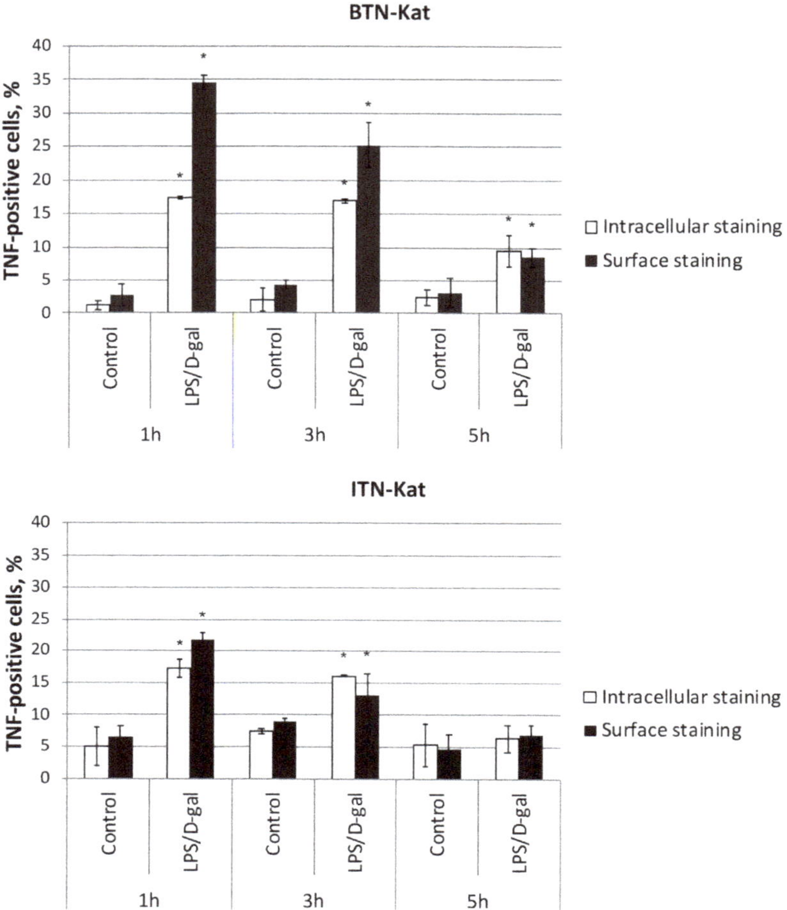


**Figure 6.** F4/80-positive cells are the main source of TNF. Levels of TNF-positive F4/80-positive cells 1 h, 3 h, and 5 h after LPS/D-gal injection were measured using BTN-Kat and ITN-Kat by flow cytometry surface and intracellular staining protocols. Mean levels of TNF-positive cells ±SD are plotted. Data are representative of five independent experiments. (* $p < 0.05$).

## 4. Discussion

Molecular imaging is a promising approach to address the role of TNF in various inflammatory pathologies. Recent studies highlighted applications of red fluorescent proteins fused with VHH antibodies for bioimaging and theranostics [31–33]. The small size and the absence of Fc-fragment in VHH modules may reduce side effects caused by interactions with receptors of immunocompetent cells and with complement system, as it usually happens with classical antibodies [34]. Also, the earlier study [35] demonstrated that red fluorescent proteins with maximum of emission spectra >600 nm are more effective, because the emitted light is not absorbed by the tissues.

We attempted to utilize far-red-emitting nanobody-based fusion proteins for imaging both systemic TNF as well as TNF at the sites of its local expression in vivo. VHH domains from *Camelidae* antibodies VHH41 [23] and ahTNF-VHH [24] were used as targeting modules. It has been shown that those module had a similar affinity to hTNF, however VHH41's target region of hTNF was not involved in interaction with the TNF receptor [23]. Fluorescent protein Katushka chosen as the imaging molecule emits in the range of 620–660 nm that fits the window of biological tissue transparency in which case the absorption is minimal [35]. Moreover, Katushka was characterized by a very high pH and photostability, and by intensity of the signal 7- to 10-fold brighter than the spectrally close HcRed [36] or mPlum [37] proteins. Additionally, it readily forms tetrameric structures at pH 5.5 to 8.5, but can dissociate into the dimer at pH below 5.0 [30]. We noticed that proteins fused with Katushka also showed a tendency to form oligomeric structures (Figure 1C). The other advantage of the fluorescent complexes "VHH-Katushka" is their relatively simple and sufficiently effective expression in prokaryotic systems (Method S1, Result S1).

In human TNF knock-in (huTNFKI) mice [26] used here, human TNF was expressed in vivo instead of murine TNF. It mediated normal and pathological functions of this cytokine which can be neutralized by clinically used anti-hTNF drugs. One of the studied nanobodies, ITN-Kat, was able to bind to human TNF (Figure 3) and neutralize its activity in vitro (Figure 2) as well as in vivo (Figure 4B). The protective effect of ITN-Kat was confirmed by liver histology (Result S2, Figure S4E). On the other hand, although BTN-Kat can bind both human and murine TNF (Figure 3), it lacks blocking ability (Figure 2) and therefore could not prevent the development of acute hepatotoxicity (Figures 4A and S4D). Moreover, ITN-Kat showed higher sensitivity to hTNF than BTN-Kat. We observed a higher level of fluorescence in activated macrophages stained with ITN-Kat compared to BTN-Kat in a similar concentration range using flow cytometry (Figure 3A).

As a systemic reaction, septic shock affects all body systems, including immune cells. TNF is known to play a critical role in the process of liver injury induced by LPS/D-Gal. Soluble TNF is the main hepatotoxic mediator in this toxicity model [38]. Using fluorescent sensors targeting TNF, we confirmed that the main source of TNF during the development of LPS/D-gal-induced acute hepatotoxicity is the myeloid cell compartment (Figure 6) [39]. More specifically, our results indicate that TNF expressed by F4/80-positive cells plays the key role in the development of pathology, while T-cells and B-cells do not contribute to the increase of serum TNF levels in the process of acute inflammation (Figure S5).

The inflammatory processes in the abdominal cavity of humanized mice after LPS/D-gal administration were studied using BTN-Kat and ITN-Kat in whole-body and ex vivo imaging mode. The peak of fluorescence at the time of maximal concentration of TNF in the blood occurred 1 h after LPS/D-gal. intraperitoneal injection, consistent with the data obtained by ELISA (Figure 5D). Furthermore, the level of BTN-Kat fluorescence gradually decreased, and this paralleled the level of TNF in the blood. The elevated expression of TNF in joints in the murine model of collagen-induced arthritis was successfully visualized by BTN-Kat [40]. These data confirmed the ability of BTN-Kat to bind both human and murine TNF, with subsequent successful visualization using bioimaging methods. The level of ITN-Kat fluorescence did not correlate with the level of TNF in the blood during acute hepatotoxicity. This may indicate that this TNF inhibitor affected regulation of TNF expression since cytokine gene regulation may include positive feedback loops. This hypothesis requires additional experimental evaluation.

In summary, we developed and characterized TNF fluorescent sensor (BTN-Kat) and fluorescent sensor-inhibitor (ITN-Kat), utilizing two single-domain anti-TNF antibodies. We evaluated their ability to bind and neutralize TNF in vitro, and to serve as imaging labels in vitro and in the whole body non-invasive analysis. We concluded that BTN-Kat is a convenient tool for studying the dynamics of TNF expression without interfering with its biological functions, while ITN-Kat is a prototype theranostic agent for TNF-dependent autoimmune diseases.

**Supplementary Materials:** The following are available online at http://www.mdpi.com/2073-4468/7/4/43/s1, Figure S1: Protein sequences of BTN-Kat and ITN-Kat proteins; Figure S2: The yields of the BTN-Kat and ITN-Kat proteins purified from the various *E. coli* strains; Figure S3: Native electrophoresis of BTN-Kat and ITN-Kat; Figure S4: Liver histology (H&E stain) of mice; Figure S5: Levels of TNF-positive CD3 and CD45R-positive cells; Table S1: The *E. coli* strains used in this study; Method S1: Expression and purification of BTN-Kat and ITN-Kat recombinant proteins; Method S2: Preparation of Liver Tissues for Histological Analysis; Result S1:

Fluorescent antibodies expressed in BL21(DE3) strain of *E. coli;* Result S2: The therapeutic effect of BTN-Kat and ITN-Kat proteins.

**Author Contributions:** Conceptualization: E.N.G., G.A.E., and I.V.A.; Methodology: E.A.V., D.V.Y., K.D.E., G.A.E., M.V.S., and S.V.T.; Software: E.N.G.; Validation, I.V.A. and V.V.M.; Formal Analysis: E.N.G. and V.V.M.; Investigation: E.N.G., K.D.E., E.A.V., V.V.M., and I.V.A.; Data Curation: E.N.G. and I.V.A.; Writing—Original Draft Preparation: E.N.G., V.V.M., and I.V.A.; Writing—Review & Editing, G.A.E., M.V.S., S.V.T., and S.A.N.; Visualization: E.N.G.; Supervision: I.V.A. and S.A.N.

**Funding:** The research was supported by Russian Foundation of Basic Research- No 17-04-01478, in part it was supported by the State Contract Nos. 20.6156.2017/9.10, 20.6445.2017/9.10, and 20.6159.2017/9.10 and by the Russian Science Foundation, project No 14-50-00060.

**Acknowledgments:** We thank L. Istomin for help with design of illustrations and E. Bolshakova, O. Shkola for help with the experiments.

**Conflicts of Interest:** The authors declare no conflict of interest.

## References

1. Davis, J.; van der Heijde, D.M.; Braun, J.; Dougados, M.; Cush, J.; Clegg, D.; Inman, R.D.; Kivitz, A.; Zhou, L.; Solinger, A.; et al. Sustained durability and tolerability of etanercept in ankylosing spondylitis for 96 weeks. *Ann. Rheum. Dis.* **2005**, *64*, 1557–1562. [CrossRef] [PubMed]
2. Van der Heijde, D.; Dijkmans, B.; Geusens, P.; Sieper, J.; DeWoody, K.; Williamson, P.; Braun, J.; Ankylosing Spondylitis Study for the Evaluation of Recombinant Infliximab Therapy Study Group. Efficacy and safety of infliximab in patients with ankylosing spondylitis: Results of a randomized, placebo-controlled trial (ASSERT). *Arthritis Rheum.* **2005**, *52*, 582–591. [CrossRef] [PubMed]
3. Van der Heijde, D.; Kivitz, A.; Schiff, M.H.; Sieper, J.; Dijkmans, B.A.; Braun, J.; Dougados, M.; Reveille, J.D.; Wong, R.L.; Kupper, H.; Davis, J.C., Jr.; ATLAS Study Group. Efficacy and safety of adalimumab in patients with ankylosing spondylitis: Results of a multicenter, randomized, double-blind, placebo-controlled trial. *Arthritis Rheum.* **2006**, *54*, 2136–2146. [CrossRef] [PubMed]
4. Inman, R.D.; Davis, J.C., Jr.; Heijde, Dv.; Diekman, L.; Sieper, J.; Kim, S.I.; Mack, M.; Han, J.; Visvanathan, S.; Xu, Z.; Hsu, B.; et al. Efficacy and safety of golimumab in patients with ankylosing spondylitis: Results of a randomized, double-blind, placebo-controlled, phase III trial. *Arthritis Rheum.* **2008**, *58*, 3402–3412. [CrossRef] [PubMed]
5. Kuchmiy, A.A.; Efimov, G.A.; Nedospasov, S.A. Methods for in vivo molecular imaging. *Biochemistry* **2012**, *77*, 1339–1353. [CrossRef] [PubMed]
6. Palframan, R.; Airey, M.; Moore, A.; Vugler, A.; Nesbitt, A. Use of biofluorescence imaging to compare the distribution of certolizumab pegol, adalimumab, and infliximab in the inflamed paws of mice with collagen-induced arthritis. *J. Immunol. Methods* **2009**, *348*, 36–41. [CrossRef]
7. Malviya, G.; Conti, F.; Chianelli, M.; Scopinaro, F.; Dierckx, R.A.; Signore, A. Molecular imaging of rheumatoid arthritis by radiolabelled monoclonal antibodies: new imaging strategies to guide molecular therapies. *Eur. J. Nucl. Med. Mol. Imaging* **2010**, *37*, 386–398. [CrossRef]
8. Lambert, B.; Carron, P.; D'Asseler, Y.; Bacher, K.; van den Bosch, F.; Elewaut, D.; Verbruggen, G.; Beyaert, R.; Dumolyn, C.; De Vos, F. (99m)Tc-labelled S-HYNIC certolizumab pegol in rheumatoid arthritis and spondyloarthritis patients: a biodistribution and dosimetry study. *EJNMMI Res.* **2016**, *6*, 88. [CrossRef]
9. Carron, P.; Lambert, B.; Van Praet, L.; De Vos, F.; Varkas, G.; Jans, L.; Elewaut, D.; Van den Bosch, F. Scintigraphic detection of TNF-driven inflammation by radiolabelled certolizumab pegol in patients with rheumatoid arthritis and spondyloarthritis. *RMD Open* **2016**, *2*, e000265. [CrossRef]
10. Put, S.; Westhovens, R.; Lahoutte, T.; Matthys, P. Molecular imaging of rheumatoid arthritis: emerging markers, tools, and techniques. *Arthritis Res. Ther.* **2014**, *16*, 208. [CrossRef]
11. Hoffman, R.M. Use of fluorescent proteins and color-coded imaging to visualize cancer cells with different genetic properties. *Cancer Metast. Rev.* **2016**, *35*, 5–19. [CrossRef] [PubMed]
12. Shcherbo, D.; Merzlyak, E.M.; Chepurnykh, T.V.; Fradkov, A.F.; Ermakova, G.V.; Solovieva, E.A.; Lukyanov, K.A.; Bogdanova, E.A.; Zaraisky, A.G.; Lukyanov, S.; et al. Bright far-red fluorescent protein for whole-body imaging. *Nat. Methods* **2007**, *4*, 741–747. [CrossRef] [PubMed]

13. Kaiser, P.D.; Maier, J.; Traenkle, B.; Emele, F.; Rothbauer, U. Recent progress in generating intracellular functional antibody fragments to target and trace cellular components in living cells. *Biochim. Biophys. Acta* **2014**, *1844*, 1933–1942. [CrossRef] [PubMed]
14. Hamers-Casterman, C.; Atarhouch, T.; Muyldermans, S.; Robinson, G.; Hamers, C.; Songa, E.B.; Bendahman, N.; Hamers, R. Naturally occurring antibodies devoid of light chains. *Nature* **1993**, *363*, 446. [CrossRef] [PubMed]
15. Olichon, A.; Surrey, T. Selection of Genetically Encoded Fluorescent Single Domain Antibodies Engineered for Efficient Expression in Escherichia coli. *J. Biol. Chem.* **2007**, *282*, 36314–36320. [CrossRef] [PubMed]
16. Desmyter, A.; Transue, T.R.; Ghahroudi, M.A.; Thi, M.H.; Poortmans, F.; Hamers, R.; Muyldermans, S.; Wyns, L. Crystal structure of a camel single-domain VH antibody fragment in complex with lysozyme. *Nat. Struct. Biol.* **1996**, *3*, 803–811. [CrossRef] [PubMed]
17. Muyldermans, S.; Cambillau, C.; Wyns, L. Recognition of antigens by single-domain antibody fragments: the superfluous luxury of paired domains. *Trends Biochem. Sci.* **2001**, *26*, 230–235. [CrossRef]
18. van der Linden, R.H.J.; Frenken, L.G.; de Geus, B.; Harmsen, M.M.; Ruuls, R.C.; Stok, W.; de Ron, L.; Wilson, S.; Davis, P.; Verrips, C.T. Comparison of physical chemical properties of llama VHH antibody fragments and mouse monoclonal antibodies. *Biochim. Biophys. Acta* **1999**, *1431*, 37–46. [CrossRef]
19. Chakravarty, R.; Goel, S.; Cai, W. Nanobody: The "Magic Bullet" for Molecular Imaging? *Theranostics* **2014**, *4*, 386–398. [CrossRef]
20. Rashidian, M.; Keliher, E.J.; Bilate, A.M.; Duarte, J.N.; Wojtkiewicz, G.R.; Jacobsen, J.T.; Cragnolini, J.; Swee, L.K.; Victora, G.D.; Weissleder, R.; Ploegh, H.L. Noninvasive imaging of immune responses. *Proc. Natl. Acad. Sci. USA* **2015**, *112*, 6146–6151. [CrossRef]
21. Rothbauer, U.; Zolghadr, K.; Tillib, S.; Nowak, D.; Schermelleh, L.; Gahl, A.; Backmann, N.; Conrath, K.; Muyldermans, S.; Cardoso, M.C.; et al. Targeting and tracing antigens in live cells with fluorescent nanobodies. *Nat. Methods* **2006**, *3*, 887. [CrossRef]
22. Tillib, S.V.; Vyatchanin, A.S.; Muyldermans, S. Molecular analysis of heavy chain-only antibodies of Camelus bactrianus. *Biochemistry* **2014**, *79*, 1382–1390. [CrossRef] [PubMed]
23. Efimov, G.A.; Khlopchatnikova, Z.V.; Sazykin, A.Yu.; Drutskaya, M.C.; Kruglov, A.A.; Shilov, E.C.; Kuchmii, A.A.; Nedospasov, S.A.; Tillib, S.V. Isolation and characteristics of a new recombinant single domain antibody that specifically binds to human TNF. *Russ. J. Immunol.* **2012**, *6*, 337–345. (In Russian)
24. Coppieters, K.; Dreier, T.; Silence, K.; de Haard, H.; Lauwereys, M.; Casteels, P.; Beirnaert, E.; Jonckheere, H.; Van de Wiele, C.; Staelens, L.; Hostens, J. Formatted anti-tumor necrosis factor alpha VHH proteins derived from camelids show superior potency and targeting to inflamed joints in a murine model of collagen-induced arthritis. *Arthritis Rheum.* **2006**, *54*, 1856–1866. [CrossRef] [PubMed]
25. Plagmann, I.; Chalaris, A.; Kruglov, A.A.; Nedospasov, S.; Rosenstiel, P.; Rose-John, S.; Scheller, J. Transglutaminase-catalyzed covalent multimerization of Camelidae anti-human TNF single domain antibodies improves neutralizing activity. *J. Biotechnol.* **2009**, *142*, 170–178. [CrossRef]
26. Olleros, M.L.; Chavez-Galan, L.; Segueni, N.; Bourigault, M.L.; Vesin, D.; Kruglov, A.A.; Drutskaya, M.S.; Bisig, R.; Ehlers, S.; Aly, S.; Walter, K.; et al. Control of Mycobacterial Infections in Mice Expressing Human Tumor Necrosis Factor (TNF) but Not Mouse TNF. *Infect Immun.* **2015**, *83*, 3612–3623. [CrossRef] [PubMed]
27. Kruglov, A.A.; Tumanov, A.V.; Grivennikov, S.I.; Shebzukhov, Y.V.; Kuchmiy, A.A.; Efimov, G.A.; Drutskaya, M.S.; Scheller, J.; Kuprash, D.V.; Nedospasov, S.A. *Modalities of Experimental TNF Blockade In Vivo: Mouse Models*; Spring: New York, NY, USA, 2011.
28. Espevik, T.; Nissen-Meyer, J. A highly sensitive cell line, WEHI 164 clone 13, for measuring cytotoxic factor/tumor necrosis factor from human monocytes. *J. Immunol. Methods* **1986**, *95*, 99–105. [CrossRef]
29. Yuzhakova, D.V.; Shirmanova, M.V.; Bocharov, A.A.; Astrakhantseva, I.V.; Vasilenko, E.A.; Gorshkova, E.N.; Drutskaya, M.S.; Zagaynova, E.V.; Nedospasov, S.A.; Kruglov, A.A. Microbiota Induces Expression of Tumor Necrosis Factor in Postnatal Mouse Skin. *Biochemistry* **2016**, *81*, 1303–1308. [CrossRef]
30. Pletneva, N.V.; Pletnev, V.Z.; Shemiakina, I.I.; Chudakov, D.M.; Artemyev, I.; Wlodawer, A.; Dauter, Z.; Pletnev, S. Crystallographic study of red fluorescent protein eqFP578 and its far-red variant Katushka reveals opposite pH-induced isomerization of chromophore. *Protein Sci.* **2011**, *20*, 1265–1274. [CrossRef]
31. Kijanka, M.; Dorresteijn, B.; Oliveira, S.; van Bergen en Henegouwen, P.M. Nanobody-based cancer therapy of solid tumors. *Nanomedicine* **2015**, *10*, 161–174. [CrossRef]

32. Albert, S.; Arndt, C.; Koristka, S.; Berndt, N.; Bergmann, R.; Feldmann, A.; Schmitz, M.; Pietzsch, J.; Steinbach, J.; Bachmann, M. From mono- to bivalent: improving theranostic properties of target modules for redirection of UniCAR T cells against EGFR-expressing tumor cells in vitro and in vivo. *Oncotarget* **2018**, *9*, 25597–25616. [CrossRef] [PubMed]
33. Van Brussel, A.S.A.; Adams, A.; Oliveira, S.; Dorresteijn, B.; El Khattabi, M.; Vermeulen, J.F.; van der Wall, E.; Mali, W.P.; Derksen, P.W.; van Diest, P.J.; van Bergen En Henegouwen, P.M. Hypoxia-Targeting Fluorescent Nanobodies for Optical Molecular Imaging of Pre-Invasive Breast Cancer. *Mol. Imaging Biol.* **2016**, *18*, 535–544. [CrossRef] [PubMed]
34. Muyldermans, S. Nanobodies: Natural Single-Domain Antibodies. *Ann. Rev. Biochem.* **2013**, *82*, 775–797. [CrossRef] [PubMed]
35. Deliolanis, N.C.; Kasmieh, R.; Wurdinger, T.; Tannous, B.A.; Shah, K.; Ntziachristos, V. Performance of the Red-shifted Fluorescent Proteins in deep-tissue molecular imaging applications. *J. Biomed. Opt.* **2008**, *13*, 044008. [CrossRef] [PubMed]
36. Gurskaya, N.G.; Fradkov, A.F.; Terskikh, A.; Matz, M.V.; Labas, Y.A.; Martynov, V.I.; Yanushevich, Y.G.; Lukyanov, K.A.; Lukyanov, S.A. GFP-like chromoproteins as a source of far-red fluorescent proteins. *FEBS Lett.* **2001**, *507*, 16–20. [CrossRef]
37. Wang, L.; Jackson, W.C.; Steinbach, P.A.; Tsien, R.Y. Evolution of new nonantibody proteins via iterative somatic hypermutation. *Proc. Natl. Acad. Sci. USA* **2004**, *101*, 16745–16749. [CrossRef] [PubMed]
38. Olleros, M.L.; Vesin, D.; Fotio, A.L.; Santiago-Raber, M.L.; Tauzin, S.; Szymkowski, D.W.; Garcia, I. Soluble TNF, but not membrane TNF, is critical in LPS-induced hepatitis. *J. Hepatol.* **2010**, *53*, 1059–1068. [CrossRef]
39. Grivennikov, S.I.; Tumanov, A.V.; Liepinsh, D.J.; Kruglov, A.A.; Marakusha, B.I.; Shakhov, A.N.; Murakami, T.; Drutskaya, L.N.; Förster, I.; Clausen, B.E.; et al. Distinct and nonredundant in vivo functions of TNF produced by T cells and macrophages/neutrophils: protective and deleterious effects. *Immunity* **2005**, *22*, 93–104. [CrossRef]
40. Drutskaya, M.S.; Efimov, G.A.; Zvartsev, R.V.; Chashchina, A.A.; Chudakov, D.M.; Tillib, S.V.; Kruglov, A.A.; Nedospasov, S.A. Experimental models of arthritis in which pathogenesis is dependent on tnf expression. *Biochemistry* **2014**, *79*, 1650–1658. [CrossRef]

*Review*

# Using Nanobodies to Study Protein Function in Developing Organisms

**Gustavo Aguilar, Shinya Matsuda, M. Alessandra Vigano and Markus Affolter ***

Biozentrum, University of Basel, Klingelbergstrasse 70, 4056 Basel, Switzerland;
gustavo.aguilar@unibas.ch (G.A.); shinya.matsuda@unibas.ch (S.M.); alessandra.vigano@unibas.ch (M.A.V.)
* Correspondence: markus.affolter@unibas.ch

Received: 21 January 2019; Accepted: 1 February 2019; Published: 12 February 2019

**Abstract:** Polyclonal and monoclonal antibodies have been invaluable tools to study proteins over the past decades. While indispensable for most biological studies including developmental biology, antibodies have been used mostly in fixed tissues or as binding reagents in the extracellular milieu. For functional studies and for clinical applications, antibodies have been functionalized by covalently fusing them to heterologous partners (i.e., chemicals, proteins or other moieties). Such functionalized antibodies have been less widely used in developmental biology studies. In the past few years, the discovery and application of small functional binding fragments derived from single-chain antibodies, so-called nanobodies, has resulted in novel approaches to study proteins during the development of multicellular animals in vivo. Expression of functionalized nanobody fusions from integrated transgenes allows manipulating proteins of interest in the extracellular and the intracellular milieu in a tissue- and time-dependent manner in an unprecedented manner. Here, we describe how nanobodies have been used in the field of developmental biology and look into the future to imagine how else nanobody-based reagents could be further developed to study the proteome in living organisms.

**Keywords:** nanobody; GFP; *C. elegans*; development; drosophila; zebrafish

---

## 1. Introduction

Antibodies have been invaluable tools in basic biological sciences for several decades. Polyclonal and monoclonal antibodies can be used for manifold studies, for example, to detect the amount of a given protein in Western blots, to isolate proteins or protein complexes from cell lysates, or to localize proteins in fixed cells and tissues, just to name a few applications. More rarely, antibodies have been used for functional studies in cultured cells or in developing organisms, in particular via injection either into cells, into the body cavity of developing organisms, or into the blood stream of developing or adult organisms. The large size and multi-domain nature of antibodies as well as their instability in the intracellular milieu has hindered their more widespread use to manipulate protein function in vivo.

A major breakthrough in antibody research was made when Hamers and colleagues reported in 1993 that camels produce, in addition to the well-characterized antibodies containing a heavy and a light chain, an additional and distinct species of antibodies containing only a heavy chain [1]. The binding affinity and selectivity of these single-chain antibodies turn out to be comparable to that of classical antibodies, and the high-affinity antigen-recognizing region can be isolated from the single heavy chain and expressed as a single polypeptide chain. Such small antibody fragments were called VHH (from variable domain of heavy chain antibodies, also referred to as nanobodies) and they have dramatically changed the way antibodies have been used in developmental biology (and in many other research fields, such as structural biology, super resolution microscopy, etc., see other articles in this issue).

## 2. From Cultured Cells to Developing Organisms

The function and the behavior of a cell is to a large extent determined by its proteome, i.e., by the proteins that are expressed in a given cell as a function of complex transcriptional regulation in the nucleus. While CRISPR/Cas technologies have paved the way to fast, cheap, and efficient genome editing [2], it remains more difficult to acutely manipulate proteins in a direct and desired manner in time and space. Recognition of DNA sequences via the injection or expression of the CRISPR/Cas module is highly efficient and selective due to guide RNAs providing target site selectivity. In contrast, recognition of proteins by classical antibodies in the intracellular milieu is often inefficient since antibodies consist of four chains that have to be properly assembled and folded and then kept in a stable configuration via disulfide bridges. The small size of single-chain antibody fragments and the ease to express them from integrated transgenes in cells has opened up the route to many novel applications in the complex context of multicellular organisms. In this article, we summarize studies in the field of developmental biology that have made use of nanobodies in multicellular animals and refer the reader to excellent recent reviews summarizing studies done in cultured cells [3–6]. Several developmental studies have used protein binders other than nanobodies, including single-chain antibodies (scFvs), Designed Ankyrin Repeat Proteins (DARPins) [7] monobodies [8], and others; we will not describe these studies here, but refer the reader to recent reviews on the topic [9,10].

## 3. Use of Functionalized Nanobodies in Multicellular Animals in the Context of Developmental Biology Studies

While antibodies and nanobodies have often been used to mask (and thereby inhibit) a functional domain of a protein of interest (POI) [11,12], the availability of the easily cloneable binding domains of nanobodies has stimulated researchers to generate transgenes that link this binding region to specific functional domains. This allows for the expression of protein fusions that are not only able to associate with the POI, but also to manipulate its function in a desired manner. First reports to do so using the nanobody scaffold involved the fusion of a fluorescent protein to a nanobody recognizing a POI, thereby visualizing the POI within cells (such nanobody fusions are also referred to as chromobodies, see below). Further experimental setups included the fusion of degradation-inducing domains, localization domains, and enzymatic domains to nanobodies. In the next sections, we highlight studies done in multicellular model organisms using nanobodies. Rather than describing in detail the biological findings resulting from these studies, we describe the generation and application of different nanobody-based tools and how they allow to manipulate and study protein function.

### 3.1. Protein Degradation

A straightforward approach to studying the function of a protein during development is to remove it from a given cell population and investigate the molecular, developmental or physiological consequences of its absence. In most cases, such studies have been done indirectly using forward and reverse genetic approaches, as well as tissue-specific genetic manipulations in more recent years (using site-specific recombination or RNA interference) [13]. Another way to remove a POI is to target it for proteasomal degradation. To achieve this, a nanobody can be fused to a subunit of the E3 ubiquitin ligase complex (there are several protein domains such as F-boxes or SOCS-boxes that characterize such E3 ligases), which ultimately results in the recruitment of the POI to the complex, polyubiquitination of the nanobody-binding POI and its subsequent degradation via the proteasome. This approach was first reported by Kuo and colleagues [14] in cell culture and by Caussinus and colleagues (called deGradFP), who used it to degrade proteins in living drosophila embryos and larvae [15,16]. Caussinus et al. [17] made use of a nanobody which recognizes the Green Fluorescent Protein (GFP) and fused it to the N-terminal F-box domain of the drosophila Slmb protein, an adaptor protein which is part of the E3 ubiquitin ligase complex SCF and is required to mediate substrate-specific ubiquitination. This functionalization of the GFP nanobody allows GFP-, Venus- and YFP-tagged proteins to be recognized by the SCF complex and to be targeted for degradation.

deGradFP has been used in a number of different studies in drosophila to address, for example, the role of actomyosin during tissue morphogenesis [18–21], to study the function of proteins in adult memory function and maintenance [22], to degrade POIs and study their contribution to Septate Junction establishment and maintenance [23], to analyze the role of the tissue-specificity of Hox gene function [24] and the role of certain proteins during microtubule network remodeling [25].

deGradFP has been used mostly in drosophila and was also shown to be able to induce protein degradation in mammalian cells [16]. However, as it turned out more recently, expression of the Slmb F-box to the GFP nanobody does not result in very efficient degradation of GFP fusion proteins when assayed in zebrafish embryos [26]. Based on this finding, two groups have further developed the method. In one case, the GFP nanobody was fused to an auxin-induced degron (AID), which was then shown to allow efficient and reversible degradation of GFP fusion proteins in zebrafish embryos upon addition of auxin [27]. In another study, the F-box of drosophila Slmb in the deGradFP fusion protein was replaced with the homologous sequence from zebrafish Slmb to generate a system called zGrad [26]. Using tissue-specific and inducible promoters in combination with functional GFP fusion proteins, it was shown that zGrad can induce the degradation of transmembrane, cytosolic, and nuclear proteins globally, locally, and temporally-controlled in different zebrafish tissues, and that such protein degradation can generate loss-of-function phenotypes. A system for protein degradation similar to deGradFP was also developed for *Caenorhabditis elegans*. In this particular case, the GFP nanobody was fused to a SOCS-box containing ubiquitin ligase adaptor in order to target GFP-tagged proteins for degradation [28]. To deplete a POI, GFP was either inserted into the endogenous locus of interest using CRISPR-Cas9 or via the rescue of a null mutant with a GFP fusion construct. This approach allowed for efficient tissue-specific protein ablation in *C. elegans* [29–32].

Several similar strategies have been reported and used in the last few years to induce degradation of specific POIs. Shin et al. [33] reported that the fusion of the GFP nanobody to a portion of SPOP (Speckle-type POZ-domain protein), a E3 ligase adaptor protein based on Cullin 3 acting in the nucleus, can induce exclusive nuclear degradation of GFP-tagged proteins in zebrafish embryos. This is an interesting addition to the other nanobody-based degradation methods, since it targets only the nuclear fraction of a POI.

As more and more lines expressing endogenously-tagged fluorescent proteins are becoming available in the different model systems due to the widespread use of Crispr/Cas9-based genome editing technologies, these degradation systems will become extremely useful new additions to the existing toolbox for the analyses of protein function in complex multicellular animals. The advantage of using protein degradation in contrast to classical genetic approaches to study the consequences of depleting a POI are several-fold. First, mRNA and proteins might be delivered by the mother into the egg, in which case zygotic loss of function genetic analyses are complicated by the prevailing maternal contribution. As shown by several studies, such maternal proteins can efficiently be degraded by deGradFP and zGrad [34,35]. In other cases, the use of tissue-specific and/or inducible drivers expressing the nanobody-F-box chimera can lead to tissue-specific and inducible protein degradation, respectively, and allows to study a subset of functions of a POI. Alternatively, proteins might be very stable and persist for extended periods of time, despite the removal of the gene or the mRNA under study. This is particularly important to keep in mind for studies in adult organisms, in which many proteins might be rather stable and do not dilute out by cell division. Interestingly, expression of nanobody-ubiquitin ligase adaptor fusions can be controlled by temperature-controlled promoters, thus allowing reversible expression and recovery of protein levels in adult flies, as pioneered by the Hugo Bellen's lab [22], and it is to be expected that many more studies of this type will be reported in the near future.

### 3.2. Protein Relocalization and Trapping

Many proteins function in distinct cellular compartments (nucleus, cytoplasm, etc.) or are linked to specific cellular structures (different membrane compartments, surface of different organelles).

To investigate the role of such distinct localization, nanobodies have proven to be extremely useful in altering the localization of POIs and investigate the consequences thereof.

In a system called GrabFP, Harmansa et al. [36] constructed three nanobody-based GFP traps that localize to defined regions along the apico-basal axis of epithelial cells in drosophila. By fusing the GFP nanobody to a transmembrane domain such that the nanobody moiety is either exposed to the extracellular or to the intracellular milieu, the different GrabFP constructs allow to trap or localize proteins to distinct apico-basal positions and ask what developmental and molecular consequences this might have. GrabFP has been used to study myosin activation via Yorkie localization at the junctional cortex [37], to better define the role of Dishevelled activity in maintaining planar polarity complexes in epithelial tissues [38], the role of Dpp/Bone morphogenetic protein 2/4 dispersal in the basolateral compartment of the wing imaginal disc in drosophila [36], and to study the importance of plasma membrane location of apoptotic caspases for non-apoptotic functions [39]. In addition to this, transmembrane scaffolds, a lipid binding domain (PH domain) has also been proposed as membrane-tether for nanobody functionalization [40].

Such relocalization or trapping experiments might be particularly interesting when it comes to study secreted molecules that depend on their dispersal in in vivo settings. Secreted signaling molecules such as morphogens or hormones play crucial roles in animal development [41]. Being able to interfere with the extracellular distribution of such molecules in a predictable manner might allow to better understand the way and the importance of their dispersal in complex tissues. A system called Morphotrap, consisting of a GFP-binding nanobody fused to a transmembrane domain exposed on the surface of expressing cells, has been used to trap the GFP-fused ligand Dpp/Bmp2 in drosophila imaginal discs [42], to trap secreted GFP-fused Wnt in *C. elegans* [43] and to trap GFP-fusion proteins of the Nodal family in zebrafish [44]. Such studies allow to investigate the requirement for dispersal of secreted signaling molecules in vivo. Furthermore, these methods provide a means to generate gradients of different shapes across tissues and to investigate the developmental consequences of such altered gradients.

This toolbox has been recently been further expanded by the addition of a low affinity Morphotrap version [45]. In this case, the authors exchanged the previously used high-affinity GFP nanobody [17] with a low-affinity GFP nanobody [46]. This new tool allowed to finely tune extracellular GFP diffusivity in living zebrafish embryos.

In addition to the relocalization or trapping of proteins using nanobodies, Janusche and colleagues [47] made use of the modular nature of protein domains and combined the MS2 system and nanobody expression to alter the subcellular localization of mRNA molecules in drosophila neuroblasts. In this particular case, the mRNA was tagged with GFP using the MS2 system [48], while the nanobody against GFP was fused to specific subcellular localization domains, resulting in the efficient mislocalization of the GFP-decorated mRNA molecules.

### 3.3. Protein Post-Translational Modification

Several enzymes (for example kinases) have many different substrates in a given cell or in different cells during development, and it remains rather challenging to unravel the complexity of such complex networks. Nanobodies can be used to direct enzymes to specific and unique substrates through direct protein–protein interaction and thereby lead to enhanced target specificity. In a proof-of-principle study, Roubinet et al. [49] fused the constitutively active minimal kinase domain of Rho kinase to a GFP-binding nanobody and an apical localization domain. Co-expression of this fusion construct together with GFP-myosin regulatory light chain in drosophila neuroblasts resulted in the ectopic accumulation of the phosphorylated form of the myosin light chain in the apical cytoplasmic compartment. In case this approach would work well with other enzymes, it would certainly contribute to a better understanding of complex regulatory circuits in developing organisms.

### 3.4. Protein Visualization

A very interesting application of nanobodies is the visualization of endogenous proteins in living organisms using fluorescently labelled nanobodies (also referred to as chromobodies; [17]). Chromobodies as ready-to-use tools in developmental biology might be particularly useful if they are directed against proteins of general interest or against proteins that mark different cellular compartments or cell states. Rothbauer and colleagues have generated chromobodies against the major cytoskeletal component Actin and the cell cycle marker PCNA and validated their use in zebrafish embryos [50]. For this purpose, nanobodies binding directly to these proteins were isolated from camelids [51,52]. Chromobodies should be built from binders directed against functionally inert epitopes such as to avoid unwanted effects on mobility and function of the POI, and thus have to be carefully selected and validated for each POI. Due to the usefulness for a wide community of biologists, such chromobody-expressing transgenic animals will most likely become reliable and important additions for future developmental studies.

When proteins are tagged endogenously with GFP, distinguishing protein dynamics of a single cell can be difficult in crowded tissues where neighbor cells also express the tagged protein. To achieve single neuron protein dynamics, Kamiyama et al. [53] designed a chromobody against GFP that, when expressed in particular neurons by tissue-specific promoters or expression systems, was able to mark the GFP-fused POI with a red fluorescent protein, and thereby differentially label these neurons from the neighboring cells.

Given the highly dynamic protein expression in the developing drosophila embryo, some proteins are degraded faster than the fluorescent tag matures (up to >30 min for GFP in vivo), impeding protein visualization via this method. To solve this problem, a different tool has been designed, the LlamaTags. Instead of fusing the POI to GFP, the POI was fused to a nanobody recognizing GFP, and GFP itself was used as soluble cytoplasmic substrate that follows the POI by binding to it; while GFP itself distributed in the cytoplasm, the expression of the nanobody-fused transcription factor resulted in the nuclear translocation of GFP [54]. Following the same concept, mCherry nanobodies and soluble mCherry allowed to perform multicolor visualization of protein dynamics. Using LlamaTag, the dynamics of transcription factors in the early drosophila embryo was followed in time and space, allowing unprecedented insight into the mechanisms of coordinated gene expression in these syncytial embryos.

### 3.5. Protein Scaffolding and Cell–Cell Contact Reporters

Nanobodies have also been used in developmental studies in a more synthetic approach, allowing to trigger certain functions when a scaffolding protein is present. This is achieved in the cell of interest via the use of two distinct binders recognizing a scaffold in a non-overlapping fashion, bringing two different components to the same scaffold complex. The Cepko laboratory has used two GFP-binding nanobodies to assemble different activities in only those cells that express GFP. In a method called "transcription device dependent of GFP" (T-DDOG; [55]), both a DNA-binding domain and a transcription regulatory domain (resulting in activation or repression) was fused to one and the other GFP binder, respectively; these two activities are only assembled into one protein complex in those cells that express GFP, thus allowing to target the activation or repression of desired genes to those particular cells. This method was used to regulate gene expression in both mice and zebrafish. The same approach has also been used to reconstitute a split Cre recombinase [56], allowing to make recombination dependent on the presence of GFP, and can be adopted for many more applications.

While T-DDOG exploits the presence of an intracellular GFP to trigger a response, others have designed receptors to elicit transgene activation upon recognition of extracellular antigen in other cells. To achieve that, the Notch receptor was engineered by replacing the extracellular region with a protein binder and the intracellular tail by a transcriptional activator. Upon recognition of the extracellular antigen, the intracellular domain is cleaved, and the C-terminal transcriptional activator is thereby released to translocate to the nucleus and activate transgene expression [57]. This concept has been used in developing drosophila embryos to trace cell–cell contacts between cells expressing membrane-bound GFP and cells containing synthetic Notch receptors exposing GFP nanobody on the cell surface as well as a transgene expressing a fluorescent label upon activation of this synthetic Notch receptor [57].

## 4. A look Into the Future

With the exception of chromobody applications, one striking aspect of all the nanobody-based developmental studies described above is that they have relied on the almost exclusive use of one or two nanobodies binding to GFP. These nanobodies have been isolated from camels upon immunization with GFP and were well characterized in vitro and in vivo [58,59]. Since GFP was discovered [60] and its sequence cloned [61], it has been extensively used as a fusion partner to follow protein expression and dynamics in vivo. The many transgenic GFP lines available in the different model systems make the use of validated GFP nanobody-based tools rather straightforward.

Probably the most favored applications of antibodies in biomedical research is their use as blocking reagents, binding with high affinity to an active site or a site involved in essential protein–protein interactions and hence interfering with protein function upon binding. This approach has not been used much thus far in developmental biology, because nanobodies against endogenous proteins of model organisms such as *C. elegans*, drosophila, or zebrafish have not been identified and reported, with the exception of the nanobodies used in the context of chromobodies. In the last few years, several labs have generated and studied nanobodies directed against cellular proteins [5], but most of the studies reported so far have targeted human proteins and have been studied in cultured cells. It will be very interesting to use these or similar nanobodies in multicellular animals to dissect cell biological processes in vivo. However, the ease with which GFP binders can be used across species is most likely not replicated by these nanobodies isolated against cellular proteins, since it is unlikely that many nanobodies isolated against human proteins will recognize the homologous protein from *C. elegans* or drosophila. Nevertheless, the availability of recently described screening devices [62] or platforms will speed up the isolation and characterization of nanobodies against endogenous proteins in different model organisms used in the field of developmental biology. Therefore, it can be expected that nanobodies binding endogenous POIs will be used more often in the future, especially in species, in which tools for efficient genetic manipulation are less common. They might be used in the context of similar functionalization as already described (degradation, relocalization, chromobodies, etc.), to mask the function of a protein or certain subdomains thereof, to detect or interfere with post-translational modifications, or even as reagents stabilizing specific protein conformations, as already proposed for some nanobodies in cell culture [62]. In each case, however, the specificity of the nanobody has to be carefully evaluated in the context of the developmental system used.

One of the obvious limitations of the use of the GFP nanobodies is the failure to endogenously tag certain proteins with GFP due to functional interference. Recently, nanobodies that are able to bind to short linear epitopes have been isolated [63,64]. Upon the insertion of such a tag at the endogenous locus of a protein to be studied, it should be possible to manipulate the latter with the corresponding functionalized binder, thereby bypassing the isolation and validation procedures involved in obtaining POI-specific nanobodies. Other binders, such as single-chain antibodies derived from IgGs, have been shown to bind short epitopes; however, the multidomain structure of these binders is normally far from ideal in the cell cytoplasm (with some exceptions, [65,66]). Since binders against small epitopes can more easily be validated in complex multicellular animals (by showing that they do not influence developmental processes in the absence of the epitope tag), and since endogenous gene tagging has become very efficient using Crisp/Cas, such binders will probably be used extensively to manipulate protein function intracellularly in combination with the available functionalization domains.

One of the most exciting aspects of the use of nanobodies in developmental biology is that they can be fused to functionalization domains to generate novel reagents which specifically and directly target a POI and manipulate it in a given, desired manner (see Figure 1). The list of such potential functionalization is long, but it is likely that many possible manipulations have not even been thought of in the early days of these novel possibilities to investigate the proteome. The emergence of more and more complex functionalization strategies requires a tight control of their performances to avoid undesired effects. The development of nanobodies that promote their own degradation when the POI is not present [67] is among the incipient strategies to achieve this tight control. It will be interesting to follow how the use of nanobody-based tools will evolve in the future, and it is hoped that this approach, combined with many other approaches (such as optogenetics), will allow to better understand the role of the proteome in development and disease.

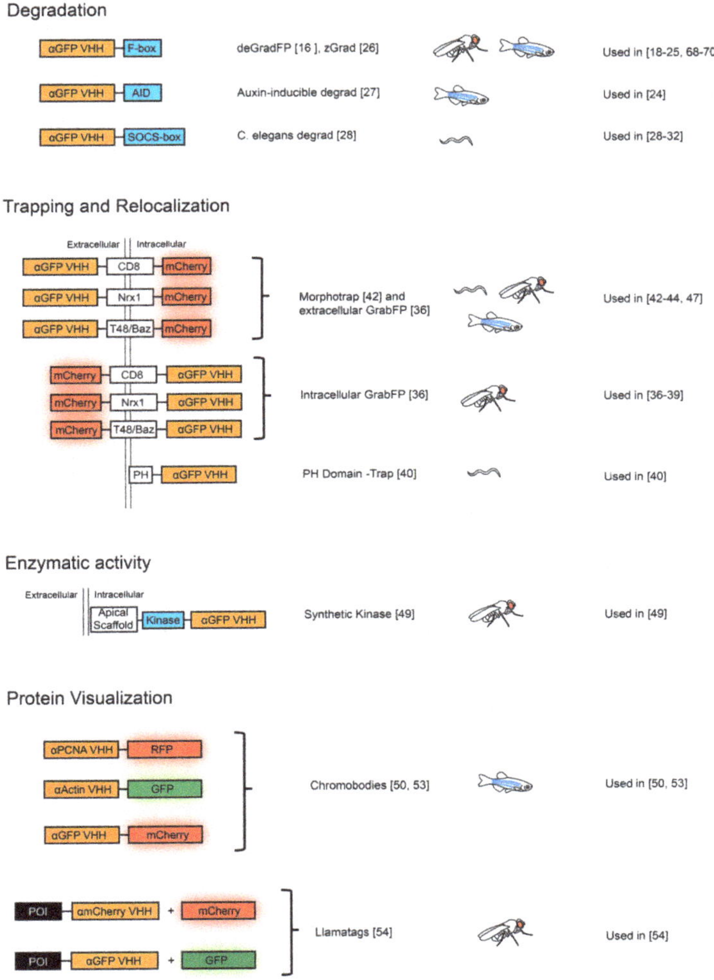


**Figure 1.** Different nanobody-based methods applied to developmental biology [16,18–32,36–40,42–44, 47,49,50,53,54,68–70].

**Author Contributions:** Conceptualization, M.A, G.A, S.M, A.V.; writing-original draft preparation, M.A., G.A.; Figure design, G.A.

**Funding:** This work was supported by the Kantons Basel-Stadt and Basel-Land and by grants from the SNSF and SystemsX (MorphogenetiX) to M.A. The "Fellowships for Excellence" International PhD Program in Molecular Life Sciences of the Biozentrum, University of Basel, supported G.A.

**Conflicts of Interest:** The authors declare no conflict of interest. The funders had no role in the design of the study; in the collection, analyses, or interpretation of data; in the writing of the manuscript, or in the decision to publish the results.

## References

1. Hamers-Casterman, C.; Atarhouch, T.; Muyldermans, S.; Robinson, G.; Hamers, C.; Songa, E.B.; Bendahman, N.; Hamers, R. Naturally occurring antibodies devoid of light chains. *Nature* **1993**, *363*, 446–448. [CrossRef] [PubMed]
2. Wang, H.; La Russa, M.; Qi, L.S. CRISPR/Cas9 in Genome Editing and Beyond. *Annu. Rev. Biochem.* **2016**, *85*, 227–264. [CrossRef] [PubMed]
3. Helma, J.; Cardoso, M.C.; Muyldermans, S.; Leonhardt, H. Nanobodies and recombinant binders in cell biology. *J. Cell. Biol.* **2015**, *209*, 633–644. [CrossRef] [PubMed]
4. Schumacher, D.; Helma, J.; Schneider, A.F.L.; Leonhardt, H.; Hackenberger, C.P.R. Nanobodies: Chemical Functionalization Strategies and Intracellular Applications. *Angew. Chem. Int. Ed. Engl.* **2018**, *57*, 2314–2333. [CrossRef] [PubMed]
5. Beghein, E.; Gettemans, J. Nanobody Technology: A Versatile Toolkit for Microscopic Imaging, Protein-Protein Interaction Analysis, and Protein Function Exploration. *Front Immunol.* **2017**, *8*, 771. [CrossRef] [PubMed]
6. Kaiser, P.D.; Maier, J.; Traenkle, B.; Emele, F.; Rothbauer, U. Recent progress in generating intracellular functional antibody fragments to target and trace cellular components in living cells. *Biochim. Biophys. Acta.* **2014**, *1844*, 1933–1942. [CrossRef] [PubMed]
7. Binz, H.K.; Stumpp, M.T.; Forrer, P.; Amstutz, P.; Plückthun, A. Designing Repeat Proteins: Well-expressed, Soluble and Stable Proteins from Combinatorial Libraries of Consensus Ankyrin Repeat Proteins. *J. Mol. Biol.* **2003**, *332*, 489–503. [CrossRef]
8. Koide, A.; Bailey, C.W.; Huang, X.; Koide, S. The fibronectin type III domain as a scaffold for novel binding proteins11Edited by J. Wells. *J. Mol. Biol.* **1998**, *284*, 1141–1151. [CrossRef] [PubMed]
9. Bieli, D.; Alborelli, I.; Harmansa, S.; Matsuda, S.; Caussinus, E.; Affolter, M. Development and Application of Functionalized Protein Binders in Multicellular Organisms. *Int. Rev. Cell Mol. Biol.* **2016**, *325*, 181–213. [CrossRef] [PubMed]
10. Harmansa, S.; Affolter, M. Protein binders and their applications in developmental biology. *Development* **2018**, *145*. [CrossRef] [PubMed]
11. Boldicke, T. Single domain antibodies for the knockdown of cytosolic and nuclear proteins. *Protein Sci.* **2017**, *26*, 925–945. [CrossRef] [PubMed]
12. Ingram, J.R.; Schmidt, F.I.; Ploegh, H.L. Exploiting Nanobodies' Singular Traits. *Annu. Rev. Immunol.* **2018**, *36*, 695–715. [CrossRef] [PubMed]
13. Housden, B.E.; Muhar, M.; Gemberling, M.; Gersbach, C.A.; Stainier, D.Y.; Seydoux, G.; Mohr, S.E.; Zuber, J.; Perrimon, N. Loss-of-function genetic tools for animal models: Cross-species and cross-platform differences. *Nat. Rev. Genet.* **2017**, *18*, 24–40. [CrossRef] [PubMed]
14. Kuo, C.L.; Oyler, G.A.; Shoemaker, C.B. Accelerated neuronal cell recovery from Botulinum neurotoxin intoxication by targeted ubiquitination. *PLoS ONE* **2011**, *6*, e20352. [CrossRef] [PubMed]
15. Caussinus, E.; Affolter, M. deGradFP: A System to Knockdown GFP-Tagged Proteins. *Methods Mol. Biol.* **2016**, *1478*, 177–187. [CrossRef] [PubMed]
16. Caussinus, E.; Kanca, O.; Affolter, M. Fluorescent fusion protein knockout mediated by anti-GFP nanobody. *Nat. Struct. Mol. Biol.* **2011**, *19*, 117–121. [CrossRef] [PubMed]
17. Rothbauer, U.; Zolghadr, K.; Tillib, S.; Nowak, D.; Schermelleh, L.; Gahl, A.; Backmann, N.; Conrath, K.; Muyldermans, S.; Cardoso, M.C.; et al. Targeting and tracing antigens in live cells with fluorescent nanobodies. *Nat. Methods* **2006**, *3*, 887–889. [CrossRef] [PubMed]
18. Ochoa-Espinosa, A.; Harmansa, S.; Caussinus, E.; Affolter, M. Myosin II is not required for Drosophila tracheal branch elongation and cell intercalation. *Development* **2017**, *144*, 2961–2968. [CrossRef] [PubMed]

19. Pasakarnis, L.; Frei, E.; Caussinus, E.; Affolter, M.; Brunner, D. Amnioserosa cell constriction but not epidermal actin cable tension autonomously drives dorsal closure. *Nat. Cell Biol.* **2016**, *18*, 1161–1172. [CrossRef] [PubMed]
20. Córdoba, S.; Estella, C. The transcription factor Dysfusion promotes fold and joint morphogenesis through regulation of Rho1. *PLoS Genetics* **2018**, *14*, e1007584. [CrossRef] [PubMed]
21. Urbano, J.M.; Naylor, H.W.; Scarpa, E.; Muresan, L.; Sanson, B. Suppression of epithelial folding at actomyosin-enriched compartment boundaries downstream of Wingless signalling in Drosophila. *Development* **2018**, *145*. [CrossRef] [PubMed]
22. Lee, P.T.; Lin, G.; Lin, W.W.; Diao, F.; White, B.H.; Bellen, H.J. A kinase-dependent feedforward loop affects CREBB stability and long term memory formation. *Elife* **2018**, *7*. [CrossRef] [PubMed]
23. Batz, T.; Forster, D.; Luschnig, S. The transmembrane protein Macroglobulin complement-related is essential for septate junction formation and epithelial barrier function in Drosophila. *Development* **2014**, *141*, 899–908. [CrossRef] [PubMed]
24. Domsch, K.; Carnesecchi, J.; Disela, V.; Friedrich, J.; Trost, N.; Ermakova, O.; Polychronidou, M.; Lohmann, I. The Hox Transcription Factor Ubx stabilizes Lineage Commitment by Suppressing Cellular Plasticity. *bioRxiv* **2018**. [CrossRef]
25. Takeda, M.; Sami, M.M.; Wang, Y.C. A homeostatic apical microtubule network shortens cells for epithelial folding via a basal polarity shift. *Nat. Cell Biol.* **2018**, *20*, 36–45. [CrossRef] [PubMed]
26. Yamaguchi, N.; Colak-Champollion, T.; Knaut, H. zGrad: A nanobody-based degron system to inactivate proteins in zebrafish. *bioRxiv* **2019**, 518944. [CrossRef] [PubMed]
27. Daniel, K.; Icha, J.; Horenburg, C.; Muller, D.; Norden, C.; Mansfeld, J. Conditional control of fluorescent protein degradation by an auxin-dependent nanobody. *Nat. Commun.* **2018**, *9*, 3297. [CrossRef] [PubMed]
28. Wang, S.; Tang, N.H.; Lara-Gonzalez, P.; Zhao, Z.; Cheerambathur, D.K.; Prevo, B.; Chisholm, A.D.; Desai, A.; Oegema, K. A toolkit for GFP-mediated tissue-specific protein degradation in C. elegans. *Development* **2017**, *144*, 2694–2701. [CrossRef] [PubMed]
29. Sallee, M.D.; Zonka, J.C.; Skokan, T.D.; Raftrey, B.C.; Feldman, J.L. Tissue-specific degradation of essential centrosome components reveals distinct microtubule populations at microtubule organizing centers. *PLoS Biology* **2018**, *16*, e2005189. [CrossRef] [PubMed]
30. Kim, K.W.; Tang, N.H.; Andrusiak, M.G.; Wu, Z.; Chisholm, A.D.; Jin, Y. A Neuronal piRNA Pathway Inhibits Axon Regeneration in C. elegans. *Neuron* **2018**, *97*, 511–519.e6. [CrossRef] [PubMed]
31. Kurup, N.; Li, Y.; Goncharov, A.; Jin, Y. Intermediate filament accumulation can stabilize microtubules in *Caenorhabditis elegans* motor neurons. *Proc. Natl. Acad. Sci. USA* **2018**, *115*, 3114–3119. [CrossRef] [PubMed]
32. Wang, S.; Wu, D.; Quintin, S.; Green, R.A.; Cheerambathur, D.K.; Ochoa, S.D.; Desai, A.; Oegema, K. NOCA-1 functions with gamma-tubulin and in parallel to Patronin to assemble non-centrosomal microtubule arrays in *C. elegans*. *Elife* **2015**, *4*, e08649. [CrossRef] [PubMed]
33. Shin, Y.J.; Park, S.K.; Jung, Y.J.; Kim, Y.N.; Kim, K.S.; Park, O.K.; Kwon, S.H.; Jeon, S.H.; Trinh le, A.; Fraser, S.E.; et al. Nanobody-targeted E3-ubiquitin ligase complex degrades nuclear proteins. *Sci. Rep.* **2015**, *5*, 14269. [CrossRef] [PubMed]
34. Perez-Mockus, G.; Mazouni, K.; Roca, V.; Corradi, G.; Conte, V.; Schweisguth, F. Spatial regulation of contractility by Neuralized and Bearded during furrow invagination in Drosophila. *Nat. Commun.* **2017**, *8*, 1594. [CrossRef] [PubMed]
35. Urban, E.; Nagarkar-Jaiswal, S.; Lehner, C.F.; Heidmann, S.K. The Cohesin Subunit Rad21 Is Required for Synaptonemal Complex Maintenance, but Not Sister Chromatid Cohesion, during Drosophila Female Meiosis. *PLoS Genetics* **2014**, *10*, e1004540. [CrossRef] [PubMed]
36. Harmansa, S.; Alborelli, I.; Bieli, D.; Caussinus, E.; Affolter, M. A nanobody-based toolset to investigate the role of protein localization and dispersal in Drosophila. *Elife* **2017**, *6*. [CrossRef] [PubMed]
37. Xu, J.; Vanderzalm, P.J.; Ludwig, M.; Su, T.; Tokamov, S.A.; Fehon, R.G. Yorkie Functions at the Cell Cortex to Promote Myosin Activation in a Non-transcriptional Manner. *Dev. Cell* **2018**, *46*, 271–284.e5. [CrossRef] [PubMed]
38. Ressurreicao, M.; Warrington, S.; Strutt, D. Rapid Disruption of Dishevelled Activity Uncovers an Intercellular Role in Maintenance of Prickle in Core Planar Polarity Protein Complexes. *Cell Rep.* **2018**, *25*, 1415–1424.e6. [CrossRef] [PubMed]

39. Amcheslavsky, A.; Wang, S.; Fogarty, C.E.; Lindblad, J.L.; Fan, Y.; Bergmann, A. Plasma Membrane Localization of Apoptotic Caspases for Non-apoptotic Functions. *Dev. Cell* **2018**, *45*, 450–464.e3. [CrossRef] [PubMed]
40. Rodriguez, J.; Peglion, F.; Martin, J.; Hubatsch, L.; Reich, J.; Hirani, N.; Gubieda, A.G.; Roffey, J.; Fernandes, A.R.; St Johnston, D.; et al. aPKC Cycles between Functionally Distinct PAR Protein Assemblies to Drive Cell Polarity. *Dev. Cell* **2017**, *42*, 400–415.e9. [CrossRef] [PubMed]
41. Rogers, K.W.; Schier, A.F. Morphogen Gradients: From Generation to Interpretation. *Annu. Rev. Cell Dev. Biol.* **2011**, *27*, 377–407. [CrossRef] [PubMed]
42. Harmansa, S.; Hamaratoglu, F.; Affolter, M.; Caussinus, E. Dpp spreading is required for medial but not for lateral wing disc growth. *Nature* **2015**, *527*, 317–322. [CrossRef] [PubMed]
43. Pani, A.M.; Goldstein, B. Direct visualization of a native Wnt in vivo reveals that a long-range Wnt gradient forms by extracellular dispersal. *Elife* **2018**, *7*. [CrossRef] [PubMed]
44. Almuedo-Castillo, M.; Blassle, A.; Morsdorf, D.; Marcon, L.; Soh, G.H.; Rogers, K.W.; Schier, A.F.; Muller, P. Scale-invariant patterning by size-dependent inhibition of Nodal signalling. *Nat. Cell Biol.* **2018**, *20*, 1032–1042. [CrossRef] [PubMed]
45. Morsdorf, D.; Muller, P. Tuning protein diffusivity with membrane tethers. *Biochemistry* **2018**. [CrossRef] [PubMed]
46. Fridy, P.C.; Li, Y.; Keegan, S.; Thompson, M.K.; Nudelman, I.; Scheid, J.F.; Oeffinger, M.; Nussenzweig, M.C.; Fenyo, D.; Chait, B.T.; et al. A robust pipeline for rapid production of versatile nanobody repertoires. *Nat. Methods* **2014**, *11*, 1253–1260. [CrossRef] [PubMed]
47. Ramat, A.; Hannaford, M.; Januschke, J. Maintenance of Miranda Localization in Drosophila Neuroblasts Involves Interaction with the Cognate mRNA. *Curr. Biol.* **2017**, *27*, 2101–2111.e5. [CrossRef] [PubMed]
48. Bertrand, E.; Chartrand, P.; Schaefer, M.; Shenoy, S.M.; Singer, R.H.; Long, R.M. Localization of ASH1 mRNA particles in living yeast. *Mol. Cell* **1998**, *2*, 437–445. [CrossRef]
49. Roubinet, C.; Tsankova, A.; Pham, T.T.; Monnard, A.; Caussinus, E.; Affolter, M.; Cabernard, C. Spatio-temporally separated cortical flows and spindle geometry establish physical asymmetry in fly neural stem cells. *Nat. Commun.* **2017**, *8*, 1383. [CrossRef] [PubMed]
50. Panza, P.; Maier, J.; Schmees, C.; Rothbauer, U.; Sollner, C. Live imaging of endogenous protein dynamics in zebrafish using chromobodies. *Development* **2015**, *142*, 1879–1884. [CrossRef] [PubMed]
51. Burgess, A.; Lorca, T.; Castro, A. Quantitative Live Imaging of Endogenous DNA Replication in Mammalian Cells. *PLoS ONE* **2012**, *7*, e45726. [CrossRef] [PubMed]
52. Rocchetti, A.; Hawes, C.; Kriechbaumer, V. Fluorescent labelling of the actin cytoskeleton in plants using a cameloid antibody. *Plant Methods* **2014**, *10*, 12. [CrossRef] [PubMed]
53. Kamiyama, D.; McGorty, R.; Kamiyama, R.; Kim, M.D.; Chiba, A.; Huang, B. Specification of Dendritogenesis Site in Drosophila aCC Motoneuron by Membrane Enrichment of Pak1 through Dscam1. *Dev. Cell* **2015**, *35*, 93–106. [CrossRef] [PubMed]
54. Bothma, J.P.; Norstad, M.R.; Alamos, S.; Garcia, H.G. LlamaTags: A Versatile Tool to Image Transcription Factor Dynamics in Live Embryos. *Cell* **2018**, *173*, 1810–1822.e16. [CrossRef] [PubMed]
55. Tang, J.C.; Szikra, T.; Kozorovitskiy, Y.; Teixiera, M.; Sabatini, B.L.; Roska, B.; Cepko, C.L. A nanobody-based system using fluorescent proteins as scaffolds for cell-specific gene manipulation. *Cell* **2013**, *154*, 928–939. [CrossRef] [PubMed]
56. Tang, J.C.; Rudolph, S.; Dhande, O.S.; Abraira, V.E.; Choi, S.; Lapan, S.W.; Drew, I.R.; Drokhlyansky, E.; Huberman, A.D.; Regehr, W.G.; et al. Cell type-specific manipulation with GFP-dependent Cre recombinase. *Nat. Neurosci.* **2015**, *18*, 1334–1341. [CrossRef] [PubMed]
57. Morsut, L.; Roybal, K.T.; Xiong, X.; Gordley, R.M.; Coyle, S.M.; Thomson, M.; Lim, W.A. Engineering Customized Cell Sensing and Response Behaviors Using Synthetic Notch Receptors. *Cell* **2016**, *164*, 780–791. [CrossRef] [PubMed]
58. Rothbauer, U.; Zolghadr, K.; Muyldermans, S.; Schepers, A.; Cardoso, M.C.; Leonhardt, H. A versatile nanotrap for biochemical and functional studies with fluorescent fusion proteins. *Mol. Cell Proteomics* **2008**, *7*, 282–289. [CrossRef] [PubMed]
59. Kirchhofer, A.; Helma, J.; Schmidthals, K.; Frauer, C.; Cui, S.; Karcher, A.; Pellis, M.; Muyldermans, S.; Casas-Delucchi, C.S.; Cardoso, M.C.; et al. Modulation of protein properties in living cells using nanobodies. *Nat. Struct. Mol. Biol.* **2010**, *17*, 133–138. [CrossRef] [PubMed]

60. Shimomura, O.; Johnson, F.H.; Saiga, Y. Extraction, Purification and Properties of Aequorin, a Bioluminescent Protein from the Luminous Hydromedusan, Aequorea. *J. Cellular Comparat. Physiol.* **1962**, *59*, 223–239. [CrossRef]
61. Prasher, D.C.; Eckenrode, V.K.; Ward, W.W.; Prendergast, F.G.; Cormier, M.J. Primary structure of the Aequorea victoria green-fluorescent protein. *Gene* **1992**, *111*, 229–233. [CrossRef]
62. Moutel, S.; Bery, N.; Bernard, V.; Keller, L.; Lemesre, E.; de Marco, A.; Ligat, L.; Rain, J.C.; Favre, G.; Olichon, A.; et al. NaLi-H1: A universal synthetic library of humanized nanobodies providing highly functional antibodies and intrabodies. *Elife* **2016**, *5*. [CrossRef] [PubMed]
63. Boersma, S.; Khuperkar, D.; Verhagen, B.M.P.; Sonneveld, S.; Grimm, J.B.; Lavis, L.D.; Tanenbaum, M.E. Multi-color single molecule imaging uncovers extensive heterogeneity in mRNA decoding. *bioRxiv* **2018**. [CrossRef]
64. Braun, M.B.; Traenkle, B.; Koch, P.A.; Emele, F.; Weiss, F.; Poetz, O.; Stehle, T.; Rothbauer, U. Peptides in headlock—A novel high-affinity and versatile peptide-binding nanobody for proteomics and microscopy. *Sci. Rep.* **2016**, *6*, 19211. [CrossRef] [PubMed]
65. Tanenbaum, M.E.; Gilbert, L.A.; Qi, L.S.; Weissman, J.S.; Vale, R.D. A Protein-Tagging System for Signal Amplification in Gene Expression and Fluorescence Imaging. *Cell* **2014**, *159*, 635–646. [CrossRef] [PubMed]
66. Zhao, N.; Kamijo, K.; Fox, P.; Oda, H.; Morisaki, T.; Sato, Y.; Kimura, H.; Stasevich, T.J. A genetically encoded probe for imaging HA-tagged protein translation, localization, and dynamics in living cells and animals. *bioRxiv* **2018**, 474668. [CrossRef]
67. Tang, J.C.; Drokhlyansky, E.; Etemad, B.; Rudolph, S.; Guo, B.; Wang, S.; Ellis, E.G.; Li, J.Z.; Cepko, C.L. Detection and manipulation of live antigen-expressing cells using conditionally stable nanobodies. *Elife* **2016**, *5*. [CrossRef] [PubMed]
68. Blattner, A.C.; Chaurasia, S.; McKee, B.D.; Lehner, C.F. Separase Is Required for Homolog and Sister Disjunction during Drosophila melanogaster Male Meiosis, but Not for Biorientation of Sister Centromeres. *PLoS Genet.* **2016**, *12*, e1005996. [CrossRef] [PubMed]
69. Raychaudhuri, N.; Dubruille, R.; Orsi, G.A.; Bagheri, H.C.; Loppin, B.; Lehner, C.F. Transgenerational Propagation and Quantitative Maintenance of Paternal Centromeres Depends on Cid/Cenp-A Presence in Drosophila Sperm. *PLoS Biology* **2012**, *10*, e1001434. [CrossRef] [PubMed]
70. Lee, K.-H.; Zhang, P.; Kim, H.J.; Mitrea, D.M.; Sarkar, M.; Freibaum, B.D.; Cika, J.; Coughlin, M.; Messing, J.; Molliex, A.; et al. C9orf72 Dipeptide Repeats Impair the Assembly, Dynamics, and Function of Membrane-Less Organelles. *Cell* **2016**, *167*, 774–788. [CrossRef] [PubMed]

*Review*

# Single-Domain Antibodies and Their Formatting to Combat Viral Infections

**Dorien De Vlieger [1,2], Marlies Ballegeer [1,2], Iebe Rossey [1,2], Bert Schepens [1,3] and Xavier Saelens [1,2,*]**

1 VIB Center for Medical Biotechnology, 9052 Ghent, Belgium; dorien.devlieger@vib-ugent.be (D.D.V.); marlies.ballegeer@vib-ugent.be (M.B.); Iebe.rossey@vib-ugent.be (I.R.); bert.schepens@vib-ugent.be (B.S.)
2 Department of Biochemistry and Microbiology, Ghent University, 9052 Ghent, Belgium
3 Department of Biomedical Molecular Biology, Ghent University, 9052 Ghent, Belgium
* Correspondence: xavier.saelens@vib-ugent.be

Received: 30 November 2018; Accepted: 14 December 2018; Published: 20 December 2018

**Abstract:** Since their discovery in the 1990s, single-domain antibodies (VHHs), also known as Nanobodies®, have changed the landscape of affinity reagents. The outstanding solubility, stability, and specificity of VHHs, as well as their small size, ease of production and formatting flexibility favor VHHs over conventional antibody formats for many applications. The exceptional ease by which it is possible to fuse VHHs with different molecular modules has been particularly explored in the context of viral infections. In this review, we focus on VHH formats that have been developed to combat viruses including influenza viruses, human immunodeficiency virus-1 (HIV-1), and human respiratory syncytial virus (RSV). Such formats may significantly increase the affinity, half-life, breadth of protection of an antiviral VHH and reduce the risk of viral escape. In addition, VHHs can be equipped with effector functions, for example to guide components of the immune system with high precision to sites of viral infection.

**Keywords:** virus; nanobody; formatting; Fc-domain; half-life

---

## 1. Introduction

The discovery of heavy chain-only antibodies in the serum of camels, first reported by Hamers et al. in 1993, opened the way for a new tool box for diverse therapeutic applications [1]. Sera from camelids such as camels, dromedaries and llamas contain conventional antibodies (IgG1 isotype) and, surprisingly, also antibodies that lack the light chain component as well as the first constant domain of the heavy chain (CH1) (IgG2 and IgG3 isotype). The epitope-binding unit of these so-called heavy chain-only antibodies thus consists of a single variable domain, called a single-domain antibody (VHH) or Nanobody®. Despite their small size (~15 kDa) these single domain binding units can have exquisite affinities and antigen-binding specificities [2–5]. Similar to the variable domain of conventional antibodies, VHHs consist of four constant framework regions (FR1–4) separated by three hypervariable complementarity determining loops (CDR1, -2 and -3). A distinct feature of VHH FR2 is the presence of a hydrophilic surface exposed patch that likely evolved to compensate for the loss of light chain binding. In addition, the CDR3 loop of a VHH often folds back over the site that normally interacts with the variable light chain. Moreover, the CDR3 of VHHs is more variable in length and typically somewhat longer than the CDR3 of conventional antibodies [5]. To compensate for the higher flexibility and otherwise entropically unfavorable binding to the target antigen, the CDR3 loop often forms a disulfide bond with the CDR1, CDR2 or FR2 [6–8].

The small size, single domain build-up and the presence of hydrophilic amino acids in FR2 go together with a typically high solubility and physical stability of VHHs. As a result, these proteins can withstand relatively harsh formulations and environments, have excellent tissue penetration capacities, can be formatted in multiple ways and can be efficiently produced at low cost in microorganisms [9]. Not surprisingly, given these appealing properties, VHHs directed against a number of viruses including influenza viruses, human immunodeficiency virus-1 (HIV-1), and human respiratory syncytial virus (RSV) have been isolated from immune, naïve or synthetic VHH libraries. Immune libraries can be generated based on peripheral blood lymphocytes isolated from a camelid that has been immunized with complete virus or a viral antigen of interest in a prime-boost strategy [10]. However, VHHs with reasonable target specificity can also be isolated from naïve libraries that were generated from a camelid that was not immunized with the target viral antigen of interest. Synthetic VHH libraries do not require any experimental animal handling. Such libraries are built based on a well characterized VHH of which the conserved FRs are retained and amino acids in the CDRs are altered by saturating site specific mutagenesis [11–14]. Immune libraries are often the first choice to isolate high affinity VHHs because natural somatic antibody maturation can create an enormous diversity. Antigen-specific VHHs are then usually isolated by phage, yeast or ribosome display [14–17].

Numerous virus-neutralizing VHHs have been described and different steps in the viral life cycle can be perturbed (Figure 1). For example, VHHs that prevent virus entry by blocking the receptor binding have been described for influenza (targeting the hemagglutinin (HA) protein), HIV (targeting gp120) and Middle East respiratory syndrome coronaviruses (MERS CoV) (targeting the spike proteins) [18–21]. Furthermore, a VHH that arrests the RSV fusion protein (F) in its prefusion state could prevent virus entry by inhibiting membrane fusion between virus and host cells [22]. When expressed within the target cell, the VHHs are often referred to as intrabodies, where they can affect, for example, viral replication and nuclear transport of viral ribonucleoproteins (vRNPs), as was shown for an anti-influenza nucleoprotein VHH, while a VHH against the HIV Rev protein could inhibit nuclear export of viral mRNA [23–25].

Apart from the direct antiviral activity that VHHs may have, VHHs can be formatted to serve as a targeting moiety to bring an effector function such as a toxin, an antiviral drug or an antibody-derived Fc domain to the site of a viral infection. VHH formatting can also be used to increase their half-life, the affinity for their target or to deliver the VHH to a certain compartment. At last, VHH formatting can be explored to develop new diagnostic tools for infectious diseases. Here we focus on the multitude of formats that have been applied to VHHs that target human viruses (Table 1).

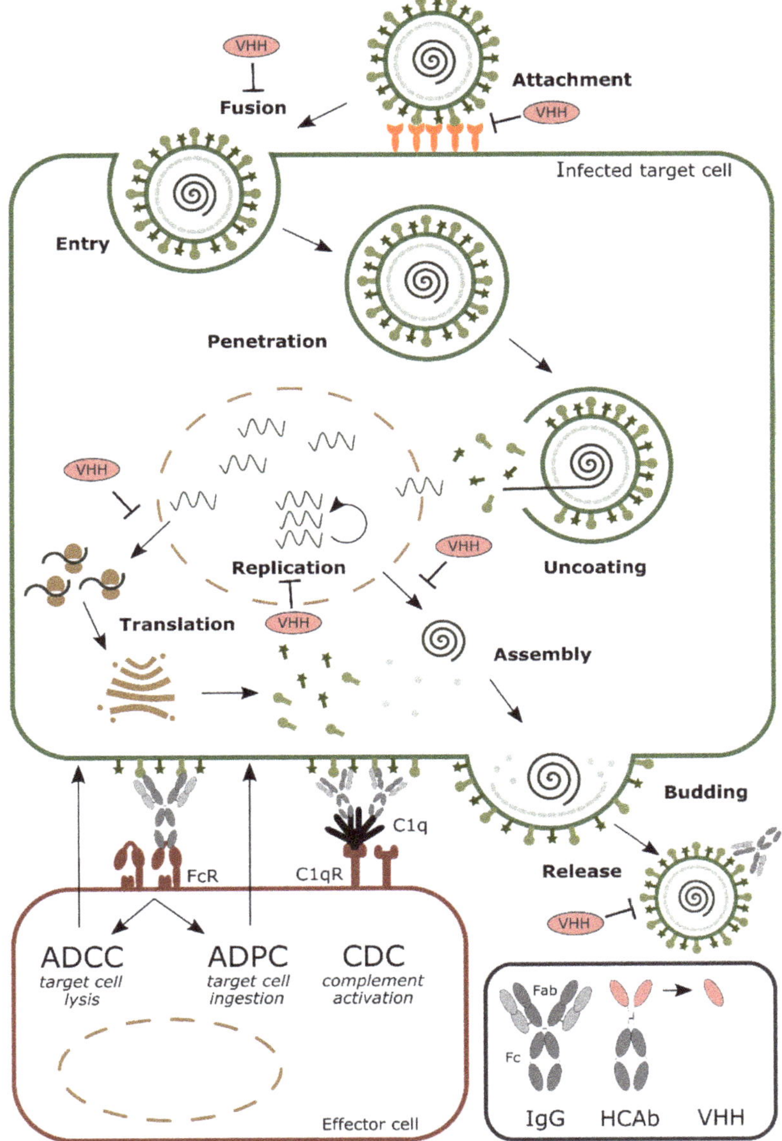


**Figure 1.** Schematic overview of the steps in a standard viral replication cycle that can be targeted by monovalent single-domain antibodies (VHHs) and antibody-mediated effector functions. General steps in viral replication include: (i) attachment and entry; (ii) penetration and uncoating; (iii) replication and translation of genomic viral RNA into proteins; (iv) assembly of virions; (v) budding and release. Antibodies (immunoglobulins or IgGs and heavy chain-only antibodies or HcAbs) employ different mechanisms to remove the infected target cells: (i) interact with Fc receptors (FcR) on effector cells to induce antibody-dependent cell-mediated cytotoxicity (ADCC) or antibody-dependent polymorphonuclear neutrophils (PMN)-mediated cytolysis (ADPC) and (ii) cell lysis through complement dependent cytotoxicity (CDC) by binding to the C1q receptor (C1qR).

## 2. Formatting of VHHs to Increase the Half-Life in Circulation

A single VHH molecule has a molecular weight of approximately 15 kDa, which is ten times smaller than a conventional IgG antibody (size around 150 kDa). Injected monomeric, dimeric and even trimeric VHHs are thus rapidly cleared from circulation by free glomerular filtration in the kidney (molecular weight cut-off 66 kDa) [26]. Such rapid removal, within a couple of hours after injection compared to 2–3 weeks for a conventional antibody, most often limits the therapeutic efficacy of a VHH [27,28]. As a result, VHHs have to be administered by infusion or repeat injection, or they are even restricted to loco-regional treatment [29]. It is important to note that there is no strict correlation between the molecular weight of a molecule and its half-life. For example, by genetic fusion of a VHH to an Fc domain the capacity to interact with the neonatal Fc receptor or FcRn is restored, which increases the retention of the VHH even more than expected based on the size [30]. It is clear that, in most cases, the half-life of the VHHs needs to be extended in order to obtain a maximal therapeutic efficacy. Prolonging the half-life will not only maintain the therapeutic threshold for a longer time but will also reduce the frequency of drug administration which will significantly benefit the patient [31]. Four major strategies, as discussed below, can be used to improve the pharmacokinetics of VHHs. In general, the goal is to limit their renal elimination and to recover the already available circulating VHH molecules.

A first frequently employed method aims to increase the size and the hydrodynamic radius of the VHH to avoid glomerular filtration. A simple way to accomplish this is by coupling two or three homologous or heterologous VHHs via a specific linker. However, even bivalent constructs are still rapidly cleared. A more complex method is the oligomerization of VHHs using a platform technology that allows to display VHHs. Different formats have been described, including the so-called fenobodies by Fan et al [32]. This display platform is based on ferritin, a spherical iron storage protein, to which, for example, broadly reactive anti-influenza VHHs have been anchored [33]. By doing so, the resulting VHH-displaying fenobody had a target affinity that was 360 times higher than the monovalent counterpart as well as a 10-fold higher half-life [32].

VHHs can also be chemically modified in order to increase the molecular weight. Such a chemical modification, for example with a polyethylene glycol (PEG) group, may also protect the VHH against proteases. The covalent attachment of PEG, a procedure that is often referred to as PEGylation and is approved by the Food and Drug Administration, to therapeutic proteins has been widely used to prolong the half-life of biopharmaceuticals. Interestingly, PEGylation of VHHs that can neutralize the highly contagious foot and mouth disease virus (FMDV, a picornavirus) did not only increase the half-life in guinea pigs but also significantly increased its neutralizing capacity [34]. The addition of a PEG molecule with a molecular mass higher than 50 kDa might result in accumulation of the VHH of interest in tissues, thereby reducing target access [35]. Next to this, chemical modification of a biotherapeutic can also result in reduced bioactivity or affinity of the molecule of interest [36,37]. Furthermore, there are safety concerns associated with the use of PEGylated protein drugs because PEG cannot be metabolized by the human body, and PEGylated biopharmaceuticals that are taken up by cells lead to vacuolation [38]. Combined with the relatively high cost of PEGylated molecules, the above-mentioned cautionary notes on the use of PEGylation to extend a drug's half-life have roused the interest of companies to explore safer and more economical alternatives. The use of recombinant PEG mimetics that can be genetically fused to the molecule of interest is one example of such an alternative (reviewed in [29,39]).

A second approach to try to avoid renal clearance of biopharmaceuticals is based on the interesting observation that negatively charged small proteins remain longer in circulation than neutral proteins. Most likely, this is due to repulsion between the negatively charged molecule and the negatively charged glomerular basement membrane in the kidney [40]. There are several ways to add negative charges to proteins including the addition of sialic acid polymers (polysialylation) or hydroxyethal starch (HESylation) and by fusion with the highly sialyated beta carboxyterminal peptide (CTP) amino acid-residue found in the human chorionic gonadotropin (hCG) hormone [39]. Unlike PEG, these naturally occurring polymers are biodegradable [41]. It would be of interest to modify VHHs with such negatively charged polymers and to assess the outcome in terms of half-life and in vivo efficacy.

Another strategy to extend the half-life of a VHH is by coupling the VHHs to long-lived serum proteins or to building blocks that target these long-lived proteins. The long serum half-life of serum albumin, for example, results from its ability to escape from catabolism after cellular uptake. Once in the acidic endosomal compartment albumin is able to bind to the FcRn, which recycles the bound albumin back to the pH neutral extracellular milieu. Human serum albumin has a terminal elimination half-life of 17–19 days in human and is a preferred target of choice for direct or indirect fusion of VHHs [42,43]. A study by Hoefman et al. reported a significant increase in half-life of different mono-, di- or trimeric VHH constructs by fusing them to an albumin binding VHH. Following a single intravenous bolus administration, these half-life extended constructs could still be detected in the serum after 10 days in mice and 35 days in monkeys while constructs fused to an irrelevant VHH were no longer detectable within one day after administration [44]. Terryn et al. also genetically fused homo-and hetero-bivalent VHHs that target the glycoprotein G of Rabies virus with an albumin-specific VHH in order to extend the half-life. The albumin-binding capacity of the construct resulted in an approximately 100-fold higher systemic exposure and improved the median survival time after Rabies virus challenge of treated mice by several weeks compared with the bivalent constructs [45]. In this respect it is important to mention that the half-life of albumin in rodents and other small mammals, is only around 1–2.5 days, which is markedly shorter than the several days to weeks observed for different immunoglobulins (Igs) [46]. An additional advantage of fusion or binding to albumin is the fact that it might help to target VHHs to specific sites in the body since it has been reported that albumin can accumulate in tumors and inflamed tissues [47].

The half-life of Igs ranges from one to almost four weeks depending on the subclass and the isotype. Therefore, IgG-based fusion constructs are also widely used to extend the half-life of therapeutic proteins including VHHs. For example, the so-called VHH2, a genetic fusion between a FMDV-neutralizing VHH and a VHH that targets the porcine IgG, was constructed by Harmsen et al. [30,34]. When binding to porcine IgG was possible, a 100-fold increase in half-life was observed. Prophylactic intramuscular treatment with VHH2 also reduced viral load and shedding. Unfortunately, the FMDV challenge-associated clinical signs and transmission of the disease could not be prevented with this molecule in the model that was used. Altogether, the data obtained by Harmsen et al. showed that prolonging the exposure time is imperative for immunotherapy efficacy but additional fine tuning of other parameters is also important [34].

A fourth approach of extending the VHH half-life is to fuse the VHH of interest to the Fc region or other Fc-binding moieties of an IgG molecule. The extended half-life of such a chimeric antibody construct was demonstrated by Raj et al., who C-terminally extended a VHH that can neutralize the MERS-CoV with the Fc part of a human IgG2. The protein, that resembles a heavy chain-only antibody, showed an enhanced MERS-CoV neutralizing capacity and protection that correlated with sustained high levels of the VHH-Fc in circulation [48]. In parallel, Zhao et al. showed stable binding of a MERS-CoV neutralizing VHH fused with a human IgG1 Fc to recombinant MERS-CoV S1 spike protein in serum collected 10 days post-injection. This is in contrast to the monovalent format of the MERS-CoV VHH where no binding could be detected by day 10 [20]. In both cases, the improved protection against MERS-CoV infection with the Fc fused constructs was attributed to the increase in size by dimerization and thus the extended half-life. The effect of binding to the FcRn was not reported in these studies.

## 3. Increasing Valency to Improve Potency

In some instances, the affinity of an antiviral VHH picked up by an initial selection step still needs to be improved in order to enhance and broaden its neutralizing capacity. In addition, increasing the affinity might be important to reduce viral escape. To this end, one could change the affinity between the VHH and its target antigen by in vitro affinity maturation in which a second-generation library is constructed by introducing random mutations in a selected part of the VHH scaffold, usually the CDRs. For example, the affinity of a parathyroid hormone (PTH)-derived peptide-specific VHH that was isolated from a phage display library that was generated from a naive llama, could be enhanced 30-fold by selecting VHH derivatives from a library of VHHs with mutations that were randomly introduced in the CDR2 and CDR3 codons [49]. Another way to increase affinity is to introduce an avidity effect by joining two or more VHHs into multivalent constructs using flexible linkers. Several studies have demonstrated the beneficial outcome of that type of VHH formatting to improve affinity through avidity effects. In addition to increasing the binding avidity, formatting could also increase neutralizing activity by increased structural restriction when 2 sites on different protomers or proteins are linked by the VHH construct. Three types of multivalent VHH formats can be distinguished: monospecific fusions in which two identical VHHs recognizing the same epitope are joined, bispecific fusions in which two VHHs each recognizing a different epitope are connected, and formats where the VHH is fused to a protein moiety that has the tendency to dimerize or multimerize [50].

The strength of introducing avidity on the affinity of the VHHs for their target antigens was extensively examined by Hultberg et al. [21]. In this study, multivalent constructs were generated against the RSV F fusion protein, influenza H5N1 HA and Rabies G protein. Remarkably, linking two identical anti-F VHHs increased the RSV neutralizing capacity by approximately 4000. In addition, bivalent and trivalent anti-HA and bivalent anti- vesicular stomatitis virus (VSV) G VHHs had significantly higher neutralizing potency than their monovalent counterparts. Flexible $Gly_4Ser$ (GS) linkers ranging from 9 to 35 amino acid residues were tested. The optimal linker length needed to achieve an avidity effect most likely depends on the availability and spacing between the viral epitopes and needs to be empirically tested for each VHH combination. Trivalent anti-influenza HA VHHs with a 20 amino acid residue linker had lower neutralizing activity compared with constructs with a 10 amino acid residue linker. In contrast bivalent RSV F specific VHH with either a 10 or 20 amino acidlinker were equally potent in neutralizing RSV [21]. Further, Hulsik et al. developed a VHH (VHH 2H10) directed against the membrane proximal external region (MPER) epitope on the HIV gp41, the viral fusion protein [51]. Bivalent constructs in which two VHH 2H10 molecules were linked by a 15 or 17 GS linker, had a 20-fold higher binding efficacy compared with monomeric VHH 2H10. This increased affinity was associated with an increased breadth of neutralization: various HIV-1 strains that were resistant to monovalent VHH 2H10 could now be neutralized. In addition to using a GS linker, Cardoso et al. used the llama IgG2c hinge as a flexible linker to fuse two identical anti-influenza neuraminidase (NA) (H5N1) VHHs [52]. Similar to the above-mentioned studies, the antiviral potency of the bivalent format was significantly enhanced. Another example of a VHH that acquired increased virus neutralizing breadth by increasing its avidity, was reported for influenza A viruses as a target. A bivalent version of the VHH R1a-B6, generated by fusing two identical VHHs directed against the stem of HA of the 2009 H1N1 pandemic virus using a 30 amino acid GS linker, had a similar neutralizing potency against H1N1 and H5N1 viruses as monovalent R1a-B6. However, neutralization against a more divergent H9N2 virus strain was increased and the bivalent format even gained the ability to neutralize a H2N2 strain [18]. The authors proposed that this increase in antiviral breadth is the result of a slower off rate which can rescue low affinity interactions. Similarly, a broader antiviral spectrum was observed for bivalent formats of VHHs directed against the CD4-binding site or CD4-binding-induced site of the HIV-1 surface envelope glycoprotein (gp120) while no significant improvement of the neutralizing potency of these multivalent proteins in terms of $IC_{50}$ was observed compared to those of the parental monovalent VHHs [53]. Apart from the in vitro increase in antiviral potency, the impact of the avidity effect was also observed in in vivo experiments. Ibañez et al. showed that an H5N1-neutralizing VHH in a bivalent format protected mice against a lethal H5N1 virus challenge and was at least 60-fold more effective in doing so than the monovalent counterpart based on lung viral titers in a dose-response comparison experiment [54]. Similarly, the intranasal administration of a bivalent RSV F-specific VHH and a bivalent rabies G-specific VHH showed increased in vivo potency compared to the monovalent counterpart [45,55]. Furthermore, ALX-0171, a homotrimeric

VHH with high binding affinity for RSV F and potent RSV neutralizing activity, is currently being evaluated in a phase IIb clinical trial for the treatment of disease caused by RSV infections in infants. Interestingly, in in vitro experiments, the trivalent VHH outperformed palivizumab, a monoclonal humanized IgG1 antibody directed against the same epitope of the RSV F protein and currently used as a prophylactic anti-RSV treatment [56]. This could possibly be explained by the enhanced accessibility of the epitope due to the small size of the VHH, a more flexible GS linker compared with the antibody hinge or the presence of a third F-binding module in ALX-0171. To explore the impact of the avidity effect on the emergence of escape mutants in vitro, Palomo et al. serially passaged the RSV long strain in the presence of the monovalent or trivalent VHH [57]. This revealed that it was much more difficult for the virus to escape from the trivalent compared to the monovalent VHH. In addition, the trivalent format was still able to neutralize the RSV escape mutants.

Another way to overcome viral escape and broaden the neutralizing activity is to use a bispecific format, comprised of two VHHs that each target a different viral epitope. Hultberg et al. demonstrated that cross-neutralization was significantly improved with both RSV- and Rabies-neutralizing bispecific formats compared to the corresponding monovalent and bivalent monospecific constructs [21]. The generation of a single molecule (a head-to-tail fusion with a GS linker) with bispecific target recognition was key for this improvement because mere mixing of the monovalent VHHs in an equimolar ratio hardly increased the potency compared with the individual monovalent VHHs. Interestingly, the position of the VHHs relative to each other in the bispecific fusion construct can influence the potency [17]. Such an influence was also observed for bispecific VHHs (JM2x5 or JM5x2) directed against the CD4-binding site or CD4-binding induced site of HIV-1 gp120. The bispecific construct with JM5 VHH at the N-terminal and JM2 VHH at the C-terminal position outperformed the construct with the VHHs in the opposite position [53].

Finally, it is also possible to increase affinity and neutralizing potency by fusing a VHH with a protein moiety that has a tendency to oligomerize. For example, Tillib et al. fused a trimerizing variant of the leucine zipper domain derived from the yeast transcription factor GCN4 to the C-terminus of an anti-influenza HA (H5N2) VHH [58]. The formatted VHHs adopted a trimeric conformation and were approximately 100 times more active compared with the unformatted VHHs. Likewise, Boruah et al. used another type of coiled-coil forming domain to generate multivalent VHHs [59]. It is even possible to generate so called combodies, constructs in which the VHH is fused with the coiled-coil domain derived from the human cartilage oligomeric matrix protein (COMP48), which results in a pentavalent VHH format. Compared with the monovalent format, the combodies were able to neutralize an 85-fold higher input of Rabies virus pseudotypes in vitro. Since the pentamerizing building block in combodies is of human origin, they have a reduced risk to become immunogenic upon administration in humans compared with the yeast leucine zipper domain. Another oligomerization approach is the development of the so-called fenobodies, as described before [32]. A very attractive way to generate bivalent VHHs is by fusion to the Fc domain of a conventional antibody (an IgG or IgA). For example, Cardoso et al. fused influenza H5N1 NA-specific VHHs to a mouse IgG2a-derived Fc domain [52]. The Fc formatted VHHs showed a 250-fold higher NA inhibition and approximately 50-fold higher in vitro antiviral activity compared to the monovalent VHHs. An important advantage of a VHH-Fc fusion construct is the introduction of Fc receptor-dependent effector functions such as complement activation and the possibility to engage effector functions such as antibody-dependent phagocytosis activity (discussed in the next paragraph).

## 4. Arming VHHs with Effector Functions

Next to direct virus neutralization, conventional antibodies can also employ Fc-mediated effector functions to control viral infections. These effector functions include antibody-dependent cell-mediated cytotoxicity (ADCC), antibody-dependent cell-mediated phagocytosis (ADCP) and complement-dependent cytotoxicity (CDC); mechanisms that can eliminate infected cells or virions [60]. The importance of these effector functions is increasingly recognized for antibody mediated protection against infections with e.g., influenza virus, RSV, Zika virus and HIV [61–67]. VHHs lack the Fc region and therefore cannot facilitate such effector functions as such. However, thanks to their single domain nature, VHHs can be readily equipped with various effector functions through molecular or biochemical engineering. As such VHHs can be linked to Fc regions but also to toxins, liposomes, and other ligand binding scaffolds.

Fc regions (CH2-CH3) of different human and mouse antibody subclasses have been genetically fused to VHHs that target different viruses. As outlined above, an Fc fusion greatly increases the serum half-life and enhances target antigen binding through an avidity effect (Table 1). For some antiviral VHH-Fcs the importance of the acquired Fc-associated effector functions has been investigated, e.g., in in vitro assays. VHH VUN400 targets the second extracellular loop of the CXCR4 chemokine receptor and can prevent entry and replication of an HIV-1 strain that uses CXCR4 as a co-receptor [68]. Fusing VUN400 to the CH2-CH3 domains of a human IgG1 antibody strongly increased the potency of this VHH to bind CXCR4 on the surface of T cells and to prevent HIV-1 infection. In addition, VUN400-Fc enabled ADCC of CXCR4-expressing target cells as evidenced by activation of FcγRIIIa, induction of natural killer (NK) cell degranulation and selective killing of target cells by NKs. VUN400-Fc could also mediate complement-dependent killing of cells that expressed high levels of CXCR4. Whether those acquired effector functions can also contribute to protection (both in vitro and in vivo) by VUN400-Fc remains to be determined.

The potential impact of VHH-Fc effector functions in vivo has been well illustrated for VHHs targeting the rotavirus VP6 protein and the influenza NA and HA proteins. The anti-rotavirus protein 1 (ARP1) is a VHH directed against the conserved rotavirus inner capsid protein VP6. Whereas conventional antibodies directed against the outer capsid protein VP7 can typically mediate serotype specific neutralization, antibodies directed against VP6 generally fail to do so. Remarkably, some non-neutralizing VP6 specific IgA monoclonal antibodies do protect mice against rotavirus infection when given systemically but not when delivered at the luminal side of the intestine [69]. When delivered intracellularly by protein lipofection, these antibodies were able to block rotavirus transcription by interfering with the rotavirus capsid structure [70]. As such non-neutralizing VP6-specific IgA could control rotavirus replication in vivo during transcytosis [69]. In sharp contrast to conventional VP6-specific antibodies, VHHs such as ARP1 that target VP6 can potently neutralize rotaviruses of diverse serotypes and control rotavirus infection in neonatal mice when fed orally [71,72]. A recently reported clinical trial demonstrated that therapeutic oral administration of yeast produced ARP1 was safe and effective in reducing diarrhea in male infants with severe rotavirus-associated diarrhea [73]. To try to further improve its protective capacity, the ARP1 VHH was fused to a mouse IgG1 Fc region and evaluated in vivo. Compared to monovalent ARP1 VHH, oral prophylactic treatment of Fc-ARP1 was far more effective in reducing the prevalence, duration and severity of diarrhea in rotavirus-infected neonatal mice pups. The observation that the (Fab′)$_2$ fragment of Fc-ARP1 was much less effective illustrates that the capacity of Fc-ARP is not due to its bivalent nature, suggesting the importance of the Fc-associated effector functions. The role of Fc effector functions was further demonstrated by the reduced protective capacity of the $Fc_{N434D}$-ARP1 variant. In contrast to Fc-ARP1, $Fc_{N434D}$-ARP1 cannot interact with FcRn or the intracellular antibody receptor TRIM21 but is expected to have unaffected interaction with the Fcγ receptors. This suggests that next to direct neutralization, Fc-ARP may additionally mediate intracellular neutralization after internalization via FcRn and binding to TRIM21. In the cell, the intracellular antibody receptor TRIM21 can recognize virus–antibody complexes and target these for proteasomal degradation and initiate antiviral innate immune responses [74]. As the $Fc_{N434D}$-ARP1 was still more potent than the ARP1 VHH, other Fc effector functions might also contribute to the protection provided by Fc-ARP1.

**Table 1.** Different VHH formats and their functionality used to tackle various viral infections.

| VHH Format | Functionality | Virus | Reference |
|---|---|---|---|
| Homobivalent VHHs | - Enhance and broaden antiviral activity | RSV<br>Influenza<br>Rabies<br>HIV | Hultberg et al. [21]<br>Schepens et al. [55]<br>Detalle et al. [56]<br>Palomo et al. [57]<br>Hultberg et al. [21]<br>Cardoso et al. [52]<br>Hufton et al. [18]<br>Ibanez et al. [54]<br>Hultberg et al. [21]<br>Terryn et al. [45]<br>Hulsik et al. [51]<br>Matz et al. [53] |
| Bispecific VHHs<br>- VHH linked to anti albumin VHH<br>- VHH linked to anti IgG VHH<br>- VHH linked to VHH which binds different epitopes on same target | - Half-life extension<br>- Half-life extension<br>- Enhance and broaden antiviral activity | Rabies<br>FMDV<br>RSV<br>Rabies<br>HIV | Terryn et al. [45]<br>Harmsen et al. [30]<br>Hultberg et al. [21]<br>Hultberg et al. [21]<br>Matz et al. [53] |

**Table 1.** *Cont.*

| VHH Format | Functionality | Virus | Reference |
| --- | --- | --- | --- |
| PEGylation | - Half-life extension | FMDV | Harmsen et al. [34] |
| VHH linked to IgG Fc region | - Half-life extension<br>- Enhance and broaden antiviral activity<br>- Effector function | MERS<br>Influenza<br>HIV<br>Rotavirus | Raj et al. [48]<br>Zhao et al. [20]<br>Cardoso et al. [52]<br>Laursen et al. [75]<br>Bobkov et al. [68]<br>Günaydın et al. [76] |
| VHH linked to ferritin | - Half-life extension | Influenza | Fan et al. [32] |
| VHH linked to GCN4 | - Enhance and broaden antiviral activity | Influenza | Tillib et al. [58] |

**Table 1.** *Cont.*

| VHH Format | Functionality | Virus | Reference |
| --- | --- | --- | --- |
| VHH linked to COMP48 | - Enhance and broaden antiviral activity | Rabies | Boruah et al. [59] |
| VHH linked to IgG | - Effector function<br>- Targeting | HIV | Sun et al. [77] |
| VHH linked to cytotoxic domain | - Effector function | HSV-2 | Geoghegan et al. [78] |
| VHH linked to liposome | - Effector function<br>- Targeting | HIV | Wang et al. [79] |

**Table 1.** *Cont.*

| VHH Format | Functionality | Virus | Reference |
| --- | --- | --- | --- |
| VHH linked to bacteria | - Targeting | Rotavirus | Pant et al. [80]<br>Günaydın et al. [76]<br>Alvarez et al. [81] |
| VHH linked to GPI | - Targeting | HIV<br>RSV | Liu et al. [82]<br>Tiwari et al. [83] |
| VHH linked to cell-penetrating peptide | - Targeting | HCV | Thueng-in et al. [84]<br>Phalaphol et al. [85]<br>Glab-ampai et al. [86]<br>Tarr et al. [87] |

VHHs, single-domain antibodies; RSV: human respiratory syncytial virus; HIV: human immunodeficiency virus; IgG: immunoglobulin G; PEG: Polyethylene glycol; FMDV: foot-and mouth disease virus; MERS CoV: Middle East respiratory syndrome coronavirus; GCN4: amino acid starvation-responsive transcription factor; COMP48: human cartilage oligomeric matrix protein; HSV-2: Herpes Simplex Virus-2; GPI: glycosylphosphatidylinositol; HCV: hepatitis C virus.

The potential contribution of Fc effector functions in VHH-Fc mediated protection was also observed for a non-neutralizing Fc-VHH directed against the H5N1 NA protein [52]. We isolated H5N1 NA binding VHHs that can block the NA activity and consequently also in vitro viral replication. Bivalent formats of VHHs with NA inhibiting activity generated by head-to-tail or Fc fusion were at least 200-fold more potent than their monovalent counterparts at inhibiting NA activity and at least 30-fold in a plaque size reduction assay. In contrast, VHHN1-7-VHHm and its Fc fusion format bind H5N1 NA but fail to inhibit NA activity and viral replication in vitro. However, despite the lack of detectable anti-viral activity in vitro, prophylactic treatment of mice with N1-7-VHH in its Fc fusion format but not as a monovalent VHH could partially protect against an otherwise lethal H5N1 influenza virus infection. This suggests that the protective activity of pN1-7-VHH-Fc is dependent on its Fc effector functions.

More direct evidence for the importance of the Fc effector functions was recently reported for a multivalent VHH-Fc protein that can provide very broad protection against influenza A and B virus infection [75]. In this study, the experimental molecule MD3606 is comprised of a head-to-tail fusion of four different VHHs linked to a human IgG Fc region. Separately, these VHH can potently neutralize group 1 influenza A viruses, group 2 influenza A viruses, Victoria lineage influenza B or Yamagata lineage influenza B viruses. Prophylactic treatment of mice with MD3606 could protect them from an infection with diverse influenza A and B viruses at a challenge dose that was lethal for control treated animals. Leucine to alanine mutations in the Fc regions (LALA for human IgG1 and IgG2$\sigma$ for mouse IgG2a MD3606 variants) that abrogate the interactions with Fc$\gamma$ receptors and C1q (thereby eliminating ADCC, ADCP and CDC activity) substantially decreased the protective capacity of MD3606. This clearly demonstrates the importance of the Fc effector functions for the antiviral activity of VHH-Fc proteins.

Next to fusion with Fc regions also other approaches can be used to arm VHHs with Fc effector functions. For example, Sun et al. fused a conventional anti-CD4 antibody to a VHH that targets a HIV-1 gp120 co-receptor binding site that is exposed upon CD4 binding [77]. This antibody-VHH chimera was more efficient in blocking HIV-1 infections of T cells than the VHH as such. If the Fc region in this antibody-VHH construct could also engage effector functions was however not investigated. Gray et al. linked a HER-2-specific VHH to dinitrophenyl (DNP) which can recruit anti-DNP antibodies that are omnipresent in human sera. These DNP-fused VHHs could engage peripheral blood mononuclear cells (PBMCs) to kill HER-2 expressing target cells by ADCC [88]. Alternatively, a VHH that specifically targets the human Fc$\gamma$R III (CD16) fused to a tumor antigen-specific VHH could kill tumor cells by inducing ADCC and reduce tumor growth in mice that were xenografted with tumor cells and human PBMCs [89]. These last approaches might be of interest for arming virus-specific VHHs with various Fc effector functions.

Next to exploiting host Fc effector functions, VHHs can also be equipped with foreign effector functions. A VHH targeting the herpes simplex type 2 (HSV-2) glycoprotein D can bind to cells that express this protein but fails to neutralize HSV-2 infection in vitro. However once fused to the cytotoxic domain of *Pseudomonas aeruginosa* exotoxin A, this VHH (R33ExoA) can efficiently kill HSV-2 infected cells and as such reduce viral replication [78]. Fusion of CD7- and CD38-specific VHHs with toxins has been explored as a strategy to respectively control T-cell acute lymphoblastic leukemia and multiple myeloma [90,91]. VHH-based immunotoxins could represent a valuable approach for the control of latent viral infections such as those caused by HSV-2. Virus-specific VHHs could also be used to bring a cargo comprised of e.g., liposomes loaded with an antiviral drug selectively to infected cells. Shielding a toxic drug in a liposome may reduce systemic toxicity while increasing the efficacy of the delivered antiviral at the site of infection. Wang et al. conjugated the neutralizing HIV-1 gp120 specific VHH J3 to liposomes. Whereas non-covalent conjugation did not interfere with the neutralization by VHH J3, covalent conjugation of J3 to liposomes did. Although the reverse transcriptase inhibitor dapivirine encapsulated in liposomes had higher antiviral activity than free dapivirine, conjugation of J3 to dapivirine loaded liposomes did not further increase the antiviral activity [79]. Further optimization could potentially lead to a more pronounced antiviral therapy.

## 5. Targeting and Delivery of VHHs

In some cases, VHHs are only effective or needed in a certain organ, cell type or cell compartment. Multiple formatting options are available to specifically target or deliver VHHs to one of these regions, e.g., linkage to antibodies or peptides, provide posttranslational modifications or in situ expression by bacterial vectors.

Delivery of VHHs to an organ system such as the gastrointestinal tract can be easily achieved by the use of bacterial vectors. This approach was used to develop a therapy against rotavirus infections. Direct oral administration of the anti-rotavirus VHH ARP1 reduced rotavirus-induced diarrhea with 22.5% in a clinical study [71,73,92]. Delivering the VHHs with the use of a bacterial vehicle could result in enhanced and sustained protection. VHH expression by a bacterial vector can be accomplished by directing the VHH to the surface of the bacteria or by secretion of the VHH. *Lactobacillus paracasei* was engineered in two ways, resulting in either expression of mono- or bivalent ARP1 (or ARP3, another rotavirus-neutralizing VHH) on the bacterial surface, or the production of the monovalent ARP1 and ARP3 as secreted and/or surface displayed proteins [76,80]. The first strategy led to *Lactobacilli* which could reduce the rate of diarrhea in mice in both a prophylactic and therapeutic setting with a slightly improved efficacy for the bivalent VHH- compared to the monovalent VHH-expressing bacterium. Using the second, combined strategy, the virus could be captured by a VHH on the surface of the *Lactobacillus* whereas the secreted VHH could bind to a distinct epitope. The efficacy of this strategy has not yet been tested in vivo. Another group made use of the probiotic strain *Lactobacillus rhamnosus* GG, which was modified to display ARP1 on its surface [81]. Display of ARP1 on this *Lactobacillus* strain, which has intrinsic anti-rotavirus activity, resulted in ameliorated clinical parameters upon rotavirus challenge infection in mice compared to the original strain. None of these modified bacteria have been tested yet in clinical studies. Such studies are somewhat complicated from a regulatory perspective because it concerns the use of genetically modified micro-organisms. Also, the use of bacterial vectors is restricted to organs with a microbiome such as the gut, the skin, the nasal cavity and the vagina. Other organs can be targeted by linking the VHH of interest to a VHH that specifically binds to an organ-specific marker protein, which can enrich the VHH in the organ of choice upon systemic administration.

Next to targeting a complete organ, it can be advantageous to target only one cell type. This is the case with the anti-HIV iMabM36 antibody format, consisting of the HIV-neutralizing Ibalizumab (iMab) monoclonal antibody that was linked to two copies of the broadly HIV-neutralizing VHH m36 [77]. iMab is specific for CD4, the receptor of HIV. By linking the VHH to this antibody, the concentration of the broadly neutralizing VHH is enhanced at the site of infection resulting in a synergistic 10-fold antiviral effect compared to a mixture of the separate iMab and m36 VHHs.

At last, VHHs can also be targeted to a specific cell compartment to exert their function. Liu et al., for example, described a strategy to target HIV-neutralizing VHHs towards the lipid rafts on the cell membrane [82]. For entry and viral release, HIV relies on lipid rafts on the host cell membrane. Targeting anti-HIV VHHs to this site of infection could therefore increase the effectiveness of the VHHs. This is possible with the use of a glycosylphosphatidylinositol (GPI) signal, a hydrophobic moiety which guides proteins to the lipid rafts in the plasma membrane. A gene construct was made linking the coding information of the HIV CD4-receptor gp120-binding JM4 with the GPI signal sequence [53]. Transduction of $CD4^+$ T cells with this construct resulted in broad and potent neutralization of HIV-1, while transduction with a construct coding for secreted JM4 did not neutralize any of the HIV-1 subtypes at all. A limitation of this technique is that the GPI-VHHs need to be produced by the $CD4^+$ cells for them to be functional. This could be achieved by modifying patient-derived $CD4^+$ cell with lentiviral vectors carrying the GPI-VHH sequence and then transfusing these cells back to the patient.

Another example of a GPI-linked VHH was described by Tiwari et al. who developed an anti-RSV strategy based on mRNA coding for an (GPI-linked) RSV-neutralizing VHH (F-VHH-4 described in Rossey et al.) [22,83]. In this strategy, in vitro produced VHH-coding mRNA is directly delivered to the lung resulting in a transient expression of the therapeutic VHH in the respiratory epithelial cells. By including a GPI anchor in the construct, the VHHs are directed to the epithelial cell surface, the site of RSV entry. Compared to a construct where the RSV-VHH is secreted, GPI-anchored RSV-neutralizing VHH could further reduce RSV infection in mice, even when administered seven days before infection. This is mainly due to the enhanced lung retention thanks to the GPI anchor.

Some viral targets may be located inside the infected cell and thus hard to reach by antiviral VHHs. This problem can however be overcome by linking the VHH to a cell-penetrating peptide such as penetratin. This approach has been applied for the development of anti-hepatitis C virus (HCV) VHHs. Several intracellular viral proteins were considered as targets: the viral RNA-dependent RNA polymerase, the NS3 helicase/NTPase and the multifunctional HCV NS4B protein [84–87]. In vitro, penetratin-linked VHHs directed against these targets could suppress HCV replication. These VHHs have not been tested yet in vivo.

## 6. Antiviral Single Domain Antibodies as Tools for Diagnostic and Antigen Display

The sturdy nature, small size and ease of production make single domain antibodies very well suited as building blocks that can be used in applications such as diagnostics. However, in the context of virus-targeting VHHs, we came across very few examples in the literature that exploited VHH for the detection of viral antigens. One explanation for this paucity of studies is probably the tremendous sensitivity of nucleic acid-based detection methods for the diagnosis of a viral infection. However, in terms of speed, it is clear that rapid antibody-based detection methods outcompete DNA- and RNA-based detection methods. For some viruses that are known to manifest antigenic diversity in their structural proteins, a VHH-based detection method is a challenge unless a broadly reactive, yet highly specific VHH is available. Human noroviruses are an example of antigenically variable viruses, with more than 40 different genotypes that are classified into two main genogroups [93]. Furthermore, norovirus outbreaks can spread at an astonishing speed, warranting the availability of a rapid and robust diagnosis method, which could help to contain an outbreak as soon as possible after the first patients fall ill. One VHH that recognizes an epitope in the lower region of the protruding domains of the norovirus capsid that is conserved among genotype II noroviruses was developed into a tool for rapid diagnostics [94]. The norovirus capsid-binding VHH was modified with biotin and conjugated to gold, modifications that allowed its use in a rapid lateral flow immunoassay. Interestingly, compared with a commercially available ELISA for the analysis of human stool samples, the VHH lateral flow immunoassay had a higher specificity (86% compared with 67%), although one out of five cases were missed with the VHH set up (sensitivity of 80% for the lateral flow immunoassay compared with 100% for the commercial ELISA) [94]. The use of multiple VHHs, that ideally also can recognize genotype I noroviruses, would likely contribute to a higher sensitivity. Influenza viruses likewise display a high antigenic variability. Here also, a rapid detection method based on single domain antibodies would require those to be broadly reactive. A sandwich type ELISA has been proposed based on two single domain antibodies derived from a camel that had been immunized with inactivated semi-purified A/Texas/1/1977 H3N2 virus. These single domain antibodies were specific for the hemagglutinin of an H3N2 virus. In a sandwich ELISA, in which one of the single domain antibodies was tethered to magnetic beads and the second one was coupled with a reporter enzyme, the semi-purified H3N2 virus was detectable up to a minimal concentration of 50 ng/mL. It is unclear from this report if the single domain antibody pair was able to recognize multiple H3N2 strains [95].

A second VHH-based tool in the context of a virus has been developed by the group of David Rowlands. This group has a long standing in the design of hepatitis B core-based virus-like particles (VLPs) that can be expressed in robust expression systems such as those based on *E. coli* and plants. Recently, the group reported a so-called tandem fusion hepatitis B core VLP assembly method. When a single-domain antibody directed against green fluorescent proteins (GFP) was inserted in the major immunodominant region of the second hepatitis B core moiety of the tandem construct, recombinant VLPs were produced and purified that could capture GFP on their surface [96]. A similar VLP with a VHH directed against a nondisclosed virus was also generated by transient expression in *Nicotiana benthamiana* leaves. These so-called tandibodies could be used to display antigens for vaccination purposes. It will be interesting to see further reports on this technology for vaccine design purposes. Most likely, the hepatitis B core as well as the VHH will also be immunogenic. In addition, the VHH might shield an epitope on the displayed antigen.

## 7. Conclusions

Due to their unique properties such as high solubility, stability, ease of production and formatting flexibility, VHHs seem to be very well suited to develop high affinity reagents to fight human viral diseases. Currently eight VHHs are in clinical development (all from the company Ablynx), of which one, ALX-0171, a trivalent anti-RSV VHH currently tested in a phase IIb clinal trial, is directed against a viral target. Their single-domain build-up allows formatting in multiple ways to obtain "best-in-class" molecules. Head-to-tail fusion of identical VHHs or VHHs recognizing different epitopes or fusion to multimerizing protein moieties have successfully been demonstrated to enhance and broaden neutralization activity. The first strategy being exemplified by the superior antiviral activity and strain coverage of the trivalent VHH ALX-0171 compared with its monovalent counterpart. A key component in the generation of long-lasting antiviral therapeutics is the implementation of half-life extension techniques such as PEGylation, fusion to a serum albumin-binding VHH or fusion to an IgG Fc-domain. Clinical proof-of-concept of the extended half-life that is obtained in this way has been achieved for an anti-IL-6R and anti-TNF VHH fused to a serum albumin-binding VHH, used in the treatment of rheumatoid arthritis. PEGylated VHHs or VHHs fused to an Fc tail have not yet been tested in clinical trials. The fusion to an Fc tail however seems a promising approach since not only half-life is extended but also avidity and Fc effector functions are introduced. Moreover, several Fc fusion-based therapeutics are already on the market [97]. Alternatively, immune effector functions can also be added by fusion to an anti-CD3 VHH, resulting in the recruitment of T-cells. Next to improving affinity, extending the half-life and introducing effector functions, formatting can also improve VHH activity by targeting a certain organ, cell type or cell compartment. In this context, using a bacterial vector to deliver the VHH to specific organs seems an interesting approach to explore further. Finally, although VHHs are also well suited for the development of diagnostic tests, this is still a poorly explored area in the context of viral infections.

**Author Contributions:** Writing—Original Draft Preparation, writing—Review & Editing: D.D.V., M.B., I.R., B.S., X.S.; Visualization; D.V.V. and M.B.

**Funding:** This research was funded by FWO (projects F043515N and G0B1917N). D.D.V. is a predoctoral and I.R. a postdoctoral fellow supported by FWO. B.S. is a doctoral assistant at the Department of Biomedical Molecular Biology.

**Conflicts of Interest:** The authors declare no conflict of interest.

## References

1. Hamers-Casterman, C.; Atarhouch, T.; Muyldermans, S.; Robinson, G.; Hammers, C.; Songa, E.B.; Bendahman, N.; Hammers, R. Naturally occurring antibodies devoid of light chains. *Nature* **1993**, *363*, 446–448. [CrossRef] [PubMed]
2. Gonzalez-Sapienza, G.; Rossotti, M.A.; Tabares-da Rosa, S. Single-Domain Antibodies as Versatile Affinity Reagents for Analytical and Diagnostic Applications. *Front. Immunol.* **2017**, *8*. [CrossRef] [PubMed]
3. Mitchell, L.S.; Colwell, L.J. Analysis of nanobody paratopes reveals greater diversity than classical antibodies. *Protein Eng. Des. Sel.* **2018**, *31*, 267–275. [CrossRef]
4. Mitchell, L.S.; Colwell, L.J. Comparative analysis of nanobody sequence and structure data. *Proteins Struct. Funct. Bioinform.* **2018**, *86*, 697–706. [CrossRef] [PubMed]

5. Henry, K.A.; MacKenzie, C.R. Antigen recognition by single-domain antibodies: Structural latitudes and constraints. *MAbs* **2018**, *10*, 815–826. [CrossRef] [PubMed]
6. Sequence and Structure of VH Domain from Naturally Occurring Camel Heavy Chain Immunoglobulins Lacking Light Chains. Available online: https://www.ncbi.nlm.nih.gov/pubmed/7831284 (accessed on 2 November 2018).
7. Vu, K.B.; Ghahroudi, M.A.; Wyns, L.; Muyldermans, S. Comparison of llama VH sequences from conventional and heavy chain antibodies. *Mol. Immunol.* **1997**, *34*, 1121–1131. [CrossRef]
8. Harmsen, M.M.; Ruuls, R.C.; Nijman, I.J.; Niewold, T.A.; Frenken, L.G.; de Geus, B. Llama heavy-chain V regions consist of at least four distinct subfamilies revealing novel sequence features. *Mol. Immunol.* **2000**, *37*, 579–590. [CrossRef]
9. Muyldermans, S.; Baral, T.N.; Retamozzo, V.C.; De Baetselier, P.; De Genst, E.; Kinne, J.; Leonhardt, H.; Magez, S.; Nguyen, V.K.; Revets, H.; et al. Camelid immunoglobulins and nanobody technology. *Vet. Immunol. Immunopathol.* **2009**, *128*, 178–183. [CrossRef]
10. Holliger, P.; Hudson, P.J. Engineered antibody fragments and the rise of single domains. *Nat. Biotechnol.* **2005**, *23*, 1126–1136. [CrossRef]
11. Sabir, J.S.M.; Atef, A.; El-Domyati, F.M.; Edris, S.; Hajrah, N.; Alzohairy, A.M.; Bahieldin, A. Construction of naïve camelids VHH repertoire in phage display-based library. *C. R. Biol.* **2014**, *337*, 244–249. [CrossRef]
12. Goldman, E.R.; Anderson, G.P.; Liu, J.L.; Delehanty, J.B.; Sherwood, L.J.; Osborn, L.E.; Cummins, L.B.; Hayhurst, A. Facile generation of heat-stable antiviral and antitoxin single domain antibodies from a semisynthetic llama library. *Anal. Chem.* **2006**, *78*, 8245–8255. [CrossRef] [PubMed]
13. Wang, Y.; Fan, Z.; Shao, L.; Kong, X.; Hou, X.; Tian, D.; Sun, Y.; Xiao, Y.; Yu, L. Nanobody-derived nanobiotechnology tool kits for diverse biomedical and biotechnology applications. *Int. J. Nanomed.* **2016**, *11*, 3287–3303. [CrossRef] [PubMed]
14. McMahon, C.; Baier, A.S.; Pascolutti, R.; Wegrecki, M.; Zheng, S.; Ong, J.X.; Erlandson, S.C.; Hilger, D.; Rasmussen, S.G.F.; Ring, A.M.; et al. Yeast surface display platform for rapid discovery of conformationally selective nanobodies. *Nat. Struct. Mol. Biol.* **2018**, *25*, 289–296. [CrossRef] [PubMed]
15. Pardon, E.; Laeremans, T.; Triest, S.; Rasmussen, S.G.F.; Wohlkönig, A.; Ruf, A.; Muyldermans, S.; Hol, W.G.J.; Kobilka, B.K.; Steyaert, J. A general protocol for the generation of Nanobodies for structural biology. *Nat. Protoc.* **2014**, *9*, 674–693. [CrossRef] [PubMed]
16. Ledsgaard, L.; Kilstrup, M.; Karatt-Vellatt, A.; McCafferty, J.; Laustsen, A.H. Basics of Antibody Phage Display Technology. *Toxins* **2018**, *10*, 236. [CrossRef] [PubMed]
17. Bencurova, E.; Pulzova, L.; Flachbartova, Z.; Bhide, M. A rapid and simple pipeline for synthesis of mRNA-ribosome-V(H)H complexes used in single-domain antibody ribosome display. *Mol. Biosyst.* **2015**, *11*, 1515–1524. [CrossRef]
18. Hufton, S.E.; Risley, P.; Ball, C.R.; Major, D.; Engelhardt, O.G.; Poole, S. The breadth of cross sub-type neutralisation activity of a single domain antibody to influenza hemagglutinin can be increased by antibody valency. *PLoS ONE* **2014**, *9*, e103294. [CrossRef]
19. Forsman, A.; Beirnaert, E.; Aasa-Chapman, M.M.I.; Hoorelbeke, B.; Hijazi, K.; Koh, W.; Tack, V.; Szynol, A.; Kelly, C.; McKnight, A.; et al. Llama antibody fragments with cross-subtype human immunodeficiency virus type 1 (HIV-1)-neutralizing properties and high affinity for HIV-1 gp120. *J. Virol.* **2008**, *82*, 12069–12081. [CrossRef]
20. Zhao, G.; He, L.; Sun, S.; Qiu, H.; Tai, W.; Chen, J.; Li, J.; Chen, Y.; Guo, Y.; Wang, Y.; et al. A Novel Nanobody Targeting Middle East Respiratory Syndrome Coronavirus (MERS-CoV) Receptor-Binding Domain Has Potent Cross-Neutralizing Activity and Protective Efficacy against MERS-CoV. *J. Virol.* **2018**, *92*. [CrossRef]
21. Hultberg, A.; Temperton, N.J.; Rosseels, V.; Koenders, M.; Gonzalez-Pajuelo, M.; Schepens, B.; Ibañez, L.I.; Vanlandschoot, P.; Schillemans, J.; Saunders, M.; et al. Llama-derived single domain antibodies to build multivalent, superpotent and broadened neutralizing anti-viral molecules. *PLoS ONE* **2011**, *6*, e17665. [CrossRef]
22. Rossey, I.; Gilman, M.S.A.; Kabeche, S.C.; Sedeyn, K.; Wrapp, D.; Kanekiyo, M.; Chen, M.; Mas, V.; Spitaels, J.; Melero, J.A.; et al. Potent single-domain antibodies that arrest respiratory syncytial virus fusion protein in its prefusion state. *Nat. Commun.* **2017**, *8*, 14158. [CrossRef] [PubMed]

23. Ashour, J.; Schmidt, F.I.; Hanke, L.; Cragnolini, J.; Cavallari, M.; Altenburg, A.; Brewer, R.; Ingram, J.; Shoemaker, C.; Ploegh, H.L. Intracellular Expression of Camelid Single-Domain Antibodies Specific for Influenza Virus Nucleoprotein Uncovers Distinct Features of Its Nuclear Localization. *J. Virol.* **2014**, *89*, 2792–2800. [CrossRef] [PubMed]
24. Vercruysse, T.; Pardon, E.; Vanstreels, E.; Steyaert, J.; Daelemans, D. An Intrabody Based on a Llama Single-Domain Antibody Targeting the N-terminal α-Helical Multimerization Domain of HIV-1 Rev Prevents Viral Production. *J. Biol. Chem.* **2010**, *285*, 21768–21780. [CrossRef] [PubMed]
25. Boons, E.; Li, G.; Vanstreels, E.; Vercruysse, T.; Pannecouque, C.; Vandamme, A.-M.; Daelemans, D. A stably expressed llama single-domain intrabody targeting Rev displays broad-spectrum anti-HIV activity. *Antivir. Res.* **2014**, *112*, 91–102. [CrossRef] [PubMed]
26. Cortez-Retamozo, V.; Lauwereys, M.; Hassanzadeh Gh., G.; Gobert, M.; Conrath, K.; Muyldermans, S.; De Baetselier, P.; Revets, H. Efficient tumor targeting by single-domain antibody fragments of camels. *Int. J. Cancer* **2002**. [CrossRef] [PubMed]
27. Huston, J.S.; George, A.J.; Adams, G.P.; Stafford, W.F.; Jamar, F.; Tai, M.S.; McCartney, J.E.; Oppermann, H.; Heelan, B.T.; Peters, A.M.; et al. Single-chain Fv radioimmunotargeting. *Q. J. Nucl. Med.* **1996**, *40*, 320–333. [CrossRef]
28. Batra, S.K.; Jain, M.; Wittel, U.A.; Chauhan, S.C.; Colcher, D. Pharmacokinetics and biodistribution of genetically engineered antibodies. *Curr. Opin. Biotechnol.* **2002**, *13*, 603–608. [CrossRef]
29. Kontermann, R.E. Strategies to extend plasma half-lives of recombinant antibodies. *BioDrugs* **2009**, *23*, 93–109. [CrossRef]
30. Harmsen, M.M.; Solt, C.B.; Fijten, H.; Setten, M.C. Prolonged in vivo residence times of llama single-domain antibody fragments in pigs by binding to porcine immunoglobulins. *Vaccine* **2005**, *23*, 4926–4934. [CrossRef]
31. Kuo, T.T.; Aveson, V.G. Neonatal Fc receptor and IgG-based therapeutics. *MAbs* **2011**, *3*, 422–430. [CrossRef]
32. Fan, K.; Jiang, B.; Guan, Z.; He, J.; Yang, D.; Xie, N.; Nie, G.; Xie, C.; Yan, X. Fenobody: A {Ferritin-Displayed} Nanobody with High Apparent Affinity and {Half-Life} Extension. *Anal. Chem.* **2018**, *90*, 5671–5677. [CrossRef] [PubMed]
33. Mu, B.; Huang, X.; Bu, P.; Zhuang, J.; Cheng, Z.; Feng, J.; Yang, D.; Dong, C.; Zhang, J.; Yan, X. Influenza virus detection with pentabody-activated nanoparticles. *J. Virol. Methods* **2010**, *169*, 282–289. [CrossRef] [PubMed]
34. Harmsen, M.M.; van Solt, C.B.; Fijten, H.P.D.; van Keulen, L.; Rosalia, R.A.; Weerdmeester, K.; Cornelissen, A.H.M.; Bruin, M.G.M.; Eblé, P.L.; Dekker, A. Passive immunization of guinea pigs with llama single-domain antibody fragments against foot-and-mouth disease. *Vet. Microbiol.* **2007**, *120*, 193–206. [CrossRef] [PubMed]
35. Yamaoka, T.; Tabata, Y.; Ikada, Y. Distribution and tissue uptake of poly(ethylene glycol) with different molecular weights after intravenous administration to mice. *J. Pharm. Sci.* **1994**, *83*, 601–606. [CrossRef] [PubMed]
36. Gaberc-Porekar, V.; Zore, I.; Podobnik, B.; Menart, V. Obstacles and pitfalls in the PEGylation of therapeutic proteins. *Curr. Opin. Drug Discov. Dev.* **2008**, *11*, 242–250.
37. Kubetzko, S. Protein PEGylation Decreases Observed Target Association Rates via a Dual Blocking Mechanism. *Mol. Pharmacol.* **2005**, *68*, 1439–1454. [CrossRef] [PubMed]
38. Bendele, A.; Seely, J.; Richey, C.; Sennello, G.; Shopp, G. Short communication: Renal tubular vacuolation in animals treated with polyethylene-glycol-conjugated proteins. *Toxicol. Sci.* **1998**, *42*, 152–157. [CrossRef]
39. Strohl, W.R. Fusion Proteins for {Half-Life} Extension of Biologics as a Strategy to Make Biobetters. *BioDrugs* **2015**, *29*, 215–239. [CrossRef]
40. Tang, L.; Persky, A.M.; Hochhaus, G.; Meibohm, B. Pharmacokinetic aspects of biotechnology products. *J. Pharm. Sci.* **2004**, *93*, 2184–2204. [CrossRef]
41. Constantinou, A.; Epenetos, A.A.; Hreczuk-Hirst, D.; Jain, S.; Wright, M.; Chester, K.A.; Deonarain, M.P. Site-specific polysialylation of an antitumor single-chain Fv fragment. *Bioconjug. Chem.* **2009**, *20*, 924–931. [CrossRef]
42. Berson, S.A.; Yalow, R.S.; Schreiber, S.S.; Post, J. Tracer experiments with I131 labeled human serum albumin: Distribution and degradation studies. *J. Clin. Investig.* **1953**, *32*, 746–768. [CrossRef] [PubMed]
43. Peters, T. Serum Albumin. *Adv. Protein Chem.* **1985**, *37*, 161–245. [CrossRef] [PubMed]

44. Hoefman, S.; Ottevaere, I.; Baumeister, J.; Sargentini-Maier, M. Pre-Clinical Intravenous Serum Pharmacokinetics of Albumin Binding and Non-Half-Life Extended Nanobodies®. *Antibodies* **2015**, *4*, 141–156. [CrossRef]
45. Terryn, S.; Francart, A.; Lamoral, S.; Hultberg, A.; Rommelaere, H.; Wittelsberger, A.; Callewaert, F.; Stohr, T.; Meerschaert, K.; Ottevaere, I.; et al. Protective Effect of Different {Anti-Rabies} Virus {VHH} Constructs against Rabies Disease in Mice. *PLoS ONE* **2014**, *9*, e109367. [CrossRef]
46. Kim, J. Albumin turnover: FcRn-mediated recycling saves as much albumin from degradation as the liver produces. *AJP Gastrointest. Liver Physiol.* **2006**, *290*, G352–G360. [CrossRef]
47. Kratz, F. Albumin as a drug carrier: Design of prodrugs, drug conjugates and nanoparticles. *J. Control. Release* **2008**. [CrossRef] [PubMed]
48. Raj, V.S.; Okba, N.M.A.; Gutierrez-Alvarez, J.; Drabek, D.; van Dieren, B.; Widagdo, W.; Lamers, M.M.; Widjaja, I.; Fernandez-Delgado, R.; Sola, I.; et al. Chimeric camel/human heavy-chain antibodies protect against MERS-CoV infection. *Sci. Adv.* **2018**, *4*, eaas9667. [CrossRef]
49. Yau, K.Y.F.; Dubuc, G.; Li, S.; Hirama, T.; Mackenzie, C.R.; Jermutus, L.; Hall, J.C.; Tanha, J. Affinity maturation of a V(H)H by mutational hotspot randomization. *J. Immunol. Methods* **2005**, *297*, 213–224. [CrossRef] [PubMed]
50. Saerens, D.; Ghassabeh, G.H.; Muyldermans, S. Single-domain antibodies as building blocks for novel therapeutics. *Curr. Opin. Pharmacol.* **2008**, *8*, 600–608. [CrossRef] [PubMed]
51. Lutje Hulsik, D.; Liu, Y.; Strokappe, N.M.; Battella, S.; El Khattabi, M.; McCoy, L.E.; Sabin, C.; Hinz, A.; Hock, M.; Macheboeuf, P.; et al. A gp41 MPER-specific Llama VHH Requires a Hydrophobic CDR3 for Neutralization but not for Antigen Recognition. *PLoS Pathog.* **2013**, *9*. [CrossRef] [PubMed]
52. Cardoso, F.M.; Ibañez, L.I.; Van den Hoecke, S.; De Baets, S.; Smet, A.; Roose, K.; Schepens, B.; Descamps, F.J.; Fiers, W.; Muyldermans, S.; et al. Single-domain antibodies targeting neuraminidase protect against an H5N1 influenza virus challenge. *J. Virol.* **2014**, *88*, 8278–8296. [CrossRef] [PubMed]
53. Matz, J.; Kessler, P.; Bouchet, J.; Combes, O.; Ramos, O.H.P.; Barin, F.; Baty, D.; Martin, L.; Benichou, S.; Chames, P. Straightforward selection of broadly neutralizing single-domain antibodies targeting the conserved CD4 and coreceptor binding sites of HIV-1 gp120. *J. Virol.* **2013**, *87*, 1137–1149. [CrossRef] [PubMed]
54. Ibañez, L.I.; De Filette, M.; Hultberg, A.; Verrips, T.; Temperton, N.; Weiss, R.A.; Vandevelde, W.; Schepens, B.; Vanlandschoot, P.; Saelens, X. Nanobodies with in vitro neutralizing activity protect mice against H5N1 influenza virus infection. *J. Infect. Dis.* **2011**, *203*, 1063–1072. [CrossRef] [PubMed]
55. Schepens, B.; Ibañez, L.I.; De Baets, S.; Hultberg, A.; Bogaert, P.; De Bleser, P.; Vervalle, F.; Verrips, T.; Melero, J.; Vandevelde, W.; et al. Nanobodies® specific for respiratory syncytial virus fusion protein protect against infection by inhibition of fusion. *J. Infect. Dis.* **2011**, *204*, 1692–1701. [CrossRef] [PubMed]
56. Detalle, L.; Stohr, T.; Palomo, C.; Piedra, P.A.; Gilbert, B.E.; Mas, V.; Millar, A.; Power, U.F.; Stortelers, C.; Allosery, K.; et al. Generation and Characterization of ALX-0171, a Potent Novel Therapeutic Nanobody for the Treatment of Respiratory Syncytial Virus Infection. *Antimicrob. Agents Chemother.* **2015**, *60*, 6–13. [CrossRef] [PubMed]
57. Palomo, C.; Mas, V.; Detalle, L.; Depla, E.; Cano, O.; Vázquez, M.; Stortelers, C.; Melero, J.A. Trivalency of a Nanobody Specific for the Human Respiratory Syncytial Virus Fusion Glycoprotein Drastically Enhances Virus Neutralization and Impacts Escape Mutant Selection. *Antimicrob. Agents Chemother.* **2016**, *60*, 6498–6509. [CrossRef] [PubMed]
58. Tillib, S.V.; Ivanova, T.I.; Vasilev, L.A.; Rutovskaya, M.V.; Saakyan, S.A.; Gribova, I.Y.; Tutykhina, I.L.; Sedova, E.S.; Lysenko, A.A.; Shmarov, M.M.; et al. Formatted single-domain antibodies can protect mice against infection with influenza virus (H5N2). *Antivir. Res.* **2013**, *97*, 245–254. [CrossRef] [PubMed]
59. Boruah, B.M.; Liu, D.; Ye, D.; Gu, T.; Jiang, C.; Qu, M.; Wright, E.; Wang, W.; He, W.; Liu, C.; et al. Single Domain Antibody Multimers Confer Protection against Rabies Infection. *PLoS ONE* **2013**, *8*, e71383. [CrossRef] [PubMed]
60. Nimmerjahn, F.; Ravetch, J.V. Fc$\gamma$ receptors as regulators of immune responses. *Nat. Rev. Immunol.* **2008**, *8*, 34–47. [CrossRef]
61. Van den Hoecke, S.; Ehrhardt, K.; Kolpe, A.; El Bakkouri, K.; Deng, L.; Grootaert, H.; Schoonooghe, S.; Smet, A.; Bentahir, M.; Roose, K.; et al. Hierarchical and Redundant Roles of Activating Fc$\gamma$Rs in Protection against Influenza Disease by M2e-Specific IgG1 and IgG2a Antibodies. *J. Virol.* **2017**, *91*. [CrossRef]

62. El Bakkouri, K.; Descamps, F.; De Filette, M.; Smet, A.; Festjens, E.; Birkett, A.; Van Rooijen, N.; Verbeek, S.; Fiers, W.; Saelens, X. Universal vaccine based on ectodomain of matrix protein 2 of influenza A: Fc receptors and alveolar macrophages mediate protection. *J. Immunol.* **2011**, *186*, 1022–1031. [CrossRef] [PubMed]
63. Jans, J.; Vissers, M.; Heldens, J.G.M.; de Jonge, M.I.; Levy, O.; Ferwerda, G. Fc gamma receptors in respiratory syncytial virus infections: implications for innate immunity. *Rev. Med. Virol.* **2014**, *24*, 55–70. [CrossRef] [PubMed]
64. Vogt, M.R.; Dowd, K.A.; Engle, M.; Tesh, R.B.; Johnson, S.; Pierson, T.C.; Diamond, M.S. Poorly neutralizing cross-reactive antibodies against the fusion loop of West Nile virus envelope protein protect in vivo via Fcgamma receptor and complement-dependent effector mechanisms. *J. Virol.* **2011**, *85*, 11567–11580. [CrossRef] [PubMed]
65. Chung, K.M.; Nybakken, G.E.; Thompson, B.S.; Engle, M.J.; Marri, A.; Fremont, D.H.; Diamond, M.S. Antibodies against West Nile Virus Nonstructural Protein NS1 Prevent Lethal Infection through Fc $\gamma$ Receptor-Dependent and -Independent Mechanisms. *J. Virol.* **2006**, *80*, 1340–1351. [CrossRef] [PubMed]
66. Cocklin, S.L.; Schmitz, J.E. The role of Fc receptors in HIV infection and vaccine efficacy. *Curr. Opin. HIV AIDS* **2014**, *9*, 257–262. [CrossRef] [PubMed]
67. Boesch, A.W.; Brown, E.; Ackerman, M.E. The role of Fc Receptors in HIV Prevention and Therapy. *Immunol. Rev.* **2015**, *268*, 296–310. [CrossRef] [PubMed]
68. Bobkov, V.; Zarca, A.M.; Van Hout, A.; Arimont, M.; Doijen, J.; Bialkowska, M.; Toffoli, E.; Klarenbeek, A.; van der Woning, B.; van der Vliet, H.J.; et al. Nanobody-Fc constructs targeting chemokine receptor CXCR4 potently inhibit signaling and CXCR4-mediated HIV-entry and induce antibody effector functions. *Biochem. Pharmacol.* **2018**. [CrossRef] [PubMed]
69. Burns, J.W.; Siadat-Pajouh, M.; Krishnaney, A.A.; Greenberg, H.B. Protective effect of rotavirus VP6-specific IgA monoclonal antibodies that lack neutralizing activity. *Science* **1996**, *272*, 104–107. [CrossRef]
70. Feng, N.; Lawton, J.A.; Gilbert, J.; Kuklin, N.; Vo, P.; Prasad, B.V.V.; Greenberg, H.B. Inhibition of rotavirus replication by a non-neutralizing, rotavirus VP6-specific IgA mAb. *J. Clin. Investig.* **2002**, *109*, 1203–1213. [CrossRef]
71. Aladin, F.; Einerhand, A.W.C.; Bouma, J.; Bezemer, S.; Hermans, P.; Wolvers, D.; Bellamy, K.; Frenken, L.G.J.; Gray, J.; Iturriza-Gómara, M. In vitro neutralisation of rotavirus infection by two broadly specific recombinant monovalent llama-derived antibody fragments. *PLoS ONE* **2012**, *7*, e32949. [CrossRef]
72. Garaicoechea, L.; Olichon, A.; Marcoppido, G.; Wigdorovitz, A.; Mozgovoj, M.; Saif, L.; Surrey, T.; Parreño, V. Llama-derived single-chain antibody fragments directed to rotavirus VP6 protein possess broad neutralizing activity in vitro and confer protection against diarrhea in mice. *J. Virol.* **2008**, *82*, 9753–9764. [CrossRef] [PubMed]
73. Sarker, S.A.; Jäkel, M.; Sultana, S.; Alam, N.H.; Bardhan, P.K.; Chisti, M.J.; Salam, M.A.; Theis, W.; Hammarström, L.; Frenken, L.G.J. Anti-Rotavirus Protein Reduces Stool Output in Infants With Diarrhea: A Randomized Placebo-Controlled Trial. *Gastroenterology* **2013**, *145*, 740–748.e8. [CrossRef] [PubMed]
74. McEwan, W.A.; Tam, J.C.H.; Watkinson, R.E.; Bidgood, S.R.; Mallery, D.L.; James, L.C. Intracellular antibody-bound pathogens stimulate immune signaling via the Fc receptor TRIM21. *Nat. Immunol.* **2013**, *14*, 327–336. [CrossRef] [PubMed]
75. Laursen, N.S.; Friesen, R.H.E.; Zhu, X.; Jongeneelen, M.; Blokland, S.; Vermond, J.; van Eijgen, A.; Tang, C.; van Diepen, H.; Obmolova, G.; et al. Universal protection against influenza infection by a multidomain antibody to influenza hemagglutinin. *Science* **2018**, *362*, 598–602. [CrossRef] [PubMed]
76. Günaydın, G.; Alvarez, B.; Lin, Y.; Hammarström, L.; Marcotte, H. Co-expression of anti-rotavirus proteins (llama VHH antibody fragments) in Lactobacillus: Development and functionality of vectors containing two expression cassettes in tandem. *PLoS ONE* **2014**, *9*, e96409. [CrossRef] [PubMed]
77. Sun, M.; Pace, C.S.; Yao, X.; Yu, F.; Padte, N.N.; Huang, Y.; Seaman, M.S.; Li, Q.; Ho, D.D. Rational design and characterization of the novel, broad and potent bispecific HIV-1 neutralizing antibody iMabm36. *J. Acquir. Immune Defic. Syndr.* **2014**, *66*, 473–483. [CrossRef] [PubMed]
78. Geoghegan, E.M.; Zhang, H.; Desai, P.J.; Biragyn, A.; Markham, R.B. Antiviral activity of a single-domain antibody immunotoxin binding to glycoprotein D of herpes simplex virus 2. *Antimicrob. Agents Chemother.* **2015**, *59*, 527–535. [CrossRef]

79. Wang, S.X.; Michiels, J.; Ariën, K.K.; New, R.; Vanham, G.; Roitt, I. Inhibition of HIV Virus by Neutralizing Vhh Attached to Dual Functional Liposomes Encapsulating Dapivirine. *Nanoscale Res. Lett.* **2016**, *11*, 350. [CrossRef]
80. Pant, N.; Marcotte, H.; Hermans, P.; Bezemer, S.; Frenken, L.; Johansen, K.; Hammarström, L. Lactobacilli producing bispecific llama-derived anti-rotavirus proteins in vivo for rotavirus-induced diarrhea. *Future Microbiol.* **2011**, *6*, 583–593. [CrossRef]
81. Álvarez, B.; Krogh-Andersen, K.; Tellgren-Roth, C.; Martínez, N.; Günaydın, G.; Lin, Y.; Martín, M.C.; Álvarez, M.A.; Hammarström, L.; Marcotte, H. An Exopolysaccharide-Deficient Mutant of Lactobacillus rhamnosus GG Efficiently Displays a Protective Llama Antibody Fragment against Rotavirus on Its Surface. *Appl. Environ. Microbiol.* **2015**, *81*, 5784–5793. [CrossRef]
82. Liu, L.; Wang, W.; Matz, J.; Ye, C.; Bracq, L.; Delon, J.; Kimata, J.T.; Chen, Z.; Benichou, S.; Zhou, P. The Glycosylphosphatidylinositol-Anchored Variable Region of Llama Heavy Chain-Only Antibody JM4 Efficiently Blocks both Cell-Free and T Cell-T Cell Transmission of Human Immunodeficiency Virus Type 1. *J. Virol.* **2016**, *90*, 10642–10659. [CrossRef] [PubMed]
83. Tiwari, P.M.; Vanover, D.; Lindsay, K.E.; Bawage, S.S.; Kirschman, J.L.; Bhosle, S.; Lifland, A.W.; Zurla, C.; Santangelo, P.J. Engineered mRNA-expressed antibodies prevent respiratory syncytial virus infection. *Nat. Commun.* **2018**, *9*, 3999. [CrossRef] [PubMed]
84. Thueng-in, K.; Thanongsaksrikul, J.; Srimanote, P.; Bangphoomi, K.; Poungpair, O.; Maneewatch, S.; Choowongkomon, K.; Chaicumpa, W. Cell penetrable humanized-VH/V(H)H that inhibit RNA dependent RNA polymerase (NS5B) of HCV. *PLoS ONE* **2012**, *7*, e49254. [CrossRef] [PubMed]
85. Phalaphol, A.; Thueng-In, K.; Thanongsaksrikul, J.; Poungpair, O.; Bangphoomi, K.; Sookrung, N.; Srimanote, P.; Chaicumpa, W. Humanized-VH/VHH that inhibit HCV replication by interfering with the virus helicase activity. *J. Virol. Methods* **2013**, *194*, 289–299. [CrossRef]
86. Glab-Ampai, K.; Malik, A.A.; Chulanetra, M.; Thanongsaksrikul, J.; Thueng-In, K.; Srimanote, P.; Tongtawe, P.; Chaicumpa, W. Inhibition of HCV replication by humanized-single domain transbodies to NS4B. *Biochem. Biophys. Res. Commun.* **2016**, *476*, 654–664. [CrossRef] [PubMed]
87. Tarr, A.W.; Lafaye, P.; Meredith, L.; Damier-Piolle, L.; Urbanowicz, R.A.; Meola, A.; Jestin, J.-L.; Brown, R.J.P.; McKeating, J.A.; Rey, F.A.; et al. An alpaca nanobody inhibits hepatitis C virus entry and cell-to-cell transmission. *Hepatology* **2013**, *58*, 932–939. [CrossRef] [PubMed]
88. Gray, M.A.; Tao, R.N.; DePorter, S.M.; Spiegel, D.A.; McNaughton, B.R. A Nanobody Activation Immunotherapeutic That Selectively Destroys HER2-Positive Breast Cancer Cells. *ChemBioChem* **2016**, *17*, 155–158. [CrossRef] [PubMed]
89. Rozan, C.; Cornillon, A.; Pétiard, C.; Chartier, M.; Behar, G.; Boix, C.; Kerfelec, B.; Robert, B.; Pèlegrin, A.; Chames, P.; et al. Single-domain antibody-based and linker-free bispecific antibodies targeting FcγRIII induce potent antitumor activity without recruiting regulatory T cells. *Mol. Cancer Ther.* **2013**, *12*, 1481–1491. [CrossRef] [PubMed]
90. Yu, Y.; Li, J.; Zhu, X.; Tang, X.; Bao, Y.; Sun, X.; Huang, Y.; Tian, F.; Liu, X.; Yang, L. Humanized CD7 nanobody-based immunotoxins exhibit promising anti-T-cell acute lymphoblastic leukemia potential. *Int. J. Nanomed.* **2017**, *12*, 1969–1983. [CrossRef]
91. Li, T.; Qi, S.; Unger, M.; Hou, Y.N.; Deng, Q.W.; Liu, J.; Lam, C.M.C.; Wang, X.W.; Xin, D.; Zhang, P.; et al. Immuno-targeting the multifunctional CD38 using nanobody. *Sci. Rep.* **2016**, *6*, 27055. [CrossRef]
92. Van der Vaart, J.M.; Pant, N.; Wolvers, D.; Bezemer, S.; Hermans, P.W.; Bellamy, K.; Sarker, S.A.; van der Logt, C.P.E.; Svensson, L.; Verrips, C.T.; et al. Reduction in morbidity of rotavirus induced diarrhoea in mice by yeast produced monovalent llama-derived antibody fragments. *Vaccine* **2006**, *24*, 4130–4137. [CrossRef] [PubMed]
93. Vinjé, J. Advances in laboratory methods for detection and typing of norovirus. *J. Clin. Microbiol.* **2015**, *53*, 373–381. [CrossRef] [PubMed]
94. Doerflinger, S.Y.; Tabatabai, J.; Schnitzler, P.; Farah, C.; Rameil, S.; Sander, P.; Koromyslova, A.; Hansman, G.S. Development of a Nanobody-Based Lateral Flow Immunoassay for Detection of Human Norovirus. *mSphere* **2016**, *1*. [CrossRef] [PubMed]
95. Zhu, M.; Hu, Y.; Li, G.; Ou, W.; Mao, P.; Xin, S.; Wan, Y. Combining magnetic nanoparticle with biotinylated nanobodies for rapid and sensitive detection of influenza H3N2. *Nanoscale Res. Lett.* **2014**, *9*, 528. [CrossRef] [PubMed]

96. Peyret, H.; Gehin, A.; Thuenemann, E.C.; Blond, D.; El Turabi, A.; Beales, L.; Clarke, D.; Gilbert, R.J.C.; Fry, E.E.; Stuart, D.I.; et al. Tandem fusion of hepatitis B core antigen allows assembly of virus-like particles in bacteria and plants with enhanced capacity to accommodate foreign proteins. *PLoS ONE* **2015**, *10*, e0120751. [CrossRef] [PubMed]
97. Czajkowsky, D.M.; Hu, J.; Shao, Z.; Pleass, R.J. Fc-fusion proteins: New developments and future perspectives. *EMBO Mol. Med.* **2012**, *4*, 1015–1028. [CrossRef]

*Review*

# Current Approaches and Future Perspectives for Nanobodies in Stroke Diagnostic and Therapy

**Larissa Jank [1], Carolina Pinto-Espinoza [2], Yinghui Duan [1], Friedrich Koch-Nolte [2], Tim Magnus [1] and Björn Rissiek [1,*]**

1 Department of Neurology, University Medical Center Hamburg-Eppendorf, 20246 Hamburg, Germany; larissajank@gmail.com (L.J.); yinghui.duan@stud.uke.uni-hamburg.de (Y.D.); t.magnus@uke.de (T.M.)

2 Institute of Immunology, University Medical Center Hamburg-Eppendorf, 20246 Hamburg, Germany; c.pinto@uke.de (C.P.-E.); nolte@uke.de (F.K.-N.)

* Correspondence: b.rissiek@uke.de; Tel.: +49-40-7410-58168

Received: 30 November 2018; Accepted: 27 December 2018; Published: 3 January 2019

**Abstract:** Antibody-based biologics are the corner stone of modern immunomodulatory therapy. Though highly effective in dampening systemic inflammatory processes, their large size and Fc-fragment mediated effects hamper crossing of the blood brain barrier (BBB). Nanobodies (Nbs) are single domain antibodies derived from llama or shark heavy-chain antibodies and represent a new generation of biologics. Due to their small size, they display excellent tissue penetration capacities and can be easily modified to adjust their vivo half-life for short-term diagnostic or long-term therapeutic purposes or to facilitate crossing of the BBB. Furthermore, owing to their characteristic binding mode, they are capable of antagonizing receptors involved in immune signaling and of neutralizing proinflammatory mediators, such as cytokines. These qualities combined make Nbs well-suited for down-modulating neuroinflammatory processes that occur in the context of brain ischemia. In this review, we summarize recent findings on Nbs in preclinical stroke models and how they can be used as diagnostic and therapeutic reagents. We further provide a perspective on the design of innovative Nb-based treatment protocols to complement and improve stroke therapy.

**Keywords:** nanobodies; ischemia; stroke; MCAO; single domain antibodies

---

## 1. Stroke and Post-Stroke Inflammation

According to the WHO Global Health, strokes are the second leading cause of death worldwide (10.2% of all deaths in 2016) and the second leading cause for loss of healthy years (5.2% of all disability-adjusted life years in 2016). In the future, these numbers are expected to further increase. In upper-middle income countries, prevalence is increasing due to the aging population [1], while in low-income countries, stroke incidence is rising due to changes in lifestyle and lack of adequate risk factor management [2]. Ischemic stroke is characterized by a reduced blood supply to the brain parenchyma. The following four underlying causes each account for about 25% of the ischemic strokes: (1) Embolization of a cardiac thrombus, (2) occlusion of a large vessel with atherosclerotic lesions, (3) small vasculature pathology usually leading to lacunar infarcts, and (4) other causes [3,4]. Due to the reduced blood flow, there is an energy deficit in neuron as well as a build-up of cellular waste products, such as lactate. This causes ionic disbalance, inducing the release of neurotransmitters, notably glutamate [5]. Glutamate binds to ionotropic glutamate receptors on neurons and calcium accumulates intracellularly. The calcium overload activates enzymatic cascades involved in neuron necrosis and apoptosis. These enzymes include phospholipases compromising membrane integrity as well as proteases mediating cell death and mitochondrial reactive oxygen species (ROS) production [6,7]. Furthermore, lack of adenosine triphosphate (ATP) reduces the activity of $Na^+/K^+$ ATPase, inducing neuronal edema [8].

Ischemic damage to neurons and tissue necrosis in the infarct core involves the release of damage associated molecular patterns (DAMPs) into the extracellular space, such as high mobility group protein B1 [9], ATP [10], heat shock protein 70 [11,12], histones, and DNA [13]. Extracellular DAMPs can bind to pattern recognition receptors (including the receptor for advanced glycation end products (RAGE), P2X7, and Toll-like receptors) on brain resident innate immune cells such as microglia, initiating an innate immune response within the first minutes after vessel occlusion [14]. In the first hours following stroke onset, microglia activation orchestrates the infiltration of other mononuclear cells in the peri-infarcted region, the penumbra [15]. The main functions of microglia include initiation and amplification of sterile inflammation by releasing proinflammatory cytokines (tumor necrosis factor $\alpha$ (TNF$\alpha$), IL-1$\beta$ and IL-6), generating ROS and nitric oxide (NO), phagocytosis to clear cell debris, and attracting peripheral immune cells to the penumbra with cytokines and chemokines, including monocyte chemoattractant protein 1 (MCP-1), macrophages inflammatory protein 1 $\alpha$ (MIP-1$\alpha$), and CXCL-8 [16–19]. Three days post-ischemia, the influx of peripheral immune cells is at its maximum [15]. Neutrophils are the most abundant peripheral immune cell population in the ischemic brain, which further enhance the sterile inflammation and contribute to infarct size growth [14]. At the peak of peripheral immune cell infiltration, T-cells are also attracted to the penumbra. $CD4^+$ and $CD8^+$ T-cells are involved in a major histocompatibility complex (MHC) dependent, i.e., antigen specific adaptive immune response, while more innate-like lymphocyte populations, such as $\gamma\delta$ T-cells, NKT cells, and NK cells are activated by cytokines and other molecules of the inflammatory milieu. This heterogeneous population of cells can contribute to infarct size growth either directly by cell-cell interactions, or indirectly through the induction of a humoral immune response or the release of cytotoxic substances [14,20].

To reach the penumbra, the attracted leukocytes need to cross the blood brain barrier (BBB). This structure consists of a monolayer of brain endothelial cells (ECs) surrounded by a basal membrane, pericytes, and astrocytes [21]. Proinflammatory cytokines released during cerebral ischemia activate ECs, leading to an increase in vesicles for transcellular transport and an increase in cell surface molecules associated with leukocyte recruitment [22] e.g., P-selectin and intercellular adhesion molecule 1 (ICAM-1), which mediate leukocyte rolling and adhesion, respectively [23,24]. Furthermore, matrix metalloproteases (MMPs) released in the penumbra change the tight junction conformation, enabling paracellular transport across the BBB [25].

## 2. Nanobodies—Single Domain Antibodies

### 2.1. Structure of Nanobodies and Conventional Antibody Fragments

Camelids, nurse sharks, and spotted ratfish exhibit naturally occurring heavy-chain-only antibodies (HcAbs) (Figure 1A). Interestingly, in camelids, HcAbs have evolved from conventional antibodies (cAbs), suggesting that they exhibit certain functional characteristics that are missing in cAbs [26]. This might be attributed to their smaller size and unique structure. Immunoglobulin G (IgG), the most abundant serum antibodies isotype in humans, consists of two heavy chains with three constant (CH1-3) and one variable domains (VH) each and two light chains with one constant (CL) and one variable domain (VL) each. In contrast, HcAbs only contain two heavy chains with two constant and one variable heavy-chain domain (VHH). Therefore, antigen binding by HcAbs is reduced to the VHH domain [27,28].

Both cAbs and HcAbs can be fragmented into smaller antigen-binding subunits in order to improve their tissue penetration. Common IgG modifications include: Cleavage of the Fc region to obtain Fab fragments, fusion of VH and VL domains with linker peptides to obtain single chain variable fragments (scFv), and the generation of autonomous human heavy-chain variable fragments (VH) [29,30].

VHH domains derived from HcAbs can be expressed as recombinant proteins, termed "Nanobodies" (Nbs), since they are one-tenth of the molecular size of an IgG molecule (Nbs: 15 kDa and IgG: 150 kDa). Nbs consists of four conserved framework regions and three antigen-binding loops, known as the complementarity determining regions (CDRs). The particularly long CDR3 rends the paratope its convex shape, building protrusions that can reach cryptic epitopes often not accessible to cAbs [27,28,31,32].

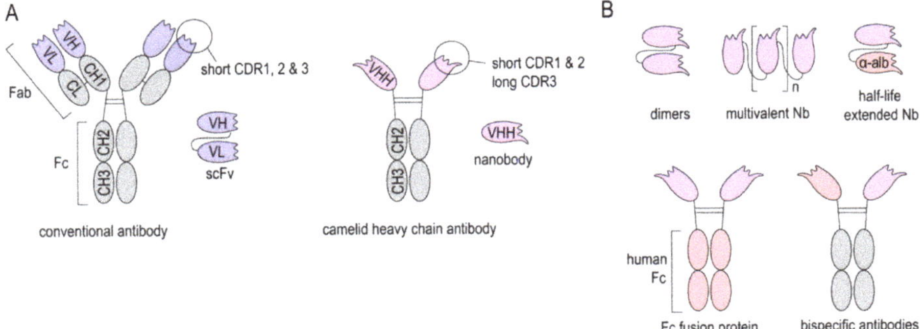


**Figure 1.** Structure of Nanobodies. (**A**) Structure of conventional antibodies (cAb), single-chain variable fragments (scFv), heavy-chain antibody (HcAb), and Nanobodies (Nbs). (**B**) Nbs can be produced as dimers and multimers to improve binding to their target or linked to an anti-albumin Nb to increase their in vivo half-life. Latter can be also achieved by fusing an Fc region of e.g., human IgG. Further, bispecific Nb-Fc-fusion proteins can also be expressed.

The first step in Nb generation is usually the immunization of HcAb-bearing large animals, such as llamas, alpacas, or sharks, followed by multiple boost immunizations in order to achieve an enrichment of high-affinity binders [33,34]. To overcome some of the logistical and financial limitations associated with immunization of large camelids, mice producing heavy-chain antibodies are being generated [35].

## 2.2. Advantages and Limitations of Nanobodies

One major advantage of Nbs is their small molecular size, which enables good tissue penetration and distribution. Furthermore, Nbs can refold after certain denaturation processes. This makes them very stable at extreme temperatures, low pH, and in the presence of proteases [36–38]. Additionally, Nbs are highly soluble in aqueous solutions, even at high concentrations [39,40]. These properties facilitate different routes of administration (e.g., intravenous, intraperitoneal, intrathecal, etc.), as well as various sites of action, such as pathological micro milieus. Due to their relatively simple structure, Nbs can be optimized by genetic engineering to obtain desired properties. They can be genetically linked to anti-albumin Nbs (Figure 1B) to extend their in vivo half-life. Further, Nb-Fc-fusion proteins allow binding to Fc receptors [41]. Several of these genetic modifications are aiming at facilitating crossing of the blood-brain-barrier (BBB) and are discussed below.

The small size of Nbs allows good tissue penetration, and also accounts for their short in vivo half-life when injected into experimental animals or into humans, since monovalent Nbs ($\approx$15 kDa) are rapidly eliminated via the kidney (70 kDa cut-off for renal filtration in humans) [42]. Though this might be beneficial for short-term applications, such as molecular imaging, it also is considered to be detrimental in long-term therapeutic applications. Increasing the size of Nbs through dimerization/multimerization, fusion to an anti-albumin Nb [43,44] or Fc engineering [45,46] can increase their serum half-life (Figure 1B). However, an increase in size and change in structure may also affect tissue penetration, affinity, stability, and solubility of Nbs. Another limitation of Nbs is that they are potentially immunogenic in humans since they originate from camelid species. Though recombinant Nbs lack an Fc region and share a large sequence identity with the human VH of family 3 [47], the risk of eliciting an anti-Nb adoptive immune response increases upon repeated application. Humanization of

Nbs is a strategy to address this problem, but it does not always sufficiently prevent antidrug antibody responses. Repeated injections of humanized Nbs (Caplacizumab) against von Willebrand Factor (vWF) resulted in a low incidence (9%) of antidrug antibody responses in the TITAN phase II study [48,49], while a clinical trial with a humanized anti-DR5 Nb (TAS266) had to be terminated because the applied Nbs evoked adverse host immune responses [50]. To this end, human scFvs or mutated human IgG lacking Fc-mediated effector functions have to be considered as a nonimmunogenic alternative to Nbs. However, for stroke diagnostic and therapy e.g., modulation of post-stroke inflammation, a single application of Nbs early after stroke onset might not reach the threshold for inducing an anti-Nb adaptive immune response. Since this threshold is highly dependent on the individual Nb, future studies on the application of Nbs in stroke should also address the issue of immunogenicity.

### 2.3. Nanobodies at the BBB

A major challenge for brain-targeting biologics is crossing of the BBB (Figure 2A). Under physiological conditions, only a very small fraction of intravenously injected cAbs cross the BBB (IgG CNS/plasma ratio: 0.1–1%), and once reaching the brain parenchyma, they are rapidly cleared by FcRn mediated efflux [51,52]. Nbs, on the other hand, lack an Fc region, are smaller in size, and more stable, promising facilitated delivery to the brain. However, when administered under non-pathological conditions, monovalent Nbs do not reach sufficient concentrations for in vivo brain imaging [53] and therapeutic purposes [42]. Pierre Lafaye's group was able to show that the brain penetration of Nbs can be improved by exploiting the process of adsorptive mediated transcytosis [54,55]. This transcytosis mechanism has been earlier identified to shuttle basic proteins and peptides across the BBB [56]. The basicity of a Nb can be increased by exchanging the carboxyl groups of the Nb with positively charged amino groups, thereby increasing the isoelectric point. The same group developed Nb-fluorochrome constructs (with pI = 8.3 and 9.5) that successfully label targets in an Alzheimer's disease model after being administered intravenously [57]. Nevertheless, high Nb concentrations (10–50 mg/kg) are required for detection and Nbs were only detectable for 4 h post-injection, suggesting a half-life too short for therapeutic purposes. The latter issue can be addressed by extending the half-life of Nbs. However, there is some controversy about the benefit of half-life extended Nbs for brain targeting. Iqbal et al. showed that fusion of an anti-EGFR Nb to a human Fc fragment improved the imaging of brain tumors [58], while another study of the kinetics of Nb-Fc fusion proteins showed that despite the extended serum half-life, the modification did not improve delivery across the BBB [59].

An alternative approach to deliver drugs to the brain is by receptor mediated transport (Figure 2B). Therapeutics are linked to ligands of or antibodies against receptors that are highly expressed on the BBB, such as the transferrin receptor [60,61], the insulin receptor [62,63], or the low-density lipoprotein receptor-related protein [64]. This antibody-mediated delivery of therapeutic proteins or peptides was studied in various neurological diseases, including stroke [65,66]. However, to date, there is only a limited number of studies in which therapeutic Nbs are delivered across the BBB via receptor mediated transcytosis. Rotman et al. loaded anti-amyloid Nbs into glutathione PEGylated liposomes. Glutathione can bind to receptors on cerebral endothelial cells and by this the liposomes are transported across the BBB [59].

Furthermore, Nbs that facilitate receptor mediated transport of biologics have been generated. The Nb clone FC5 was generated by phage-display in order to select Nbs that transmigrate across human cerebromicrovascular endothelial cells [67]. Later, it was discovered that FC5 targets a luminal sialoglycoprotein receptor (TMEM-30A), which induces the formation of clathrin vesicles and ultimately transcytosis [68]. By this mechanism, FC5 can act as a Trojan horse, transporting attached molecules across the BBB. Webster et al., for instance, generated FC5-Fc fusion proteins and conjugated these with the analgesic peptides dalargin and neuropeptide Y to deliver them across the BBB. Brain penetration of the FC5-Fc proteins was up to 30-fold higher compared to Fc protein alone [46]. The same group created bispecific antibodies with one FC5-arm and one arm targeting the metabotropic glutamate receptor-1 (mGluR1). These bispecific constructs showed a 20-fold higher brain penetration than unmodified anti-mGluR1 IgG [69].

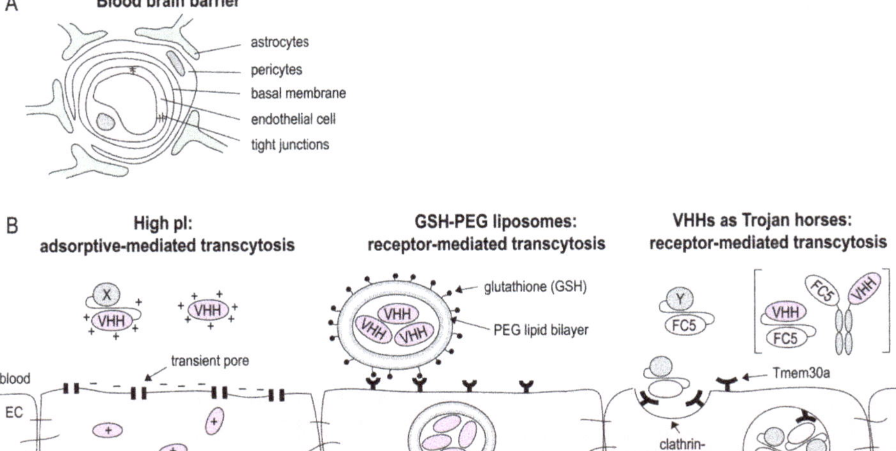


**Figure 2.** Blood-brain-barrier (BBB) crossing Nanobodies. (**A**) The BBB is built by the neurovascular unit consisting of endothelial cells connected via tight junctions, a basal membrane, pericytes and astrocytes foot processes. (**B**) Various strategies have been applied to shuttle Nbs (VHH) across the BBB: increasing the isoelectric point (pI) to facilitate uptake by endothelial cells (EC); package of Nbs in glutathione coated liposomes and receptor-mediated uptake into EC. Nbs such as the Tmem30a-specific Nb FC5 that target EC membrane receptors can act as Trojan horse to shuttle other Nbs or peptides across the BBB.

## 3. Stroke Imaging—New Job Opportunities for Nanobodies?

### 3.1. Principles of Stroke Imaging

In stroke therapy, early intervention by thrombolysis or mechanical thrombectomy is essential to save hypoxic tissue. However, the mere assessment of clinical signs and symptoms of stroke are not sufficient for diagnosis. Hence, imaging plays an important role in stroke diagnostics and management.

Acute imaging has to be fast and rule out other possible diagnoses, such as intracerebral bleeding, or so-called stroke mimics (e.g., epileptic seizures or migraine) [70]. The current standard procedure is computed tomography (CT) or magnetic resonance imaging (MRI) (if applicable with angiograms) within the first 4.5 h of stroke onset [71]. In the subacute stage, imaging reveals risk factors of cerebrovascular events, such as atherosclerotic plaques or dissections, in order to initiate adequate secondary prevention. To this end, imaging of the extracranial and intracranial arteries, the aorta and the heart is performed with CTA, MRA, carotid Doppler ultrasonography and echocardiography.

Anatomical imaging modalities, such as conventional CT and MRI, detect the secondary consequences of post-ischemic inflammation e.g., changes in diffusion and edema. Molecular and cellular imaging techniques, on the other hand, can be used to visualize and quantify distinct molecules, cell populations and processes. Here, we will focus on antibody- and Nb-assisted molecular imaging.

### 3.2. Imaging Endothelial Activation

Most studies on antibody-mediated molecular imaging in stroke target endothelial markers. These molecules are upregulated directly after occlusion and the antibody can bind these epitopes without crossing the BBB [72]. For example, Quenault et al. used microparticle of iron oxide (MPIOs) coated with P-selectin-targeting antibodies to identify transient ischemic attacks and to exclude other differential diagnosis, such as epilepsy and migraine, in MRI scans [73]. Other known endothelial activation markers used for antibody-mediated MRI stroke imaging include vascular cellular adhesion molecule 1 (VCAM-1) [74,75], platelet and endothelial cell adhesion molecule 1 (PECAM-1) [76], and ICAM-1 [77].

MRI is the modality of choice because it combines desirable properties, including relatively fast acquisition times, easy accessibility, and high safety. Nevertheless, molecular, nuclear, and optical imaging are important alternatives due to their high sensitivity. However, each imaging modality has its own drawbacks, including radiation for nuclear and CT imaging, possible tissue accumulation of MRI contrast agents, and limited imaging depth for optical imaging. Besides imaging-based limitations, cAb-mediated imaging may cause further difficulties in clinical application, including their long serum half-life (1–3 weeks), and therefore, strong background signal [78]. This could be addressed by replacing cAbs with Nbs. Devoogdt's group, for instance, created Nb-based imaging probes for positron-emission tomography (PET)/CT [79] and single photon emission computed tomography (SPECT) [80] targeting VCAM-1 for atherosclerosis plaque risk assessment. It is worth noting that Nbs unite several beneficial characteristics for endovascular imaging, including a high affinity to withstand shear forces in the vascular lumen and short serum half-life, which is essential, since imaging is preformed after the unbound contrast agent has been cleared from the blood [81].

As mentioned above, antibody-based molecular imaging of the brain is restricted to extracerebral markers, since antibodies usually do not spontaneously cross the BBB [72]. However, under brain pathophysiological conditions, such as stroke, multiple sclerosis, or Alzheimer's disease, the integrity of the BBB is impaired [82], allowing antigen-binding constructs facilitated access to the brain. Several studies have shown that Nbs labelled with fluorochromes or radioligands can access the brain in mouse models of Alzheimer disease [57], glioblastoma [58], and sleeping sickness [83], visualizing intracerebral processes, such as amyloid deposition, tumor-marker (EGFR) expression, and cerebral *Trypanosoma* invasion. Interestingly, Vandesquille et al. could even show that an MRI contrast agent (gadoterate meglumine), which alone does not cross the BBB, is able to pass once bound to a Nb [84]. Further, Li et al. could show that intravenously injected fluorochrome-labelled Nbs can be used to visualize brain amyloid plaques in an Alzheimer's disease mouse model [57]. However, to date, no Nbs have been utilized to image stroke-induced cerebral inflammation.

## 4. Nanobodies as New Thrombolytic Agents

The only FDA-approved treatment for acute cerebral ischemia is thrombolysis, i.e., the pharmaceutical resolution of occluding blood clots with recombinant tissue plasmin activator (rt-PA). However, in 2009, only 3.4–5.2% of acute stroke patients received this treatment in the USA [85]. Despite recent efforts to extend the therapeutic window with MRI imaging [86], the indications for rt-PA remain limited because of the high risk of bleeding.

Nbs-based thrombolysis may be a promising alternative to rt-PA or might improve its efficacy while simultaneously reducing adverse effects of thrombolysis, such as bleeding [87]. Interestingly, in August 2018, Caplacizumab, a Nb directed against von Willebrand Factor (vWF), was EMA-approved for acquired thrombotic thrombocytopenic purpura (aTTP) [48,88]. Caplacizumab inhibits the interaction of vWF with platelet glycoprotein Ib$\alpha$ (GPIb$\alpha$) receptors by binding the vWF A1 domain. This reduces platelet adhesion to damaged vessels and thrombus growth without increasing the risk for intracerebral hemorrhage [89]. Momi et al. showed that Caplacizumab is an effective therapy in a guinea pig stroke model. When given up to 15 min after occlusion, Caplacizumab prevented both clot formation and induced reperfusion, thereby reducing brain damage. In contrast to tirofiban (GP-IIb/IIIa-antagonist) and rt-PA, Caplacizumab did not increase intracerebral hemorrhage [90]. Furthermore, vWF inhibition also dampens thrombo-inflammatory processes including leukocyte infiltration [91].

Another potential target for pharmaceutical thrombolysis in stroke is thrombin-activatable fibrinolysis inhibitor (TAFI). TAFI is activated by thrombin or thrombin-thrombomodulin complexes during fibrinolysis. It acts as a negative feedback regulator, i.e., inhibits fibrinolysis. In stroke patients TAFI is elevated in the acute phase of ischemia [92] and is associated with a poor outcome [93]. Furthermore, studies on murine stroke models have shown that anti-TAFI monoclonal antibodies (MA-TCK26D6) reduce fibrinogen deposition, hence improving reperfusion [94]. Nbs against TAFI have been developed. They induce fibrinolysis in vitro and in vivo in a mouse model of thromboembolism [87]. The advantage of Nbs over conventional anti-TAFI antibodies is that Nbs can target different activation states of TAFI [95]. Hence, Nbs not only counteract TAFI activation, but can additionally inhibit already activated TAFI. It remains to be tested if this therapeutic Nb has beneficial effects in stroke.

## 5. Nanobodies to Modulate Post-Stroke Inflammation

The concept of post-stroke inflammation was established a decade ago. However, so far, no studies on Nanobody-based therapy for post-ischemic inflammation have been carried out. Therefore, in this section we will point out possible options to use already existing Nbs as modulators of post-stroke sterile inflammatory processes. The therapeutic approaches discussed include: DAMP inactivation, cytokine neutralization, and inhibition of cell migration.

### 5.1. *Targeting DAMP Signaling*

Within the first few minutes after stroke onset, DAMPs such as high mobility group protein B1 [9], ATP [10], heat shock protein 70 [11,12], histones, and DNA [13] are released. These molecules play a central role in initiating a sterile innate immune response by binding to corresponding DAMP receptors (including RAGE, P2X7, and Toll-like receptors) [14]. Inhibition of DAMPs and their receptors is a promising therapeutic strategy in stroke. Interestingly, Nbs inhibiting ATP/P2X7 signaling have been generated and successfully tested in two different inflammatory mouse models [96]. During inflammation, ATP is released into the extracellular space by damaged neurons and glial cells. Binding of ATP to P2X7 ion channel induces gating leading to $Na^+/Ca^{2+}$ influx and $K^+$ efflux. This activates the inflammasome, a multiprotein complex that cleaves inactive pro-IL1$\beta$ into its active form. In stroke patients the release of proinflammatory cytokine IL-1$\beta$ is associated with poor outcome [97]. Hence, inhibiting P2X7 may be a successful therapeutic approach. However, in vivo preclinical studies show opposing results when it comes to evaluating P2X7 as therapeutic target in stroke. In some studies, P2X7 inhibitors such as Reactive Blue 2 [98], Brilliant Blue G [99], adenosine 5′-triphosphate-2′,3′-dialdehyde (oATP), and A438079 [100] reduced ischemic brain damage in rat stroke models. Conversely, other studies suggest that P2X7 has neuroprotective effects. Kaiser et al. for instance found that P2X7 knockout mice develop worse cerebral edema after experimental stroke [101] and Yanagisawa et al. observed increased brain damage after P2X7 inhibitor (oATP) treatment [102]. Using P2X7 knockout mice and P2X7 inhibitors (oATP, PPADS, and KN62), another group observed that P2X7 had no significant effect on brain damage in experimental stroke [103]. These contradictory results may be attributed to differences in the stroke model, dosage, as well as starting time and duration of P2X7 inhibitor treatment. Furthermore, many of the used inhibitors have a poor specificity for P2X7 [104]. The latter issue may be addressed by using the P2X7-blocking Nbs developed by Danquah et al. [96], since they are highly specific and potent with an $IC_{50}$ in the subnanomolar range. Therefore, they represent valuable tools to further study the role of P2X7 in post-ischemic inflammation.

### 5.2. *Inflammatory Cytokine Neutralization*

Cytokines are major regulators of post-ischemic sterile inflammation. The main proinflammatory cytokines in stroke are TNF$\alpha$, IL-1$\beta$, and IL-6. In stroke patients, these cytokines rise after occlusion and correlate with neurological outcome [105]. In the following section we will discuss TNF$\alpha$ and IL-1$\beta$ as two potential targets for Nb-based therapy. In contrast to TNF$\alpha$ and IL-1$\beta$, IL-6 mainly has neuroprotective effects [106] and plays a major role in body temperature regulation in stroke patients [107]. Therefore, benefits of interfering with the IL-6 signaling pathway, especially with neutralizing anti-/nanobodies, are of potential negative outcome.

5.2.1. TNFα

In stroke patients, serum TNFα is elevated, peaking at day seven post-ischemia [108] and early TNFα levels in the cerebrospinal fluid (CSF) correlate with neurological outcome [109]. TNFα is mainly produced in macrophages and microglia [110] and binds to TNFα receptors 1 or 2 (TNFR1/2). These receptors initiate several different signaling cascades, e.g., MAPK, NF-κB and caspase 8/10 pathways. Hence, depending on the target cell and the micromilieu, TNFα ligation can lead to inflammation, apoptosis or proliferation [111]. Neutralizing TNFα has different effects depending on the nature (transitory/permanent) and time point of inhibition. Several authors have described that post-ischemic TNFα neutralization significantly reduces the infarct volume in both transient and permanent ischemia models [112–114]. On the contrary, Lambertsen et al. showed that TNFα and TNFR knockout mice had larger infarct volumes compared to wild-type mice [115]. Nawashiro et al. demonstrated that low-dose TNFα pretreatment increases the ischemic tolerance [116], suggesting protective effects of TNFα. Targeting TNFα in stroke, therefore, demands careful planning of the time of administration and, ideally, specific inhibitors that only target certain TNFα signaling pathways.

TNFα inhibitors approved by the FDA and EMA are etanercept (TNFR2-Fc-fusion protein), infliximab, adalimumab, golimumab (anti-TNFα monoclonal antibodies), and certizumab (anti-TNFα Fab fragment). Although these inhibitors are currently used to treat autoimmune diseases such as rheumatoid arthritis, they can have severe side effects, such as increased risk for infectious diseases and malignancies, notably lymphomas [117,118]. Another complication during treatment with these biologics is secondary non-response due to the formation of antidrug antibodies [119].

TNFα-targeting Nbs represent a promising alternative that addresses several limitations of the therapeutics listed above. When engineered into dimers or bispecific constructs, TNFα-specific Nbs show a higher potency to neutralize TNFα both in vitro and in mouse models of RA when compared to monoclonal antibodies [118]. Several groups have developed Nbs that inhibit TNFα signaling more selectively, thereby reducing possible adverse effects. Efimor et al. for instance linked antihuman TNFα Nbs to cell lineage marker, such as F4/80 for myeloid cells to neutralize TNFα in a targeted manner [120]. Steeland et al. generated anti-TNFR1 Nbs that selectively inhibit TNFα-TNFR1 interaction, hence sustaining beneficial effects (promoting Tregs) mediated by TNFR2 [121]. This group also tested these Nbs in the EAE mouse model for multiple sclerosis and observed an increase of Nbs in the brain of EAE-induced mice compared to healthy controls [122]. These studies suggest that targeting TNFα signaling with Nbs might be a promising therapeutic approach to dampen post-stroke inflammation.

5.2.2. IL-1β

IL-1β is one of three cytokines in the IL-1 superfamily: IL-1α, IL-1β, and IL-1Ra (IL-1 receptor antagonist). After stroke, the expression of IL-1β, IL-1Ra, and the IL-1 receptors increases [123–125]. The two main IL-1 signaling molecules studied in stroke are IL-1β and IL1Ra. IL1Ra competes with IL-1α/β for the IL-1 receptors, thereby inhibiting downstream pro-inflammatory effects [126]. In murine models of stroke, Anakinra, a modified recombinant IL1Ra, improves the neurological outcome, even when administered up to 3 h after onset [127]. Anakinra is FDA-approved for rheumatoid arthritis [128] and phase II trials in stroke patients were successful [129]. In contrast to IL1Ra, IL-1β has detrimental effects in stroke. Preclinical studies with IL-1α/β knockout mice [130] and intrathecal injection of recombinant IL-1β [131] show that IL-1β worsens the neurological outcome. Liberale et al. showed that inhibiting IL-1β with a monoclonal antibody dampens post-ischemic inflammation and reduces infarct size [132]. Canakinumab, a human anti-IL-1β monoclonal antibody, is FDA-approved for arthritis, but remains to be tested in stroke patients. To the best of our knowledge, no Nbs targeting IL-1β or its receptors have been developed so far.

### 5.3. Cell Migration

After stroke onset, leukocytes from the periphery migrate to the ischemic lesion and contribute to post-ischemic inflammation [15]. The most abundant cell type recruited are neutrophil granulocytes, which promote both neuroinflammatory and neuroprotective effects [133]. To this end, regulating cell migration, in particular neutrophil influx, may be a new therapeutic approach to control post-ischemic inflammation.

#### 5.3.1. CXCR2

The CXCR2/CXCL8 axis is involved in chemotaxis of granulocytes and NK cells to the infarcted area after stroke [134]. Targeting this signaling pathway may therefore be beneficial in reducing post-ischemic inflammation. In a rat model of stroke, Connell et al. showed that CXCR1/2 antagonists Repertaxin [135] and G31P [136] significantly reduce ischemic brain damage. Similarly, He et al. suggest that targeting CXCR2 may be beneficial in humans [134]. Interestingly, Brait et al. also found that treatment with a CXCR2 antagonist (SB225002) reduced neutrophil infiltration, but had no effect on neurological outcome [137]. Therefore, the benefits in terms of disease outcome and the optimum time of CXCR2 inhibition to dampen inflammatory effects without limiting regenerative effects of neutrophils still need to be studied. Hereby, Nbs may be a useful instrument, due to their short serum half-life and high specificity. Antihuman CXCR2 Nbs have been developed [138].

#### 5.3.2. CXCR4

Another well-studied chemokine axis in stroke involves CXCR4 and CXCL12. It is associated with both proinflammatory and regenerative processes, including angiogenesis, and the recruitment of neural stem cells and various immune cells to the infarct zone [139]. Inhibiting this pathway with CX549 [140] and AMD3100 [141,142] reduces ischemic brain damage and improves neurological outcome. As for CXCR2, anti-CXCR4 Nbs have been developed [143,144], but have not yet been tested in stroke.

## 6. Conclusions

According to WHO Global Health estimates, strokes are the second leading cause of death worldwide (10.2% of all deaths in 2016). In the future, the prevalence is expected to rise further. This scenario has led to an extensive amount of research in this field. Today, there are many different clinical and preclinical studies evaluating the causes of stroke, diagnostic tools, and possible therapeutic targets. However, despite the extensive research, particularly on post-stroke inflammation, the only treatment for acute ischemic stroke is rt-PA. In part, this can be attributed to the difficult delivery of therapeutics across the BBB. Here, Nbs may be of great benefit. They are small in size, but have a high specificity and can be modified to facilitate crossing of the BBB. Nevertheless, further research has to be undertaken to fully understand which Nb modifications optimize brain penetration and which targets are best suited for Nb-based therapy of stroke.

**Author Contributions:** Conceptualization, B.R. and L.J.; Writing—original draft preparation, B.R., L.J., C.P.-E., and Y.D.; Writing—review and editing, F.K.-N., T.M.; Visualization, L.J.

**Funding:** This research was funded by grants from the "Hermann und Lilly Schilling Stiftung für Medizinische Forschung" to T.M. and the Deutsche Forschungsgemeinschaft to T.M. (SFB1328/A13) and F.K.-N. (SFB1328/Z2).

**Conflicts of Interest:** The authors declare no conflict of interest.

## References

1. Truelsen, T.; Piechowski-Jozwiak, B.; Bonita, R.; Mathers, C.; Bogousslavsky, J.; Boysen, G. Stroke incidence and prevalence in europe: A review of available data. *Eur. J. Neurol.* **2006**, *13*, 581–598. [CrossRef] [PubMed]

2. Pandian, J.D.; William, A.G.; Kate, M.P.; Norrving, B.; Mensah, G.A.; Davis, S.; Roth, G.A.; Thrift, A.G.; Kengne, A.P.; Kissela, B.M.; et al. Strategies to improve stroke care services in low- and middle-income countries: A systematic review. *Neuroepidemiology* **2017**, *49*, 45–61. [CrossRef] [PubMed]
3. Ay, H.; Furie, K.L.; Singhal, A.; Smith, W.S.; Sorensen, A.G.; Koroshetz, W.J. An evidence-based causative classification system for acute ischemic stroke. *Ann. Neurol.* **2005**, *58*, 688–697. [CrossRef] [PubMed]
4. Amarenco, P.; Bogousslavsky, J.; Caplan, L.R.; Donnan, G.A.; Hennerici, M.G. Classification of stroke subtypes. *Cerebrovasc. Dis.* **2009**, *27*, 493–501. [CrossRef] [PubMed]
5. Lee, J.M.; Grabb, M.C.; Zipfel, G.J.; Choi, D.W. Brain tissue responses to ischemia. *J. Clin. Investig.* **2000**, *106*, 723–731. [CrossRef] [PubMed]
6. Lipton, P. Ischemic cell death in brain neurons. *Physiol. Rev.* **1999**, *79*, 1431–1568. [CrossRef] [PubMed]
7. Xing, C.; Arai, K.; Lo, E.H.; Hommel, M. Pathophysiologic cascades in ischemic stroke. *Int. J. Stroke* **2012**, *7*, 378–385. [CrossRef]
8. Liang, D.; Bhatta, S.; Gerzanich, V.; Simard, J.M. Cytotoxic edema: Mechanisms of pathological cell swelling. *Neurosurg. Focus* **2007**, *22*, E2. [CrossRef]
9. Liesz, A.; Dalpke, A.; Mracsko, E.; Antoine, D.J.; Roth, S.; Zhou, W.; Yang, H.; Na, S.Y.; Akhisaroglu, M.; Fleming, T.; et al. Damp signaling is a key pathway inducing immune modulation after brain injury. *J. Neurosci.* **2015**, *35*, 583–598. [CrossRef]
10. Pedata, F.; Dettori, I.; Coppi, E.; Melani, A.; Fusco, I.; Corradetti, R.; Pugliese, A.M. Purinergic signalling in brain ischemia. *Neuropharmacology* **2016**, *104*, 105–130. [CrossRef]
11. Giffard, R.G.; Han, R.Q.; Emery, J.F.; Duan, M.; Pittet, J.F. Regulation of apoptotic and inflammatory cell signaling in cerebral ischemia: The complex roles of heat shock protein 70. *Anesthesiology* **2008**, *109*, 339–348. [CrossRef] [PubMed]
12. Sharp, F.R.; Lu, A.; Tang, Y.; Millhorn, D.E. Multiple molecular penumbras after focal cerebral ischemia. *J. Cereb. Blood Flow Metab.* **2000**, *20*, 1011–1032. [CrossRef] [PubMed]
13. Marsman, G.; Zeerleder, S.; Luken, B.M. Extracellular histones, cell-free DNA, or nucleosomes: Differences in immunostimulation. *Cell. Death Dis.* **2016**, *7*, e2518. [CrossRef] [PubMed]
14. Iadecola, C.; Anrather, J. The immunology of stroke: From mechanisms to translation. *Nat. Med.* **2011**, *17*, 796–808. [CrossRef] [PubMed]
15. Gelderblom, M.; Leypoldt, F.; Steinbach, K.; Behrens, D.; Choe, C.U.; Siler, D.A.; Arumugam, T.V.; Orthey, E.; Gerloff, C.; Tolosa, E.; et al. Temporal and spatial dynamics of cerebral immune cell accumulation in stroke. *Stroke* **2009**, *40*, 1849–1857. [CrossRef] [PubMed]
16. Patel, A.R.; Ritzel, R.; McCullough, L.D.; Liu, F. Microglia and ischemic stroke: A double-edged sword. *Int. J. Physiol. Pathophysiol. Pharmacol.* **2013**, *5*, 73–90.
17. Yenari, M.A.; Kauppinen, T.M.; Swanson, R.A. Microglial activation in stroke: Therapeutic targets. *Neurotherapeutics* **2010**, *7*, 378–391. [CrossRef]
18. Wang, Q.; Tang, X.N.; Yenari, M.A. The inflammatory response in stroke. *J. Neuroimmunol.* **2007**, *184*, 53–68. [CrossRef]
19. Wang, L.; Li, Y.; Chen, X.; Chen, J.; Gautam, S.C.; Xu, Y.; Chopp, M. Mcp-1, mip-1, il-8 and ischemic cerebral tissue enhance human bone marrow stromal cell migration in interface culture. *Hematology* **2002**, *7*, 113–117. [CrossRef]
20. Chamorro, A.; Meisel, A.; Planas, A.M.; Urra, X.; van de Beek, D.; Veltkamp, R. The immunology of acute stroke. *Nat. Rev. Neurol.* **2012**, *8*, 401–410. [CrossRef]
21. Ballabh, P.; Braun, A.; Nedergaard, M. The blood-brain barrier: An overview: Structure, regulation, and clinical implications. *Neurobiol. Dis.* **2004**, *16*, 1–13. [CrossRef] [PubMed]
22. Haley, M.J.; Lawrence, C.B. The blood-brain barrier after stroke: Structural studies and the role of transcytotic vesicles. *J. Cereb. Blood Flow Metab.* **2017**, *37*, 456–470. [CrossRef] [PubMed]
23. Nourshargh, S.; Alon, R. Leukocyte migration into inflamed tissues. *Immunity* **2014**, *41*, 694–707. [CrossRef] [PubMed]
24. Yilmaz, G.; Granger, D.N. Leukocyte recruitment and ischemic brain injury. *Neuromol. Med.* **2010**, *12*, 193–204. [CrossRef] [PubMed]
25. Knowland, D.; Arac, A.; Sekiguchi, K.J.; Hsu, M.; Lutz, S.E.; Perrino, J.; Steinberg, G.K.; Barres, B.A.; Nimmerjahn, A.; Agalliu, D. Stepwise recruitment of transcellular and paracellular pathways underlies blood-brain barrier breakdown in stroke. *Neuron* **2014**, *82*, 603–617. [CrossRef] [PubMed]

26. Nguyen, V.K.; Su, C.; Muyldermans, S.; van der Loo, W. Heavy-chain antibodies in camelidae; a case of evolutionary innovation. *Immunogenetics* **2002**, *54*, 39–47.
27. Beghein, E.; Gettemans, J. Nanobody technology: A versatile toolkit for microscopic imaging, protein-protein interaction analysis, and protein function exploration. *Front. Immunol.* **2017**, *8*, 771. [CrossRef]
28. Steeland, S.; Vandenbroucke, R.E.; Libert, C. Nanobodies as therapeutics: Big opportunities for small antibodies. *Drug Discov. Today* **2016**, *21*, 1076–1113. [CrossRef]
29. Barthelemy, P.A.; Raab, H.; Appleton, B.A.; Bond, C.J.; Wu, P.; Wiesmann, C.; Sidhu, S.S. Comprehensive analysis of the factors contributing to the stability and solubility of autonomous human vh domains. *J. Biol. Chem.* **2008**, *283*, 3639–3654. [CrossRef]
30. Tanha, J.; Nguyen, T.D.; Ng, A.; Ryan, S.; Ni, F.; Mackenzie, R. Improving solubility and refolding efficiency of human v(h)s by a novel mutational approach. *Protein Eng. Des. Sel.* **2006**, *19*, 503–509. [CrossRef]
31. Muyldermans, S. Nanobodies: Natural single-domain antibodies. *Annu. Rev. Biochem.* **2013**, *82*, 775–797. [CrossRef] [PubMed]
32. Pirez-Schirmer, M.; Rossotti, M.; Badagian, N.; Leizagoyen, C.; Brena, B.M.; Gonzalez-Sapienza, G. Comparison of three antihapten vhh selection strategies for the development of highly sensitive immunoassays for microcystins. *Anal. Chem.* **2017**, *89*, 6800–6806. [CrossRef] [PubMed]
33. Arbabi-Ghahroudi, M. Camelid single-domain antibodies: Historical perspective and future outlook. *Front. Immunol.* **2017**, *8*, 1589. [CrossRef] [PubMed]
34. Eden, T.; Menzel, S.; Wesolowski, J.; Bergmann, P.; Nissen, M.; Dubberke, G.; Seyfried, F.; Albrecht, B.; Haag, F.; Koch-Nolte, F. A cdna immunization strategy to generate nanobodies against membrane proteins in native conformation. *Front. Immunol.* **2017**, *8*, 1989. [CrossRef] [PubMed]
35. Janssens, R.; Dekker, S.; Hendriks, R.W.; Panayotou, G.; van Remoortere, A.; San, J.K.; Grosveld, F.; Drabek, D. Generation of heavy-chain-only antibodies in mice. *Proc. Natl. Acad. Sci. USA* **2006**, *103*, 15130–15135. [CrossRef] [PubMed]
36. Kunz, P.; Zinner, K.; Mucke, N.; Bartoschik, T.; Muyldermans, S.; Hoheisel, J.D. The structural basis of nanobody unfolding reversibility and thermoresistance. *Sci. Rep.* **2018**, *8*, 7934. [CrossRef] [PubMed]
37. Hussack, G.; Hirama, T.; Ding, W.; Mackenzie, R.; Tanha, J. Engineered single-domain antibodies with high protease resistance and thermal stability. *PLoS ONE* **2011**, *6*, e28218. [CrossRef]
38. Davies, J.; Riechmann, L. Single antibody domains as small recognition units: Design and in vitro antigen selection of camelized, human vh domains with improved protein stability. *Protein Eng.* **1996**, *9*, 531–537. [CrossRef]
39. Conrath, K.; Vincke, C.; Stijlemans, B.; Schymkowitz, J.; Decanniere, K.; Wyns, L.; Muyldermans, S.; Loris, R. Antigen binding and solubility effects upon the veneering of a camel vhh in framework-2 to mimic a vh. *J. Mol. Biol.* **2005**, *350*, 112–125. [CrossRef]
40. Ewert, S.; Cambillau, C.; Conrath, K.; Pluckthun, A. Biophysical properties of camelid v(hh) domains compared to those of human v(h)3 domains. *Biochemistry* **2002**, *41*, 3628–3636. [CrossRef]
41. Schumacher, D.; Helma, J.; Schneider, A.F.L.; Leonhardt, H.; Hackenberger, C.P.R. Nanobodies: Chemical functionalization strategies and intracellular applications. *Angew. Chem. Int. Ed. Engl.* **2018**, *57*, 2314–2333. [CrossRef] [PubMed]
42. Caljon, G.; Caveliers, V.; Lahoutte, T.; Stijlemans, B.; Ghassabeh, G.H.; Van Den Abbeele, J.; Smolders, I.; De Baetselier, P.; Michotte, Y.; Muyldermans, S.; et al. Using microdialysis to analyse the passage of monovalent nanobodies through the blood-brain barrier. *Br. J. Pharmacol.* **2012**, *165*, 2341–2353. [CrossRef] [PubMed]
43. Coppieters, K.; Dreier, T.; Silence, K.; de Haard, H.; Lauwereys, M.; Casteels, P.; Beirnaert, E.; Jonckheere, H.; Van de Wiele, C.; Staelens, L.; et al. Formatted anti-tumor necrosis factor alpha vhh proteins derived from camelids show superior potency and targeting to inflamed joints in a murine model of collagen-induced arthritis. *Arthr. Rheumatol.* **2006**, *54*, 1856–1866. [CrossRef] [PubMed]

44. Van Roy, M.; Ververken, C.; Beirnaert, E.; Hoefman, S.; Kolkman, J.; Vierboom, M.; Breedveld, E.; Poelmans, S.; Bontinck, L.; Hemeryck, A.; et al. The preclinical pharmacology of the high affinity anti-il-6r nanobody(r) alx-0061 supports its clinical development in rheumatoid arthritis. *Arthr. Res. Ther.* **2015**, *17*, 135. [CrossRef] [PubMed]
45. Bell, A.; Wang, Z.J.; Arbabi-Ghahroudi, M.; Chang, T.A.; Durocher, Y.; Trojahn, U.; Baardsnes, J.; Jaramillo, M.L.; Li, S.; Baral, T.N.; et al. Differential tumor-targeting abilities of three single-domain antibody formats. *Cancer Lett.* **2010**, *289*, 81–90. [CrossRef] [PubMed]
46. Farrington, G.K.; Caram-Salas, N.; Haqqani, A.S.; Brunette, E.; Eldredge, J.; Pepinsky, B.; Antognetti, G.; Baumann, E.; Ding, W.; Garber, E.; et al. A novel platform for engineering blood-brain barrier-crossing bispecific biologics. *FASEB J.* **2014**, *28*, 4764–4778. [CrossRef]
47. Nguyen, V.K.; Hamers, R.; Wyns, L.; Muyldermans, S. Camel heavy-chain antibodies: Diverse germline v(h)h and specific mechanisms enlarge the antigen-binding repertoire. *EMBO J.* **2000**, *19*, 921–930. [CrossRef]
48. Peyvandi, F.; Scully, M.; Kremer Hovinga, J.A.; Knobl, P.; Cataland, S.; De Beuf, K.; Callewaert, F.; De Winter, H.; Zeldin, R.K. Caplacizumab reduces the frequency of major thromboembolic events, exacerbations and death in patients with acquired thrombotic thrombocytopenic purpura. *J. Thromb. Haemost.* **2017**, *15*, 1448–1452. [CrossRef]
49. Peyvandi, F.; Scully, M.; Kremer Hovinga, J.A.; Cataland, S.; Knobl, P.; Wu, H.; Artoni, A.; Westwood, J.P.; Mansouri Taleghani, M.; Jilma, B.; et al. Caplacizumab for acquired thrombotic thrombocytopenic purpura. *N. Engl. J. Med.* **2016**, *374*, 511–522. [CrossRef]
50. Papadopoulos, K.P.; Isaacs, R.; Bilic, S.; Kentsch, K.; Huet, H.A.; Hofmann, M.; Rasco, D.; Kundamal, N.; Tang, Z.; Cooksey, J.; et al. Unexpected hepatotoxicity in a phase i study of tas266, a novel tetravalent agonistic nanobody(r) targeting the dr5 receptor. *Cancer Chemother. Pharmacol.* **2015**, *75*, 887–895. [CrossRef]
51. Cooper, P.R.; Ciambrone, G.J.; Kliwinski, C.M.; Maze, E.; Johnson, L.; Li, Q.; Feng, Y.; Hornby, P.J. Efflux of monoclonal antibodies from rat brain by neonatal fc receptor, fcrn. *Brain Res.* **2013**, *1534*, 13–21. [CrossRef] [PubMed]
52. Abuqayyas, L.; Balthasar, J.P. Investigation of the role of fcgammar and fcrn in mab distribution to the brain. *Mol. Pharm.* **2013**, *10*, 1505–1513. [CrossRef]
53. Nabuurs, R.J.; Rutgers, K.S.; Welling, M.M.; Metaxas, A.; de Backer, M.E.; Rotman, M.; Bacskai, B.J.; van Buchem, M.A.; van der Maarel, S.M.; van der Weerd, L. In vivo detection of amyloid-beta deposits using heavy chain antibody fragments in a transgenic mouse model for alzheimer's disease. *PLoS ONE* **2012**, *7*, e38284. [CrossRef] [PubMed]
54. Li, T.; Bourgeois, J.P.; Celli, S.; Glacial, F.; Le Sourd, A.M.; Mecheri, S.; Weksler, B.; Romero, I.; Couraud, P.O.; Rougeon, F.; et al. Cell-penetrating anti-gfap vhh and corresponding fluorescent fusion protein vhh-gfp spontaneously cross the blood-brain barrier and specifically recognize astrocytes: Application to brain imaging. *FASEB J.* **2012**, *26*, 3969–3979. [CrossRef]
55. Herce, H.D.; Garcia, A.E. Molecular dynamics simulations suggest a mechanism for translocation of the hiv-1 tat peptide across lipid membranes. *Proc. Natl. Acad. Sci. USA* **2007**, *104*, 20805–20810. [CrossRef] [PubMed]
56. Tamai, I.; Sai, Y.; Kobayashi, H.; Kamata, M.; Wakamiya, T.; Tsuji, A. Structure-internalization relationship for adsorptive-mediated endocytosis of basic peptides at the blood-brain barrier. *J. Pharmacol. Exp. Ther.* **1997**, *280*, 410–415.
57. Li, T.; Vandesquille, M.; Koukouli, F.; Dudeffant, C.; Youssef, I.; Lenormand, P.; Ganneau, C.; Maskos, U.; Czech, C.; Grueninger, F.; et al. Camelid single-domain antibodies: A versatile tool for in vivo imaging of extracellular and intracellular brain targets. *J. Control. Release* **2016**, *243*, 1–10. [CrossRef]
58. Iqbal, U.; Trojahn, U.; Albaghdadi, H.; Zhang, J.; O'Connor-McCourt, M.; Stanimirovic, D.; Tomanek, B.; Sutherland, G.; Abulrob, A. Kinetic analysis of novel mono- and multivalent vhh-fragments and their application for molecular imaging of brain tumours. *Br. J. Pharmacol.* **2010**, *160*, 1016–1028. [CrossRef]
59. Rotman, M.; Welling, M.M.; Bunschoten, A.; de Backer, M.E.; Rip, J.; Nabuurs, R.J.; Gaillard, P.J.; van Buchem, M.A.; van der Maarel, S.M.; van der Weerd, L. Enhanced glutathione pegylated liposomal brain delivery of an anti-amyloid single domain antibody fragment in a mouse model for alzheimer's disease. *J. Control. Release* **2015**, *203*, 40–50. [CrossRef]
60. Yoshikawa, T.; Pardridge, W.M. Biotin delivery to brain with a covalent conjugate of avidin and a monoclonal antibody to the transferrin receptor. *J. Pharmacol. Exp. Ther.* **1992**, *263*, 897–903.

61. Thom, G.; Burrell, M.; Haqqani, A.S.; Yogi, A.; Lessard, E.; Brunette, E.; Delaney, C.; Baumann, E.; Callaghan, D.; Rodrigo, N.; et al. Enhanced delivery of galanin conjugates to the brain through bioengineering of the anti-transferrin receptor antibody ox26. *Mol. Pharm.* **2018**, *15*, 1420–1431. [CrossRef] [PubMed]
62. Pardridge, W.M.; Kang, Y.S.; Buciak, J.L.; Yang, J. Human insulin receptor monoclonal antibody undergoes high affinity binding to human brain capillaries in vitro and rapid transcytosis through the blood-brain barrier in vivo in the primate. *Pharm Res.* **1995**, *12*, 807–816. [CrossRef] [PubMed]
63. Boado, R.J.; Pardridge, W.M. Brain and organ uptake in the rhesus monkey in vivo of recombinant iduronidase compared to an insulin receptor antibody-iduronidase fusion protein. *Mol. Pharm.* **2017**, *14*, 1271–1277. [CrossRef] [PubMed]
64. Michaelis, K.; Hoffmann, M.M.; Dreis, S.; Herbert, E.; Alyautdin, R.N.; Michaelis, M.; Kreuter, J.; Langer, K. Covalent linkage of apolipoprotein e to albumin nanoparticles strongly enhances drug transport into the brain. *J. Pharmacol. Exp. Ther.* **2006**, *317*, 1246–1253. [CrossRef] [PubMed]
65. Zhang, Y.; Pardridge, W.M. Blood-brain barrier targeting of bdnf improves motor function in rats with middle cerebral artery occlusion. *Brain Res.* **2006**, *1111*, 227–229. [CrossRef] [PubMed]
66. Boado, R.J.; Zhang, Y.; Zhang, Y.; Wang, Y.; Pardridge, W.M. Gdnf fusion protein for targeted-drug delivery across the human blood-brain barrier. *Biotechnol. Bioeng.* **2008**, *100*, 387–396. [CrossRef] [PubMed]
67. Muruganandam, A.; Tanha, J.; Narang, S.; Stanimirovic, D. Selection of phage-displayed llama single-domain antibodies that transmigrate across human blood-brain barrier endothelium. *FASEB J.* **2002**, *16*, 240–242. [CrossRef] [PubMed]
68. Abulrob, A.; Sprong, H.; Van Bergen en Henegouwen, P.; Stanimirovic, D. The blood-brain barrier transmigrating single domain antibody: Mechanisms of transport and antigenic epitopes in human brain endothelial cells. *J. Neurochem.* **2005**, *95*, 1201–1214. [CrossRef]
69. Webster, C.I.; Caram-Salas, N.; Haqqani, A.S.; Thom, G.; Brown, L.; Rennie, K.; Yogi, A.; Costain, W.; Brunette, E.; Stanimirovic, D.B. Brain penetration, target engagement, and disposition of the blood-brain barrier-crossing bispecific antibody antagonist of metabotropic glutamate receptor type 1. *FASEB J.* **2016**, *30*, 1927–1940. [CrossRef]
70. Forster, A.; Gass, A.; Kern, R.; Ay, H.; Chatzikonstantinou, A.; Hennerici, M.G.; Szabo, K. Brain imaging in patients with transient ischemic attack: A comparison of computed tomography and magnetic resonance imaging. *Eur. Neurol.* **2012**, *67*, 136–141. [CrossRef]
71. Thomalla, G.; Cheng, B.; Ebinger, M.; Hao, Q.; Tourdias, T.; Wu, O.; Kim, J.S.; Breuer, L.; Singer, O.C.; Warach, S.; et al. Dwi-flair mismatch for the identification of patients with acute ischaemic stroke within 4.5 h of symptom onset (pre-flair): A multicentre observational study. *Lancet Neurol.* **2011**, *10*, 978–986. [CrossRef]
72. Kim, D.E. Chapter 70-principles and methods of molecular imaging in stroke a2-caplan, louis r. In *Primer on Cerebrovascular Diseases*, 2nd ed.; Biller, J., Leary, M.C., Lo, E.H., Thomas, A.J., Yenari, M., Zhang, J.H., Eds.; Academic Press: San Diego, CA, USA, 2017; pp. 332–338.
73. Quenault, A.; Martinez de Lizarrondo, S.; Etard, O.; Gauberti, M.; Orset, C.; Haelewyn, B.; Segal, H.C.; Rothwell, P.M.; Vivien, D.; Touze, E.; et al. Molecular magnetic resonance imaging discloses endothelial activation after transient ischaemic attack. *Brain* **2017**, *140*, 146–157. [CrossRef] [PubMed]
74. Frechou, M.; Beray-Berthat, V.; Raynaud, J.S.; Meriaux, S.; Gombert, F.; Lancelot, E.; Plotkine, M.; Marchand-Leroux, C.; Ballet, S.; Robert, P.; et al. Detection of vascular cell adhesion molecule-1 expression with uspio-enhanced molecular mri in a mouse model of cerebral ischemia. *Contrast Media Mol. Imaging* **2013**, *8*, 157–164. [CrossRef] [PubMed]
75. Gauberti, M.; Montagne, A.; Marcos-Contreras, O.A.; Le Behot, A.; Maubert, E.; Vivien, D. Ultra-sensitive molecular mri of vascular cell adhesion molecule-1 reveals a dynamic inflammatory penumbra after strokes. *Stroke* **2013**, *44*, 1988–1996. [CrossRef] [PubMed]
76. Deddens, L.H.; van Tilborg, G.A.; van der Toorn, A.; de Vries, H.E.; Dijkhuizen, R.M. Pecam-1-targeted micron-sized particles of iron oxide as mri contrast agent for detection of vascular remodeling after cerebral ischemia. *Contrast Media Mol. Imaging* **2013**, *8*, 393–401. [CrossRef]
77. Deddens, L.H.; van Tilborg, G.A.; van der Toorn, A.; van der Marel, K.; Paulis, L.E.; van Bloois, L.; Storm, G.; Strijkers, G.J.; Mulder, W.J.; de Vries, H.E.; et al. Mri of icam-1 upregulation after stroke: The importance of choosing the appropriate target-specific particulate contrast agent. *Mol. Imaging Biol.* **2013**, *15*, 411–422. [CrossRef] [PubMed]
78. Olafsen, T.; Wu, A.M. Antibody vectors for imaging. *Semin. Nucl. Med.* **2010**, *40*, 167–181. [CrossRef]

79. Bala, G.; Blykers, A.; Xavier, C.; Descamps, B.; Broisat, A.; Ghezzi, C.; Fagret, D.; Van Camp, G.; Caveliers, V.; Vanhove, C.; et al. Targeting of vascular cell adhesion molecule-1 by 18f-labelled nanobodies for pet/ct imaging of inflamed atherosclerotic plaques. *Eur. Heart J. Cardiovasc. Imaging* **2016**, *17*, 1001–1008. [CrossRef]
80. Broisat, A.; Hernot, S.; Toczek, J.; De Vos, J.; Riou, L.M.; Martin, S.; Ahmadi, M.; Thielens, N.; Wernery, U.; Caveliers, V.; et al. Nanobodies targeting mouse/human vcam1 for the nuclear imaging of atherosclerotic lesions. *Circ. Res.* **2012**, *110*, 927–937. [CrossRef]
81. Gauberti, M.; Montagne, A.; Quenault, A.; Vivien, D. Molecular magnetic resonance imaging of brain-immune interactions. *Front. Cell. Neurosci.* **2014**, *8*, 389. [CrossRef]
82. Sweeney, M.D.; Sagare, A.P.; Zlokovic, B.V. Blood-brain barrier breakdown in alzheimer disease and other neurodegenerative disorders. *Nat. Rev. Neurol.* **2018**, *14*, 133–150. [CrossRef]
83. Caljon, G.; Stijlemans, B.; Saerens, D.; Van Den Abbeele, J.; Muyldermans, S.; Magez, S.; De Baetselier, P. Affinity is an important determinant of the anti-trypanosome activity of nanobodies. *PLoS Negl. Trop. Dis.* **2012**, *6*, e1902. [CrossRef]
84. Vandesquille, M.; Li, T.; Po, C.; Ganneau, C.; Lenormand, P.; Dudeffant, C.; Czech, C.; Grueninger, F.; Duyckaerts, C.; Delatour, B.; et al. Chemically-defined camelid antibody bioconjugate for the magnetic resonance imaging of alzheimer's disease. *MAbs* **2017**, *9*, 1016–1027. [CrossRef] [PubMed]
85. Adeoye, O.; Hornung, R.; Khatri, P.; Kleindorfer, D. Recombinant tissue-type plasminogen activator use for ischemic stroke in the united states: A doubling of treatment rates over the course of 5 years. *Stroke* **2011**, *42*, 1952–1955. [CrossRef] [PubMed]
86. Thomalla, G.; Simonsen, C.Z.; Boutitie, F.; Andersen, G.; Berthezene, Y.; Cheng, B.; Cheripelli, B.; Cho, T.H.; Fazekas, F.; Fiehler, J.; et al. Mri-guided thrombolysis for stroke with unknown time of onset. *N. Engl. J. Med.* **2018**, *379*, 611–622. [CrossRef] [PubMed]
87. Hendrickx, M.L.; Zatloukalova, M.; Hassanzadeh-Ghassabeh, G.; Muyldermans, S.; Gils, A.; Declerck, P.J. In vitro and in vivo characterisation of the profibrinolytic effect of an inhibitory anti-rat tafi nanobody. *Thromb. Haemost.* **2014**, *111*, 824–832. [CrossRef] [PubMed]
88. Duggan, S. Caplacizumab: First global approval. *Drugs* **2018**, *78*, 1639–1642. [CrossRef]
89. De Meyer, S.F.; Stoll, G.; Wagner, D.D.; Kleinschnitz, C. Von willebrand factor: An emerging target in stroke therapy. *Stroke* **2012**, *43*, 599–606. [CrossRef]
90. Momi, S.; Tantucci, M.; Van Roy, M.; Ulrichts, H.; Ricci, G.; Gresele, P. Reperfusion of cerebral artery thrombosis by the gpib-vwf blockade with the nanobody alx-0081 reduces brain infarct size in guinea pigs. *Blood* **2013**, *121*, 5088–5097. [CrossRef]
91. Nieswandt, B.; Stoll, G. (Dis)solving the stroke problem by vWF inhibition? *Blood* **2013**, *121*, 4972–4974. [CrossRef]
92. Montaner, J.; Ribo, M.; Monasterio, J.; Molina, C.A.; Alvarez-Sabin, J. Thrombin-activable fibrinolysis inhibitor levels in the acute phase of ischemic stroke. *Stroke* **2003**, *34*, 1038–1040. [CrossRef] [PubMed]
93. Mertens, J.C.; Leenaerts, D.; Brouns, R.; Engelborghs, S.; Ieven, M.; De Deyn, P.P.; Lambeir, A.M.; Hendriks, D. Procarboxypeptidase u (procpu, tafi, procpb2) in cerebrospinal fluid during ischemic stroke is associated with stroke progression, outcome and blood-brain barrier dysfunction. *J. Thromb. Haemost.* **2018**, *16*, 342–348. [CrossRef] [PubMed]
94. Denorme, F.; Wyseure, T.; Peeters, M.; Vandeputte, N.; Gils, A.; Deckmyn, H.; Vanhoorelbeke, K.; Declerck, P.J.; De Meyer, S.F. Inhibition of thrombin-activatable fibrinolysis inhibitor and plasminogen activator inhibitor-1 reduces ischemic brain damage in mice. *Stroke* **2016**, *47*, 2419–2422. [CrossRef] [PubMed]
95. Zhou, X.; Weeks, S.D.; Ameloot, P.; Callewaert, N.; Strelkov, S.V.; Declerck, P.J. Elucidation of the molecular mechanisms of two nanobodies that inhibit thrombin-activatable fibrinolysis inhibitor activation and activated thrombin-activatable fibrinolysis inhibitor activity. *J. Thromb. Haemost.* **2016**, *14*, 1629–1638. [CrossRef]
96. Danquah, W.; Meyer-Schwesinger, C.; Rissiek, B.; Pinto, C.; Serracant-Prat, A.; Amadi, M.; Iacenda, D.; Knop, J.H.; Hammel, A.; Bergmann, P.; et al. Nanobodies that block gating of the p2x7 ion channel ameliorate inflammation. *Sci. Transl. Med.* **2016**, *8*, 366ra162. [CrossRef] [PubMed]
97. Mazzotta, G.; Sarchielli, P.; Caso, V.; Paciaroni, M.; Floridi, A.; Floridi, A.; Gallai, V. Different cytokine levels in thrombolysis patients as predictors for clinical outcome. *Eur. J. Neurol.* **2004**, *11*, 377–381. [CrossRef] [PubMed]

98. Melani, A.; Amadio, S.; Gianfriddo, M.; Vannucchi, M.G.; Volonte, C.; Bernardi, G.; Pedata, F.; Sancesario, G. P2x7 receptor modulation on microglial cells and reduction of brain infarct caused by middle cerebral artery occlusion in rat. *J. Cereb. Blood Flow Metab.* **2006**, *26*, 974–982. [CrossRef]
99. Arbeloa, J.; Perez-Samartin, A.; Gottlieb, M.; Matute, C. P2x$_7$ receptor blockade prevents atp excitotoxicity in neurons and reduces brain damage after ischemia. *Neurobiol. Dis.* **2012**, *45*, 954–961. [CrossRef]
100. Chu, K.; Yin, B.; Wang, J.; Peng, G.; Liang, H.; Xu, Z.; Du, Y.; Fang, M.; Xia, Q.; Luo, B. Inhibition of p2x7 receptor ameliorates transient global cerebral ischemia/reperfusion injury via modulating inflammatory responses in the rat hippocampus. *J. Neuroinflamm.* **2012**, *9*, 69. [CrossRef]
101. Kaiser, M.; Penk, A.; Franke, H.; Krugel, U.; Norenberg, W.; Huster, D.; Schaefer, M. Lack of functional p2x7 receptor aggravates brain edema development after middle cerebral artery occlusion. *Purinergic Signal.* **2016**, *12*, 453–463. [CrossRef]
102. Yanagisawa, D.; Kitamura, Y.; Takata, K.; Hide, I.; Nakata, Y.; Taniguchi, T. Possible involvement of p2x7 receptor activation in microglial neuroprotection against focal cerebral ischemia in rats. *Biol. Pharm. Bull.* **2008**, *31*, 1121–1130. [CrossRef] [PubMed]
103. Le Feuvre, R.A.; Brough, D.; Touzani, O.; Rothwell, N.J. Role of p2x7 receptors in ischemic and excitotoxic brain injury in vivo. *J. Cereb. Blood Flow Metab.* **2003**, *23*, 381–384. [CrossRef] [PubMed]
104. Bartlett, R.; Stokes, L.; Sluyter, R. The p2x7 receptor channel: Recent developments and the use of p2x7 antagonists in models of disease. *Pharmacol. Rev.* **2014**, *66*, 638–675. [CrossRef] [PubMed]
105. Lambertsen, K.L.; Biber, K.; Finsen, B. Inflammatory cytokines in experimental and human stroke. *J. Cereb. Blood Flow Metab.* **2012**, *32*, 1677–1698. [CrossRef] [PubMed]
106. Yamashita, T.; Sawamoto, K.; Suzuki, S.; Suzuki, N.; Adachi, K.; Kawase, T.; Mihara, M.; Ohsugi, Y.; Abe, K.; Okano, H. Blockade of interleukin-6 signaling aggravates ischemic cerebral damage in mice: Possible involvement of stat3 activation in the protection of neurons. *J. Neurochem.* **2005**, *94*, 459–468. [CrossRef] [PubMed]
107. Herrmann, O.; Tarabin, V.; Suzuki, S.; Attigah, N.; Coserea, I.; Schneider, A.; Vogel, J.; Prinz, S.; Schwab, S.; Monyer, H.; et al. Regulation of body temperature and neuroprotection by endogenous interleukin-6 in cerebral ischemia. *J. Cereb. Blood Flow Metab.* **2003**, *23*, 406–415. [CrossRef] [PubMed]
108. Intiso, D.; Zarrelli, M.M.; Lagioia, G.; Di Rienzo, F.; Checchia De Ambrosio, C.; Simone, P.; Tonali, P.; Cioffi Dagger, R.P. Tumor necrosis factor alpha serum levels and inflammatory response in acute ischemic stroke patients. *Neurol. Sci.* **2004**, *24*, 390–396. [CrossRef] [PubMed]
109. Zaremba, J.; Losy, J. Early tnf-alpha levels correlate with ischaemic stroke severity. *Acta Neurol. Scand.* **2001**, *104*, 288–295. [CrossRef]
110. Dziewulska, D.; Mossakowski, M.J. Cellular expression of tumor necrosis factor a and its receptors in human ischemic stroke. *Clin. Neuropathol.* **2003**, *22*, 35–40. [PubMed]
111. Zelova, H.; Hosek, J. Tnf-alpha signalling and inflammation: Interactions between old acquaintances. *Inflamm. Res.* **2013**, *62*, 641–651. [CrossRef] [PubMed]
112. Nawashiro, H.; Martin, D.; Hallenbeck, J.M. Inhibition of tumor necrosis factor and amelioration of brain infarction in mice. *J. Cereb. Blood Flow Metab.* **1997**, *17*, 229–232. [CrossRef] [PubMed]
113. Yang, G.Y.; Gong, C.; Qin, Z.; Liu, X.H.; Lorris Betz, A. Tumor necrosis factor alpha expression produces increased blood-brain barrier permeability following temporary focal cerebral ischemia in mice. *Brain Res. Mol. Brain Res.* **1999**, *69*, 135–143. [CrossRef]
114. Meistrell, M.E., 3rd; Botchkina, G.I.; Wang, H.; Di Santo, E.; Cockroft, K.M.; Bloom, O.; Vishnubhakat, J.M.; Ghezzi, P.; Tracey, K.J. Tumor necrosis factor is a brain damaging cytokine in cerebral ischemia. *Shock* **1997**, *8*, 341–348. [CrossRef] [PubMed]
115. Lambertsen, K.L.; Clausen, B.H.; Babcock, A.A.; Gregersen, R.; Fenger, C.; Nielsen, H.H.; Haugaard, L.S.; Wirenfeldt, M.; Nielsen, M.; Dagnaes-Hansen, F.; et al. Microglia protect neurons against ischemia by synthesis of tumor necrosis factor. *J. Neurosci.* **2009**, *29*, 1319–1330. [CrossRef] [PubMed]
116. Nawashiro, H.; Tasaki, K.; Ruetzler, C.A.; Hallenbeck, J.M. Tnf-alpha pretreatment induces protective effects against focal cerebral ischemia in mice. *J. Cereb. Blood Flow Metab.* **1997**, *17*, 483–490. [CrossRef]
117. Lis, K.; Kuzawinska, O.; Balkowiec-Iskra, E. Tumor necrosis factor inhibitors—State of knowledge. *Arch. Med. Sci.* **2014**, *10*, 1175–1185. [CrossRef]

118. Beirnaert, E.; Desmyter, A.; Spinelli, S.; Lauwereys, M.; Aarden, L.; Dreier, T.; Loris, R.; Silence, K.; Pollet, C.; Cambillau, C.; et al. Bivalent llama single-domain antibody fragments against tumor necrosis factor have picomolar potencies due to intramolecular interactions. *Front. Immunol.* **2017**, *8*, 867. [CrossRef]
119. Kalden, J.R.; Schulze-Koops, H. Immunogenicity and loss of response to tnf inhibitors: Implications for rheumatoid arthritis treatment. *Nat. Rev. Rheumatol.* **2017**, *13*, 707–718. [CrossRef]
120. Efimov, G.A.; Kruglov, A.A.; Khlopchatnikova, Z.V.; Rozov, F.N.; Mokhonov, V.V.; Rose-John, S.; Scheller, J.; Gordon, S.; Stacey, M.; Drutskaya, M.S.; et al. Cell-type-restricted anti-cytokine therapy: Tnf inhibition from one pathogenic source. *Proc. Natl. Acad. Sci. USA* **2016**, *113*, 3006–3011. [CrossRef]
121. Steeland, S.; Puimege, L.; Vandenbroucke, R.E.; Van Hauwermeiren, F.; Haustraete, J.; Devoogdt, N.; Hulpiau, P.; Leroux-Roels, G.; Laukens, D.; Meuleman, P.; et al. Generation and characterization of small single domain antibodies inhibiting human tumor necrosis factor receptor 1. *J. Biol. Chem.* **2015**, *290*, 4022–4037. [CrossRef]
122. Steeland, S.; Van Ryckeghem, S.; Van Imschoot, G.; De Rycke, R.; Toussaint, W.; Vanhoutte, L.; Vanhove, C.; De Vos, F.; Vandenbroucke, R.E.; Libert, C. Tnfr1 inhibition with a nanobody protects against EAE development in mice. *Sci. Rep.* **2017**, *7*, 13646. [CrossRef] [PubMed]
123. Murray, K.N.; Parry-Jones, A.R.; Allan, S.M. Interleukin-1 and acute brain injury. *Front. Cell Neurosci.* **2015**, *9*, 18. [CrossRef]
124. Liu, T.; McDonnell, P.C.; Young, P.R.; White, R.F.; Siren, A.L.; Hallenbeck, J.M.; Barone, F.C.; Feurestein, G.Z. Interleukin-1 beta mRNA expression in ischemic rat cortex. *Stroke* **1993**, *24*, 1746–1750; discussion 1750–1741. [CrossRef] [PubMed]
125. Wang, X.; Barone, F.C.; Aiyar, N.V.; Feuerstein, G.Z. Interleukin-1 receptor and receptor antagonist gene expression after focal stroke in rats. *Stroke* **1997**, *28*, 155–161; discussion 161–152. [CrossRef] [PubMed]
126. Pradillo, J.M.; Murray, K.N.; Coutts, G.A.; Moraga, A.; Oroz-Gonjar, F.; Boutin, H.; Moro, M.A.; Lizasoain, I.; Rothwell, N.J.; Allan, S.M. Reparative effects of interleukin-1 receptor antagonist in young and aged/co-morbid rodents after cerebral ischemia. *Brain Behav. Immun.* **2017**, *61*, 117–126. [CrossRef] [PubMed]
127. Mulcahy, N.J.; Ross, J.; Rothwell, N.J.; Loddick, S.A. Delayed administration of interleukin-1 receptor antagonist protects against transient cerebral ischaemia in the rat. *Br. J. Pharmacol.* **2003**, *140*, 471–476. [CrossRef] [PubMed]
128. Mertens, M.; Singh, J.A. Anakinra for rheumatoid arthritis. *Cochrane Database Syst. Rev.* **2009**, CD005121. [CrossRef] [PubMed]
129. Emsley, H.C.; Smith, C.J.; Georgiou, R.F.; Vail, A.; Hopkins, S.J.; Rothwell, N.J.; Tyrrell, P.J.; Acute Stroke, I. A randomised phase ii study of interleukin-1 receptor antagonist in acute stroke patients. *J. Neurol. Neurosurg. Psychiatry* **2005**, *76*, 1366–1372. [CrossRef]
130. Boutin, H.; LeFeuvre, R.A.; Horai, R.; Asano, M.; Iwakura, Y.; Rothwell, N.J. Role of il-1alpha and il-1beta in ischemic brain damage. *J. Neurosci.* **2001**, *21*, 5528–5534. [CrossRef]
131. Stroemer, R.P.; Rothwell, N.J. Exacerbation of ischemic brain damage by localized striatal injection of interleukin-1beta in the rat. *J. Cereb. Blood Flow Metab.* **1998**, *18*, 833–839. [CrossRef]
132. Liberale, L.; Diaz-Canestro, C.; Bonetti, N.R.; Paneni, F.; Akhmedov, A.; Beer, J.H.; Montecucco, F.; Luscher, T.F.; Camici, G.G. Post-ischaemic administration of the murine canakinumab-surrogate antibody improves outcome in experimental stroke. *Eur. Heart J.* **2018**, *39*, 3511–3517. [CrossRef] [PubMed]
133. Jickling, G.C.; Liu, D.; Ander, B.P.; Stamova, B.; Zhan, X.; Sharp, F.R. Targeting neutrophils in ischemic stroke: Translational insights from experimental studies. *J. Cereb. Blood Flow Metab.* **2015**, *35*, 888–901. [CrossRef] [PubMed]
134. He, Q.; Shi, X.; Zhou, B.; Teng, J.; Zhang, C.; Liu, S.; Lian, J.; Luo, B.; Zhao, G.; Lu, H.; et al. Interleukin 8 (cxcl8)-cxc chemokine receptor 2 (cxcr2) axis contributes to mir-4437-associated recruitment of granulocytes and natural killer cells in ischemic stroke. *Mol. Immunol.* **2018**, *101*, 440–449. [CrossRef] [PubMed]
135. Garau, A.; Bertini, R.; Colotta, F.; Casilli, F.; Bigini, P.; Cagnotto, A.; Mennini, T.; Ghezzi, P.; Villa, P. Neuroprotection with the cxcl8 inhibitor repertaxin in transient brain ischemia. *Cytokine* **2005**, *30*, 125–131. [CrossRef]
136. Connell, B.J.; Gordon, J.R.; Saleh, T.M. Elr-cxc chemokine antagonism is neuroprotective in a rat model of ischemic stroke. *Neurosci. Lett.* **2015**, *606*, 117–122. [CrossRef]

137. Brait, V.H.; Rivera, J.; Broughton, B.R.; Lee, S.; Drummond, G.R.; Sobey, C.G. Chemokine-related gene expression in the brain following ischemic stroke: No role for cxcr2 in outcome. *Brain Res.* **2011**, *1372*, 169–179. [CrossRef] [PubMed]
138. Bradley, M.E.; Dombrecht, B.; Manini, J.; Willis, J.; Vlerick, D.; De Taeye, S.; Van den Heede, K.; Roobrouck, A.; Grot, E.; Kent, T.C.; et al. Potent and efficacious inhibition of cxcr2 signaling by biparatopic nanobodies combining two distinct modes of action. *Mol. Pharmacol.* **2015**, *87*, 251–262. [CrossRef]
139. Wang, Y.; Huang, J.; Li, Y.; Yang, G.Y. Roles of chemokine cxcl12 and its receptors in ischemic stroke. *Curr. Drug Targets* **2012**, *13*, 166–172. [CrossRef] [PubMed]
140. Wu, K.J.; Yu, S.J.; Shia, K.S.; Wu, C.H.; Song, J.S.; Kuan, H.H.; Yeh, K.C.; Chen, C.T.; Bae, E.; Wang, Y. A novel cxcr4 antagonist cx549 induces neuroprotection in stroke brain. *Cell. Transpl.* **2017**, *26*, 571–583. [CrossRef]
141. Walter, H.L.; van der Maten, G.; Antunes, A.R.; Wieloch, T.; Ruscher, K. Treatment with amd3100 attenuates the microglial response and improves outcome after experimental stroke. *J. Neuroinflamm.* **2015**, *12*, 24. [CrossRef]
142. Ruscher, K.; Kuric, E.; Liu, Y.; Walter, H.L.; Issazadeh-Navikas, S.; Englund, E.; Wieloch, T. Inhibition of cxcl12 signaling attenuates the postischemic immune response and improves functional recovery after stroke. *J. Cereb. Blood Flow Metab.* **2013**, *33*, 1225–1234. [CrossRef] [PubMed]
143. de Wit, R.H.; Heukers, R.; Brink, H.J.; Arsova, A.; Maussang, D.; Cutolo, P.; Strubbe, B.; Vischer, H.F.; Bachelerie, F.; Smit, M.J. Cxcr4-specific nanobodies as potential therapeutics for whim syndrome. *J. Pharmacol. Exp. Ther.* **2017**, *363*, 35–44. [CrossRef] [PubMed]
144. Jahnichen, S.; Blanchetot, C.; Maussang, D.; Gonzalez-Pajuelo, M.; Chow, K.Y.; Bosch, L.; De Vrieze, S.; Serruys, B.; Ulrichts, H.; Vandevelde, W.; et al. Cxcr4 nanobodies (vhh-based single variable domains) potently inhibit chemotaxis and hiv-1 replication and mobilize stem cells. *Proc. Natl. Acad. Sci. USA* **2010**, *107*, 20565–20570. [CrossRef] [PubMed]

*Review*

# Single-Domain Antibodies as Therapeutic and Imaging Agents for the Treatment of CNS Diseases

**Kasandra Bélanger [1,*], Umar Iqbal [1], Jamshid Tanha [1,2], Roger MacKenzie [1], Maria Moreno [1] and Danica Stanimirovic [1]**

1 Human Health Therapeutics Research Centre, National Research Council Canada, Ottawa, ON K1A 0R6, Canada; Umar.Iqbal@nrc-cnrc.gc.ca (U.I.); Jamshid.Tanha@nrc-cnrc.gc.ca (J.T.); Colin.MacKenzie@nrc-cnrc.gc.ca (R.M.); Maria.Moreno@nrc-cnrc.gc.ca (M.M.); Danica.Stanimirovic@nrc-cnrc.gc.ca (D.S.)

2 Department of Biochemistry, Microbiology and Immunology, University of Ottawa, Ottawa, ON K1H 8M5, Canada

* Correspondence: Kasandra.belanger@nrc-cnrc.gc.ca; Tel.: +1-613-993-7464

Received: 1 March 2019; Accepted: 28 March 2019; Published: 5 April 2019

**Abstract:** Antibodies have become one of the most successful therapeutics for a number of oncology and inflammatory diseases. So far, central nervous system (CNS) indications have missed out on the antibody revolution, while they remain 'hidden' behind several hard to breach barriers. Among the various antibody modalities, single-domain antibodies (sdAbs) may hold the 'key' to unlocking the access of antibody therapies to CNS diseases. The unique structural features of sdAbs make them the smallest monomeric antibody fragments suitable for molecular targeting. These features are of particular importance when developing antibodies as modular building blocks for engineering CNS-targeting therapeutics and imaging agents. In this review, we first introduce the characteristic properties of sdAbs compared to traditional antibodies. We then present recent advances in the development of sdAbs as potential therapeutics across brain barriers, including their use for the delivery of biologics across the blood–brain and blood–cerebrospinal fluid (CSF) barriers, treatment of neurodegenerative diseases and molecular imaging of brain targets.

**Keywords:** single-domain antibodies; neurodegenerative diseases; brain imaging; blood–brain barrier; delivery

---

## 1. Introduction to sdAbs

### 1.1. Structure and Characteristics

The concept of single-domain antibodies (sdAbs) originated in the 90's, with the proof-of-concept experiments demonstrating sdAbs as *bone fide* antigen binding fragments [1], and the discovery of camelid [2] and shark [3] heavy chain-only antibodies (HCAbs). Single-domain antibodies can be derived from the antigen binding variable domains of homodimeric, light-chain lacking immunoglobulins, such as camelid HCAbs [4] and shark immunoglobulin new antigen receptors (IgNARs) [5], variable light chain ($V_L$) or variable heavy chain ($V_H$) domains of tetrameric—typically human—conventional immunoglobulins [6] (Figure 1). The variable domains of camelid HCAbs and shark IgNARs are referred to as $V_H$Hs (or nanobodies) and $V_{NAR}$s, respectively. While $V_H$Hs and $V_{NAR}$s are almost without exception non-aggregating and highly soluble, the opposite is true for $V_H$s and $V_L$s. However, various strategies have been developed to successfully obtain aggregation resistant and soluble human $V_H$s and $V_L$s ([7,8] and references therein) including transgenic mice technology [9,10]. Human $V_H$ and $V_L$ domains are of interest because of their human nature, a property that presumably makes them less immunogenic in humans compared to camelid $V_H$Hs or nurse shark $V_{NAR}$s.

The desirable biophysical, biochemical, and structural properties of sdAbs, particularly those from natural repertoires are generally well known, and have been described in several reviews [4–6,11,12]. Despite their significantly smaller combining site, consisting of only three complementarity-determining regions (CDRs) or hypervariable loops—as opposed to six for conventional antibodies such as monoclonal antibodies (mAbs)—sdAbs demonstrate comparable antigen binding affinities. Interrelated properties such as small size (12–15 kDa vs. 150 kDa for mAbs); strict monomericity; high solubility, including at therapeutic doses; aggregation resistance; chemical, physical and protease stability; efficient folding /refolding; good recombinant expression, notably in economic microbial expression systems such as yeast and *E. coli*; excellent shelf life; excellent manufacturability; and low cost of production make sdAbs an attractive alternative to other antibody formats such as mAbs, Fabs (fragments antigen binding), and scFvs (single chain variable fragments) as therapeutic and diagnostic agents. Resistance to aggregation is particularly noteworthy as it significantly reduces the risk of immunogenicity. Furthermore, their small size and frequently extended CDR3 make sdAbs the antibody of choice when targeting recessed epitopes of proteins such as enzymes' active sites or receptors' cavities. Longer CDR3s also increase the combining site's surface area, and to a significant degree compensate for the absence of $V_L$ CDRs. In addition, their fast blood clearance and effective tissue penetration, attributed to their small size, make sdAbs ideal imaging agents, e.g., against tumors. In this respect, the high stability and folding properties of sdAbs provide flexibility for labeling reactions with optimal outcomes. Modularity is another hallmark of sdAbs, and becomes a key property when engineering sdAb-based multimeric and multispecific constructs as CNS diagnostics and therapeutics (see Section 1.3).

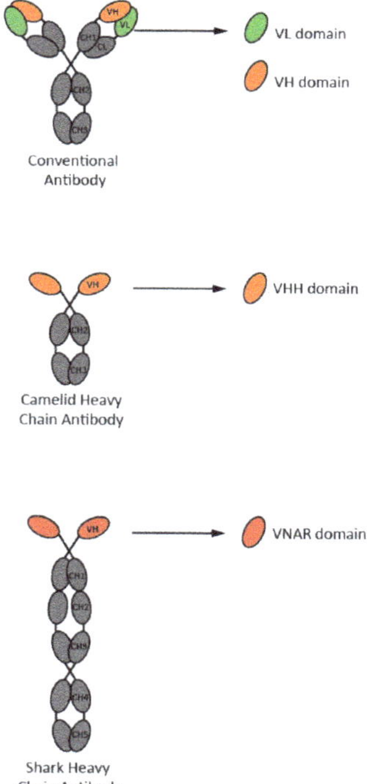


**Figure 1.** Schematic representation of the four types of sdAbs described in the current review. Antibody constant domains are in grey, whereas antibody variable domains from which sdAbs are derived are in color.

### 1.2. Single-Domain Antibody Libraries and Selection

Single-domain antibodies have been typically isolated from display libraries mostly phage-displayed, although other display platforms also exist, such as yeast and ribosome display [13]. While $V_HHs$ and $V_{NAR}s$ have been obtained from all types of libraries including immune, non-immune, semi-synthetic/synthetic libraries [4,14,15], human $V_Hs$ and $V_Ls$ have been most commonly obtained from synthetic libraries [6]. Unlike immune $V_{NAR}s$ and $V_HHs$ that have high affinities as a result of in vivo somatic hypermutation, sdAbs obtained from non-immune or synthetic/semi-synthetic sdAb libraries are of low affinity, and often require further mutation for improved affinity and function. However, more recently, human $V_Hs$ have also been isolated from immune $V_H$ display libraries derived from HCAb-producing transgenic mice that are immunized with a target antigen of interest [9,10]. While these $V_Hs$ display high solubility, stability and affinity of immune $V_HHs$, they are advantageously expected to be less immunogenic.

Constructing natural repertoire sdAb libraries is well established and relatively straightforward [4,15,16]. Human $V_H$ and $V_L$ synthetic libraries are typically built on a single scaffold with demonstrated good biophysical properties, such as high thermostability, solubility, and expression ([6,8] and references therein, [17,18]). Library diversity generation entails introducing random or specific amino acids at all or selected positions in the three CDRs. Owing to their small size and single-domain nature, in contrast to more complex, multidomain scFvs and Fabs, sdAbs lend themselves to a facile and straightforward library construction, and are not associated with $V_H/V_L$ mispairing phenomenon that occurs during the construction of scFv and Fab libraries, which adversely affects the library quality.

In its simplest and most commonly practiced format, the selection, or panning, the process for the isolation of sdAbs from phage-displayed libraries, involves selecting for a single property—affinity for the target antigen. Most commonly, this involves exposure of a library to an antigen immobilized on a microtitre plate, washing away unbound phage, and eluting the bound phage molecules, which are amplified for another round of panning. Depending on the type of library, between two and four rounds of panning are typically sufficient to obtain around half a dozen sdAbs with affinity for the target antigen. With human sdAb libraries, affinity selection may be coupled with selection for stability for a more efficient isolation of aggregation-resistant binders [6].

One of the great advantages of antibody library display technologies over hybridoma technology for the isolation of mAbs is the capability to drive, in some measure, the selection process towards isolating antibodies with specific properties. For example, by panning in the presence of proteases it has been possible to isolate sdAbs with enhanced protease resistance [19]. In the context of this review, it is especially noteworthy that sdAbs that transmigrate across an in vitro human blood–brain barrier (BBB) model have been isolated [20].

### 1.3. Modular Building of Multispecific Molecules

Their small size and monomeric nature make sdAbs ideal building blocks for the construction of multivalent and multispecific therapeutic and imaging molecules of improved function and potency (compared to monomeric versions) with good development capacity and manufacturability [11,21,22]. For example, bivalent or bispecific sdAbs have been generated by linking two identical or two different sdAbs using a short spacer sequence [23–32]. Successful generation of trivalent bispecific and tetravalent bispecific sdAbs—where sdAb moieties are linked through short linker sequences—have also been reported [28,29,33–36]. Monospecific pentavalent sdAbs have been constructed by fusing sdAbs to the N- or C-terminus of the verotoxin 1B (VT1B) subunit [37]. Similarly, fusing different sdAbs to the N- and C-terminus of VT1B has yielded bispecific decavalent molecules [38]. Bivalent monospecific or tetravalent bispecific sdAbs can also be made by fusing sdAbs to an antibody Fc fragment [8,39–43]; this has the added advantage of greatly extending the serum half-life of sdAbs [44] and imparting effector functions such as antibody-dependent cellular cytotoxicity (ADCC) [42] or complement-dependent cytotoxicity (CDC) [45]. Single-domain antibodies should also be ideal molecules for constructing bivalent and bispecific antibodies incorporating a heterodimeric Fc region [46]. More complex constructs such IgG-sdAb fusions have also been reported [47]. For therapeutic applications, sdAbs have been linked to enzymes or toxins, either by cloning or by chemical conjugation [48–53].

### 1.4. Developing sdAbs as CNS Diagnostics or Therapeutics

Treating CNS disorders remains one of the greatest challenges in modern medicine. Although several promising therapeutics are developed every year, their failure to reach brain target prevents their advancement to the clinic. This is mainly due to the presence of the BBB acting as a gatekeeper to maintain brain homeostasis and protect neurological capabilities [54]. The BBB is composed of specialized endothelial cells sealed together by tight junctions to form a physical barrier lining the brain blood vessels. These cells differ from endothelial cells lining peripheral vessels by their lack of fenestrations and limited pinocytic activity thereby restricting transcellular transport. The brain endothelial layer is surrounded by pericytes and astrocyte end-feet, which are essential for maintaining the integrity of the BBB. In addition, several efflux transporters are present at the BBB and function to remove unwanted molecules from the brain. Although this restrictive physiology is necessary to prevent undesirable blood-borne material from penetrating the brain, it also limits the effective delivery of CNS therapeutics. Therefore, agents designed for use as CNS diagnostics or therapeutics must be delivered to sites of action via administration routes that circumvent the BBB, such as intrathecal/intraventricular, intracerebral administration, or combined with delivery technologies that increase their penetration across the BBB upon systemic administration.

The brain neuropil is packed with interacting cells, including neurons, neuronal processes, and various types of glial cells. The brain extracellular space (ECS), filled with brain extracellular fluid, is tight and very convoluted—modeling studies estimate its width between 35 and 60 nm [55]. Any compound administered directly into the neuropil will diffuse through the ECS to distances inversely proportional to the size of the molecule. Monoclonal antibodies exhibit limited diffusion in the brain ECS due to their large size and interactions with the extracellular matrix (ECM). Single-domain antibodies have a distinct advantage as intracerebrally administered reagents/therapeutics, achieving diffusion across longer distances from the site of injection [55]. In addition, the lack of a Fc fragment reduces their interactions with the ECM and brain efflux via an FcRn-mediated reverse transcytosis. A recent study on the brain biodistribution of antibodies via perivascular transport after intrathecal infusion in rodents [56] demonstrated both deeper brain penetration and broader brain exposure of a smaller $V_HH$ fragment compared to a full mAb. This study demonstrated that sdAbs are advantageous as a CNS therapeutic antibody modality developed for intracerebral (local) or intrathecal administration. This is particularly relevant for brain diseases originating from or confined to a specific brain area, such as Parkinson's disease.

However, the majority of CNS diseases can be considered 'whole-brain' diseases, even when they initially affect more localized brain regions. The brain's vascular network is particularly dense, and thus each brain capillary supplies only few neurons, the diffusion distance of compounds, including antibodies, delivered across the BBB to their neuronal targets is only ~25 µm. Transvascular (cross-BBB) brain delivery would therefore achieve a more global brain distribution of antibodies, regardless of their size, since these diffusion distances are readily achievable even by mAbs. Systemic delivery of therapeutic antibodies targeting the CNS could be improved using 'carrier' molecules selected or engineered for the ability to traverse the BBB.

In the following sections, sdAbs that have been developed as delivery agents across the BBB, as treatments against the most common neurodegenerative diseases and as neuroimaging tools are descried and summarized in Table 1.

## 2. Single-Domain Antibodies as Delivery Agents across the BBB

The most investigated method to deliver macromolecules into the brain is via receptor-mediated transcytosis (RMT) [57]. This process is crucial for the proper delivery of macromolecules essential for brain function such as vitamins, proteins and nutrients. They use naturally occurring transport systems to shuttle between the blood and the brain. Such systems can potentially be 'hi-jacked' to facilitate the delivery of therapeutics into brain.

The process of RMT is initiated by ligand binding to its receptor expressed at the luminal face of endothelial cells to trigger internalization of the receptor-ligand complex into endosomal vesicles (Figure 2). These vesicles then travel inside the cytoplasm of the cell via a complex vesicular sorting pathway to finally fuse with the abluminal surface of endothelial cells and deliver their cargo in the brain parenchyma. The receptor is then recycled at the luminal cell surface. Currently, the main RMT receptors that have been studied are the transferrin receptor (TfR) and insulin receptor (IR) [57]. Ligands against these receptors, including different antibody formats, have been used as carriers to deliver their therapeutic cargo inside the brain [7,58–65]. Single-domain antibodies present numerous advantages over conventional antibodies as potential transvascular brain delivery vectors including small size, low non-specific interactions with tissues expressing high levels of Fc receptors (e.g., liver, spleen), remarkable stability against harsh conditions and low immunogenicity (see Section 1). In the case of IR, although there are examples of mAbs and peptides specific for this receptor, no IR-specific sdAbs have been described to date. The only example of a sdAb targeting TfR is a $V_{NAR}$, termed TBX4, which was obtained from a synthetic library following a combination of in vitro and in vivo phage display techniques [66]. When fused to an immunoglobulin Fc backbone, this antibody was enriched in the brain parenchyma of mice following vein tail injection. Furthermore, bispecific variants of this antibody fused to a CD20 targeting agent were able to reach aberrant B cells in the brain and induced cell toxicity. The use of this $V_{NAR}$ for the delivery of a variety of biologics to the brain is currently under investigation.

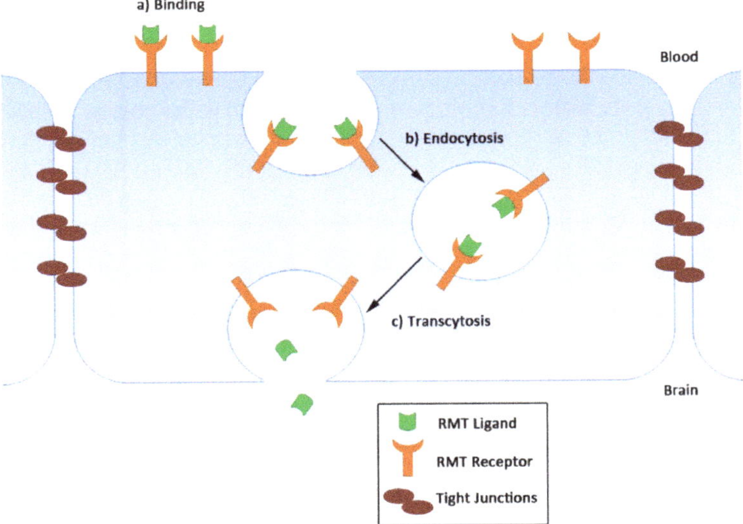


**Figure 2.** Representation of the receptor-mediated transcytosis (RMT) process. (**a**) Initially, an RMT ligand binds to a specific RMT receptor on the luminal cell membrane, which (**b**) leads to the internalization of both receptor and ligand in intracellular vesicles via endocytosis. (**c**) These vesicles then travel within the cell cytoplasm to reach the abluminal membrane where fusion of endosomes with the cell membrane releases the vesicular cargo inside the brain.

Although there is now growing evidence that the use of the aforementioned receptors for therapeutic delivery into the brain is a promising avenue, recent studies suggest that these may not be ideal RMT targets [67–69]. Major drawbacks associated with these receptors include their ubiquitous expression in numerous peripheral organs and their involvement in essential physiological functions, which may raise significant safety concerns. Therefore, the chase for alternative RMT receptors with more optimal BBB crossing properties continues.

FC5 and FC44, two camelid $V_H$Hs of a non-immune phage-displayed library, were isolated by phenotypic panning for their ability to interact and internalize into the human brain cerebromicrovascular endothelial cells (HBECs) [20]. The two antibodies were found to transmigrate in an in vitro rat BBB model, and to accumulate in the brain at high levels following tail vein injection in rodents [20,70]. The investigation of their mechanism of action led to the finding that it was RMT-mediated [71], and in the case of FC5, the receptor was later identified as transmembrane domain protein 30A (TMEM-30A) [72]. Engineered fusions of FC5 to the human IgG1 Fc in a monovalent or bivalent format showed increased migration across the BBB in vitro, and achieved a significantly higher brain exposure in vivo compared to the control antibody-Fc [73]. FC5-Fc constructs were also detected in brain vessels and in the brain parenchyma in rat brain sections. Furthermore, the conjugation of FC5 antibodies with impermeable analgesic peptides, dalargin or neuropeptide Y, induced an important analgesic effect in a thermal hyperalgesia model whereas the systemic administration of the neuropeptides alone had no suppressive effect. Although all antibody formats were able to reduce hyperalgesia, the bivalent and monovalent Fc fusions showed a pronounced increase in the response at equal dose compared to the $V_H$H suggesting that improving serum pharmacokinetics plays a determining role in the pharmacological potency of FC5. In addition to peptides, the CNS delivery of a monoclonal antibody antagonist of metabotropic glutamate receptor 1 (mGluR1) was successfully achieved using the BBB-crossing $V_H$H FC5 [74]. Following intravenous injection in a rat model of persistent inflammatory pain, the BBB-mGluR1 bispecific antibody co-localized with thalamic neurons involved in mGluR1-mediated pain processing, and subsequently inhibited hyperalgesia.

Insulin growth factor 1 receptor (IGF1R) has been identified as a potential RMT candidate based on observation that its ligand, IGF-1 is transported across the BBB. A panel of $V_H$Hs, targeting the ectodomain of this receptor was isolated from a phage-displayed immune library [75]. Following humanization, the ability of several IGF1R-binding $V_H$Hs to transmigrate in rat and human BBB models in vitro was confirmed [75,76]. When expressed in fusion with a murine antibody fragment, the resultant IGF1R-specific $V_H$Hs were found to significantly increase brain and CSF exposure in mice and rats compared to the control [75]. IGF1R-targeting $V_H$Hs conjugated to galanin—A systemically restricted neuroactive peptide that produces analgesia by binding GalR1 and GalR2 receptors expressed in the brain [77]—Induced a strong analgesic effect in a rat model of inflammatory hyperalgesia following a single dose injection [75], suggesting the ability of these IGF1R-binding $V_H$Hs to act as delivery carriers across the BBB.

The above studies provided evidence for the feasibility of using sdAbs as carriers targeting a new generation of RMT receptors for the development of CNS therapeutics.

## 3. Single-Domain Antibodies as Treatments against Neurodegenerative Diseases

### *3.1. Protein-Misfolding Diseases (PMDs)*

A vast majority of neurodegenerative diseases are associated with misfolded proteins that interact with each other to form large aggregates referred to as amyloid fibrils [78]. These complexes are insoluble, highly organized and extremely stable, and their accumulation is toxic to the cell. Although all PMDs share a common mechanism of action, the nature of the misfolded proteins differs between each disorder and dictates the identity of the disease. Alzheimer's disease is caused by the accumulation of amyloid $\beta$ (A$\beta$) peptides and Tau proteins, whereas the aggregation of $\alpha$-synucleins ($\alpha$Syn) is at the origin of Parkinson's disease. Similarly, aggregates formed by huntingtin (Htt) proteins lead to Huntington's disease while Prion disease is associated with the conversion of the normal, cellular prion protein ($PrP^c$) into its disordered scrapie isoform ($PrP^{sc}$), which accumulates into large oligomers. The formation of fibrils is a complex phenomenon involving several intermediate distinct structures [78,79]. The identification and characterization of the various species formed during the process is essential for the development of early diagnostic tools and new therapeutic strategies. This is, however, an extremely difficult task due to the insolubility and heterogeneity of the different forms involved in the process of fibril formation. This explains the lack of effective treatments against the devastating PMDs to date.

Single-domain antibodies represent a promising asset for the treatment of PMDs since they possess unique characteristics allowing them to access unprecedented epitopes (see Section 1). In addition, their high specificity and stability ensured the targeting of specific species under harsh solubilizing conditions along the process of fibril maturation. In this next section, we will review recent advances in the use of sdAbs for the diagnostic and treatment of the main PMDs affecting the CNS.

#### 3.1.1. Alzheimer's Disease (AD)

The first description of AD goes back to 1906 [80]. However, it took several decades until it was finally established as a major neurodegenerative disorder. It is now considered the main cause of dementia accounting for up to 80% of all cases [81]. Patients affected with the disease suffer several symptoms including the loss of memory and cognitive functions. This is believed to be due to two main phenomena. First, the extracellular accumulation of Aβ peptides made of 39–42 amino acids to form amyloid plaques in the CNS [82] and second, the aggregation of hyperphosphorylated Tau proteins into Tau tangles inside neurons [83]. The presence of these aggregates or their precursor forms severely affects the normal function of neurons leading to cell death. In recent years, several efforts have been deployed to generate antibody fragments against Aβ and Tau aggregates in view of developing novel therapeutics for AD.

$V_HH$ B10 emerged from a synthetic phage-displayed library panned against biotinylated Aβ (1-40) fibrils [84]. This antibody was shown to bind specifically to mature amyloid fibrils as well as to protofibrils which were defined as the aggregated species forming prior to the assembly of more stable mature fibrils [85]. $V_HH$ did not interact with disaggregated peptides or other non-fibrillar Aβ oligomers. In addition, the authors demonstrated the antibody's ability to stabilize protofibrils upon interacting with it, thus inhibiting mature fibril formation. However, B10 did not have the ability to disintegrate preformed fibrils. Similarly, another $V_HH$ isolated from a synthetic phage-displayed library, was shown to interact specifically with non-fibrillar Aβ (1-40) oligomers and prevent the formation of mature fibrils [86]. The antibody could not induce the disaggregation of already formed fibrils. In their report, the authors immobilized biotinylated Aβ (1-40) oligomers to select a conformation-specific binder that they named KW1. They demonstrated that the addition of KW1 to preformed Aβ oligomers prevented their synaptotoxic effect. Nevertheless, another report published two years later showed that Aβ oligomers formed in the presence of the same KW1 antibody were highly toxic [87], which seems to indicate a time-sensitive beneficial effect by the $V_HH$. ni3A is a $V_HH$ that was isolated from a non-immune phage-displayed library using Aβ (1-42) as antigen [88]. This $V_HH$ bound to its target with high specificity and affinity and showed BBB-crossing abilities in vitro [89]. When tested in vivo [90], ni3A successfully detected Aβ deposits in a transgenic mouse model of AD, suggesting its potential as a diagnostic tool.

In contrast to the $V_HH$s described above, three additional ones were isolated from a phage-displayed library made from the blood of a llama immunized with a mixture of Aβ (1-42) monomers, small oligomers and fibrils [91]. These antibodies bound specifically to monomers and small oligomers formed exclusively by Aβ (1-42) but not to higher molecular-mass aggregates or fibrils or to Aβ (1-40)-originating species. One $V_HH$ in particular, V31-1, was found to inhibit the formation of amyloid fibrils and to prevent the toxic cellular effect of Aβ oligomers [91]. Another immunization campaign—this time using brain homogenates from an AD patient as immunogen in alpacas—Led to the identification of three $V_HHs$, PrioAD12, PrioAD13 and PrioAD120 targeting Aβ (1-40), Aβ (1-42) or Tau (1-16) peptides, respectively [92]. PrioAD12 had the ability to detect Aβ plaques in brain sections from an individual affected with AD while no detection (staining) was observed on sections from a normal brain. Finally, Aβ-specific $V_Hs$ were isolated following immunization of a mouse with Aβ (1-42) peptides and construction of a phage-displayed $V_H$ library [93]. Selected antibodies were found to interact with different regions of the full-length peptide and inhibit its cell toxicity. Moreover, one $V_H$ (VH1.27), when tested for its ability to clear amyloid deposits in a mouse model of AD following intracranial injection was shown to significantly reduce the amyloid burden compared to the control.

Perchiacca and colleagues developed a new strategy to generate a series of sdAbs against disordered proteins [94–97]. They used defined algorithms to select motifs within disordered proteins that are predicted to participate in amyloid formation based on charge, hydrophobicity and propensity to form β-sheets [94]. They subsequently grafted peptides corresponding to the selected motifs into the CDR3 of a human $V_H$ with good solubility characteristics. By using their technique, the authors generated a pool of antibodies against amyloidogenic epitopes within Aβ (1-42) peptides. The $V_Hs$ demonstrated specific and sensitive recognition of Aβ monomers, soluble oligomers or fibrillar intermediates depending on the region covered by the grafted peptide and prevented toxicity induced by the targeted conformers [94]. It was later demonstrated that binding of the $V_Hs$ with their amyloidogenic target led to the assembly of Aβ-$V_Hs$ non-toxic complexes thereby preventing the formation of mature amyloid fibrils [95]. This technique was extended to construct one $V_H$ specific for an aggregation-prone epitope within αSyn with the ability to inhibit fibrillization by the protein.

A similar grafting method was used to generate additional $V_Hs$ targeting Aβ (1-42) or αSyn [98]. In this case, complementary peptides to the target sequence were designed based on interactions between amino acid sequences in the Protein Data Bank (PDB) and inserted into the CDR3 of a human $V_H$. Resulting antibodies all showed specific binding to their respective target. In addition, one anti-αSyn $V_H$ was tested for its neutralization potency in in vitro assays and demonstrated the ability to significantly reduce the aggregation of the targeted protein [98].

#### 3.1.2. Parkinson's Disease (PD)

PD represents the second most common neurological disorder affecting approximately 10 million people worldwide, and this number is predicted to increase over the coming years due to population aging [99]. Hallmarks of the disease include the loss or degeneration of dopamine producing neurons leading to severe motor control impairment. At the molecular level, PD is associated with the appearance of intracellular fibrillar aggregates known as Lewy bodies (LB) or Lewy neurites composed mostly of αSyn [100]. These large inclusions are responsible for neuronal cell death. Therefore, antibodies targeting the small, αSyn protein represent a promising treatment against PD.

Three sdAbs recognizing αSyn have been described in addition to the ones mentioned in the previous section [94,101–103]. First, following immunization of a dromedary with monomeric αSyn and subsequent construction and screening of a phage-displayed library, a $V_HH$ (NbSyn2) interacting with the soluble form of the protein was identified [101]. Based on nuclear magnetic resonance (NMR) spectroscopy and X-ray crystallography, the epitope of NbSyn2 was mapped to the C-terminus of αSyn within the last four residues [104]. Interestingly, $V_HH$ also interacted with amyloid fibrils formed by αSyn suggesting that this region of the protein remains exposed following its aggregation. In this same report, the authors demonstrated that the binding of NbSyn2 to αSyn did not induce any structural changes nor did it have any effect on the kinetics of formation of fibrils [101]. However, the affinity of the binding decreased as the process of fibril formation progressed suggesting that there might be conformational rearrangements of the C-terminal region of αSyn upon fibril maturation.

The same group isolated a second $V_HH$ (NbSyn87) from a phage-displayed library generated from the blood of an immunized llama this time using a mutant of αSyn (A53T) [102], which has been associated with early onset of PD [105]. This antibody also interacted with a region encompassing the C-terminus of the monomeric αSyn distinct from the NbSyn2 epitope and had the ability to bind to amyloid fibrils without structural consequences. As was observed for NbSyn2, there was a time-dependent decrease in the affinity of the antibody for its amyloid target. Further characterization of NbSyn2 and NbSyn87 led to the observation that both $V_H$Hs could inhibit the formation of mature fibrils in vitro [103]. They also had the ability to induce the conversion of αSyn from more stable oligomers into less stable oligomers significantly reducing the cellular toxicity caused by the protein.

The last αSyn-specific sdAb is a human $V_H$ (VH14) against the non-amyloid component (NAC) region of monomeric αSyn, which was selected from a non-immune yeast-displayed scFv library [106]. Although it was shown to have the highest affinity for its target, this antibody failed to rescue the cytotoxicity induced by αSyn. Nevertheless, fusion of this domain antibody to a proteosomal targeting PEST motif increased its solubility and conferred the ability to induce αSyn clearance thereby reducing the toxic effect associated with protein aggregation both in situ [107] and in vivo [108]. When compared to NbSyn87-PEST, the VH14-PEST fusion demonstrated a more pronounced effect suggesting that the NAC region of αSyn is a preferable therapeutic target.

#### 3.1.3. Huntington's Disease (HD)

HD is caused by an autosomal dominantly inherited CAG trinucleotide repeat expansion in the Htt gene [109]. Due to its high tendency to aggregate, the resulting mutant protein is at the origin of neuronal anomalies leading to cell death. In patients, this translates into numerous psychiatric and motor dysfunctions. There appears to be an inverse correlation between the length of expansion and age of onset. Although there are currently no curative treatments for HD, it certainly represents one of the most treatable neurological disorder since the molecular triggers are clearly defined. In this regard, sdAbs binding to mutant Htt have the potential to reduce its associated toxicity.

The use of a non-immune yeast-displayed scFv library led to the isolation of a human $V_L$ sdAb targeting the first 20 amino acids of the Htt protein [110]. The $V_L$ showed the same affinity for its target compared to its precursor scFv while achieving higher levels of cytoplasmic expression. However, inhibition of Htt aggregation demonstrated in a cell-free in vitro assay as well as in mammalian cells was only modest, requiring high amounts of the sdAb. In view of increasing the potency of $V_L$, the same group submitted it to mutagenesis to remove its disulfide bond for efficient expression of natively folded sdAbs in the cytoplasm and subsequently increase its binding affinity [111]. The mutant, $V_L$12.3, was able to strongly inhibit the formation of Htt aggregates and rescue cell toxicity in rat and yeast HD models. Adenoviral-delivery of this sdAb was shown to significantly improve behavior and neuropathology in a lentiviral mouse model of HD [112]. In contrast, when injected in transgenic HD mice, the antibody was found to increase the severity of the disease leading to a higher mortality rate. This was later attributed to a higher nuclear retention of Htt in the presence of $V_L$12.3 in the transgenic HD mouse model [113].

Similarly, two more $V_L$ domain antibodies (Happ1 and Happ3) targeting the proline-rich region of Htt were selected from a non-immune phage-displayed human scFv library [113]. Their capacity to reduce Htt-induced toxicity in cell culture increased compared to their scFv predecessors. Furthermore, they both had a greater ability to prevent neurodegeneration in a brain slice model of HD. Their mechanism of action involved an increased turnover rate of mutant Htt. Adenoviral-delivery of Happ1 demonstrated its efficacy in vivo in different mouse models of HD in which marked reduction of the disease-associated symptoms was observed following administration of the sdAb [112].

The first $V_H$Hs (i$V_H$H1–i$V_H$H4) against the N-terminal region of Htt have been isolated from an immunized llama using phage display technologies [114]. Although the functionality of these sdAbs remains to be examined, they were found to interact with purified human wild-type and mutant Htt and also co-immunoprecipitated with both species following incubation with human HD brain lysates.

#### 3.1.4. Prion Diseases

Prion diseases, also known as transmissible spongiform encephalopathies (TSEs) comprise of a group of fatal transmissible neurodegenerative diseases caused by the misfolding of the cellular prion protein ($PrP^c$) into the abnormally shaped scrapie prion protein ($PrP^{sc}$). The emergence of the diseased state of the prion protein (PrP) can be spontaneous, genetic, or acquired [115]. In all cases, each newly formed $PrP^{sc}$ acts as a template and promotes the conversion of more $PrP^c$ leading to the assembly of large insoluble amyloid fibrils associated with neurotoxicity and spongiform change in the brain parenchyma. The most common TSE is the Creutzfeldt-Jakob disease with a new incidence rate of about 1–2 cases per million of population worldwide [116]. People suffering from this disorder show a wide variety of psychiatric symptoms that rapidly progress leading ultimately to death. Despite the severity of prion diseases, only two sdAbs targeting PrPs have been generated to date.

The first one, PrioV3 was isolated from a phage-displayed $V_HH$ library generated from the blood of dromedaries immunized with brain homogenates from scrapie-infected mice adsorbed on magnetic beads [117]. The $V_HH$ showed high affinity binding to a linear epitope at the C-terminus of both $PrP^c$ and $PrP^{sc}$. PrioV3 was shown to cross the BBB in vitro in rat and human brain endothelial cell lines via RMT [92,117]. Moreover, when injected intravenously in rats, the $V_HH$ was detected in the brain parenchyma suggesting its ability to cross the BBB in vivo. It also had the capacity to reduce $PrP^c$ expression and $PrP^{sc}$ accumulation in prion-permissive cells following its addition to the culture medium. When the treatment was prolonged over four days, $PrP^{sc}$ was undetectable by Western blot suggesting complete and permanent inhibition of its replication by the antibody. Similar results were obtained in vivo in mice inoculated with scrapie-infected brain homogenates receiving a weekly dose of PrioV3 [92,117]. This treatment severely abrogated the accumulation of $PrP^{sc}$ in the spleen of the animals. Finally, PrioV3 showed no sign of neurotoxicity in vitro.

Nb484 was selected from a pool of 14 $V_HHs$ identified following llama immunization with murine PrPs and construction of phage-displayed $V_HH$ libraries [118]. This specific $V_HH$ showed the highest affinity for human PrPs. Assessment of its neutralizing properties revealed that the antibody could delay the formation of fibrils and abrogate the expression of $PrP^{sc}$ in scrapie-infected murine cells. In addition, Nb484 was used as a crystallization chaperone, allowing the solution of the first crystal structures of the full length human $PrP^c$ and a C-terminal truncated version of the protein, revealing novel structural insights on the early events of the conversion of $PrP^c$ into $PrP^{sc}$.

### 3.2. Glioblastoma Multiforme (GBM)

GBM is the most common type of brain tumors with the emergence of approximately 1000 new cases every year worldwide [119]. It is a highly aggressive malignancy showing rapid growth, intensive vascularization and predominant necrosis. Current treatments generally consist of maximal surgical resection followed by radiotherapy and chemotherapy. However, even with the use of these interventions, the prognosis remains extremely low and patients usually succumb to the disease within the first two years following diagnosis. This is in part due to the highly invasive nature of GBM and the difficulty of surgically removing all tumor cells. In addition, there is now growing evidence that the presence of chemotherapy and radiotherapy resistant stem-like cells within the tumor contributes to the resilience and recurrence of GBM [120]. Since early stages of the disease are mostly asymptomatic, current therapeutic strategies also suffer from late diagnosis. Alternative tools for the treatment and diagnosis of GBM are therefore urgently needed. Here we will review the different applications for sdAbs to improve current therapeutic modalities against GBM.

In view of identifying novel biomarkers for GBM, Jocevzka and colleagues prepared a phage-displayed $V_HH$ library from the blood of a llama immunized with a human GBM cell line enriched in stem-like cells [121,122]. Following several rounds of selection using protein extracts from diverse biological samples, three GBM-specific $V_HHs$ were designated for further characterization. Identification of their antigen by mass spectrometry revealed two proteins, Trim28 and β-actin, which showed enrichment in GBM compared to control samples. The relevance of these proteins as GBM biomarkers remains to be determined. Using a similar approach, the same group isolated seven additional $V_HHs$ specifically interacting with GBM antigens [123]. Initial Western blot and qPCR analyses complemented with bioinformatics demonstrated differential expression of some of the identified proteins in GBM compared to low grade gliomas suggesting their potential application as glioma class differentiation markers. Moreover, one antigen, mitochondrial translation elongation factor (TUFM) was isolated in a second independent screening by the same authors in which the selection of a $V_HH$ specific for GBM stem cells (GSCs) was achieved [124]. The specificity of the $V_HH$ for its target was confirmed by immunocytochemistry, and cytotoxicity assays demonstrated its profound effect on GSC growth.

VH-9.7 is a GSC-binding human $V_H$ that emerged from a non-immune yeast-displayed human scFv library using a patient-derived GSC line for selection [125]. This sdAb showed selective binding to five GSC lines and successfully identified GSCs in mouse brain xenografts by flow cytometry. Its ability to detect and localize to GSCs was also demonstrated in vivo in mice harboring orthotopic GSC xenografts following intravenous injection of a fluorophore-conjugated VH-9.7.

The following study aimed to develop novel strategies to target GBM vasculature using an in vivo panning technique to isolate camelid phage-displayed sdAbs specifically accumulating in tumor vessels [126]. This led to the identification of the C-C7 $V_HH$, which was later shown to target a distinct population of tumor vessels in mice xenografts as well as in GBM patient samples. The antibody also had the capacity to accumulate in the tumor vasculature following injection in mice harboring orthotopic xenografts while no antibodies were detected in normal brain vessels. Using a yeast-two-hybrid method, the antigen of C-C7 was identified as Dynactin-1-p150Glued, which was expressed exclusively on activated endothelial cells and may represent a valuable tumor vessel target. The antibody presented here could be used to assess the level of angiogenesis in GBM patients and determine the severity of the disease.

Finally, $V_HHs$ targeting the epidermal growth factor receptor (EGFR) have been investigated as GBM therapeutic agents [127]. EGFR is well known to be overexpressed and mutated in a wide variety of tumors including GBM and has been extensively studied as an anti-cancer target [128,129]. $V_HHs$ targeting this receptor were isolated from an immune phage-displayed library following llama immunization with overexpressing cell preparations [130]. These antibodies (ENb1 and ENb2) were selected specifically for their ability to prevent binding of the EGF ligand to the receptor via competitive elution strategies. When engineered for sustained on-site delivery by neural-stem cells, the $V_HHs$ were shown to localize specifically in the tumor environment and inhibit EGFR signaling in vitro and to significantly reduce tumor growth in mouse models of malignant and invasive GBMs [127].

## 4. Single-Domain Antibodies as Neuroimaging Tools

### *4.1. Single-Domain Antibodies as Targeted Molecular Imaging Agents*

Molecular imaging using advance and hybrid imaging modalities such as computed tomography (CT), positron emission tomography (PET), single photon emission computed tomography (SPECT), magnetic resonance imaging (MRI), and optical imaging, provide noninvasive means to characterize physiological processes and correlate molecular alterations with clinical outcomes. These technologies are improving early disease diagnosis, surgical guidance, patient stratification, and treatment monitoring [131]. Molecular imaging has advanced significantly during the last few decades through the identification of novel molecular targets and the development of multifunctional contrast agents along with new imaging instrumentation and analysis tools to extract quantitative data.

Targeted molecular imaging consists of an imaging probe linked to an agent that targets a specific biomarker of clinical relevance. Targeted molecular imaging agents have unique requirements that often differ from those of targeted therapeutic agents. In both cases, a high expression of the target antigen in the diseased versus normal tissue is required. However, for a targeted molecular imaging agent, a short half-life in circulation is preferable. Standard mAbs have a long half-life with slow liver clearance, which is a major hindrance for imaging applications, where a high contrast at early time points is critical for clinical applications [132]. The small size of sdAbs enables good tissue penetration and a fast clearance of the unbound fraction primarily via renal filtration (~60 kDa cutoff) [133–135]. This also allows the use of short-lived radionuclides, such as $^{68}$Ga ($t_{1/2}$ = 68 min) or $^{18}$F ($t_{1/2}$ = 109.8 min) for PET imaging which significantly reduces the patient's exposure to radiation. A first-in human PET study using an anti-HER2 sdAb labeled with $^{68}$Ga-NOTA in female patients with metastatic breast cancer showed that imaging at 60–90 min provides suitable contrast to detect small and large tumor lesions with a fast blood clearance of the sdAb, such that only 7.2% of initial activity was remaining at 90 min [131].

The unique modularity of sdAbs to be engineered in different multivalent formats, including monomers, dimers, and pentamers provide additional flexibility to fulfill their antigen binding and pharmacokinetic characteristics to specific applications [136]. For instance, anti-EGFR sdAbs in monomer and pentamer formats showed to be particularly suitable for molecular optical imaging of glioblastoma tumors due to their respective short half-lives of 40 min and 80 min, while the same sdAbs engineered into a bivalent format fused with human IgG Fc have better potential to be exploited for therapeutic applications due to their extended half-lives (12.5 h) and enhanced avidity [136].

### 4.2. Single-Domain Antibodies for Imaging Brain Tumor Vasculature

The brain tumor vasculature represents a readily reachable target for molecular imaging due to its direct access via blood perfusion after intravenous administration. For brain tumors, such as GBM, assessment of tumor angiogenesis can provide information on the severity of the disease and guide appropriate treatment regimens [137,138]. Various tumor vascular targets that are overexpressed in the disease brain tissue and not in normal brain have been previously exploited by molecular targeted moieties for the non-invasive assessment of tumor angiogenesis using PET, optical imaging, and MRI. These include vascular endothelial growth factor receptor 2 (VEGFR2) [139], endothelial cell adhesion molecules ($\alpha v\beta 3$ and $\alpha v\beta 5$ integrins) [140], and insulin-like growth factor binding protein 7 (IGFBP7) [141,142]. IGFBP7, in particular, is a secreted protein that accumulates in the basement membrane of tumor endothelial cells, and its expression is believed to be associated with higher-grade gliomas [141–143]. Since the tumor's malignancy is highly correlated with the degree of angiogenesis [137,138], the use of anti-IGFBP7 sdAb linked to a contrast agent for tumor vascular imaging could aid in the diagnosis and clinical management of brain tumors. In preclinical mouse models of GBM, an anti-IGFBP7 sdAb linked to a fluorophore was capable of non-invasively imaging the degree of angiogenesis [142]. Furthermore, bimodal optical-MRI contrast agents were developed by the bio-conjugation of anti-IGFBP7 sdAbs and the near-infrared fluorophore Cy5.5 to the surface of two types of nanoparticles, gadolinium-coated lipid particles for T1-weighted MRI imaging [144] and PEG functionalized-iron oxide nanoparticles for T2-weighted MRI imaging [145]. In both cases, after intravenous administration, the agents elicited an increased MRI contrast enhancement and fluorescent signal in a xenograft GBM tumor compared to a non-targeted nanoparticle. The molecular localization of the anti-IGFBP7 sdAb in the tumor brain vessels was further demonstrated by fluorescence microscopy [142].

### 4.3. Single-Domain Antibodies for Imaging Brain Targets

Due to the presence of the BBB, which limits the access of most biologics (i.e., proteins, peptides, antibodies) to the brain, radioligands used for PET imaging of CNS targets have been based on small molecular weight (<500 Da) molecules [146]. Targeted radioligands utilizing antibodies, antibody fragments, or sdAbs have been mainly developed for peripheral targets and used in different applications, including the detection of tumor markers, monitoring inflammatory processes, and visualization of antitumor immune responses [147]. However, a variety of strategies have been employed to allow the delivery of protein molecules across the BBB using both disruptive and non-disruptive methods.

As previously described, transmigration of antibodies across the BBB via RMT is a non-disruptive method for gaining access to brain targets (see Section 2). Using this strategy molecular imaging probes coupled to BBB carriers can be shuttled to the brain. For instance, taking advantage of the modularity of the BBB-transmigrating FC5 sdAb, a lipid-based nanoparticle was designed to encapsulate the anti-cancer drug doxorubicin and to display on its surface both a near-infrared (NIR) imaging agent and FC5 sdAb [148]. Upon intravenous injection in mice, in vivo optical imaging indicated increased brain delivery of the FC5-targeted versus non-targeted doxorubicin-containing liposomes. The optical fluorescent signal detected in vivo in the brain parenchyma correlated with the amount of doxorubicin delivered in the brain and measured ex vivo. Thus, this method allows for a non-invasive estimation of drug delivery into the brain.

BBB transmigration of macromolecules may also be achieved via adsorptive-mediated endocytosis through non-specific, charge-based interactions with the endothelial cell surface [149]. Endothelial cells are characterized by the presentation of negatively-charged clathrin-coated pits at the luminal surface, which can bind cationic proteins and facilitate their penetration through the BBB [150]. In an AD in vivo two-photon imaging study, sdAbs, selected for a basic isoelectric point (i.e., due to cationic amino acids) and for their binding to brain Aβ deposits or Tau inclusions were able to penetrate the BBB and bind to their respective brain target in vivo [135]. Interestingly, it was suggested that, in addition to the basic isoelectric pI, the molecular size of the sdAb was an important factor in the BBB penetration capability, as larger constructs (i.e., sdAb dimers) demonstrated reduced BBB penetration [151].

Some sdAbs have also been shown to interact with intracellular targets (i.e., penetrate cells) [152]. For instance, $V_HHs$ against the astrocyte marker glial fibrillary acidic protein (GFAP) has been shown to cross the BBB, reach the brain tissue, and penetrate into astrocytes, as demonstrated by immunofluorescence studies on injected animal tissue sections [151]. This feature of sdAbs has the potential to open up "difficult to access" intracellular targets in the brain or within brain cell subtypes.

In summary, sdAbs hold promise for dynamic imaging compared to other antibody-based agents due to their small size that allows better tissue penetration, rapid and homogeneous tumor/brain accumulation and fast blood clearance, which results in high tissue-to-background noise ratios. Single-domain antibodies are versatile, stable in very harsh conditions (pH, temperature), easy to conjugate to different imaging probes, and relatively safe due to their high specificity.

**Table 1.** Overview of single-domain antibodies developed for central nervous system applications.

| Product Name | Target | sdAb Type | Source | References |
|---|---|---|---|---|
| | | **BBB Shuttles** | | |
| TBX4 | TfR1 | $V_{NAR}$ | Synthetic phage-displayed $V_{NAR}$ library | [66] |
| IGF1R-3 | IGF1R | $V_HH$ | Immune phage-displayed $V_HH$ library | [75,76] |
| FC5 | TMEM-30A | $V_HH$ | Non-immune phage-displayed $V_HH$ library | [20,70–74] |
| FC44 | Unknown | $V_HH$ | Non-immune phage-displayed $V_HH$ library | [20,70,71] |

**Table 1.** *Cont.*

| Product Name | Target | sdAb Type | Source | References |
|---|---|---|---|---|
| | | **Neurodegenerative diseases** | | |
| | | **Alzheimer's disease** | | |
| B10 | Mature Aβ (1-40) fribrils and protofibrils | $V_HH$ | Synthetic phage-displayed $V_HH$ library | [84] |
| KW1 | Non-fibrillar Aβ (1-40) oligomers | $V_HH$ | Synthetic phage-displayed $V_HH$ library | [86,87] |
| ni3A | Aβ (1-42) deposits | $V_HH$ | Non-immune phage-displayed $V_HH$ library | [88–90] |
| V31-1 | Monomers and small Aβ (1-42) oligomers | $V_HH$ | Immune phage-displayed $V_HH$ library | [91] |
| PrioAD12 | Aβ (1-40) peptide | $V_HH$ | Immune phage-displayed $V_HH$ library | [92] |
| PrioAD13 | Aβ (1-42) peptide | $V_HH$ | Immune phage-displayed $V_HH$ library | [92] |
| PrioAD120 | Tau (1-16) peptide | $V_HH$ | Immune phage-displayed $V_HH$ library | [92] |
| VH1.27, VH1.28, VH2.8 | Aβ (1-42) peptide | $V_H$ | Immune phage-displayed mouse $V_H$ library | [93] |
| Aβ (1-10), Aβ (3-12), Aβ (6-15), Aβ (9-18), Aβ (12-21), Aβ (15-24), Aβ (18-27), Aβ (21-30), Aβ (24-33), Aβ (27-36), Aβ (30-39), Aβ (33-42) | AB monomers, soluble oligomers or fibrils | $V_H$ | Grafted amyloid-motif antibodies (Gammabody) | [94–97] |
| DesAb-Aβ | Aβ (15-21) peptide | $V_H$ | Gammabody | [98] |
| | | **Parkinson's disease** | | |
| αSyn (69-78) | αSyn fibrils | $V_H$ | Gammabody | [95] |
| DesAb-D, DesAb-E, DesAb-F | αSyn (61-67) or αSyn (70-76) peptide | $V_H$ | Gammabody | [98] |
| NbSyn2 | Monomeric αSyn and mature fibrils | $V_HH$ | Immune phage-displayed $V_HH$ library | [101,102,104] |

**Table 1.** *Cont.*

| Product Name | Target | sdAb Type | Source | References |
| --- | --- | --- | --- | --- |
| | | **Parkinson's disease** | | |
| NbSyn87 | Monomeric αSyn(A53T) and mature fibrils | $V_HH$ | Immune phage-displayed $V_HH$ library | [102,103,107,108] |
| VH14 | Monomeric αSyn | $V_H$ | Non-immune yeast-displayed human scFv library | [106–108] |
| | | **Huntington's disease** | | |
| $V_L$12.3 | Htt protein | $V_L$ | Non-immune yeast-displayed human scFv library | [110–113] |
| Happ1, Happ3 | Htt protein | $V_L$ | Non-immune phage-displayed human scFv library | [112,113] |
| i$V_HH$1, i$V_HH$2, i$V_HH$3, i$V_HH$4 | Htt protein | $V_HH$ | Immune phage-displayed $V_HH$ library | [114] |
| | | **Prion diseases** | | |
| PrioV3 | $PrP^c$ and $PrP^{sc}$ | $V_HH$ | Immune phage-displayed $V_HH$ library | [92,117] |
| Nb484 | MoPrP (23-230) | $V_HH$ | Immune phage-displayed $V_HH$ library | [118] |
| | | **Glioblastoma multiforme** | | |
| Nb237 | TRIM28 | $V_HH$ | Immune phage-displayed $V_HH$ library | [122] |
| Nb141 | β-actin | $V_HH$ | Immune phage-displayed $V_HH$ library | [122] |
| Nb10 | ACTB/NUCL complex | $V_HH$ | Immune phage-displayed $V_HH$ library | [123] |
| Nb79 | VIM | $V_HH$ | Immune phage-displayed $V_HH$ library | [123] |
| Nb179 | NAP1L1 | $V_HH$ | Immune phage-displayed $V_HH$ library | [123] |
| Nb225 | TUFM | $V_HH$ | Immune phage-displayed $V_HH$ library | [123] |
| Nb314 | DPYSL2 and MTHFD1 | $V_HH$ | Immune phage-displayed $V_HH$ library | [123] |

**Table 1.** *Cont.*

| Product Name | Target | sdAb Type | Source | References |
|---|---|---|---|---|
| **Glioblastoma multiforme** | | | | |
| Nb394 | CRMP1 | $V_HH$ | Immune phage-displayed $V_HH$ library | [123] |
| Nb395 | ALYREF | $V_HH$ | Immune phage-displayed $V_HH$ library | [123] |
| Nb206 | TUFM | $V_HH$ | Immune phage-displayed $V_HH$ library | [124] |
| VH-9.7 | GSC | $V_H$ | Non-immune yeast-displayed human scFv library | [125] |
| C-C7 | Dynactin-1-p150Glued | $V_HH$ | Non-immune phage-displayed $V_HH$ library | [126] |
| ENb1, ENb2 | EGFR | $V_HH$ | Immune phage-displayed $V_HH$ library | [127,130] |
| **Neuroimaging** | | | | |
| EG(2) | EGFR | $V_HH$ | Immune phage-displayed $V_HH$ library | [136] |
| sdAb 4.43 | IGFBP7 | $V_HH$ | Immune phage-displayed $V_HH$ library | [142,144,145] |
| FC5 | TMEM-30A | $V_HH$ | Non-immune phage-displayed $V_HH$ library | [148] |
| R3VQ | Aβ (1-42) peptide | $V_HH$ | Immune phage-displayed $V_HH$ library | [135] |
| A2 | Phospho-Tau protein | $V_HH$ | Immune phage-displayed $V_HH$ library | [135] |
| mVHH A10, mVHH E9, mVHH E3 | GFAP | $V_HH$ | Immune ribosome-displayed $V_HH$ library | [151] |

## 5. Conclusions

Singe-domain antibody technologies are 'coming of age' with many being tested in clinical trials. Several notable advantages of this compact antibody format, including ease of engineering, stability, recognition of unusual epitopes, and versatility for creating bi- and multifunctional molecules, have resulted in sdAbs being poised to address some of the most difficult target and disease spaces, most notably those of the CNS. CNS diseases are among the most difficult to treat not only because therapeutic targets (e.g., misfolded proteins, ion channels and G-protein coupled receptors) are very complex, but also because they are 'hidden' behind brain barriers and are thus difficult to access systemically. Selectivity of targeting of receptor/channel subtypes, often in specific activation states, specific targeting of point mutations, or epitopes 'embedded' in misfolded proteins present unique challenges, often difficult to address by either synthetic molecules or mAbs. While 'precision' targeting of desired epitopes is achievable by both sdAbs and mAbs, compact sdAb format could improve access to hidden epitopes. One distinct advantage of this format is improved diffusion in brain tissue after direct intracerebral administration, and enhanced brain tissue penetration after intrathecal infusion via perivascular flow. Furthermore, sdAbs are proving to be a versatile format for designing BBB carriers that could be easily combined in various display linkages (mono-, bi-, multivalent) with therapeutic monoclonal antibodies and other therapeutic cargos (peptides, proteins, nanocarriers, and imaging agents). The pipeline of sdAbs, both camelid and human, raised against CNS targets from naïve or immune libraries and tested in preclinical models is growing with prospects for entry into clinical testing in the near future. With parallel and significant progress in the development of BBB-delivery technologies based on sdAbs, the field of CNS, so far dominated by small molecule therapeutics, is slowly but steadily progressing into a new era of biological treatments, most notably antibody therapies for chronic neurodegenerative diseases.

**Author Contributions:** K.B., U.I., M.M., J.T., R.M. and D.S. wrote and proof-read the manuscript. K.B. assembled the manuscript.

**Funding:** This research received no external funding.

**Conflicts of Interest:** The authors declare no conflict of interest.

## References

1. Ward, E.S.; Gussow, D.; Griffiths, A.D.; Jones, P.T.; Winter, G. Binding activities of a repertoire of single immunoglobulin variable domains secreted from *Escherichia coli*. *Nature* **1989**, *341*, 544–546. [CrossRef] [PubMed]
2. Hamers-Casterman, C.; Atarhouch, T.; Muyldermans, S.; Robinson, G.; Hamers, C.; Songa, E.B.; Bendahman, N.; Hamers, R. Naturally occurring antibodies devoid of light chains. *Nature* **1993**, *363*, 446–448. [CrossRef] [PubMed]
3. Greenberg, A.S.; Avila, D.; Hughes, M.; Hughes, A.; McKinney, E.C.; Flajnik, M.F. A new antigen receptor gene family that undergoes rearrangement and extensive somatic diversification in sharks. *Nature* **1995**, *374*, 168–173. [CrossRef] [PubMed]
4. Muyldermans, S. Nanobodies: Natural single-domain antibodies. *Annu. Rev. Biochem.* **2013**, *82*, 775–797. [CrossRef] [PubMed]
5. Konning, D.; Zielonka, S.; Grzeschik, J.; Empting, M.; Valldorf, B.; Krah, S.; Schroter, C.; Sellmann, C.; Hock, B.; Kolmar, H. Camelid and shark single domain antibodies: Structural features and therapeutic potential. *Curr. Opin. Struct. Boil.* **2017**, *45*, 10–16. [CrossRef] [PubMed]
6. Kim, D.Y.; Hussack, G.; Kandalaft, H.; Tanha, J. Mutational approaches to improve the biophysical properties of human single-domain antibodies. *Biochim. Biophys. Acta* **2014**, *1844*, 1983–2001. [CrossRef]
7. Yu, Y.J.; Atwal, J.K.; Zhang, Y.; Tong, R.K.; Wildsmith, K.R.; Tan, C.; Bien-Ly, N.; Hersom, M.; Maloney, J.A.; Meilandt, W.J.; et al. Therapeutic bispecific antibodies cross the blood-brain barrier in nonhuman primates. *Sci. Transl. Med.* **2014**, *6*, 261ra154. [CrossRef] [PubMed]
8. Henry, K.A.; Kim, D.Y.; Kandalaft, H.; Lowden, M.J.; Yang, Q.; Schrag, J.D.; Hussack, G.; MacKenzie, C.R.; Tanha, J. Stability-Diversity Tradeoffs Impose Fundamental Constraints on Selection of Synthetic Human VH/VL Single-Domain Antibodies from In Vitro Display Libraries. *Front. Immunol.* **2017**, *8*, 1759. [CrossRef] [PubMed]
9. Janssens, R.; Dekker, S.; Hendriks, R.W.; Panayotou, G.; van Remoortere, A.; San, J.K.; Grosveld, F.; Drabek, D. Generation of heavy-chain-only antibodies in mice. *Proc. Natl. Acad. Sci. USA* **2006**, *103*, 15130–15135. [CrossRef] [PubMed]
10. Drabek, D.; Janssens, R.; de Boer, E.; Rademaker, R.; Kloess, J.; Skehel, J.; Grosveld, F. Expression Cloning and Production of Human Heavy-Chain-Only Antibodies from Murine Transgenic Plasma Cells. *Front. Immunol.* **2016**, *7*, 619. [CrossRef] [PubMed]
11. Steeland, S.; Vandenbroucke, R.E.; Libert, C. Nanobodies as therapeutics: Big opportunities for small antibodies. *Drug Discov. Today* **2016**, *21*, 1076–1113. [CrossRef] [PubMed]

12. Porter, C.B.a.A. VNARS: An ancient and unique repertoire of small molecules that deliver small, soluble, stable and high affinity binders of proteins. *Antibodies* **2015**, *4*, 240–258. [CrossRef]
13. Hoogenboom, H.R. Selecting and screening recombinant antibody libraries. *Nat. Biotechnol.* **2005**, *23*, 1105–1116. [CrossRef] [PubMed]
14. Zielonka, S.; Empting, M.; Grzeschik, J.; Konning, D.; Barelle, C.J.; Kolmar, H. Structural insights and biomedical potential of IgNAR scaffolds from sharks. *mAbs* **2015**, *7*, 15–25. [CrossRef] [PubMed]
15. Kovaleva, M.; Ferguson, L.; Steven, J.; Porter, A.; Barelle, C. Shark variable new antigen receptor biologics—A novel technology platform for therapeutic drug development. *Expert Opin. Boil. Ther.* **2014**, *14*, 1527–1539. [CrossRef] [PubMed]
16. Baral, T.N.; MacKenzie, R.; Arbabi Ghahroudi, M. Single-domain antibodies and their utility. *Curr. Protoc. Immunol.* **2013**, *103*, 2–17. [CrossRef]
17. Schneider, D.; Xiong, Y.; Hu, P.; Wu, D.; Chen, W.; Ying, T.; Zhu, Z.; Dimitrov, D.S.; Dropulic, B.; Orentas, R.J. A Unique Human Immunoglobulin Heavy Chain Variable Domain-Only CD33 CAR for the Treatment of Acute Myeloid Leukemia. *Front. Oncol.* **2018**, *8*, 539. [CrossRef]
18. Henry, K.A.; Tanha, J. Performance evaluation of phage-displayed synthetic human single-domain antibody libraries: A retrospective analysis. *J. Immunol. Methods* **2018**, *456*, 81–86. [CrossRef] [PubMed]
19. Hussack, G.; Riazi, A.; Ryan, S.; van Faassen, H.; MacKenzie, R.; Tanha, J.; Arbabi-Ghahroudi, M. Protease-resistant single-domain antibodies inhibit Campylobacter jejuni motility. *Protein Eng. Des. Sel. PEDS* **2014**, *27*, 191–198. [CrossRef]
20. Muruganandam, A.; Tanha, J.; Narang, S.; Stanimirovic, D. Selection of phage-displayed llama single-domain antibodies that transmigrate across human blood-brain barrier endothelium. *FASEB J.* **2002**, *16*, 240–242. [CrossRef]
21. Iezzi, M.E.; Policastro, L.; Werbajh, S.; Podhajcer, O.; Canziani, G.A. Single-Domain Antibodies and the Promise of Modular Targeting in Cancer Imaging and Treatment. *Front. Immunol.* **2018**, *9*, 273. [CrossRef] [PubMed]
22. Ubah, O.C.; Buschhaus, M.J.; Ferguson, L.; Kovaleva, M.; Steven, J.; Porter, A.J.; Barelle, C.J. Next-generation flexible formats of VNAR domains expand the drug platform's utility and developability. *Biochem. Soc. Trans.* **2018**, *46*, 1559–1565. [CrossRef]
23. Els Conrath, K.; Lauwereys, M.; Wyns, L.; Muyldermans, S. Camel single-domain antibodies as modular building units in bispecific and bivalent antibody constructs. *J. Boil. Chem.* **2001**, *276*, 7346–7350. [CrossRef]
24. Holt, L.J.; Basran, A.; Jones, K.; Chorlton, J.; Jespers, L.S.; Brewis, N.D.; Tomlinson, I.M. Anti-serum albumin domain antibodies for extending the half-lives of short lived drugs. *Protein Eng. Des. Sel. PEDS* **2008**, *21*, 283–288. [CrossRef] [PubMed]
25. Coppieters, K.; Dreier, T.; Silence, K.; de Haard, H.; Lauwereys, M.; Casteels, P.; Beirnaert, E.; Jonckheere, H.; Van de Wiele, C.; Staelens, L.; et al. Formatted anti-tumor necrosis factor alpha VHH proteins derived from camelids show superior potency and targeting to inflamed joints in a murine model of collagen-induced arthritis. *Arthritis Rheum.* **2006**, *54*, 1856–1866. [CrossRef] [PubMed]
26. Simmons, D.P.; Abregu, F.A.; Krishnan, U.V.; Proll, D.F.; Streltsov, V.A.; Doughty, L.; Hattarki, M.K.; Nuttall, S.D. Dimerisation strategies for shark IgNAR single domain antibody fragments. *J. Immunol. Methods* **2006**, *315*, 171–184. [CrossRef] [PubMed]
27. De Bernardis, F.; Liu, H.; O'Mahony, R.; La Valle, R.; Bartollino, S.; Sandini, S.; Grant, S.; Brewis, N.; Tomlinson, I.; Basset, R.C.; et al. Human domain antibodies against virulence traits of Candida albicans inhibit fungus adherence to vaginal epithelium and protect against experimental vaginal candidiasis. *J. Infect. Dis.* **2007**, *195*, 149–157. [CrossRef] [PubMed]
28. Muller, M.R.; Saunders, K.; Grace, C.; Jin, M.; Piche-Nicholas, N.; Steven, J.; O'Dwyer, R.; Wu, L.; Khetemenee, L.; Vugmeyster, Y.; et al. Improving the pharmacokinetic properties of biologics by fusion to an anti-HSA shark VNAR domain. *mAbs* **2012**, *4*, 673–685. [CrossRef]
29. Steven, J.; Muller, M.R.; Carvalho, M.F.; Ubah, O.C.; Kovaleva, M.; Donohoe, G.; Baddeley, T.; Cornock, D.; Saunders, K.; Porter, A.J.; et al. In Vitro Maturation of a Humanized Shark VNAR Domain to Improve Its Biophysical Properties to Facilitate Clinical Development. *Front. Immunol.* **2017**, *8*, 1361. [CrossRef]
30. Nosenko, M.A.; Atretkhany, K.N.; Mokhonov, V.V.; Efimov, G.A.; Kruglov, A.A.; Tillib, S.V.; Drutskaya, M.S.; Nedospasov, S.A. VHH-Based Bispecific Antibodies Targeting Cytokine Production. *Front. Immunol.* **2017**, *8*, 1073. [CrossRef]

31. Beirnaert, E.; Desmyter, A.; Spinelli, S.; Lauwereys, M.; Aarden, L.; Dreier, T.; Loris, R.; Silence, K.; Pollet, C.; Cambillau, C.; et al. Bivalent Llama Single-Domain Antibody Fragments against Tumor Necrosis Factor Have Picomolar Potencies due to Intramolecular Interactions. *Front. Immunol.* **2017**, *8*, 867. [CrossRef] [PubMed]
32. Darling, T.L.; Sherwood, L.J.; Hayhurst, A. Intracellular Crosslinking of Filoviral Nucleoproteins with Xintrabodies Restricts Viral Packaging. *Front. Immunol.* **2017**, *8*, 1197. [CrossRef] [PubMed]
33. Yang, Z.; Schmidt, D.; Liu, W.; Li, S.; Shi, L.; Sheng, J.; Chen, K.; Yu, H.; Tremblay, J.M.; Chen, X.; et al. A novel multivalent, single-domain antibody targeting TcdA and TcdB prevents fulminant Clostridium difficile infection in mice. *J. Infect. Dis.* **2014**, *210*, 964–972. [CrossRef]
34. Van Roy, M.; Ververken, C.; Beirnaert, E.; Hoefman, S.; Kolkman, J.; Vierboom, M.; Breedveld, E.; t Hart, B.; Poelmans, S.; Bontinck, L.; et al. The preclinical pharmacology of the high affinity anti-IL-6R Nanobody(R) ALX-0061 supports its clinical development in rheumatoid arthritis. *Arthritis Res. Ther.* **2015**, *17*, 135. [CrossRef] [PubMed]
35. Desmyter, A.; Spinelli, S.; Boutton, C.; Saunders, M.; Blachetot, C.; de Haard, H.; Denecker, G.; Van Roy, M.; Cambillau, C.; Rommelaere, H. Neutralization of Human Interleukin 23 by Multivalent Nanobodies Explained by the Structure of Cytokine-Nanobody Complex. *Front. Immunol.* **2017**, *8*, 884. [CrossRef]
36. Harmsen, M.M.; Fijten, H.P.; Dekker, A.; Eble, P.L. Passive immunization of pigs with bispecific llama single-domain antibody fragments against foot-and-mouth disease and porcine immunoglobulin. *Vet. Microbiol.* **2008**, *132*, 56–64. [CrossRef]
37. Zhang, J.; Tanha, J.; Hirama, T.; Khieu, N.H.; To, R.; Tong-Sevinc, H.; Stone, E.; Brisson, J.R.; MacKenzie, C.R. Pentamerization of single-domain antibodies from phage libraries: A novel strategy for the rapid generation of high-avidity antibody reagents. *J. Mol. Boil.* **2004**, *335*, 49–56. [CrossRef]
38. Stone, E.; Hirama, T.; Tanha, J.; Tong-Sevinc, H.; Li, S.; MacKenzie, C.R.; Zhang, J. The assembly of single domain antibodies into bispecific decavalent molecules. *J. Immunol. Methods* **2007**, *318*, 88–94. [CrossRef] [PubMed]
39. Zhang, J.; Liu, X.; Bell, A.; To, R.; Baral, T.N.; Azizi, A.; Li, J.; Cass, B.; Durocher, Y. Transient expression and purification of chimeric heavy chain antibodies. *Protein Expr. Purif.* **2009**, *65*, 77–82. [CrossRef] [PubMed]
40. Ubah, O.C.; Steven, J.; Kovaleva, M.; Ferguson, L.; Barelle, C.; Porter, A.J.R.; Barelle, C.J. Novel, Anti-hTNF-alpha Variable New Antigen Receptor Formats with Enhanced Neutralizing Potency and Multifunctionality, Generated for Therapeutic Development. *Front. Immunol.* **2017**, *8*, 1780. [CrossRef] [PubMed]
41. Kovaleva, M.; Johnson, K.; Steven, J.; Barelle, C.J.; Porter, A. Therapeutic Potential of Shark Anti-ICOSL VNAR Domains is Exemplified in a Murine Model of Autoimmune Non-Infectious Uveitis. *Front. Immunol.* **2017**, *8*, 1121. [CrossRef]
42. D'Eall, C.; Pon, R.A.; Rossotti, M.A.; Krahn, N.; Spearman, M.; Callaghan, D.; van Faassen, H.; Hussack, G.; Stetefeld, J.; Butler, M.; et al. Modulating antibody-dependent cellular cytotoxicity of epidermal growth factor receptor-specific heavy-chain antibodies through hinge engineering. *Immunol. Cell Boil.* **2019**, 1–12. [CrossRef] [PubMed]
43. Rossotti, M.A.; Henry, K.A.; van Faassen, H.; Tanha, J.; Callaghan, D.; Hussack, G.; Arbabi-Ghahroudi, M.; MacKenzie, C.R. Camelid single-domain antibodies raised by DNA immunization are potent inhibitors of EGFR signaling. *Biochem. J.* **2019**, *476*, 39–50. [CrossRef] [PubMed]
44. Henry, K.A.; Kandalaft, H.; Lowden, M.J.; Rossotti, M.A.; van Faassen, H.; Hussack, G.; Durocher, Y.; Kim, D.Y.; Tanha, J. A disulfide-stabilized human VL single-domain antibody library is a source of soluble and highly thermostable binders. *Mol. Immunol.* **2017**, *90*, 190–196. [CrossRef] [PubMed]
45. Schutze, K.; Petry, K.; Hambach, J.; Schuster, N.; Fumey, W.; Schriewer, L.; Rockendorf, J.; Menzel, S.; Albrecht, B.; Haag, F.; et al. CD38-Specific Biparatopic Heavy Chain Antibodies Display Potent Complement-Dependent Cytotoxicity Against Multiple Myeloma Cells. *Front. Immunol.* **2018**, *9*, 2553. [CrossRef] [PubMed]
46. Von Kreudenstein, T.S.; Escobar-Carbrera, E.; Lario, P.I.; D'Angelo, I.; Brault, K.; Kelly, J.; Durocher, Y.; Baardsnes, J.; Woods, R.J.; Xie, M.H.; et al. Improving biophysical properties of a bispecific antibody scaffold to aid developability: Quality by molecular design. *mAbs* **2013**, *5*, 646–654. [CrossRef] [PubMed]
47. Brinkmann, U.; Kontermann, R.E. The making of bispecific antibodies. *mAbs* **2017**, *9*, 182–212. [CrossRef]

48. Cortez-Retamozo, V.; Backmann, N.; Senter, P.D.; Wernery, U.; De Baetselier, P.; Muyldermans, S.; Revets, H. Efficient cancer therapy with a nanobody-based conjugate. *Cancer Res.* **2004**, *64*, 2853–2857. [CrossRef] [PubMed]
49. Baral, T.N.; Magez, S.; Stijlemans, B.; Conrath, K.; Vanhollebeke, B.; Pays, E.; Muyldermans, S.; De Baetselier, P. Experimental therapy of African trypanosomiasis with a nanobody-conjugated human trypanolytic factor. *Nat. Med.* **2006**, *12*, 580–584. [CrossRef] [PubMed]
50. Yu, Y.; Li, J.; Zhu, X.; Tang, X.; Bao, Y.; Sun, X.; Huang, Y.; Tian, F.; Liu, X.; Yang, L. Humanized CD7 nanobody-based immunotoxins exhibit promising anti-T-cell acute lymphoblastic leukemia potential. *Int. J. Nanomed.* **2017**, *12*, 1969–1983. [CrossRef] [PubMed]
51. Behdani, M.; Zeinali, S.; Karimipour, M.; Khanahmad, H.; Schoonooghe, S.; Aslemarz, A.; Seyed, N.; Moazami-Godarzi, R.; Baniahmad, F.; Habibi-Anbouhi, M.; et al. Development of VEGFR2-specific Nanobody Pseudomonas exotoxin A conjugated to provide efficient inhibition of tumor cell growth. *New Biotechnol.* **2013**, *30*, 205–209. [CrossRef]
52. Tian, B.; Wong, W.Y.; Hegmann, E.; Gaspar, K.; Kumar, P.; Chao, H. Production and characterization of a camelid single domain antibody-urease enzyme conjugate for the treatment of cancer. *Bioconj. Chem.* **2015**, *26*, 1144–1155. [CrossRef] [PubMed]
53. Tian, B.; Wong, W.Y.; Uger, M.D.; Wisniewski, P.; Chao, H. Development and Characterization of a Camelid Single Domain Antibody-Urease Conjugate That Targets Vascular Endothelial Growth Factor Receptor 2. *Front. Immunol.* **2017**, *8*, 956. [CrossRef] [PubMed]
54. Abbott, N.J.; Patabendige, A.A.; Dolman, D.E.; Yusof, S.R.; Begley, D.J. Structure and function of the blood-brain barrier. *Neurobiol. Dis.* **2010**, *37*, 13–25. [CrossRef] [PubMed]
55. Wolak, D.J.; Thorne, R.G. Diffusion of macromolecules in the brain: Implications for drug delivery. *Mol. Pharm.* **2013**, *10*, 1492–1504. [CrossRef] [PubMed]
56. Pizzo, M.E.; Wolak, D.J.; Kumar, N.N.; Brunette, E.; Brunnquell, C.L.; Hannocks, M.J.; Abbott, N.J.; Meyerand, M.E.; Sorokin, L.; Stanimirovic, D.B.; et al. Intrathecal antibody distribution in the rat brain: Surface diffusion, perivascular transport and osmotic enhancement of delivery. *J. Physiol.* **2018**, *596*, 445–475. [CrossRef] [PubMed]
57. Jones, A.R.; Shusta, E.V. Blood-brain barrier transport of therapeutics via receptor-mediation. *Pharm. Res.* **2007**, *24*, 1759–1771. [CrossRef] [PubMed]
58. Pardridge, W.M.; Buciak, J.L.; Friden, P.M. Selective transport of an anti-transferrin receptor antibody through the blood-brain barrier in vivo. *J. Pharmacol. Exp. Ther.* **1991**, *259*, 66–70. [PubMed]
59. Yu, Y.J.; Zhang, Y.; Kenrick, M.; Hoyte, K.; Luk, W.; Lu, Y.; Atwal, J.; Elliott, J.M.; Prabhu, S.; Watts, R.J.; et al. Boosting brain uptake of a therapeutic antibody by reducing its affinity for a transcytosis target. *Sci. Transl. Med.* **2011**, *3*, 84ra44. [CrossRef] [PubMed]
60. Niewoehner, J.; Bohrmann, B.; Collin, L.; Urich, E.; Sade, H.; Maier, P.; Rueger, P.; Stracke, J.O.; Lau, W.; Tissot, A.C.; et al. Increased brain penetration and potency of a therapeutic antibody using a monovalent molecular shuttle. *Neuron* **2014**, *81*, 49–60. [CrossRef]
61. Haqqani, A.S.; Delaney, C.E.; Brunette, E.; Baumann, E.; Farrington, G.K.; Sisk, W.; Eldredge, J.; Ding, W.; Tremblay, T.L.; Stanimirovic, D.B. Endosomal trafficking regulates receptor-mediated transcytosis of antibodies across the blood brain barrier. *J. Cereb. Blood Flow Metab.* **2018**, *38*, 727–740. [CrossRef] [PubMed]
62. Coloma, M.J.; Lee, H.J.; Kurihara, A.; Landaw, E.M.; Boado, R.J.; Morrison, S.L.; Pardridge, W.M. Transport across the primate blood-brain barrier of a genetically engineered chimeric monoclonal antibody to the human insulin receptor. *Pharm. Res.* **2000**, *17*, 266–274. [CrossRef] [PubMed]
63. Boado, R.J.; Zhou, Q.H.; Lu, J.Z.; Hui, E.K.; Pardridge, W.M. Pharmacokinetics and brain uptake of a genetically engineered bifunctional fusion antibody targeting the mouse transferrin receptor. *Mol. Pharm.* **2010**, *7*, 237–244. [CrossRef] [PubMed]
64. Giugliani, R.; Giugliani, L.; de Oliveira Poswar, F.; Donis, K.C.; Corte, A.D.; Schmidt, M.; Boado, R.J.; Nestrasil, I.; Nguyen, C.; Chen, S.; et al. Neurocognitive and somatic stabilization in pediatric patients with severe Mucopolysaccharidosis Type I after 52 weeks of intravenous brain-penetrating insulin receptor antibody-iduronidase fusion protein (valanafusp alpha): An open label phase 1-2 trial. *Orphanet J. Rare Dis.* **2018**, *13*, 110. [CrossRef] [PubMed]

65. Pardridge, W.M.; Boado, R.J.; Patrick, D.J.; Ka-Wai Hui, E.; Lu, J.Z. Blood-Brain Barrier Transport, Plasma Pharmacokinetics, and Neuropathology Following Chronic Treatment of the Rhesus Monkey with a Brain Penetrating Humanized Monoclonal Antibody Against the Human Transferrin Receptor. *Mol. Pharm.* **2018**. [CrossRef] [PubMed]
66. Walsh, F.S.; Wicher, K.; Szary, J.; Stocki, P.; Demydchuk, M.; Rutkowski, L. Delivery of a CD20 transferrin receptor VNAR bispecific antibody to the brain for CNS lymphoma. In Proceedings of the AACR Annual Meeting, Washington, DC, USA, 1–5 April 2017.
67. Bahney, J.; von Bartheld, C.S. The Cellular Composition and Glia-Neuron Ratio in the Spinal Cord of a Human and a Nonhuman Primate: Comparison with Other Species and Brain Regions. *Anat. Rec.* **2018**, *301*, 697–710. [CrossRef]
68. Couch, J.A.; Yu, Y.J.; Zhang, Y.; Tarrant, J.M.; Fuji, R.N.; Meilandt, W.J.; Solanoy, H.; Tong, R.K.; Hoyte, K.; Luk, W.; et al. Addressing safety liabilities of TfR bispecific antibodies that cross the blood-brain barrier. *Sci. Transl. Med.* **2013**, *5*, 183ra57. [CrossRef] [PubMed]
69. Webster, C.I.; Stanimirovic, D.B. A gateway to the brain: Shuttles for brain delivery of macromolecules. *Ther. Deliv.* **2015**, *6*, 1321–1324. [CrossRef]
70. Haqqani, A.S.; Caram-Salas, N.; Ding, W.; Brunette, E.; Delaney, C.E.; Baumann, E.; Boileau, E.; Stanimirovic, D. Multiplexed evaluation of serum and CSF pharmacokinetics of brain-targeting single-domain antibodies using a NanoLC-SRM-ILIS method. *Mol. Pharm.* **2013**, *10*, 1542–1556. [CrossRef]
71. Abulrob, A.; Sprong, H.; Van Bergen en Henegouwen, P.; Stanimirovic, D. The blood-brain barrier transmigrating single domain antibody: Mechanisms of transport and antigenic epitopes in human brain endothelial cells. *J. Neurochem.* **2005**, *95*, 1201–1214. [CrossRef]
72. Albulrob, A.; Stanimirovic, D.; Muruganandam, A. Blood-Brain Barrier Epitopes and Uses Thereof. Patent WO2007036021A1, 5 April 2007.
73. Farrington, G.K.; Caram-Salas, N.; Haqqani, A.S.; Brunette, E.; Eldredge, J.; Pepinsky, B.; Antognetti, G.; Baumann, E.; Ding, W.; Garber, E.; et al. A novel platform for engineering blood-brain barrier-crossing bispecific biologics. *FASEB J.* **2014**, *28*, 4764–4778. [CrossRef] [PubMed]
74. Webster, C.I.; Caram-Salas, N.; Haqqani, A.S.; Thom, G.; Brown, L.; Rennie, K.; Yogi, A.; Costain, W.; Brunette, E.; Stanimirovic, D.B. Brain penetration, target engagement, and disposition of the blood-brain barrier-crossing bispecific antibody antagonist of metabotropic glutamate receptor type 1. *FASEB J.* **2016**, *30*, 1927–1940. [CrossRef] [PubMed]
75. Stanimirovic, D.; Kemmerich, K.; Haqqani, A.S.; Sulea, T.; Arbabi-Ghahroudi, M.; Massie, B.; Gilbert, R. Insulin-Like Growth Factor 1 Receptor-Specific Antibodies and Uses Thereof. U.S. Patent US20170015748A1, 19 January 2017.
76. Ribecco-Lutkiewicz, M.; Sodja, C.; Haukenfrers, J.; Haqqani, A.S.; Ly, D.; Zachar, P.; Baumann, E.; Ball, M.; Huang, J.; Rukhlova, M.; et al. A novel human induced pluripotent stem cell blood-brain barrier model: Applicability to study antibody-triggered receptor-mediated transcytosis. *Sci. Rep.* **2018**, *8*, 1873. [CrossRef] [PubMed]
77. Robertson, C.S.; Cherian, L.; Shah, M.; Garcia, R.; Navarro, J.C.; Grill, R.J.; Hand, C.C.; Tian, T.S.; Hannay, H.J. Neuroprotection with an erythropoietin mimetic peptide (pHBSP) in a model of mild traumatic brain injury complicated by hemorrhagic shock. *J. Neurotrauma* **2012**, *29*, 1156–1166. [CrossRef]
78. Chiti, F.; Dobson, C.M. Protein misfolding, functional amyloid, and human disease. *Annu. Rev. Biochem.* **2006**, *75*, 333–366. [CrossRef] [PubMed]
79. Dobson, C.M. Principles of protein folding, misfolding and aggregation. *Semin. Cell Dev. Boil.* **2004**, *15*, 3–16. [CrossRef] [PubMed]
80. Katzman, R. Editorial: The prevalence and malignancy of Alzheimer disease. A major killer. *Arch. Neurol.* **1976**, *33*, 217–218. [CrossRef] [PubMed]
81. Barker, W.W.; Luis, C.A.; Kashuba, A.; Luis, M.; Harwood, D.G.; Loewenstein, D.; Waters, C.; Jimison, P.; Shepherd, E.; Sevush, S.; et al. Relative frequencies of Alzheimer disease, Lewy body, vascular and frontotemporal dementia, and hippocampal sclerosis in the State of Florida Brain Bank. *Alzheimer Dis. Assoc. Disord.* **2002**, *16*, 203–212. [CrossRef] [PubMed]
82. Hardy, J.; Selkoe, D.J. The amyloid hypothesis of Alzheimer's disease: Progress and problems on the road to therapeutics. *Science* **2002**, *297*, 353–356. [CrossRef]

83. Wang, D.S.; Dickson, D.W.; Malter, J.S. Tissue transglutaminase, protein cross-linking and Alzheimer's disease: Review and views. *Int. J. Clin. Exp. Pathol.* **2008**, *1*, 5–18.
84. Habicht, G.; Haupt, C.; Friedrich, R.P.; Hortschansky, P.; Sachse, C.; Meinhardt, J.; Wieligmann, K.; Gellermann, G.P.; Brodhun, M.; Gotz, J.; et al. Directed selection of a conformational antibody domain that prevents mature amyloid fibril formation by stabilizing Abeta protofibrils. *Proc. Natl. Acad. Sci. USA* **2007**, *104*, 19232–19237. [CrossRef] [PubMed]
85. Walsh, D.M.; Lomakin, A.; Benedek, G.B.; Condron, M.M.; Teplow, D.B. Amyloid beta-protein fibrillogenesis. Detection of a protofibrillar intermediate. *J. Boil. Chem.* **1997**, *272*, 22364–22372. [CrossRef]
86. Morgado, I.; Wieligmann, K.; Bereza, M.; Ronicke, R.; Meinhardt, K.; Annamalai, K.; Baumann, M.; Wacker, J.; Hortschansky, P.; Malesevic, M.; et al. Molecular basis of beta-amyloid oligomer recognition with a conformational antibody fragment. *Proc. Natl. Acad. Sci. USA* **2012**, *109*, 12503–12508. [CrossRef] [PubMed]
87. Wacker, J.; Ronicke, R.; Westermann, M.; Wulff, M.; Reymann, K.G.; Dobson, C.M.; Horn, U.; Crowther, D.C.; Luheshi, L.M.; Fandrich, M. Oligomer-targeting with a conformational antibody fragment promotes toxicity in Abeta-expressing flies. *Acta Neuropathol. Commun.* **2014**, *2*, 43. [CrossRef] [PubMed]
88. Rutgers, K.S.; van Remoortere, A.; van Buchem, M.A.; Verrips, C.T.; Greenberg, S.M.; Bacskai, B.J.; Frosch, M.P.; van Duinen, S.G.; Maat-Schieman, M.L.; van der Maarel, S.M. Differential recognition of vascular and parenchymal beta amyloid deposition. *Neurobiol. Aging* **2011**, *32*, 1774–1783. [CrossRef] [PubMed]
89. Rutgers, K.S.; Nabuurs, R.J.; van den Berg, S.A.; Schenk, G.J.; Rotman, M.; Verrips, C.T.; van Duinen, S.G.; Maat-Schieman, M.L.; van Buchem, M.A.; de Boer, A.G.; et al. Transmigration of beta amyloid specific heavy chain antibody fragments across the in vitro blood-brain barrier. *Neuroscience* **2011**, *190*, 37–42. [CrossRef]
90. Nabuurs, R.J.; Rutgers, K.S.; Welling, M.M.; Metaxas, A.; de Backer, M.E.; Rotman, M.; Bacskai, B.J.; van Buchem, M.A.; van der Maarel, S.M.; van der Weerd, L. In vivo detection of amyloid-beta deposits using heavy chain antibody fragments in a transgenic mouse model for Alzheimer's disease. *PLoS ONE* **2012**, *7*, e38284. [CrossRef] [PubMed]
91. Lafaye, P.; Achour, I.; England, P.; Duyckaerts, C.; Rougeon, F. Single-domain antibodies recognize selectively small oligomeric forms of amyloid beta, prevent Abeta-induced neurotoxicity and inhibit fibril formation. *Mol. Immunol.* **2009**, *46*, 695–704. [CrossRef]
92. David, M.A.; Jones, D.R.; Tayebi, M. Potential candidate camelid antibodies for the treatment of protein-misfolding diseases. *J. Neuroimmunol.* **2014**, *272*, 76–85. [CrossRef] [PubMed]
93. Medecigo, M.; Manoutcharian, K.; Vasilevko, V.; Govezensky, T.; Munguia, M.E.; Becerril, B.; Luz-Madrigal, A.; Vaca, L.; Cribbs, D.H.; Gevorkian, G. Novel amyloid-beta specific scFv and VH antibody fragments from human and mouse phage display antibody libraries. *J. Neuroimmunol.* **2010**, *223*, 104–114. [CrossRef] [PubMed]
94. Perchiacca, J.M.; Ladiwala, A.R.; Bhattacharya, M.; Tessier, P.M. Structure-based design of conformation- and sequence-specific antibodies against amyloid beta. *Proc. Natl. Acad. Sci. USA* **2012**, *109*, 84–89. [CrossRef] [PubMed]
95. Ladiwala, A.R.; Bhattacharya, M.; Perchiacca, J.M.; Cao, P.; Raleigh, D.P.; Abedini, A.; Schmidt, A.M.; Varkey, J.; Langen, R.; Tessier, P.M. Rational design of potent domain antibody inhibitors of amyloid fibril assembly. *Proc. Natl. Acad. Sci. USA* **2012**, *109*, 19965–19970. [CrossRef] [PubMed]
96. Julian, M.C.; Lee, C.C.; Tiller, K.E.; Rabia, L.A.; Day, E.K.; Schick, A.J., 3rd; Tessier, P.M. Co-evolution of affinity and stability of grafted amyloid-motif domain antibodies. *Protein Eng. Des. Sel. PEDS* **2015**, *28*, 339–350. [CrossRef] [PubMed]
97. Lee, C.C.; Julian, M.C.; Tiller, K.E.; Meng, F.; DuConge, S.E.; Akter, R.; Raleigh, D.P.; Tessier, P.M. Design and Optimization of Anti-amyloid Domain Antibodies Specific for beta-Amyloid and Islet Amyloid Polypeptide. *J. Boil. Chem.* **2016**, *291*, 2858–2873. [CrossRef] [PubMed]
98. Sormanni, P.; Aprile, F.A.; Vendruscolo, M. Rational design of antibodies targeting specific epitopes within intrinsically disordered proteins. *Proc. Natl. Acad. Sci. USA* **2015**, *112*, 9902–9907. [CrossRef] [PubMed]
99. Hoehn, M.M.; Yahr, M.D. Parkinsonism: Onset, progression, and mortality. *Neurology* **1998**, *50*, 318. [CrossRef] [PubMed]
100. Malek, N.; Swallow, D.; Grosset, K.A.; Anichtchik, O.; Spillantini, M.; Grosset, D.G. Alpha-synuclein in peripheral tissues and body fluids as a biomarker for Parkinson's disease—A systematic review. *Acta Neurol. Scand.* **2014**, *130*, 59–72. [CrossRef]

101. De Genst, E.J.; Guilliams, T.; Wellens, J.; O'Day, E.M.; Waudby, C.A.; Meehan, S.; Dumoulin, M.; Hsu, S.T.; Cremades, N.; Verschueren, K.H.; et al. Structure and properties of a complex of alpha-synuclein and a single-domain camelid antibody. *J. Mol. Boil.* **2010**, *402*, 326–343. [CrossRef]
102. Guilliams, T.; El-Turk, F.; Buell, A.K.; O'Day, E.M.; Aprile, F.A.; Esbjorner, E.K.; Vendruscolo, M.; Cremades, N.; Pardon, E.; Wyns, L.; et al. Nanobodies raised against monomeric alpha-synuclein distinguish between fibrils at different maturation stages. *J. Mol. Boil.* **2013**, *425*, 2397–2411. [CrossRef]
103. Iljina, M.; Hong, L.; Horrocks, M.H.; Ludtmann, M.H.; Choi, M.L.; Hughes, C.D.; Ruggeri, F.S.; Guilliams, T.; Buell, A.K.; Lee, J.E.; et al. Nanobodies raised against monomeric a-synuclein inhibit fibril formation and destabilize toxic oligomeric species. *BMC Boil.* **2017**, *15*, 57. [CrossRef]
104. Vuchelen, A.; O'Day, E.; De Genst, E.; Pardon, E.; Wyns, L.; Dumoulin, M.; Dobson, C.M.; Christodoulou, J.; Hsu, S.T. $^{1}$H, $^{13}$C and $^{15}$N assignments of a camelid nanobody directed against human alpha-synuclein. *Biomol. NMR Assign.* **2009**, *3*, 231–233. [CrossRef] [PubMed]
105. Polymeropoulos, M.H.; Lavedan, C.; Leroy, E.; Ide, S.E.; Dehejia, A.; Dutra, A.; Pike, B.; Root, H.; Rubenstein, J.; Boyer, R.; et al. Mutation in the alpha-synuclein gene identified in families with Parkinson's disease. *Science* **1997**, *276*, 2045–2047. [CrossRef] [PubMed]
106. Lynch, S.M.; Zhou, C.; Messer, A. An scFv intrabody against the nonamyloid component of alpha-synuclein reduces intracellular aggregation and toxicity. *J. Mol. Boil.* **2008**, *377*, 136–147. [CrossRef] [PubMed]
107. Butler, D.C.; Joshi, S.N.; Genst, E.; Baghel, A.S.; Dobson, C.M.; Messer, A. Bifunctional Anti-Non-Amyloid Component alpha-Synuclein Nanobodies Are Protective In Situ. *PLoS ONE* **2016**, *11*, e0165964. [CrossRef]
108. Chatterjee, D.; Bhatt, M.; Butler, D.; De Genst, E.; Dobson, C.M.; Messer, A.; Kordower, J.H. Proteasome-targeted nanobodies alleviate pathology and functional decline in an alpha-synuclein-based Parkinson's disease model. *NPJ Park. Dis.* **2018**, *4*, 25. [CrossRef] [PubMed]
109. Butler, D.C.; McLear, J.A.; Messer, A. Engineered antibody therapies to counteract mutant huntingtin and related toxic intracellular proteins. *Prog. Neurobiol.* **2012**, *97*, 190–204. [CrossRef] [PubMed]
110. Colby, D.W.; Garg, P.; Holden, T.; Chao, G.; Webster, J.M.; Messer, A.; Ingram, V.M.; Wittrup, K.D. Development of a human light chain variable domain (V(L)) intracellular antibody specific for the amino terminus of huntingtin via yeast surface display. *J. Mol. Boil.* **2004**, *342*, 901–912. [CrossRef] [PubMed]
111. Colby, D.W.; Chu, Y.; Cassady, J.P.; Duennwald, M.; Zazulak, H.; Webster, J.M.; Messer, A.; Lindquist, S.; Ingram, V.M.; Wittrup, K.D. Potent inhibition of huntingtin aggregation and cytotoxicity by a disulfide bond-free single-domain intracellular antibody. *Proc. Natl. Acad. Sci. USA* **2004**, *101*, 17616–17621. [CrossRef]
112. Southwell, A.L.; Ko, J.; Patterson, P.H. Intrabody gene therapy ameliorates motor, cognitive, and neuropathological symptoms in multiple mouse models of Huntington's disease. *J. Neurosci.* **2009**, *29*, 13589–13602. [CrossRef] [PubMed]
113. Southwell, A.L.; Khoshnan, A.; Dunn, D.E.; Bugg, C.W.; Lo, D.C.; Patterson, P.H. Intrabodies binding the proline-rich domains of mutant huntingtin increase its turnover and reduce neurotoxicity. *J. Neurosci.* **2008**, *28*, 9013–9020. [CrossRef] [PubMed]
114. Schut, M.H.; Pepers, B.A.; Klooster, R.; van der Maarel, S.M.; El Khatabi, M.; Verrips, T.; den Dunnen, J.T.; van Ommen, G.J.; van Roon-Mom, W.M. Selection and characterization of llama single domain antibodies against N-terminal huntingtin. *Neurol. Sci.* **2015**, *36*, 429–434. [CrossRef] [PubMed]
115. Harris, D.A. Cellular biology of prion diseases. *Clin. Microbiol. Rev.* **1999**, *12*, 429–444. [CrossRef] [PubMed]
116. Ladogana, A.; Puopolo, M.; Croes, E.A.; Budka, H.; Jarius, C.; Collins, S.; Klug, G.M.; Sutcliffe, T.; Giulivi, A.; Alperovitch, A.; et al. Mortality from Creutzfeldt-Jakob disease and related disorders in Europe, Australia, and Canada. *Neurology* **2005**, *64*, 1586–1591. [CrossRef]
117. Jones, D.R.; Taylor, W.A.; Bate, C.; David, M.; Tayebi, M. A camelid anti-PrP antibody abrogates PrP replication in prion-permissive neuroblastoma cell lines. *PLoS ONE* **2010**, *5*, e9804. [CrossRef] [PubMed]
118. Abskharon, R.N.; Giachin, G.; Wohlkonig, A.; Soror, S.H.; Pardon, E.; Legname, G.; Steyaert, J. Probing the N-terminal beta-sheet conversion in the crystal structure of the human prion protein bound to a nanobody. *J. Am. Chem. Soc.* **2014**, *136*, 937–944. [CrossRef] [PubMed]
119. Veliz, I.; Loo, Y.; Castillo, O.; Karachaliou, N.; Nigro, O.; Rosell, R. Advances and challenges in the molecular biology and treatment of glioblastoma-is there any hope for the future? *Ann. Transl. Med.* **2015**, *3*, 7. [CrossRef] [PubMed]
120. Shibata, M.; Shen, M.M. The roots of cancer: Stem cells and the basis for tumor heterogeneity. *BioEssays News Rev. Mol. Cell. Dev. Boil.* **2013**, *35*, 253–260. [CrossRef] [PubMed]

121. Bourkoula, E.; Mangoni, D.; Ius, T.; Pucer, A.; Isola, M.; Musiello, D.; Marzinotto, S.; Toffoletto, B.; Sorrentino, M.; Palma, A.; et al. Glioma-associated stem cells: A novel class of tumor-supporting cells able to predict prognosis of human low-grade gliomas. *Stem Cells* **2014**, *32*, 1239–1253. [CrossRef]
122. Jovcevska, I.; Zupanec, N.; Kocevar, N.; Cesselli, D.; Podergajs, N.; Stokin, C.L.; Myers, M.P.; Muyldermans, S.; Ghassabeh, G.H.; Motaln, H.; et al. TRIM28 and beta-actin identified via nanobody-based reverse proteomics approach as possible human glioblastoma biomarkers. *PLoS ONE* **2014**, *9*, e113688. [CrossRef]
123. Jovcevska, I.; Zupanec, N.; Urlep, Z.; Vranic, A.; Matos, B.; Stokin, C.L.; Muyldermans, S.; Myers, M.P.; Buzdin, A.A.; Petrov, I.; et al. Differentially expressed proteins in glioblastoma multiforme identified with a nanobody-based anti-proteome approach and confirmed by OncoFinder as possible tumor-class predictive biomarker candidates. *Oncotarget* **2017**, *8*, 44141–44158. [CrossRef]
124. Samec, N.; Jovcevska, I.; Stojan, J.; Zottel, A.; Liovic, M.; Myers, M.P.; Muyldermans, S.; Sribar, J.; Krizaj, I.; Komel, R. Glioblastoma-specific anti-TUFM nanobody for in-vitro immunoimaging and cancer stem cell targeting. *Oncotarget* **2018**, *9*, 17282–17299. [CrossRef]
125. Zorniak, M.; Clark, P.A.; Umlauf, B.J.; Cho, Y.; Shusta, E.V.; Kuo, J.S. Yeast display biopanning identifies human antibodies targeting glioblastoma stem-like cells. *Sci. Rep.* **2017**, *7*, 15840. [CrossRef] [PubMed]
126. van Lith, S.A.; Roodink, I.; Verhoeff, J.J.; Makinen, P.I.; Lappalainen, J.P.; Yla-Herttuala, S.; Raats, J.; van Wijk, E.; Roepman, R.; Letteboer, S.J.; et al. In vivo phage display screening for tumor vascular targets in glioblastoma identifies a llama nanobody against dynactin-1-p150Glued. *Oncotarget* **2016**, *7*, 71594–71607. [CrossRef] [PubMed]
127. van de Water, J.A.; Bagci-Onder, T.; Agarwal, A.S.; Wakimoto, H.; Roovers, R.C.; Zhu, Y.; Kasmieh, R.; Bhere, D.; Van Bergen en Henegouwen, P.M.; Shah, K. Therapeutic stem cells expressing variants of EGFR-specific nanobodies have antitumor effects. *Proc. Natl. Acad. Sci. USA* **2012**, *109*, 16642–16647. [CrossRef] [PubMed]
128. Ciardiello, F.; Tortora, G. EGFR antagonists in cancer treatment. *N. Engl. J. Med.* **2008**, *358*, 1160–1174. [CrossRef] [PubMed]
129. Huang, P.H.; Xu, A.M.; White, F.M. Oncogenic EGFR signaling networks in glioma. *Sci. Signal.* **2009**, *2*, re6. [CrossRef] [PubMed]
130. Roovers, R.C.; Laeremans, T.; Huang, L.; De Taeye, S.; Verkleij, A.J.; Revets, H.; de Haard, H.J.; van Bergen en Henegouwen, P.M. Efficient inhibition of EGFR signaling and of tumour growth by antagonistic anti-EFGR Nanobodies. *Cancer Immunol. Immunother. CII* **2007**, *56*, 303–317. [CrossRef] [PubMed]
131. Keyaerts, M.; Xavier, C.; Heemskerk, J.; Devoogdt, N.; Everaert, H.; Ackaert, C.; Vanhoeij, M.; Duhoux, F.P.; Gevaert, T.; Simon, P.; et al. Phase I Study of 68Ga-HER2-Nanobody for PET/CT Assessment of HER2 Expression in Breast Carcinoma. *J. Nucl. Med.* **2016**, *57*, 27–33. [CrossRef] [PubMed]
132. Freise, A.C.; Wu, A.M. In vivo imaging with antibodies and engineered fragments. *Mol. Immunol.* **2015**, *67*, 142–152. [CrossRef] [PubMed]
133. Xavier, C.; Vaneycken, I.; D'Huyvetter, M.; Heemskerk, J.; Keyaerts, M.; Vincke, C.; Devoogdt, N.; Muyldermans, S.; Lahoutte, T.; Caveliers, V. Synthesis, preclinical validation, dosimetry, and toxicity of 68Ga-NOTA-anti-HER2 Nanobodies for iPET imaging of HER2 receptor expression in cancer. *J. Nucl. Med.* **2013**, *54*, 776–784. [CrossRef] [PubMed]
134. Li, Z.; Krippendorff, B.F.; Sharma, S.; Walz, A.C.; Lave, T.; Shah, D.K. Influence of molecular size on tissue distribution of antibody fragments. *mAbs* **2016**, *8*, 113–119. [CrossRef] [PubMed]
135. Li, T.; Vandesquille, M.; Koukouli, F.; Dudeffant, C.; Youssef, I.; Lenormand, P.; Ganneau, C.; Maskos, U.; Czech, C.; Grueninger, F.; et al. Camelid single-domain antibodies: A versatile tool for in vivo imaging of extracellular and intracellular brain targets. *J. Control. Release* **2016**, *243*, 1–10. [CrossRef] [PubMed]
136. Iqbal, U.; Trojahn, U.; Albaghdadi, H.; Zhang, J.; O'Connor-McCourt, M.; Stanimirovic, D.; Tomanek, B.; Sutherland, G.; Abulrob, A. Kinetic analysis of novel mono- and multivalent VHH-fragments and their application for molecular imaging of brain tumours. *Br. J. Pharmacol.* **2010**, *160*, 1016–1028. [CrossRef] [PubMed]
137. Daumas-Duport, C.; Varlet, P.; Tucker, M.L.; Beuvon, F.; Cervera, P.; Chodkiewicz, J.P. Oligodendrogliomas. Part I: Patterns of growth, histological diagnosis, clinical and imaging correlations: A study of 153 cases. *J. Neuro-Oncol.* **1997**, *34*, 37–59. [CrossRef]

138. Daumas-Duport, C.; Tucker, M.L.; Kolles, H.; Cervera, P.; Beuvon, F.; Varlet, P.; Udo, N.; Koziak, M.; Chodkiewicz, J.P. Oligodendrogliomas. Part II: A new grading system based on morphological and imaging criteria. *J. Neuro-Oncol.* **1997**, *34*, 61–78. [CrossRef]
139. Olsson, A.K.; Dimberg, A.; Kreuger, J.; Claesson-Welsh, L. VEGF receptor signalling—In control of vascular function. *Nat. Rev. Mol. Cell Boil.* **2006**, *7*, 359–371. [CrossRef]
140. Varner, J.A.; Cheresh, D.A. Integrins and cancer. *Curr. Opin. Cell Boil.* **1996**, *8*, 724–730. [CrossRef]
141. Pen, A.; Moreno, M.J.; Martin, J.; Stanimirovic, D.B. Molecular markers of extracellular matrix remodeling in glioblastoma vessels: Microarray study of laser-captured glioblastoma vessels. *Glia* **2007**, *55*, 559–572. [CrossRef]
142. Iqbal, U.; Albaghdadi, H.; Luo, Y.; Arbabi, M.; Desvaux, C.; Veres, T.; Stanimirovic, D.; Abulrob, A. Molecular imaging of glioblastoma multiforme using anti-insulin-like growth factor-binding protein-7 single-domain antibodies. *Br. J. Cancer* **2010**, *103*, 1606–1616. [CrossRef]
143. Nagakubo, D.; Murai, T.; Tanaka, T.; Usui, T.; Matsumoto, M.; Sekiguchi, K.; Miyasaka, M. A high endothelial venule secretory protein, mac25/angiomodulin, interacts with multiple high endothelial venule-associated molecules including chemokines. *J. Immunol.* **2003**, *171*, 553–561. [CrossRef]
144. Iqbal, U.; Albaghdadi, H.; Nieh, M.P.; Tuor, U.I.; Mester, Z.; Stanimirovic, D.; Katsaras, J.; Abulrob, A. Small unilamellar vesicles: A platform technology for molecular imaging of brain tumors. *Nanotechnology* **2011**, *22*, 195102. [CrossRef] [PubMed]
145. Tomanek, B.; Iqbal, U.; Blasiak, B.; Abulrob, A.; Albaghdadi, H.; Matyas, J.R.; Ponjevic, D.; Sutherland, G.R. Evaluation of brain tumor vessels specific contrast agents for glioblastoma imaging. *Neuro-Oncology* **2012**, *14*, 53–63. [CrossRef] [PubMed]
146. Pardridge, W.M. Blood-brain barrier drug targeting: The future of brain drug development. *Mol. Interv.* **2003**, *3*, 90–105, 151. [CrossRef] [PubMed]
147. Rashidian, M.; Keliher, E.J.; Bilate, A.M.; Duarte, J.N.; Wojtkiewicz, G.R.; Jacobsen, J.T.; Cragnolini, J.; Swee, L.K.; Victora, G.D.; Weissleder, R.; et al. Noninvasive imaging of immune responses. *Proc. Natl. Acad. Sci. USA* **2015**, *112*, 6146–6151. [CrossRef] [PubMed]
148. Iqbal, U.; Abulrob, A.; Stanimirovic, D.B. Integrated platform for brain imaging and drug delivery across the blood-brain barrier. *Methods Mol. Boil.* **2011**, *686*, 465–481. [CrossRef]
149. Herve, F.; Ghinea, N.; Scherrmann, J.M. CNS delivery via adsorptive transcytosis. *AAPS J.* **2008**, *10*, 455–472. [CrossRef] [PubMed]
150. Bickel, U.; Yoshikawa, T.; Pardridge, W.M. Delivery of peptides and proteins through the blood-brain barrier. *Adv. Drug Deliv. Rev.* **2001**, *46*, 247–279. [CrossRef]
151. Li, T.; Bourgeois, J.P.; Celli, S.; Glacial, F.; Le Sourd, A.M.; Mecheri, S.; Weksler, B.; Romero, I.; Couraud, P.O.; Rougeon, F.; et al. Cell-penetrating anti-GFAP VHH and corresponding fluorescent fusion protein VHH-GFP spontaneously cross the blood-brain barrier and specifically recognize astrocytes: Application to brain imaging. *FASEB J.* **2012**, *26*, 3969–3979. [CrossRef]
152. Traenkle, B.; Emele, F.; Anton, R.; Poetz, O.; Haeussler, R.S.; Maier, J.; Kaiser, P.D.; Scholz, A.M.; Nueske, S.; Buchfellner, A.; et al. Monitoring interactions and dynamics of endogenous beta-catenin with intracellular nanobodies in living cells. *Mol. Cell. Proteom. MCP* **2015**, *14*, 707–723. [CrossRef]

*Review*

# Targeted Nanobody-Based Molecular Tracers for Nuclear Imaging and Image-Guided Surgery

**Pieterjan Debie, Nick Devoogdt and Sophie Hernot ***

Laboratory for in vivo Cellular and Molecular Imaging, ICMI-BEFY/MIMA, Vrije Universiteit Brussel, Laarbeeklaan 103, 1090 Brussels, Belgium; pieterjan.debie@vub.be (P.D.); ndevoogd@vub.be (N.D.)

* Correspondence: sophie.hernot@gmail.com; Tel.: +32-2-477-4991

Received: 29 November 2018; Accepted: 7 January 2019; Published: 11 January 2019

**Abstract:** Molecular imaging is paving the way towards noninvasive detection, staging, and treatment follow-up of diseases such as cancer and inflammation-related conditions. Monoclonal antibodies have long been one of the staples of molecular imaging tracer design, although their long blood circulation and high nonspecific background limits their applicability. Nanobodies, unique antibody-binding fragments derived from camelid heavy-chain antibodies, have excellent properties for molecular imaging as they are able to specifically find their target early after injection, with little to no nonspecific background. Nanobody-based tracers using either nuclear or fluorescent labels have been heavily investigated preclinically and are currently making their way into the clinic. In this review, we will discuss different important factors in nanobody-tracer design, as well as the current state of the art regarding their application for nuclear and fluorescent imaging purposes. Furthermore, we will discuss how nanobodies can also be exploited for molecular therapy applications such as targeted radionuclide therapy and photodynamic therapy.

**Keywords:** single-domain antibody fragments; molecular imaging; molecular therapy; nuclear imaging; targeted fluorescence imaging; intraoperative imaging

---

## 1. Introduction

Molecular imaging has been defined as "the visualization, characterization, and measurement of biological processes at the molecular and cellular level in living subjects" [1]. The technique makes use of molecular tracers in probing for biomarkers expressed in (patho)physiological processes. Molecular tracers most often consist of a targeting moiety to direct the tracer and a signaling moiety for detection. Following administration, the molecular tracer will specifically accumulate in areas where the biomarker reveals itself, while unbound tracer will be eliminated. The bound molecular tracer is then visualized by means of an appropriate imaging modality. Radioactivity and fluorescence are particularly suited for the application because of their high detection sensitivity. Yet, the choice of detection system and associated signaling molecule is directly linked to the intended application. In the clinic, nuclear molecular imaging is mainly used as a noninvasive technique for diagnosis of diseases, treatment follow-up, or early patient stratification according to the expression of predictive biomarkers [2]. Contrarily, as fluorescence signals have limited depth penetration (several mm), the use of fluorescence imaging is restricted to imaging of the skin surface and interventional procedures, i.e., surgery, endoscopic, and intravascular imaging. Over the last decade, fluorescence-guided interventions have gained interest as a method to assist surgeons in real-time by demarcating cancerous tissues for precise and complete resection [3], by highlighting healthy tissues that should be preserved [4,5], by guiding biopsy [6,7], or by interrogating suspicious lesions in vivo at the molecular level [8].

The design of molecular tracers is driven by certain requirements: the tracer should remain stable in vivo, it must accumulate specifically and in sufficient amounts into the tissue of interest, no or low uptake in nontargeted tissues and organs should be detected in order to reach high contrast, and an appropriately fast imaging time point and sufficiently extended imaging window should be attained. These parameters are predominantly defined by the pharmacokinetics of the tracer, and thus by the choice of targeting agent. Several classes of targeting agents can be considered, including antibodies, antibody fragments, scaffolds, peptides, or small molecules [9]. Antibodies, given their natural capacity to recognize specific epitopes as part of the body's humoral immune system, have been extensively considered for in vivo targeting. However, their large size results in slow systemic clearance and hampers deep tissue penetration [10]. In consequence, molecular imaging performed using antibody tracers often yields low-contrast images with low specificity, and only several days after tracer administration [11]. This review will focus on nanobodies as a platform technology, since nanobody-based tracers have been shown to possess unique characteristics in terms of versatility, specificity, and the short time needed to attain high contrast [12]. Following the description of what nanobodies are, we will give an overview of the application of radiolabeled nanobodies in nuclear medicine. Subsequently, we will discuss how the attractive properties of nanobodies can also be exploited for fluorescence-guided surgery and photodynamic therapy. The different aspects that are important to their design and utilization as molecular tracers in these fields will be addressed in more detail.

## 2. Nanobodies and Their Unique Properties

Conventional antibodies are Y-shaped and are composed of two light and two heavy polypeptide chains. In camelids, besides these, a substantial fraction of IgG-like antibodies devoid of light chains also occur naturally [13]. Although only half of the complementarity-determining regions (CDRs) needed for antigen recognition are present, these heavy-chain-only antibodies retain a similar affinity and specificity to conventional IgGs. This interesting finding has led to the exploitation of the variable domain of heavy-chain-only antibodies for many biotechnological and medical applications. This monomeric domain of 12–15 kDa in size and with nanometer-range dimensions is often referred to as a single-domain antibody (sdAb) or the variable domain of a heavy-chain antibody ($V_HH$), and has been given the commercial name Nanobody$^{TM}$ [14].

Crystallographic studies of nanobodies revealed, as for conventional $V_H$ fragments, a classical immunoglobulin fold with nine antiparallel β-strands forming two β-sheets, connected through a conserved disulfide bridge (Figure 1A). In camel-derived nanobodies, more so than for llama-derived nanobodies, an additional disulfide bridge is often present between the CDR1 and CDR3 or the CDR2 and CDR3 loops. The three CDR loops are located near the amino (N)-terminal end of the domain and opposite to the carboxy (C)-terminal end. However, adaptations are essential to compensate for the absence of a variable light chain ($V_L$). On the one hand, four hydrophobic residues which normally interact with the $V_L$ domain are changed to more hydrophilic ones [15–17]. This results in a structure with improved water solubility and which is less prone to aggregation [18]. On the other hand, the lack of three extra loops for antigen recognition is compensated by elongated CDR1 and CDR3 loops that are able to adopt alternative canonical structures. As a consequence of the often more convex shape of the paratope, nanobodies tend to bind epitopes located within cryptic clefts [19].

An important advantage of nanobodies is their high thermal and chemical stability. They typically exhibit melting temperatures above 60 °C, and antigen-binding activity is retained even after prolonged incubation at such high temperatures [20,21]. These properties open opportunities for novel chemical modifications and labeling methods.

Despite these differences, nanobodies are considered to be very weakly immunogenic due to their high degree of homology with human variable heavy-chain ($V_H$) fragments [22]. This has since been confirmed by several clinical trials where no immunogenicity or adverse effects were detected following administration [23,24]. While preexisting anti-nanobody antibodies were found in one clinical study using a tetravalent agonistic nanobody targeting the Death Receptor 5 as an antitumor agent [25], no more preexisting anti-nanobody antibodies appear to be present than preexisting autoantibodies against conventional $V_H$ fragments [26].

Nanobodies can be readily generated against many targets through the immunization of camelids with either the antigen of interest, DNA coding for it, or cells expressing the antigen on their surface. Subsequent amplification of the nanobody gene sequences from peripheral blood mononuclear cells yields an immune cDNA library from which specific nanobodies can be selected through various affinity (e.g., phage display) and/or functional screens [27]. Alternatively, synthetic libraries can be used. These generally utilize a fixed framework, where residues in the CDR regions are randomized to obtain specificity for different targets [28,29].

Nanobodies with single-digit nanomolar affinities should preferentially be selected for further *in vivo* applications, especially when the target is only expressed at low levels. The affinity of nanobodies across species furthermore facilitates their preclinical characterization, as the uptake in organs and tissues constitutively expressing the target can be better assessed.

The selected nanobody clones can then straightforwardly be expressed in bacteria (e.g., *E. coli*) or yeast strains (e.g., *S. cerevisiae* and *P. pastoris)* at yields of several milligrams of soluble nanobodies per liter of culture. Their relatively simple and single gene form allows the engineering of nanobodies into all kinds of formats, including the generation of multimeric and multispecific compounds, creation of fusion proteins, and addition of small peptide sequences (tags) for later functionalization [30].

In the last 15 years, nanobodies have been proposed as a new class of antibody-derived agents for molecular imaging because of their unique features regarding affinity, specificity, and rapid pharmacokinetics, ensuring good uptake in the targeted tissues and high target-to-background ratios [31].

## 3. Radiolabeled Nanobodies for Same-Day, High-Contrast Nuclear Imaging and Targeted Radionuclide Therapy with Minimal Toxicity

### 3.1. Radiolabeling of Nanobodies

Nuclear molecular imaging requires the targeting moiety, in this case a nanobody, to be labeled with a diagnostic radioisotope. The latter can either be a gamma-emitting isotope for single photon emission computed tomography (SPECT) or a positron-emitting isotope for positron emission tomography (PET). Clinically, the higher resolution, sensitivity, and quantitative potential of PET/Computed Tomography (CT) imaging is driving its adoption and is expected to result in a shift towards the increasing use of this technology over SPECT/CT [32]. On the contrary, microSPECT devices specifically developed for small-animal imaging typically achieve higher spatial resolution than microPET systems. It is thus an attractive imaging technology for the initial in vivo screening and characterization of a set of nanobodies, especially since nanobodies can be easily labeled with $^{99m}$Tc. Conveniently, $^{99m}$Tc–tricarbonyl reacts site-specifically with a genetically inserted C-terminal hexahistidine tag, which can also be used for purification purposes via immobilized metal affinity chromatography [33].

Radiolabeling of proteins with other radiometals (e.g., $^{64}$Cu, $^{68}$Ga, and $^{89}$Zr for PET or $^{67}$Ga and $^{111}$In for SPECT) or radiohalogens (e.g., $^{18}$F and $^{124}$I for PET and $^{123/131}$I for SPECT) usually necessitates the use of a chelator for complexation or a prosthetic group for electrophilic substitution, respectively. For human applications, the PET isotopes $^{68}$Ga and $^{18}$F are particularly suited to imaging with nanobodies due to their short half-lives (68 and 110 min, respectively), which match up well with the nanobodies' biological half-life. For the attachment of chelators and prosthetic groups to the nanobody, different conjugation strategies are possible. These can be broadly divided into two categories: random or site-specific labeling. Random labeling typically occurs through conjugation to primary amines (lysines) in the framework. Although this is a common and straightforward method, it however results in a heterogenous mixture with varying amounts of labels per nanobody at different positions. Contrarily, site-specific strategies aim to obtain homogenous and consistent tracers through the conjugation of a single contrast agent to predetermined, specific sites. Positioning of the contrast label opposite to the antigen-binding site furthermore avoids interference with the binding capacity of nanobodies [34]. Different types of site-specific labeling methods with nanobodies have been explored for in vivo applications. For example, the incorporation of a C-terminal cysteine tag enables reaction with maleimide-functionalized agents after the prior reduction of dimeric nanobodies or nanobodies with a blocked cysteine. Importantly, the reduction reaction must be carefully titrated to prevent disruption of the nanobodies' internal disulfide bridges [35]. Although reversal of the thioether bond is known to occur in vivo [36,37], this is not expected to happen fast enough to pose a problem to nanobody probes due to their fast pharmacokinetics [34]. A more elegant method for site-specific conjugation is enzyme-mediated ligation through the transpeptidase Sortase A. Here, the enzyme catalyzes the formation of a new peptide bond between the peptide motif LPXTG expressed C-terminally on the nanobody and the label containing a N-terminal oligo-glycine motif [38,39]. Other methods under investigation for the design of site-specifically labeled nanobodies for molecular imaging are alkyne–azide click reactions and those involving the incorporation of unnatural amino acids into the nanobody structure [34,40].

### 3.2. In Vivo Biodistribution of Radiolabeled Nanobodies

Upon intravenous administration, radiolabeled nanobodies are rapidly cleared from the blood circulation. In mice, normally less than 0.5% Injected Activity (IA)/g remains present in the blood pool at 1 h post-injection [41–44]. In humans, the early- and late-phase half-life of the anti-human epidermal growth factor receptor 2 (HER2) $^{68}$Ga-labeled nanobody 2Rs15d was calculated as 2.9 and 25.5 min, respectively, and at 1 h post-injection of the tracer, only 10% of the injected activity remained in the blood pool [23]. The size of nanobodies being below 60 kDa causes them to be filtered through the glomeruli in the kidneys. However, nanobodies are subsequently reabsorbed by the proximal tubuli, resulting in their retention in the renal cortex. It has been previously shown that the endocytic receptor megalin, which is abundantly expressed in the brush border, is at least partially involved in the renal retention of nanobodies (megalin-deficient mice show 40% less renal retention of a $^{99m}$Tc-labeled nanobody than wild-type mice) [45]. The long-term renal retention of (radiolabeled) nanobodies and/or their (radio)catabolites can be an issue as it can possibly lead to undesired nephrotoxicity. Furthermore, the imaging of molecular targets in the vicinity of the kidneys is hampered due to intense renal signals. Therefore, several possible strategies to reduce the renal reabsorption of nanobodies have been investigated. Coadministration of positively charged amino acids or gelofusin, which competitively interact with megalin/cubulin receptors, have long been known to reduce the renal retention of radiometal-labeled antibody fragments and peptides [46,47], and this has also been confirmed for nanobodies [45,48]. Alternatively, the removal of charged amino-acid tags, for example, used for purification or radiolabeling purposes, has an effect on the polarity of nanobodies and consequently has an important impact on the degree of kidney retention. Indeed, Myc–His- and His-tagged nanobodies show considerably higher kidney values compared to their untagged analogues [48,49]. For clinical applications, the His tag is recommended to be removed anyway to prevent immunogenic reactions [50,51]. Finally, the degree of kidney retention for radiohalogenated (fluorinated and iodinated) nanobodies is significantly lower than for radiometal-labeled analogues. Catabolites of radiohalogenated compounds formed in the kidneys are thought to be nonresidualizing and hydrophobic and rapidly excreted via the urine [52–54].

Other than the accumulation in kidneys and urine, the uptake of radiolabeled nanobodies in nontargeted organs and tissues is very low (Figure 1). In combination with efficient penetration into and diffusion through tissues and fast targeting, this consequently results in high target-to-background ratios early after administration, allowing same-day imaging [55]. This is in stark contrast to full-length antibodies, where due to their long circulation time, optimal tumor-to-background contrast is only obtained several days after administration of the tracer and nonspecific uptake is generally much higher [11]. Other non-immunoglobulin low-molecular-weight protein scaffolds (e.g., Affibodies, DARPins, Adnectins, ADAPTs, or knottins) share many of the in vivo characteristics of nanobodies. A recent review discusses in detail the clinical application and promising preclinical developments of nanobodies and other small proteins for radionuclide-based imaging within the field of oncology [55]. The main advantage of nanobodies over these scaffolds remains the relatively simple process to generate new nanobodies with high affinity through the immunization of camelids. On the other hand, further size-reduced compounds could eventually be produced synthetically instead of via fermentation, once a lead compound has been identified [56].

### 3.3. Nuclear Medicine Applications with Nanobodies

A major application field for nanobody-based radiotracers is cancer imaging. Radiolabeled nanobodies targeting tumor biomarkers such as the epidermal growth factor receptor (EGFR), carcinoembryonic antigen (CEA), mesothelin, prostate-specific membrane antigen (PSMA), CD20, or HER2 showed high specific tumor uptake (ranging typically between 2 and 10% IA/g, depending on the expression level of the target) in preclinical tumor models as soon as 1 h after administration, with tumor-to-blood ratios of up to 10–30 (Figure 1B,C) [23,35,38,41,42,45,48,49,53,54,57–71]. Of particular interest is the anti-HER2 nanobody 2Rs15d that has been selected as a lead compound for clinical translation. In the very first clinical trial with a radiolabeled nanobody, Keyaerts et al. demonstrated that $^{68}$Ga–NOTA–2Rs15d PET/CT enabled the visualization of both primary lesions and/or local or distant metastases in HER2-positive breast cancer patients, without adverse effects (Figure 1H). Good tumor uptake was observed, with mean standard uptake values of up to 11.8 for primary tumors and 6.0 in metastases between 60 and 90 min post-injection. With the exception of the kidneys, intestines, and liver, background uptake was low (weak uptake in glandular tissues such as the salivary glands, pituitary, lacrimal glands, and axillary sweat glands was thought to be related to low levels of HER2 expression or chelator-mediated trapping mechanisms). Ultimately, 90 min post-injection was chosen as the optimal imaging time point, due to decreased liver uptake compared to 60 min post-injection. In this study, no preexisting or tracer-induced antibodies against the nanobody 2Rs15d could be detected. These findings imply the potential application of $^{68}$Ga–2Rs15d for the noninvasive assessment of the HER2 status of patients [23]. A phase II study with this tracer has since been initiated, evaluating its potential to detect brain metastasis in breast cancer patients (NCT03331601).

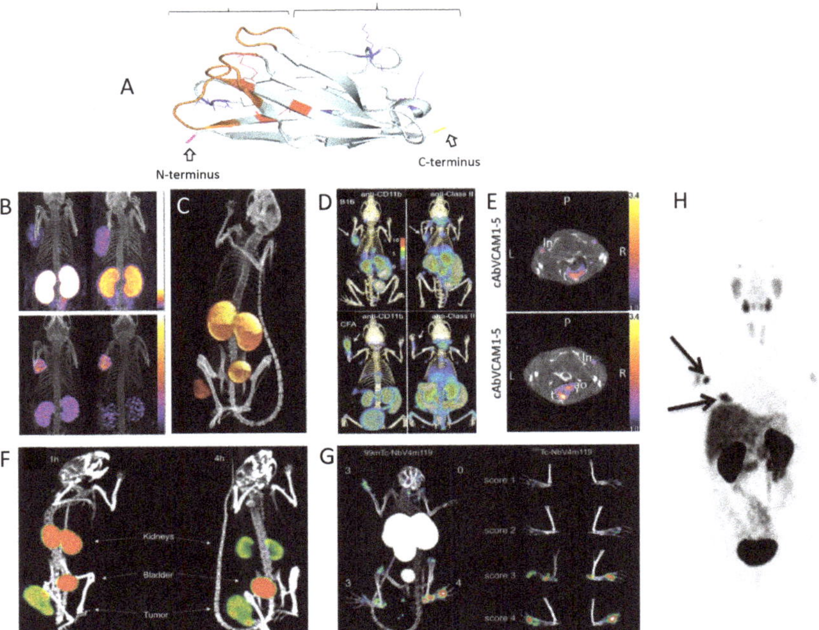


**Figure 1.** Schematic representation of the structure of a nanobody and illustrative positron emission tomography (PET) and single photon emission computed tomography (SPECT) preclinical and clinical

images obtained using nanobodies that are labeled with distinct radionuclides in diverse medical applications, from oncology, immunology, atherosclerosis, and arthritis to the theranostic imaging of a radiotherapeutic probe. (**A**) Ribbon diagram of the nanobody 2Rs15d. The complementarity-determining regions (CDRs) are shown in orange, lysines (used for random conjugation methods) in blue, and cysteines and cysteine bridges in red. The C-terminus (in yellow) can be easily genetically modified for site-specific conjugation methods. (**B**) SPECT/Computed Tomography (CT) images of the biodistribution of $^{111}$In-labeled JV7 nanobodies at 3 h post-injection in PSMA+ tumor-bearing mice (on the left shoulder). Effect on renal retention by the removal of tags (top panels: Myc–Cys-tagged nanobody, bottom: Cys-tagged nanobody) and coinjection of positively charged amino acids and gelofusin (left panels: no injection, right panels: with coinjection) is shown. Adapted from [48]. (**C**) SPECT/CT imaging of an EGFR+ tumor-bearing mouse 1 h after injection of $^{99m}$Tc-labeled 7C12 nanobody. Adapted from [64]. (**D**) PET/CT immune cell imaging 90 min after injection of $^{18}$F-labeled nanobodies against murine CD11b and major histocompatibility complex (MHC) class II. Top: C57Bl/6 mice inoculated with B16 tumor cells on the left shoulder; bottom: animals injected with complete Freund's adjuvant on the left paw. Adapted from [72]. (**E**) SPECT/CT coronal image taken at 2–3 h post-injection of $^{99m}$Tc-labeled cAbVCAM1-5 nanobody, showing uptake in atherosclerotic lesions (ao) of ApoE$^{-/-}$ mice (bottom) and absence of signals in the aortic arch of C57Bl/6J mice (top) [43]. (**F**) SPECT/CT images of the biodistribution of the $^{131}$I-labeled 2Rs15d therapeutic nanobody in a mouse model with subcutaneous HER2+ xenograft at 1 and 4 h post-injection. Adapted from [54]. (**G**) SPECT/CT imaging of arthritis in a mouse model with a VSIG4/CRIg-specific $^{99m}$Tc-labeled nanobody. Adapted from [73]. (**H**) PET/CT image of the biodistribution of $^{68}$Ga-labeled anti-HER2 nanobody in a breast cancer patient 90 min post-injection showing uptake in breast tumor lesions. Adapted from [23].

Next to imaging the tumor cells themselves for the prognosis and prediction of therapy response, a different application is the characterization and quantification of specific immune cells within the tumor environment. This approach could eventually also aid in better understanding and evaluating drug action during drug development. In mice, radiolabeled nanobodies have been proven to be able to track the infiltration of CD11b (macrophages, dendritic cells, and neutrophils) and major histocompatibility complex (MHC) class II (macrophages and dendritic cells) positive cells in both xenogeneic and syngeneic tumors, as well as after injection of complete Freund's adjuvant [72] (Figure 1D). Similarly, macrophage mannose receptor (MMR)-specific nanobodies were used to image tumor-associated macrophages in mice [44,74]. The human-MMR-specific nanobody MMR3.49 labelled with $^{68}$Ga is currently being translated to the clinic, with a phase I safety trial to be initiated soon [75]. Of growing interest is the prediction of immune-checkpoint blockade therapy outcome. Nuclear imaging with radiolabeled nanobodies against the T-cell marker CD8 or programmed death-ligand 1 (PD-L1) shows promising results as a tool to image antitumor immune responses [76–79].

Macrophage-specific nanobodies have also been found to be of interest for imaging in other inflammatory diseases, examples of which are hepatic inflammation and collagen-induced arthritis (CIA). Indeed, radiolabeled nanobodies targeting MMR or the complement receptor of the Ig superfamily (CRIg or VSIG4) were found to accumulate specifically in inflamed joints of mouse CIA models (Figure 1G). Furthermore, CRIg/VSIG4 imaging appeared to be sufficiently sensitive to detect early signs of inflammation, even before the manifestation of clinical signs [73,80–82].

Applications that heavily rely on an elevated uptake with low nonspecific surrounding background signals are the imaging of pancreatic islets after transplantation [83] and the imaging of vulnerable atherosclerotic plaques. The feasibility to noninvasively detect small, inflamed atherosclerotic lesions in the aortic arch of mice or along the aortas of atherosclerotic rabbits has been shown with radiolabeled nanobodies targeting MMR, vascular cell adhesion molecule-1 (VCAM-1) (Figure 1E), and lectin-like oxidized low-density lipoprotein receptor (LOX-1) [43,52,84–89]. However, their prognostic value for the identification of high-risk patients remains to be demonstrated in clinical trials. Within this context, the anti-VCAM-1 nanobody cAbVCAM1-5 is currently in the process of being translated into the clinic.

In analogy with molecular imaging, the incorporation of high-energy β⁻ ($^{177}$Lu, $^{131}$I) or α-emitters ($^{211}$At, $^{213}$Bi, and $^{225}$Ac) allows the use of nanobodies for targeted radionuclide therapy (TRNT). The nanobody is used as a vehicle to specifically bring damaging ionizing radiation to the tumor cells. Radiation can inflict damage to tumor cells either by direct DNA damage or by the generation of reactive oxygen species (ROS). Furthermore, the destruction of cancer cells can release antigens and immune-triggers into the environment, activating an anticancer immune response [90,91]. β-emitters, which have a lower linear energy transfer (LET), deposit their energy over a longer path length than α-emitters, and can thus be advantageous in heterogenous tumors where not all malignant cells express the molecular target. On the other hand, α-emitters deliver higher therapeutic absorbed doses, but might require internalization as these operate at very short distances. α-TRNT is mostly suggested for the treatment of micro-metastasis and minimal residual disease [92,93].

In a theranostic approach, therapy is combined with molecular imaging, the latter being used to predict susceptibility to TRNT and as a means to follow up treatment. This can be accomplished either by using the same nanobody for both the preparation of the diagnostic and therapeutic radio-analogue (e.g., pairing of $^{68}$Ga/$^{177}$Lu or $^{123/124}$I/$^{131}$I-labeled agents) or by radiolabeling the nanobody with a radioisotope with both diagnostic and therapeutic properties (e.g., $^{177}$Lu and $^{131}$I) [92,93]. This strategy has successfully been applied preclinically in breast and ovarian cancer, non-Hodgkin lymphoma, and multiple myeloma using anti-HER2 (Figure 1F), anti-CD20, and anti-idiotypic (multiple myeloma) nanobodies, respectively [49,54,68]. Efficient tumor therapy in a preclinical setting could be demonstrated through either significant improvement in overall survival compared to all controls in mice inoculated subcutaneously or intraperitoneally with HER2- or CD20-positive tumors cells, or the inhibition of disease progress in the 5T2 multiple myeloma mouse model. Furthermore, the fast pharmacokinetics and low nonspecific uptake of nanobodies lead to minimal toxicity. Contrarily, the prolonged blood residence time of monoclonal antibodies has important implications for toxicity to bone marrow and other highly perfused organs such as the spleen and liver. So far, renal toxicity due to kidney retention has not been observed in mice, even after repeated administration. Furthermore, by using $^{131}$I-labeled nanobodies and having nonresidual catabolites in the kidneys, the absorbed dose to the kidneys could even be reduced below the dose delivered to the tumor [49,54,68]. A phase I trial using a low-dose $^{131}$I-labeled anti-HER2 nanobody 2Rs15d in breast cancer patients has recently been completed (NCT02683083) [94].

While monovalent nanobodies are most often used for nuclear imaging and TRNT, the use of multimeric nanobodies has also been investigated. Dimeric monospecific nanobodies generally show similar pharmacokinetics to monovalent tracers with regard to fast blood elimination and renal clearance. Tumor uptake of bivalent nanobodies at early time points has been found to be slightly lower than that of monovalent ones, although the bivalent compound shows considerably longer tumor retention [69]. However, due to the increased avidity and size of bivalent nanobodies, they likely exhibit inferior tumor penetration properties and are more limited to the perivascular region [95,96]. Bivalent nanobodies can consequently be used as a means to cope with on-target off-tumor uptake. This was demonstrated by Movahedi et al., who used a bivalent unlabeled anti-MMR nanobody to modify the biodistribution of the radiolabeled monomeric nanobody. Impact on tumor uptake was minimal, while specific uptake in nontumor organs and tissues was almost completely blocked [44]. When extension of the blood half-life is desirable, multimeric constructs containing one or more tumor-specific domains and an albumin-binding domain can be engineered. The resulting radiolabeled compounds show vastly increased blood retention, and consequently, tumor values significantly increase over time, although so does the nonspecific uptake [69,97–99]. Half-life-extended nanobodies are thus theoretically more relevant in a therapeutic context, and less so for diagnostic applications. The negative impact on deep tumor penetration should also be considered.

An overview of nanobody-based radiopharmaceuticals with potential application in clinical nuclear medicine is provided in Table 1.

**Table 1.** Overview of preclinically and clinically tested nanobody-based radiopharmaceuticals with applications in nuclear medicine.

| Application Field | Molecular Target | Lead Compound | Radiolabel | Disease | Development Phase | References |
|---|---|---|---|---|---|---|
| Tumor cell imaging/therapy | HER2 | 2Rs15d | $^{99m}Tc$, $^{111}In$, $^{177}Lu$, $^{18}F$, $^{225}Ac$ | Breast cancer | Preclinical | [42,57–60,100] |
| | | | $^{68}Ga$ | | Phase II ongoing (NCT03331601) | [23,41] |
| | | | $^{131}I$ | | Phase I completed (NCT02683083) | [54,94] |
| | | 5F7 | $^{125}I$, $^{131}I$, $^{18}F$ | | Preclinical | [60–63] |
| | EGFR | 7C12,7D12 | $^{99m}Tc$, $^{177}Lu$, $^{68}Ga$, $^{89}Zr$ | Skin cancer | Preclinical | [45,64,65,97] |
| | | D10 | $^{99m}Tc$ | | | [66] |
| | HER3 | MSB0010853 | $^{89}Zr$ | Non-small cell lung cancer, head and neck cancer | Preclinical | [99] |
| | PSMA | PSMA30 | $^{99m}Tc$ | Prostate cancer | Preclinical | [67] |
| | | JVZ-007 | $^{111}In$ | | | [48] |
| | CEA | CEA5 | $^{99m}Tc$ | Colon cancer | Preclinical | [71] |
| | Mesothelin | A1 | $^{99m}Tc$ | Breast cancer | Preclinical | [70] |
| | CD20 | 9077, 9079 | $^{99m}Tc$, $^{111}In$, $^{177}Lu$, $^{68}Ga$ | Non-Hodgkin lymphoma | Preclinical | [68,69] |
| | HGF | 1E6-Alb8, 6E10-Alb8 | $^{89}Zr$ | Glioma | Preclinical | [98] |
| | Mouse monoclonal protein | R3b23 | $^{99m}Tc$, $^{177}Lu$ | Multiple myeloma | Preclinical | [101] |

**Table 1.** *Cont.*

| Application Field | Molecular Target | Lead Compound | Radiolabel | Disease | Development Phase | References |
|---|---|---|---|---|---|---|
| Tumor immunology and inflammatory diseases | Mouse CD8 | VHH-X118 | $^{89}Zr$ | Tumor immunology | Preclinical | [76] |
| | Mouse PD-L1 | B3 | $^{18}F$ | Immune checkpoint | Preclinical | [78] |
| | | C3,E2 | $^{99m}Tc$ | | | [77] |
| | Mouse dendritic cells | DC1.8, DC2.1 | $^{99m}Tc$ | Tumor immunology | Preclinical | [102] |
| | Mouse Cd11b | VHHDC13 | $^{18}F$, $^{64}Cu$ | Tumor immunology | Preclinical | [72] |
| | Mouse MHC class II | VHH7 | $^{18}F$, $^{64}Cu$ | Tumor immunology | Preclinical | [39,72] |
| | Human MHC class II | VHH4 | $^{64}Cu$ | Graft vs. host disease | Preclinical | [103] |
| | Mouse MMR | MMRCl1 | $^{99m}Tc$ | Tumor immunology | Preclinical | [44] |
| | | | $^{99m}Tc$ | Arthritis | Preclinical | [81] |
| | | MMR3.49 | $^{99m}Tc$, $^{18}F$, $^{68}Ga$ | Tumor immunology | Clinical translation | [74,75] |
| | Human MMR | MMR3.49 | $^{99m}Tc$, $^{64}Cu$, $^{68}Ga$ | Atherosclerosis | Preclinical | [84,86] |
| | CRIg/VSIG4 | VM119 | $^{99m}Tc$, $^{18}F$ | Arthritis, liver inflammation | Preclinical | [73,82] |
| | Clec4F | C4m22 | $^{99m}Tc$ | Liver inflammation | Preclinical | [82] |
| | VCAM-1 | cAbVCAM1-5 | $^{99m}Tc$, $^{111}In$, $^{18}F$, $^{64}Cu$, $^{68}Ga$ | Atherosclerosis | Clinical translation | [43,52,85–87, 89] |
| | LOX-1 | Lox1.14 | $^{99m}Tc$, $^{64}Cu$ | Atherosclerosis | Preclinical | [86,88] |
| Amyloidosis | Gelsolin | FAF Nb1 | $^{99m}Tc$ | Gelsolin amyloidosis | Preclinical | [104,105] |
| | B-amyloid | Ni3A, pa2H | $^{99m}Tc$ | Alzheimer's | Preclinical | [106] |
| Diabetes | DPP6 | 4hD29 | $^{99m}Tc$, $^{111}In$ | Diabetes | Preclinical | [83] |

HER2: human epidermal growth factor receptor 2, EGFR: epidermal growth factor receptor, HER3: human epidermal growth factor receptor 3, PSMA: prostate-specific membrane antigen, CEA: carcinoembryonic antigen, HGF: hepatocyte growth factor, PD-L1: programmed death-ligand 1, MMR: macrophage mannose receptor, CRIg/VSIG4: complement receptor of the immunoglobulin family/V-set and immunoglobulin domain containing 4, Clec4F: C-type lectin domain family 4 member F, VCAM-1: vascular cell adhesion molecule 1, LOX-1: lectin-like oxidized low-density lipoprotein receptor-1, DPP6: dipeptidyl peptidase like 6.

## 4. Image-Guided Surgery and Photodynamic Therapy Using Fluorescent Nanobodies

### 4.1. Design of Fluorescent Nanobody-Based Tracers

Alternatively to nuclear imaging, fluorescence imaging requires the presence of a fluorescent label for sensitive detection. While a wide range of fluorescent dyes are available for biotechnological purposes, for in vivo imaging applications, the choice of fluorophore is limited to those emitting in the near-infrared (NIR) region, and more specifically, those with a maximal excitation and emission wavelength between 650 and 900 nm. In this range, scattering, nonspecific tissue autofluorescence and absorption by endogenous chromophores are minimal. These characteristics infer intrinsically improved signal-to-background ratios and as such provide sharper contrast, better resolution, and deeper signal detection (from several mm to 1 cm) [107]. Commonly, hydrophilic (sulfonated) variants of cyanine dyes with a penta- or heptamethine chain are used, e.g., Cy5, Alexa Fluor (AF)680, IRDye680RD, or IRDye800CW. The actual choice of the dye will depend on the specifications of the camera system used for detection (the wavelength(s) the system is able to detect) and whether the tracer is intended to be used in combination with a second tracer providing complementary information, e.g., highlighting of the tumor tissue to be resected and the nerves to be preserved (multiplexing).

Analogous conjugation strategies to those described above for chelators and prosthetic groups are applicable for fluorophore conjugation. However, it is increasingly being recognized that fluorophore conjugation can have a significant impact on the pharmacokinetics of antibody tracers [108]. The impact is most likely even more pronounced for smaller fragments such as nanobodies. In fact, randomly IRDye800CW-labeled nanobodies have been demonstrated to have an atypical tissue distribution with high background signals, high liver accumulation, and low tumor contrast [109–111]. Likewise, randomly IRDye680RD- and AF680-labeled nanobodies do not exhibit such persistent background signals, but are partially excreted via the hepatobiliary route [111–113]. The chosen conjugation chemistry appears to be an additional determining factor, as the site-specific labeling of IRDye800CW and IRDye680RD via a C-terminal cysteine tag yields nanobody tracers with normal biodistribution profiles, meaning fast tumor targeting, renal excretion, and no nonspecific uptake (Figure 2A) [111]. Conjugation of more than one dye per nanobody is furthermore undesirable since a higher dye/nanobody ratio may cause quenching of the fluorescent signal due to the close proximity of the dyes.

Compared to nuclear imaging techniques, fluorescent imaging requires sufficient uptake of the tracer in the tissue of interest for sensitive detection. Because of the depth-related attenuation of fluorescent signals, high injected doses are often required in humans, at the limits of or above the microdose level (less than 30 nmol or 100 μg [114]) [115–119]. Moreover, as for equal doses, the maximal tumor signal that can be attained will be lower for nanobodies than for long-circulating antibodies, a higher molar concentration will likely be needed as well [113]. In mice, typically, 1–5 nmol of fluorescent nanobody based on the dye concentration (25–75 μg of protein) is injected. Adjusting the injected dose does not appear to significantly affect the biodistribution of nanobodies (in comparison, a higher dose of antibodies results in an increase of nonspecific signal), and dose optimization could thus lead to superior image quality in terms of signal intensity and tumor-to-background ratios [113]. The use of radio- or bimodal (radioactivity in combination with fluorescence)-labeled tracers could be an alternative approach to increase sensitivity for the intraoperative detection of deeper located lesions, but this remains to be investigated for nanobodies [120].

### 4.2. Fluorescence-Guided Surgery Using Nanobody-Based Contrast Agents

Several fluorescently labeled nanobodies have been successfully evaluated in the context of intraoperative imaging (an overview is provided in Table 2). IRDye800CW-labeled anti-EGFR nanobodies could clearly delineate orthotopic tongue tumors in mice, and even enabled the identification of a lymph node metastasis (Figure 2C) [109]. Of note, in this study the optimal imaging time point appeared to be 24 h post-injection, likely due to the random characteristic of the labeling method. The site-specifically IRDye800CW- and IRDye680RD-labeled nanobodies 11A4 and B9, respectively targeting HER2 and carbonic anhydrase IX (CAIX), showed accumulation in breast cancer lesions (DCIS) and lung metastasis in an experimental setup mimicking the surgical setting [121,122]. Furthermore, it was demonstrated that the combination of nanobodies targeting these two independent tumor markers, but labeled with the same fluorescent dye, could further improve tumor-to-background ratios and overcome tumor heterogeneity [122]. Finally, using a mouse model of intraperitoneal disseminated tumor lesions mimicking late-stage ovarian cancer, the advantage of fluorescence guidance with the anti-HER2 nanobody tracer 2Rs15d-IRDye800CW on the efficiency of debulking surgery was demonstrated. Submillimeter lesions could be visualized with high contrast at 1.5 h post-injection, leading to the excision of significantly more tumor tissue as compared to traditional surgery and resection of less false-positive tissue (Figure 2B) [123].

**Table 2.** Overview of in vivo preclinically evaluated fluorescent nanobodies with potential for clinical interventional molecular imaging and photodynamic therapy.

| Molecular Target | Lead Compound | Fluorophore | Conjugation Strategy | Intended Clinical Application | References |
|---|---|---|---|---|---|
| HER2 | 2Rs15d | IRDye800CW<br>IRDye680RD | Random<br>(Lys–NHS) | - | [111] |
| | | IRDye800CW<br>IRDye680RD | Site-specific<br>(Cys–maleimide) | Intraoperative imaging of breast/ovarian cancer | [111,123] |
| | | Cy5 | Site-specific<br>(Sortase A) | Intraoperative imaging of breast cancer | [38] |
| | 11A4 | IRDye800CW<br>IRDye680RD | Site-specific<br>(Cys–maleimide) | Intraoperative imaging of breast cancer | [122,124] |
| CAIX | B9 | IRDye800CW | Site-specific<br>(Cys–maleimide) | Intraoperative imaging of breast cancer | [121,122] |
| EGFR | 7D12 | IRDye800CW | Random<br>(Lys–NHS) | Intraoperative imaging of head and neck cancer | [109] |
| | 7D12,<br>7D12-9G6 | IRDye700DX | Random<br>(Lys–NHS) | Photodynamic therapy of head and neck cancer | [125] |

NHS: N-Hydroxysuccinimide, HER2: human epidermal growth factor receptor 2, CAIX: carbonic anhydrase 9, EGFR: epidermal growth factor receptor.

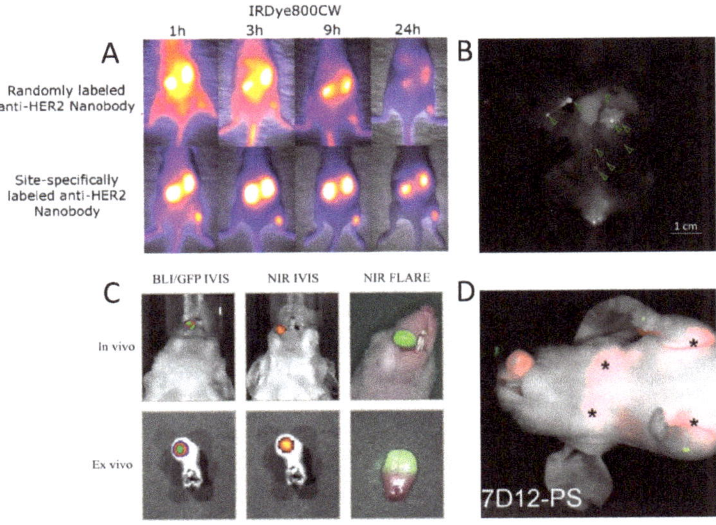


**Figure 2.** Examples of in vivo fluorescent molecular imaging with nanobody tracers in mouse tumor models. (**A**) Comparison of the biodistribution and tumor-targeting potential of the anti-HER2 nanobody

2Rs15d conjugated with IRDye800CW either randomly (top) or site-specifically (bottom). Adapted with permission from [111]. Copyright 2017 American Chemical Society. (**B**) Fluorescence image acquired during the surgical resection of intraperitoneally disseminated HER2+ tumor lesions. Site-specifically IRDye800CW-labeled 2Rs15d nanobody was injected 90 min before surgery. Fluorescent signal in tumor lesions (indicated by green arrows) is clearly discernible from background signal. Adapted with permission from [123]. (**C**) Real-time fluorescence imaging of orthotopic tongue tumor 24 h post-injection of an EGFR-specific randomly IRDye800CW-conjugated nanobody. Colocalization with bioluminescence imaging (BLI) and green fluorescent protein (GFP) signals is shown. Adapted with permission from [109]. (**D**) Fluorescent imaging of an orthotopically inoculated tongue tumor at 1 h post-injection of an EGFR-specific randomly IRDye700DX-conjugated nanobody for photodynamic therapy (PDT). Stars denote the presence of fluorescent tracer uptake in invaded lymph nodes. Adapted with permission from [125].

Similarly to the theranostic approach in nuclear medicine, where TRNT is combined with diagnostic imaging, the conjugation of a photosensitizer to a tumor-targeting nanobody enables its use for image-guided resection followed by photodynamic therapy (PDT) of residual malignant cells. In PDT, a photosensitizer is activated by incidence light to produce ROS. These ROS can damage the tumor by directly causing cell death through apoptosis and necrosis, damaging the tumor vasculature, and inducing an immune response [126]. This was investigated in a mouse orthotopic tongue tumor model with EGFR-specific nanobodies conjugated randomly to the photosensitizer IRDye700DX (Figure 2D). The effectiveness of a monovalent nanobody and bispecific variant, which binds two different sites on EGFR, was compared with that of a conventional antibody. Both nanobody photosensitizers outperformed the antibody after therapeutic illumination, with more homogenous damage to the tumor and less nontarget damage. Furthermore, despite the higher internalization seen in vitro for the bispecific variant, better in vivo therapeutic results were obtained with the monovalent nanobody. This is in accordance with the assumption that smaller, monovalent compounds diffuse more homogenously through tumor tissue [125,127].

## 5. Conclusions and Perspectives

Nanobodies, with their unique properties, show great promise as targeting moieties in molecular imaging and therapy. Their fast blood clearance, rapid and homogenous tissue penetration, and low background retention allow highly specific imaging at early time points after administration and effective therapy with minimal nonspecific toxicity. The utility of nanobody tracers is now broadly recognized thanks to the convincing preclinical data obtained so far. However, clinical data on their use in this field is still very limited. The expensive and time-consuming process required to translate nanobodies into the clinic (current Good Manufacturing Practice (cGMP) production, toxicity studies, and Investigational Medicinal Product Dossier (IMPD) filing) is probably the major limiting factor. This, however, holds true for any molecular tracer.

The first clinically translated radiolabeled nanobody, the anti-HER2 nanobody 2Rs15d, labeled either with $^{68}$Ga or $^{131}$I, has been investigated in two phase I trials as a potential tool to provide predictive and responsive information on targeted tumor therapies. Follow-up studies respectively evaluating the diagnosis and treatment of breast cancer brain metastasis are now ongoing or planned. In the next years, clinical data of two additional radiolabeled nanobodies is expected, as their clinical translation is almost completed. A nanobody targeting the inflammatory marker VCAM-1 will be evaluated for vulnerable plaque screening, and a macrophage-specific nanobody will also be investigated, which opens up opportunities to image immune cell activation and dynamics in oncology and inflammatory diseases. The latter approach is expected to be further exploited in the future to aid in the development, selection, and monitoring of (novel) immunotherapies.

In the context of intraoperative imaging, properly designed fluorescent nanobody tracers seem to be promising tools to assist and guide surgeons during complex interventions. Evaluated only in a preclinical setting so far, their feasibility and surgical benefit in humans remains to be demonstrated. The strategy to move most rapidly towards the clinic would be to fluorescently label clinical-grade nanobodies which are already available (e.g., anti-HER2 nanobody), in analogy with antibodies currently under investigation as fluorescent contrast agents. However, preferentially, novel nanobodies targeting more relevant biomarkers for the application of image-guided surgery are expected to be developed and clinically translated.

Regarding the design of nanobody tracers, further advances towards novel chemistries permitting conjugation of contrast labels in a more controlled manner are warranted, as ultimately, any labeling method that is considered for clinical translation must be evaluated in a regulatory context. Furthermore, developments of chelators, prosthetic groups, fluorescent dyes, and bimodal labels with improved effects on the pharmacokinetics of nanobodies would be of interest, especially related to kidney retention. This aspect remains a critical point for potential toxicity issues, particularly for therapeutic applications.

**Funding:** Our studies with nanobody-based probes were funded by Kom op tegen kanker, Stichting tegen Kanker, Fund for Scientific Research Flanders (FWO), Industrial Research Fund (IOF), Strategic Research Program (SRP), Scientific Fund Willy Gepts, Melanoma Research Alliance, Juvenile Diabetes Research Foundation, Innoviris, the Novo Nordisk Foundation, and Horizon2020.

**Conflicts of Interest:** N.D. is a consultant of Camel-IDS and holds ownership interest (including patents) in camelid single-domain diagnostics and therapeutics. S.H. holds a patent in camelid single-domain diagnostics and therapeutics.

## References

1. Mankoff, D.A. A definition of molecular imaging. *J. Nucl. Med.* **2007**, *48*, 18N–21N. [PubMed]
2. Weber, J.; Haberkorn, U.; Mier, W. Cancer stratification by molecular imaging. *Int. J. Mol. Sci.* **2015**, *16*, 4918–4946. [CrossRef] [PubMed]
3. Vahrmeijer, A.L.; Hutteman, M.; van der Vorst, J.R.; van de Velde, C.J.H.; Frangioni, J.V. Image-guided cancer surgery using near-infrared fluorescence. *Nat. Rev. Clin. Oncol.* **2013**, *10*, 507–518. [CrossRef] [PubMed]
4. Hingorani, D.V.; Whitney, M.A.; Friedman, B.; Kwon, J.-K.; Crisp, J.L.; Xiong, Q.; Gross, L.; Kane, C.J.; Tsien, R.Y.; Nguyen, Q.T. Nerve-targeted probes for fluorescence-guided intraoperative imaging. *Theranostics* **2018**, *8*, 4226–4237. [CrossRef] [PubMed]
5. Cha, J.; Nani, R.R.; Luciano, M.P.; Broch, A.; Kim, K.; Namgoong, J.M.; Kulkarni, R.A.; Meier, J.L.; Kim, P.; Schnermann, M.J. A chemically stable fluorescent marker of the ureter. *Bioorg. Med. Chem. Lett.* **2018**, *28*, 2741–2745. [CrossRef] [PubMed]
6. Burggraaf, J.; Kamerling, I.M.C.; Gordon, P.B.; Schrier, L.; De Kam, M.L.; Kales, A.J.; Bendiksen, R.; Indrevoll, B.; Bjerke, R.M.; Moestue, S.A.; et al. Detection of colorectal polyps in humans using an intravenously administered fluorescent peptide targeted against c-Met. *Nat. Med.* **2015**, *21*, 955–961. [CrossRef] [PubMed]
7. Nagengast, W.B.; Hartmans, E.; Garcia-Allende, P.B.; Peters, F.T.M.; Linssen, M.D.; Koch, M.; Koller, M.; Tjalma, J.J.J.; Karrenbeld, A.; Jorritsma-Smit, A.; et al. Near-infrared fluorescence molecular endoscopy detects dysplastic oesophageal lesions using topical and systemic tracer of vascular endothelial growth factor A. *Gut* **2017**, *68*, 1–4. [CrossRef]
8. Jaffer, F.A.; Calfon, M.A.; Rosenthal, A.; Mallas, G.; Razansky, N.; Mauskapf, A.; Weissleder, R.; Libby, P.; Ntziachristos, V. Two-Dimensional Intravascular Near-Infrared Fluorescence Molecular Imaging of Inflammation in Atherosclerosis and Stent-Induced Vascular Injury. *J. Am. Coll. Cardiol.* **2011**, *57*, 2516–2526. [CrossRef]
9. Chen, K.; Chen, X. Design and Development of Molecular Imaging Probes. *Curr. Top. Med. Chem.* **2010**, *10*, 1227–1236. [CrossRef]
10. Thurber, G.M.; Schmidt, M.M.; Wittrup, K.D. Factors determining antibody distribution in tumors. *Trends Pharmacol. Sci.* **2008**, *29*, 57–61. [CrossRef]

11. Wu, A.M. Engineered antibodies for molecular imaging of cancer. *Methods* **2014**, *65*, 139–147. [CrossRef] [PubMed]
12. Chakravarty, R.; Goel, S.; Cai, W. Nanobody: The "Magic Bullet" for Molecular Imaging? *Theranostics* **2014**, *4*, 386–398. [CrossRef] [PubMed]
13. Hamers-Casterman, C.; Atarhouch, T.; Muyldermans, S.; Robinson, G.; Hamers, C.; Songa, E.B.; Bendahman, N.; Hamers, R. Naturally occurring antibodies devoid of light chains. *Nature* **1993**, *363*, 446–448. [CrossRef] [PubMed]
14. Muyldermans, S.; Baral, T.N.; Retamozzo, V.C.; De Baetselier, P.; De Genst, E.; Kinne, J.; Leonhardt, H.; Magez, S.; Nguyen, V.K.; Revets, H.; et al. Camelid immunoglobulins and nanobody technology. *Vet. Immunol. Immunopathol.* **2009**, *128*, 178–183. [CrossRef] [PubMed]
15. Vu, K.B.; Ghahroudi, M.A.; Wyns, L.; Muyldermans, S. Comparison of llama V(H) sequences from conventional and heavy chain antibodies. *Mol. Immunol.* **1997**, *34*, 1121–1131. [CrossRef]
16. Muyldermans, S.; Atarhouch, T.; Saldanha, J.; Barbosa, J.A; Hamers, R. Sequence and structure of VH domain from naturally occurring camel heavy chain immunoglobulins lacking light chains. *Protein Eng.* **1994**, *7*, 1129–1135. [CrossRef]
17. Harmsen, M.M.; Ruuls, R.C.; Nijman, I.J.; Niewold, T.A.; Frenken, L.G.J.; De Geus, B. Llama heavy-chain V regions consist of at least four distinct subfamilies revealing novel sequence features. *Mol. Immunol.* **2000**, *37*, 579–590. [CrossRef]
18. Muyldermans, S.; Cambillau, C.; Wyns, L. Recognition of antigens by single-domain antibody fragments: The superfluous luxury of paired domains. *Trends Biochem. Sci.* **2001**, *26*, 230–235. [CrossRef]
19. De Genst, E.; Silence, K.; Decanniere, K.; Conrath, K.; Loris, R.; Kinne, J.; Muyldermans, S.; Wyns, L. Molecular basis for the preferential cleft recognition by dromedary heavy-chain antibodies. *Proc. Natl. Acad. Sci. USA* **2006**, *103*, 4586–4591. [CrossRef]
20. Van Der Linden, R.H.J.; Frenken, L.G.J.; De Geus, B.; Harmsen, M.M.; Ruuls, R.C.; Stok, W.; De Ron, L.; Wilson, S.; Davis, P.; Verrips, C.T. Comparison of physical chemical properties of llama V(HH) antibody fragments and mouse monoclonal antibodies. *Biochim. Biophys. Acta Protein Struct. Mol. Enzymol.* **1999**, *1431*, 37–46. [CrossRef]
21. Dumoulin, M.; Conrath, K.; Van Meirhaeghe, A.; Meersman, F.; Heremans, K.; Frenken, L.G.J.; Muyldermans, S.; Wyns, L.; Matagne, A. Single-domain antibody fragments with high conformational stability. *Protein Sci.* **2002**, *11*, 500–515. [CrossRef] [PubMed]
22. Muyldermans, S. Nanobodies: Natural Single-Domain Antibodies. *Annu. Rev. Biochem.* **2013**, *82*, 775–797. [CrossRef] [PubMed]
23. Keyaerts, M.; Xavier, C.; Heemskerk, J.; Devoogdt, N.; Everaert, H.; Ackaert, C.; Vanhoeij, M.; Duhoux, F.P.; Gevaert, T.; Simon, P.; et al. Phase I Study of 68Ga-HER2-Nanobody for PET/CT Assessment of HER2 Expression in Breast Carcinoma. *J. Nucl. Med.* **2016**, *57*, 27–33. [CrossRef] [PubMed]
24. Bartunek, J.; Barbato, E.; Heyndrickx, G.; Vanderheyden, M.; Wijns, W.; Holz, J.B. Novel antiplatelet agents: ALX-0081, a nanobody directed towards von Willebrand factor. *J. Cardiovasc. Transl. Res.* **2013**, *6*, 355–363. [CrossRef]
25. Papadopoulos, K.P.; Isaacs, R.; Bilic, S.; Kentsch, K.; Huet, H.A.; Hofmann, M.; Rasco, D.; Kundamal, N.; Tang, Z.; Cooksey, J.; et al. Unexpected hepatotoxicity in a phase I study of TAS266, a novel tetravalent agonistic Nanobody®targeting the DR5 receptor. *Cancer Chemother. Pharmacol.* **2015**, *75*, 887–895. [CrossRef]
26. Holland, M.C.; Wurthner, J.U.; Morley, P.J.; Birchler, M.A.; Lambert, J.; Albayaty, M.; Serone, A.P.; Wilson, R.; Chen, Y.; Forrest, R.M.; et al. Autoantibodies to variable heavy (VH) chain Ig sequences in humans impact the safety and clinical pharmacology of a VHdomain antibody antagonist of TNF-$\alpha$ receptor 1. *J. Clin. Immunol.* **2013**, *33*, 1192–1203. [CrossRef]
27. Arbabi Ghahroudi, M.; Desmyter, A.; Wyns, L.; Hamers, R.; Muyldermans, S. Selection and identification of single domain antibody fragments from camel heavy-chain antibodies. *FEBS Lett.* **1997**, *414*, 521–526. [CrossRef]
28. Moutel, S.; Bery, N.; Bernard, V.; Keller, L.; Lemesre, E.; De Marco, A.; Ligat, L.; Rain, J.C.; Favre, G.; Olichon, A.; et al. NaLi-H1: A universal synthetic library of humanized nanobodies providing highly functional antibodies and intrabodies. *eLife* **2016**, *5*, e16228. [CrossRef]

29. McMahon, C.; Baier, A.S.; Pascolutti, R.; Wegrecki, M.; Zheng, S.; Ong, J.X.; Erlandson, S.C.; Hilger, D.; Rasmussen, S.G.F.; Ring, A.M.; et al. Yeast surface display platform for rapid discovery of conformationally selective nanobodies. *Nat. Struct. Mol. Biol.* **2018**, *25*, 289–296. [CrossRef]
30. Bannas, P.; Hambach, J.; Koch-Nolte, F. Nanobodies and nanobody-based human heavy chain antibodies as antitumor therapeutics. *Front. Immunol.* **2017**, *8*, 1603. [CrossRef]
31. De Vos, J.; Devoogdt, N.; Lahoutte, T.; Muyldermans, S. Camelid single-domain antibody-fragment engineering for (pre)clinical in vivo molecular imaging applications: Adjusting the bullet to its target. *Expert Opin. Biol. Ther.* **2013**, *13*, 1149–1160. [CrossRef] [PubMed]
32. Hicks, R.J.; Hofman, M.S. Is there still a role for SPECT–CT in oncology in the PET–CT era? *Nat. Rev. Clin. Oncol.* **2012**, *9*, 712–720. [CrossRef] [PubMed]
33. Xavier, C.; Devoogdt, N.; Hernot, S.; Vaneycken, I.; Huyvetter, M.D.; De Vos, J.; Massa, S.; Lahoutte, T.; Caveliers, V. Site-Specific Labeling of His-Tagged Nanobodies with 99mTc: A Practical Guide. In *Single Domain Antibodies: Methods and Protocols*; Springer: Berlin, Germnay, 2012; Chapter 30, pp. 485–490, ISBN 978-1-61779-967-9.
34. Massa, S.; Xavier, C.; Muyldermans, S.; Devoogdt, N. Emerging site-specific bioconjugation strategies for radioimmunotracer development. *Expert Opin. Drug Deliv.* **2016**, *13*, 1149–1163. [CrossRef] [PubMed]
35. Massa, S.; Xavier, C.; De Vos, J.; Caveliers, V.; Lahoutte, T.; Muyldermans, S.; Devoogdt, N. Site-specific labeling of cysteine-tagged camelid single-domain antibody-fragments for use in molecular imaging. *Bioconjug. Chem.* **2014**, *25*, 979–988. [CrossRef] [PubMed]
36. Fontaine, S.D.; Reid, R.; Robinson, L.; Ashley, G.W.; Santi, D.V. Long-term stabilization of maleimide-thiol conjugates. *Bioconjug. Chem.* **2015**, *26*, 145–152. [CrossRef] [PubMed]
37. Alley, S.C.; Benjamin, D.R.; Jeffrey, S.C.; Okeley, N.M.; Meyer, D.L.; Sanderson, R.J.; Senter, P.D. Contribution of linker stability to the activities of anticancer immunoconjugates. *Bioconjug. Chem.* **2008**, *19*, 759–765. [CrossRef] [PubMed]
38. Massa, S.; Vikani, N.; Betti, C.; Ballet, S.; Vanderhaegen, S.; Steyaert, J.; Descamps, B.; Vanhove, C.; Bunschoten, A.; van Leeuwen, F.W.B.; et al. Sortase A-mediated site-specific labeling of camelid single-domain antibody-fragments: A versatile strategy for multiple molecular imaging modalities. *Contrast Media Mol. Imaging* **2016**, *11*, 328–339. [CrossRef]
39. Rashidian, M.; Wang, L.; Edens, J.G.; Jacobsen, J.T.; Hossain, I.; Wang, Q.; Victora, G.D.; Vasdev, N.; Ploegh, H.; Liang, S.H. Enzyme-Mediated Modification of Single-Domain Antibodies for Imaging Modalities with Different Characteristics. *Angew. Chem. Int. Ed.* **2016**, *55*, 528–533. [CrossRef]
40. Billen, B.; Vincke, C.; Hansen, R.; Devoogdt, N.; Muyldermans, S.; Adriaensens, P.; Guedens, W. Cytoplasmic versus periplasmic expression of site-specifically and bioorthogonally functionalized nanobodies using expressed protein ligation. *Protein Expr. Purif.* **2017**, *133*, 25–34. [CrossRef]
41. Xavier, C.; Vaneycken, I.; D'huyvetter, M.; Heemskerk, J.; Keyaerts, M.; Vincke, C.; Devoogdt, N.; Muyldermans, S.; Lahoutte, T.; Caveliers, V. Synthesis, Preclinical Validation, Dosimetry, and Toxicity of 68Ga-NOTA-Anti-HER2 Nanobodies for iPET Imaging of HER2 Receptor Expression in Cancer. *J. Nucl. Med.* **2013**, *54*, 776–784. [CrossRef]
42. Vaneycken, I.; Devoogdt, N.; Van Gassen, N.; Vincke, C.; Xavier, C.; Wernery, U.; Muyldermans, S.; Lahoutte, T.; Caveliers, V. Preclinical screening of anti-HER2 nanobodies for molecular imaging of breast cancer. *FASEB J.* **2011**, *25*, 2433–2446. [CrossRef] [PubMed]
43. Broisat, A.; Hernot, S.; Toczek, J.; De Vos, J.; Riou, L.M.; Martin, S.; Ahmadi, M.; Thielens, N.; Wernery, U.; Caveliers, V.; et al. Nanobodies Targeting Mouse/Human VCAM1 for the Nuclear Imaging of Atherosclerotic Lesions. *Circ. Res.* **2012**, *110*, 927–937. [CrossRef] [PubMed]
44. Movahedi, K.; Schoonooghe, S.; Laoui, D.; Houbracken, I.; Waelput, W.; Breckpot, K.; Bouwens, L.; Lahoutte, T.; De Baetselier, P.; Raes, G.; et al. Nanobody-Based Targeting of the Macrophage Mannose Receptor for Effective In Vivo Imaging of Tumor-Associated Macrophages. *Cancer Res.* **2012**, *72*, 4165–4177. [CrossRef] [PubMed]
45. Tchouate-gainkam, L.O.; Caveliers, V.; Devoogdt, N.; Vanhove, C.; Xavier, C.; Boerman, O.; Muyldermans, S.; Bossuyt, A.; Lahoutte, T. Localization, mechanism and reduction of renal retention of technetium-99m labeled epidermal growth factor receptor-specific nanobody in mice. *Contrast Media Mol. Imaging* **2011**, *6*, 85–92. [CrossRef]

46. De Jong, M.; Barone, R.R.; Krenning, E.; Bernard, B.; Melis, M.; Visser, T.; Gekle, M.; Willnow, T.E.; Walrand, S.; Jamar, F.; et al. Megalin Is Essential for Renal Proximal Tubule Reabsorption of 111In-DTPA-Octreotide. *J. Nucl. Med.* **2005**, *46*, 1696–1700. [PubMed]
47. Behr, T.M.; Sharkey, R.M.; Juweid, M.E.; Blumenthal, R.D.; Dunn, R.M.; Griffiths, G.L.; Bair, H.J.; Wolf, F.G.; Becker, W.S.; Goldenberg, D.M. Reduction of the Renal Uptake of Radiolabeled Monoclonal Antibody Fragments by Cationic Amino Acids and Their Derivatives. *Cancer Res.* **1995**, *55*, 3825–3834. [PubMed]
48. Chatalic, K.L.S.; Veldhoven-Zweistra, J.; Bolkestein, M.; Hoeben, S.; Koning, G.A.; Boerman, O.C.; de Jong, M.; van Weerden, W.M. A Novel 111In-Labeled Anti-Prostate-Specific Membrane Antigen Nanobody for Targeted SPECT/CT Imaging of Prostate Cancer. *J. Nucl. Med.* **2015**, *56*, 1094–1099. [CrossRef]
49. D'Huyvetter, M.; Vincke, C.; Xavier, C.; Aerts, A.; Impens, N.; Baatout, S.; De Raeve, H.; Muyldermans, S.; Caveliers, V.; Devoogdt, N.; et al. Targeted radionuclide therapy with A 177Lu-labeled anti-HER2 nanobody. *Theranostics* **2014**, *4*, 708–720. [CrossRef]
50. Khan, F.; Legler, P.M.; Mease, R.M.; Duncan, E.H.; Bergmann-Leitner, E.S.; Angov, E. Histidine affinity tags affect MSP142 structural stability and immunodominance in mice. *Biotechnol. J.* **2012**, *7*, 133–147. [CrossRef]
51. Randolph, T.W. The two faces of His-tag: Immune response versus ease of protein purification. *Biotechnol. J.* **2012**, *7*, 18–19. [CrossRef]
52. Bala, G.; Crauwels, M.; Blykers, A.; Remory, I.; Marschall, A.L.J.; Dübel, S.; Dumas, L.; Broisat, A.; Martin, C.; Ballet, S.; et al. Radiometal-labeled anti-VCAM-1 nanobodies as molecular tracers for atherosclerosis—Impact of radiochemistry on pharmacokinetics. *Biol. Chem.* **2018**. [CrossRef] [PubMed]
53. Xavier, C.; Blykers, A.; Vaneycken, I.; D'Huyvetter, M.; Heemskerk, J.; Lahoutte, T.; Devoogdt, N.; Caveliers, V. 18F-nanobody for PET imaging of HER2 overexpressing tumors. *Nucl. Med. Biol.* **2016**, *43*, 247–252. [CrossRef] [PubMed]
54. D'Huyvetter, M.; De Vos, J.; Xavier, C.; Pruszynski, M.; Sterckx, Y.G.J.; Massa, S.; Raes, G.; Caveliers, V.; Zalutsky, M.R.; Lahoutte, T.; et al. 131I-labeled anti-HER2 camelid sdAb as a theranostic tool in cancer treatment. *Clin. Cancer Res.* **2017**, *23*, 6616–6628. [CrossRef]
55. Krasniqi, A.; D'Huyvetter, M.; Devoogdt, N.; Frejd, F.Y.; Sörensen, J.; Orlova, A.; Keyaerts, M.; Tolmachev, V. Same-Day Imaging Using Small Proteins: Clinical Experience and Translational Prospects in Oncology. *J. Nucl. Med.* **2018**, *59*, 885–891. [CrossRef]
56. Richards, D.A. Exploring alternative antibody scaffolds: Antibody fragments and antibody mimics for targeted drug delivery. *Drug Discov. Today Technol.* **2018**, *30*, 35–46. [CrossRef] [PubMed]
57. D'Huyvetter, M.; Aerts, A.; Xavier, C.; Vaneycken, I.; Devoogdt, N.; Gijs, M.; Impens, N.; Baatout, S.; Ponsard, B.; Muyldermans, S.; et al. Development of 177Lu-nanobodies for radioimmunotherapy of HER2-positive breast cancer: Evaluation of different bifunctional chelators. *Contrast Media Mol. Imaging* **2012**, *7*, 254–264. [CrossRef] [PubMed]
58. Zhou, Z.; Vaidyanathan, G.; McDougald, D.; Kang, C.M.; Balyasnikova, I.; Devoogdt, N.; Ta, A.N.; McNaughton, B.R.; Zalutsky, M.R. Fluorine-18 Labeling of the HER2-Targeting Single-Domain Antibody 2Rs15d Using a Residualizing Label and Preclinical Evaluation. *Mol. Imaging Biol.* **2017**, *19*, 867–877. [CrossRef] [PubMed]
59. Zhou, Z.; Chitneni, S.K.; Devoogdt, N.; Zalutsky, M.R.; Vaidyanathan, G. Fluorine-18 labeling of an anti-HER2 VHH using a residualizing prosthetic group via a strain-promoted click reaction: Chemistry and preliminary evaluation. *Bioorganic Med. Chem.* **2018**, *26*, 1939–1949. [CrossRef]
60. Zhou, Z.; McDougald, D.L.; Devoogdt, N.; Zalutsky, M.R.; Vaidyanathan, G. Labeling Single Domain Antibody Fragments with Fluorine-18 Using 2,3,5,6-Tetrafluorophenyl 6-[$^{18}$F]Fluoronicotinate Resulting in High Tumor to Kidney Ratios. *Mol. Pharm.* **2018**. [CrossRef]
61. Pruszynski, M.; Koumarianou, E.; Vaidyanathan, G.; Revets, H.; Devoogdt, N.; Lahoutte, T.; Lyerly, H.K.; Zalutsky, M.R. Improved Tumor Targeting of Anti-HER2 Nanobody Through N-Succinimidyl 4-Guanidinomethyl-3-Iodobenzoate Radiolabeling. *J. Nucl. Med.* **2014**, *55*, 650–656. [CrossRef]
62. Vaidyanathan, G.; McDougald, D.; Choi, J.; Koumarianou, E.; Weitzel, D.; Osada, T.; Lyerly, H.K.; Zalutsky, M.R. Preclinical Evaluation of 18F-Labeled Anti-HER2 Nanobody Conjugates for Imaging HER2 Receptor Expression by Immuno-PET. *J. Nucl. Med.* **2016**, *57*, 967–973. [CrossRef] [PubMed]

63. Pruszynski, M.; Koumarianou, E.; Vaidyanathan, G.; Revets, H.; Devoogdt, N.; Lahoutte, T.; Zalutsky, M.R. Targeting breast carcinoma with radioiodinated anti-HER2 Nanobody. *Nucl. Med. Biol.* **2013**, *40*, 52–59. [CrossRef] [PubMed]
64. Gainkam, L.O.T.; Huang, L.; Caveliers, V.; Keyaerts, M.; Hernot, S.; Vaneycken, I.; Vanhove, C.; Revets, H.; De Baetselier, P.; Lahoutte, T. Comparison of the Biodistribution and Tumor Targeting of Two 99mTc-Labeled Anti-EGFR Nanobodies in Mice, Using Pinhole SPECT/Micro-CT. *J. Nucl. Med.* **2008**, *49*, 788–795. [CrossRef] [PubMed]
65. Vosjan, M.J.W.D.; Perk, L.R.; Roovers, R.C.; Visser, G.W.M.; Stigter-Van Walsum, M.; Van Bergen En Henegouwen, P.M.P.; Van Dongen, G.A.M.S. Facile labelling of an anti-epidermal growth factor receptor Nanobody with 68Ga via a novel bifunctional desferal chelate for immuno-PET. *Eur. J. Nucl. Med. Mol. Imaging* **2011**, *38*, 753–763. [CrossRef] [PubMed]
66. Krüwel, T.; Nevoltris, D.; Bode, J.; Dullin, C.; Baty, D.; Chames, P.; Alves, F. In vivo detection of small tumour lesions by multi-pinhole SPECT applying a 99mTc-labelled nanobody targeting the Epidermal Growth Factor Receptor. *Sci. Rep.* **2016**, *6*, 1–12. [CrossRef] [PubMed]
67. Evazalipour, M.; D'Huyvetter, M.; Tehrani, B.S.; Abolhassani, M.; Omidfar, K.; Abdoli, S.; Arezumand, R.; Morovvati, H.; Lahoutte, T.; Muyldermans, S.; et al. Generation and characterization of nanobodies targeting PSMA for molecular imaging of prostate cancer. *Contrast Media Mol. Imaging* **2014**, *9*, 211–220. [CrossRef] [PubMed]
68. Krasniqi, A.; D'Huyvetter, M.; Xavier, C.; Van der Jeught, K.; Muyldermans, S.; Van Der Heyden, J.; Lahoutte, T.; Tavernier, J.; Devoogdt, N. Theranostic Radiolabeled Anti-CD20 sdAb for Targeted Radionuclide Therapy of Non-Hodgkin Lymphoma. *Mol. Cancer Ther.* **2017**, *16*, 2828–2839. [CrossRef] [PubMed]
69. Krasniqi, A.; Bialkowska, M.; Xavier, C.; Van der Jeught, K.; Muyldermans, S.; Devoogdt, N.; D'Huyvetter, M. Pharmacokinetics of radiolabeled dimeric sdAbs constructs targeting human CD20. *New Biotechnol.* **2018**, *45*, 69–79. [CrossRef] [PubMed]
70. Montemagno, C.; Bacot, S.; Ahmadi, M.; Kerfelec, B.; Baty, D.; Debiossat, M.; Soubies, A.; Perret, P.; Riou, L.; Fagret, D.; et al. Preclinical Evaluation of Mesothelin-Specific Ligands for SPECT Imaging of Triple-Negative Breast Cancer. *J. Nucl. Med.* **2018**, *59*, 1056–1062. [CrossRef] [PubMed]
71. Vaneycken, I.; Govaert, J.; Vincke, C.; Caveliers, V.; Lahoutte, T.; De Baetselier, P.; Raes, G.; Bossuyt, A.; Muyldermans, S.; Devoogdt, N. In Vitro Analysis and In Vivo Tumor Targeting of a Humanized, Grafted Nanobody in Mice Using Pinhole SPECT/Micro-CT. *J. Nucl. Med.* **2010**, *51*, 1099–1106. [CrossRef] [PubMed]
72. Rashidian, M.; Keliher, E.J.; Bilate, A.M.; Duarte, J.N.; Wojtkiewicz, G.R.; Jacobsen, J.T.; Cragnolini, J.; Swee, L.K.; Victora, G.D.; Weissleder, R.; et al. Noninvasive imaging of immune responses. *Proc. Natl. Acad. Sci. USA* **2015**, *112*, 6146–6151. [CrossRef] [PubMed]
73. Zheng, F.; Put, S.; Bouwens, L.; Lahoutte, T.; Matthys, P.; Muyldermans, S.; De Baetselier, P.; Devoogdt, N.; Raes, G.; Schoonooghe, S. Molecular imaging with macrophage CRIg-targeting nanobodies for early and preclinical diagnosis in a mouse model of rheumatoid arthritis. *J. Nucl. Med.* **2014**, *55*, 824–829. [CrossRef]
74. Blykers, A.; Schoonooghe, S.; Xavier, C.; D'hoe, K.; Laoui, D.; D'Huyvetter, M.; Vaneycken, I.; Cleeren, F.; Bormans, G.; Heemskerk, J.; et al. PET Imaging of Macrophage Mannose Receptor-Expressing Macrophages in Tumor Stroma Using 18F-Radiolabeled Camelid Single-Domain Antibody Fragments. *J. Nucl. Med.* **2015**, *56*, 1265–1271. [CrossRef] [PubMed]
75. Xavier, C.; Blykers, A.; Laoui, D.; Bolli, E.; Vaneycken, I.; Bridoux, J.; Baudhuin, H.; Raes, G.; Everaert, H.; Movahedi, K.; et al. Clinical translation of [68Ga]Ga-NOTA-anti-MMR-sdAb for PET/CT imaging of protumorigenic macrophages. *Mol. Imaging Biol.* **2018**. [CrossRef]
76. Rashidian, M.; Ingram, J.R.; Dougan, M.; Dongre, A.; Whang, K.A.; LeGall, C.; Cragnolini, J.J.; Bierie, B.; Gostissa, M.; Gorman, J.; et al. Predicting the response to CTLA-4 blockade by longitudinal noninvasive monitoring of CD8 T cells. *J. Exp. Med.* **2017**, *214*, 2243–2255. [CrossRef] [PubMed]
77. Broos, K.; Keyaerts, M.; Lecocq, Q.; Renmans, D.; Nguyen, T.; Escors, D.; Liston, A.; Raes, G.; Breckpot, K.; Devoogdt, N. Non-invasive assessment of murine PD-L1 levels in syngeneic tumor models by nuclear imaging with nanobody tracers. *Oncotarget* **2017**, *8*, 41932–41946. [CrossRef] [PubMed]
78. Ingram, J.R.; Dougan, M.; Rashidian, M.; Knoll, M.; Keliher, E.J.; Garrett, S.; Garforth, S.; Blomberg, O.S.; Espinosa, C.; Bhan, A.; et al. PD-L1 is an activation-independent marker of brown adipocytes. *Nat. Commun.* **2017**, *8*, 647. [CrossRef]

79. Broos, K.; Lecocq, Q.; Raes, G.; Devoogdt, N.; Keyaerts, M.; Breckpot, K. Noninvasive imaging of the PD-1: PD-L1 immune checkpoint: Embracing nuclear medicine for the benefit of personalized immunotherapy. *Theranostics* **2018**, *8*, 3559–3570. [CrossRef]
80. Zheng, F.; Devoogdt, N.; Sparkes, A.; Morias, Y.; Abels, C.; Stijlemans, B.; Lahoutte, T.; Muyldermans, S.; De Baetselier, P.; Schoonooghe, S.; et al. Monitoring liver macrophages using nanobodies targeting Vsig4: Concanavalin A induced acute hepatitis as paradigm. *Immunobiology* **2015**, *220*, 200–209. [CrossRef]
81. Put, S.; Schoonooghe, S.; Devoogdt, N.; Schurgers, E.; Avau, A.; Mitera, T.; D'Huyvetter, M.; De Baetselier, P.; Raes, G.; Lahoutte, T.; et al. SPECT Imaging of Joint Inflammation with Nanobodies Targeting the Macrophage Mannose Receptor in a Mouse Model for Rheumatoid Arthritis. *J. Nucl. Med.* **2013**, *54*, 807–814. [CrossRef]
82. Zheng, F.; Sparkes, A.; De Baetselier, P.; Schoonooghe, S.; Stijlemans, B.; Muyldermans, S.; Flamand, V.; Van Ginderachter, J.A.; Devoogdt, N.; Raes, G.; et al. Molecular Imaging with Kupffer Cell-Targeting Nanobodies for Diagnosis and Prognosis in Mouse Models of Liver Pathogenesis. *Mol. Imaging Biol.* **2017**, *19*, 49–58. [CrossRef] [PubMed]
83. Balhuizen, A.; Massa, S.; Mathijs, I.; Turatsinze, J.V.; De Vos, J.; Demine, S.; Xavier, C.; Villate, O.; Millard, I.; Egrise, D.; et al. A nanobody-based tracer targeting DPP6 for non-invasive imaging of human pancreatic endocrine cells. *Sci. Rep.* **2017**, *7*, 15130. [CrossRef]
84. Bala, G.; Baudhuin, H.; Remory, I.; Gillis, K.; Debie, P.; Krasniqi, A.; Lahoutte, T.; Raes, G.; Devoogdt, N.; Cosyns, B.; et al. Evaluation of [99mTc]Radiolabeled Macrophage Mannose Receptor-Specific Nanobodies for Targeting of Atherosclerotic Lesions in Mice. *Mol. Imaging Biol.* **2018**, *20*, 260–267. [CrossRef] [PubMed]
85. Bala, G.; Blykers, A.; Xavier, C.; Descamps, B.; Broisat, A.; Ghezzi, C.; Fagret, D.; Van Camp, G.; Caveliers, V.; Vanhove, C.; et al. Targeting of vascular cell adhesion molecule-1 by18F-labelled nanobodies for PET/CT imaging of inflamed atherosclerotic plaques. *Eur. Heart J. Cardiovasc. Imaging* **2016**, *17*, 1001–1008. [CrossRef] [PubMed]
86. Senders, M.L.; Hernot, S.; Carlucci, G.; van de Voort, J.C.; Fay, F.; Calcagno, C.; Tang, J.; Alaarg, A.; Zhao, Y.; Ishino, S.; et al. Nanobody-Facilitated Multiparametric PET/MRI Phenotyping of Atherosclerosis. *JACC Cardiovasc. Imaging* **2018**, 2751. [CrossRef] [PubMed]
87. Broisat, A.; Toczek, J.; Dumas, L.S.; Ahmadi, M.; Bacot, S.; Perret, P.; Slimani, L.; Barone-Rochette, G.; Soubies, A.; Devoogdt, N.; et al. 99mTc-cAbVCAM1-5 imaging is a sensitive and reproducible tool for the detection of inflamed atherosclerotic lesions in mice. *J. Nucl. Med.* **2014**, *55*, 1678–1684. [CrossRef]
88. De Vos, J.; Mathijs, I.; Xavier, C.; Massa, S.; Wernery, U.; Bouwens, L.; Lahoutte, T.; Muyldermans, S.; Devoogdt, N. Specific Targeting of Atherosclerotic Plaques in ApoE−/−Mice Using a New Camelid sdAb Binding the Vulnerable Plaque Marker LOX-1. *Mol. Imaging Biol.* **2014**, *16*, 690–698. [CrossRef] [PubMed]
89. Dumas, L.S.; Briand, F.; Clerc, R.; Brousseau, E.; Montemagno, C.; Ahmadi, M.; Bacot, S.; Soubies, A.; Perret, P.; Riou, L.M.; et al. Evaluation of Antiatherogenic Properties of Ezetimibe Using $^{3}$H-Labeled Low-Density-Lipoprotein Cholesterol and $^{99m}$ Tc-cAbVCAM1-5 SPECT in ApoE $^{-/-}$ Mice Fed the Paigen Diet. *J. Nucl. Med.* **2017**, *58*, 1088–1093. [CrossRef]
90. Prise, K.M.; Schettino, G.; Folkard, M.; Held, K.D. New insights on cell death from radiation exposure. *Lancet Oncol.* **2005**, *6*, 520–528. [CrossRef]
91. Walle, T.; Monge, R.M.; Cerwenka, A.; Ajona, D.; Melero, I.; Lecanda, F. Radiation effects on antitumor immune responses: Current perspectives and challenges. *Ther. Adv. Med. Oncol.* **2018**, *10*, 1–27. [CrossRef]
92. D'Huyvetter, M.; Xavier, C.; Caveliers, V.; Lahoutte, T.; Muyldermans, S.; Devoogdt, N. Radiolabeled nanobodies as theranostic tools in targeted radionuclide therapy of cancer. *Expert Opin. Drug Deliv.* **2014**, *11*, 1939–1954. [CrossRef] [PubMed]
93. Dekempeneer, Y.; Keyaerts, M.; Krasniqi, A.; Puttemans, J.; Muyldermans, S.; Lahoutte, T.; D'huyvetter, M.; Devoogdt, N. Targeted alpha therapy using short-lived alpha-particles and the promise of nanobodies as targeting vehicle. *Expert Opin. Biol. Ther.* **2016**, *16*, 1035–1047. [CrossRef] [PubMed]
94. Keyaerts, M.; De Vos, J.; Duhoux, F.P.; Caveliers, V.; Fontaine, C.; Vanhoeij, M.; D'Huyvetter, M.; Everaert, H.; Ghykiere, P.; Devoogdt, N.; et al. Phase I results of CAM-H2 confirm excellent safety profile in human and tumor targeting in patients. In Proceedings of the ASCO Conference Chicago, Chicago, IL, USA, 1–5 June 2018.

95. Xenaki, K.T.; Oliveira, S.; van Bergen en Henegouwen, P.M.P. Antibody or Antibody Fragments: Implications for Molecular Imaging and Targeted Therapy of Solid Tumors. *Front. Immunol.* **2017**, *8*, 1287. [CrossRef] [PubMed]
96. Debie, P.; Lafont, C.; Hansen, I.; Defrise, M.; Van Willigen, D.M.; van Leeuwen, F.; Gijsbers, R.; D'huyvetter, M.; Busson, M.; Devoogdt, N.; et al. Intravital imaging of tumor targeting with nanobodies. In Proceedings of the Abstract Book EMIM 2017, Cologne, Germany, 3 January 2017.
97. Tijink, B.M.; Laeremans, T.; Budde, M.; Walsum, M.S.-V.; Dreier, T.; de Haard, H.J.; Leemans, C.R.; van Dongen, G.A.M.S. Improved tumor targeting of anti-epidermal growth factor receptor Nanobodies through albumin binding: Taking advantage of modular Nanobody technology. *Mol. Cancer Ther.* **2008**, *7*, 2288–2297. [CrossRef] [PubMed]
98. Vosjan, M.J.W.D.; Vercammen, J.; Kolkman, J.A.; Stigter-van Walsum, M.; Revets, H.; van Dongen, G.A.M.S. Nanobodies Targeting the Hepatocyte Growth Factor: Potential New Drugs for Molecular Cancer Therapy. *Mol. Cancer Ther.* **2012**, *11*, 1017–1025. [CrossRef] [PubMed]
99. Warnders, F.J.; Terwisscha van Scheltinga, A.G.T.; Knuehl, C.; van Roy, M.; de Vries, E.F.J.; Kosterink, J.G.W.; de Vries, E.G.E.; Lub-de Hooge, M.N. Human Epidermal Growth Factor Receptor 3–Specific Tumor Uptake and Biodistribution of $^{89}$ Zr-MSB0010853 Visualized by Real-Time and Noninvasive PET Imaging. *J. Nucl. Med.* **2017**, *58*, 1210–1215. [CrossRef] [PubMed]
100. Pruszynski, M.; D'Huyvetter, M.; Bruchertseifer, F.; Morgenstern, A.; Lahoutte, T. Evaluation of an Anti-HER2 Nanobody Labeled with225Ac for Targeted α-Particle Therapy of Cancer. *Mol. Pharm.* **2018**, *15*, 1457–1466. [CrossRef]
101. Lemaire, M.; D'Huyvetter, M.; Lahoutte, T.; Van Valckenborgh, E.; Menu, E.; De Bruyne, E.; Kronenberger, P.; Wernery, U.; Muyldermans, S.; Devoogdt, N.; et al. Imaging and radioimmunotherapy of multiple myeloma with anti-idiotypic Nanobodies. *Leukemia* **2014**, *28*, 444–447. [CrossRef]
102. De Groeve, K.; Deschacht, N.; De Koninck, C.; Caveliers, V.; Lahoutte, T.; Devoogdt, N.; Muyldermans, S.; De Baetselier, P.; Raes, G. Nanobodies as Tools for In Vivo Imaging of Specific Immune Cell Types. *J. Nucl. Med.* **2010**, *51*, 782–789. [CrossRef]
103. Van Elssen, C.H.M.J.; Rashidian, M.; Vrbanac, V.; Wucherpfennig, K.W.; Habre, Z.E.; Sticht, J.; Freund, C.; Jacobsen, J.T.; Cragnolini, J.; Ingram, J.; et al. Noninvasive Imaging of Human Immune Responses in a Human Xenograft Model of Graft-Versus-Host Disease. *J. Nucl. Med.* **2017**, *58*, 1003–1008. [CrossRef]
104. Verhelle, A.; Van Overbeke, W.; Peleman, C.; De Smet, R.; Zwaenepoel, O.; Lahoutte, T.; Van Dorpe, J.; Devoogdt, N.; Gettemans, J. Non-Invasive Imaging of Amyloid Deposits in a Mouse Model of AGel Using 99mTc-Modified Nanobodies and SPECT/CT. *Mol. Imaging Biol.* **2016**, *18*, 887–897. [CrossRef] [PubMed]
105. Verhelle, A.; Nair, N.; Everaert, I.; Van Overbeke, W.; Supply, L.; Zwaenepoel, O.; Peleman, C.; Van Dorpe, J.; Lahoutte, T.; Devoogdt, N.; et al. AAV9 delivered bispecific nanobody attenuates amyloid burden in the gelsolin amyloidosis mouse model. *Hum. Mol. Genet.* **2017**, *26*, 1353–1364. [CrossRef] [PubMed]
106. Nabuurs, R.J.A.; Rutgers, K.S.; Welling, M.M.; Metaxas, A.; de Backer, M.E.; Rotman, M.; Bacskai, B.J.; van Buchem, M.A.; van der Maarel, S.M.; van der Weerd, L. In vivo detection of amyloid-β deposits using heavy chain antibody fragments in a transgenic mouse model for alzheimer's disease. *PLoS ONE* **2012**, *7*, e38284. [CrossRef] [PubMed]
107. Hong, G.; Antaris, A.L.; Dai, H. Near-infrared fluorophores for biomedical imaging. *Nat. Biomed. Eng.* **2017**, *1*, 0010. [CrossRef]
108. Cilliers, C.; Nessler, I.; Christodolu, N.; Thurber, G.M. Tracking Antibody Distribution with Near-Infrared Fluorescent Dyes: Impact of Dye Structure and Degree of Labeling on Plasma Clearance. *Mol. Pharm.* **2017**, *14*, 1623–1633. [CrossRef] [PubMed]
109. Van Driel, P.B.A.A.; Van Der Vorst, J.R.; Verbeek, F.P.R.; Oliveira, S.; Snoeks, T.J.A.; Keereweer, S.; Chan, B.; Boonstra, M.C.; Frangioni, J.V.; Van Bergen En Henegouwen, P.M.P.; et al. Intraoperative fluorescence delineation of head and neck cancer with a fluorescent Anti-epidermal growth factor receptor nanobody. *Int. J. Cancer* **2014**, *134*, 2663–2673. [CrossRef]
110. Oliveira, S.; Van Dongen, G.A.M.S.; Stigter-Van Walsum, M.; Roovers, R.C.; Stam, J.C.; Mali, W.; Van Diest, P.J.; Van Bergen En Henegouwen, P.M.P. Rapid visualization of human tumor xenografts through optical imaging with a near-infrared fluorescent anti-epidermal growth factor receptor nanobody. *Mol. Imaging* **2012**, *11*, 33–46. [CrossRef]

111. Debie, P.; Van Quathem, J.; Hansen, I.; Bala, G.; Massa, S.; Devoogdt, N.; Xavier, C.; Hernot, S. Effect of dye and conjugation chemistry on the biodistribution profile of near-infrared-labeled nanobodies as tracers for image-guided surgery. *Mol. Pharm.* **2017**, *14*, 1145–1153. [CrossRef]
112. Bannas, P.; Well, L.; Lenz, A.; Rissiek, B.; Haag, F.; Schmid, J.; Hochgräfe, K.; Trepel, M.; Adam, G.; Ittrich, H.; et al. In vivo near-infrared fluorescence targeting of T cells: Comparison of nanobodies and conventional monoclonal antibodies. *Contrast Media Mol. Imaging* **2014**, *9*, 135–142. [CrossRef]
113. Bannas, P.; Lenz, A.; Kunick, V.; Well, L.; Fumey, W.; Rissiek, B.; Haag, F.; Schmid, J.; Schütze, K.; Eichhoff, A.; et al. Molecular imaging of tumors with nanobodies and antibodies: Timing and dosage are crucial factors for improved in vivo detection. *Contrast Media Mol. Imaging* **2015**, *10*, 367–378. [CrossRef]
114. Kummar, S.; Doroshow, J.H.; Tomaszewski, J.E.; Calvert, A.H. Phase 0 clinical trials: Recommendations from the task force on methodology for the development of innovative cancer therapies. *Eur. J. Cancer* **2008**, *45*, 741–746. [CrossRef] [PubMed]
115. Tummers, Q.R.J.G.; Hoogstins, C.E.S.; Gaarenstroom, K.N.; de Kroon, C.D.; van Poelgeest, M.I.E.; Vuyk, J.; Bosse, T.; Smit, V.T.H.B.M.; van de Velde, C.J.H.; Cohen, A.F.; et al. Intraoperative imaging of folate receptor alpha positive ovarian and breast cancer using the tumor specific agent EC17. *Oncotarget* **2016**, *7*, 32144–32155. [CrossRef] [PubMed]
116. Hoogstins, C.E.S.; Boogerd, L.S.F.; Sibinga Mulder, B.G.; Mieog, J.S.D.; Swijnenburg, R.J.; van de Velde, C.J.H.; Farina Sarasqueta, A.; Bonsing, B.A.; Framery, B.; Pèlegrin, A.; et al. Image-Guided Surgery in Patients with Pancreatic Cancer: First Results of a Clinical Trial Using SGM-101, a Novel Carcinoembryonic Antigen-Targeting, Near-Infrared Fluorescent Agent. *Ann. Surg. Oncol.* **2018**, *25*, 3350–3357. [CrossRef] [PubMed]
117. Gao, R.W.; Teraphongphom, N.; de Boer, E.; Van Den Berg, N.S.; Divi, V.; Kaplan, M.J.; Oberhelman, N.J.; Hong, S.S.; Capes, E.; Colevas, A.D.; et al. Safety of panitumumab-IRDye800CW and cetuximab-IRDye800CW for fluorescence-guided surgical navigation in head and neck cancers. *Theranostics* **2018**, *8*, 2488–2495. [CrossRef] [PubMed]
118. Lamberts, L.E.; Koch, M.; de Jong, J.S.; Adams, A.L.L.; Glatz, J.; Kranendonk, M.E.G.; Terwisscha van Scheltinga, A.G.T.; Jansen, L.; de Vries, J.; Lub-de Hooge, M.N.; et al. Tumor-Specific Uptake of Fluorescent Bevacizumab–IRDye800CW Microdosing in Patients with Primary Breast Cancer: A Phase I Feasibility Study. *Clin. Cancer Res.* **2017**, *23*, 2730–2742. [CrossRef] [PubMed]
119. KleinJan, G.H.; Bunschoten, A.; van den Berg, N.S.; Olmos, R.A.V.; Klop, W.M.C.; Horenblas, S.; van der Poel, H.G.; Wester, H.J.; van Leeuwen, F.W.B. Fluorescence guided surgery and tracer-dose, fact or fiction? *Eur. J. Nucl. Med. Mol. Imaging* **2016**, *43*, 1857–1867. [CrossRef]
120. van Leeuwen, F.W.B.; Valdés-Olmos, R.; Buckle, T.; Vidal-Sicart, S. Hybrid surgical guidance based on the integration of radionuclear and optical technologies. *Br. J. Radiol.* **2016**, *89*, 20150797. [CrossRef]
121. van Brussel, A.S.A.; Adams, A.; Oliveira, S.; Dorresteijn, B.; El Khattabi, M.; Vermeulen, J.F.; van der Wall, E.; Mali, W.P.T.M.; Derksen, P.W.B.; van Diest, P.J.; et al. Hypoxia-Targeting Fluorescent Nanobodies for Optical Molecular Imaging of Pre-Invasive Breast Cancer. *Mol. Imaging Biol.* **2016**, *18*, 535–544. [CrossRef]
122. Kijanka, M.M.; van Brussel, A.S.A.; van der Wall, E.; Mali, W.P.T.M.; van Diest, P.J.; van Bergen En Henegouwen, P.M.P.; Oliveira, S. Optical imaging of pre-invasive breast cancer with a combination of VHHs targeting CAIX and HER2 increases contrast and facilitates tumour characterization. *EJNMMI Res.* **2016**, *6*, 14. [CrossRef]
123. Debie, P.; Vanhoeij, M.; Poortmans, N.; Puttemans, J.; Gillis, K.; Devoogdt, N.; Lahoutte, T.; Hernot, S. Improved Debulking of Peritoneal Tumor Implants by Near-Infrared Fluorescent Nanobody Image Guidance in an Experimental Mouse Model. *Mol. Imaging Biol.* **2018**, *20*, 361–367. [CrossRef]
124. Kijanka, M.; Warnders, F.J.; El Khattabi, M.; Lub-De Hooge, M.; Van Dam, G.M.; Ntziachristos, V.; De Vries, L.; Oliveira, S.; Van Bergen En Henegouwen, P.M.P. Rapid optical imaging of human breast tumour xenografts using anti-HER2 VHHs site-directly conjugated to IRDye 800CW for image-guided surgery. *Eur. J. Nucl. Med. Mol. Imaging* **2013**, *40*, 1718–1729. [CrossRef] [PubMed]
125. Van Driel, P.B.A.A.; Boonstra, M.C.; Slooter, M.D.; Heukers, R.; Stammes, M.A.; Snoeks, T.J.A.; De Bruijn, H.S.; Van Diest, P.J.; Vahrmeijer, A.L.; Van Bergen En Henegouwen, P.M.P.; et al. EGFR targeted nanobody-photosensitizer conjugates for photodynamic therapy in a pre-clinical model of head and neck cancer. *J. Control. Release* **2016**, *229*, 93–105. [CrossRef] [PubMed]

126. Allison, R.R.; Moghissi, K. Photodynamic therapy (PDT): PDT mechanisms. *Clin. Endosc.* **2013**, *46*, 24–29. [CrossRef] [PubMed]
127. Heukers, R.; van Bergen en Henegouwen, P.M.P.; Oliveira, S. Nanobody-photosensitizer conjugates for targeted photodynamic therapy. *Nanomed. Nanotechnol. Biol. Med.* **2014**, *10*, 1441–1451. [CrossRef] [PubMed]

*Review*

# Nanobody Engineering: Toward Next Generation Immunotherapies and Immunoimaging of Cancer

**Timothée Chanier and Patrick Chames ***

Aix Marseille University, CNRS, INSERM, Institute Paoli-Calmettes, CRCM, 13009 Marseille, France; timothee.chanier@inserm.fr

* Correspondence: patrick.chames@inserm.fr; Tel.: +33-491-828-833

Received: 18 December 2018; Accepted: 17 January 2019; Published: 21 January 2019

**Abstract:** In the last decade, cancer immunotherapies have produced impressive therapeutic results. However, the potency of immunotherapy is tightly linked to immune cell infiltration within the tumor and varies from patient to patient. Thus, it is becoming increasingly important to monitor and modulate the tumor immune infiltrate for an efficient diagnosis and therapy. Various bispecific approaches are being developed to favor immune cell infiltration through specific tumor targeting. The discovery of antibodies devoid of light chains in camelids has spurred the development of single domain antibodies (also called VHH or nanobody), allowing for an increased diversity of multispecific and/or multivalent formats of relatively small sizes endowed with high tissue penetration. The small size of nanobodies is also an asset leading to high contrasts for non-invasive imaging. The approval of the first therapeutic nanobody directed against the von Willebrand factor for the treatment of acquired thrombotic thrombocypenic purpura (Caplacizumab, Ablynx), is expected to bolster the rise of these innovative molecules. In this review, we discuss the latest advances in the development of nanobodies and nanobody-derived molecules for use in cancer immunotherapy and immunoimaging.

**Keywords:** Nanobody; Single Domain Antibody; Cancer; Immunotherapy; Imaging

---

## 1. Introduction

It is now well established that tumor cells can interact with their environment to promote an immunosuppressive environment to favor their survival and proliferation. Targeting the tumor environment for therapy has become a major interest in the past decade and is now a paradigm for new cancer therapies. Success of immunotherapy in cancer treatment, particularly the use of PD-1/PD-L1 and CTLA-4 antibodies, has led to the development of treatment targeting other immunological pathways [1,2]. However, immunotherapies are only efficient in a fraction of cancer patients [3]. Combination therapies are emerging as the path to increase response rates and tackle cancer cell escape mechanisms [4]. Their success often relies on the presence of immune cell within the tumor and their interaction with immunosuppressive ligands expressed by tumor cells. Cancers are currently best classified according to the immune infiltrate as well as the tumor cell type and localization [5].

In the case of non-infiltrated ("cold") tumors resistant to checkpoint inhibitors, new immunotherapy approaches tend to use bispecific construction targeting a tumor antigen and an immune receptor to favor immune cells infiltration and tumor cell specific targeting. Two bispecific antibodies have been approved by the US food and drug administration (FDA) (catumaxomab, CD3 × EpCAM and blinatumomab, CD3 × CD19) and many more are under clinical or pre-clinical development [6]. With the rise of molecular antibody engineering, a lot of different bispecific formats combining the heavy and light variable domains (VH + VL) with different specificities are being used for various therapeutic modalities [7].

Heavy chain only antibodies (HcAbs) have been identified in camelids. These antibodies are lacking the CH1 domain compared to conventional IgGs and are devoid of light chain. The specificity of HcAbs only relies on heavy variable domains called VHH. The recombinant production of a VHH generates a fragment called single domain antibody (sdAb), or nanobody [8]. Thanks to their high degree of sequence identity with human VHs (of family 3), nanobodies are expected to exhibit a low immunogenicity in human, and are easy to humanize for therapeutic perspectives [9], as confirmed by several phase clinical trials involving nanobodies and the recent approval by the European medicines agency (EMA) of the first therapeutic nanobody, caplacizumab [10]. The CDR3 loop of nanobodies is usually longer than conventional VH, allowing the binding to non-conventional epitopes such as protein clefts [11]. Moreover, structural studies have established that nanobodies usually have greater paratope diversity, involving amino acids within variable loops and framework regions [12]. Nanobodies are also characterized by a good solubility and stability to pH and temperatures. Importantly their small size allows for a better penetration within tissue and in cell–cell interfaces like immune synapses [13]. Conversely, this can be seen as a disadvantage for therapy, due to a quick renal elimination causing a very short serum half-life (close to 30 min). Different strategies to increase their serum half-life have been developed. One of them is based on a fusion to anti-albumin nanobody, increasing the serum half-life to 4–10 days without drastically increasing the molecule size [14]. Other strategies consist in a fusion to a human Fc fragment (CH2 and CH3 domains) allowing neonatal Fc receptor-based recycling and generating a bivalent molecule with a higher apparent affinity. Nanobodies can also be used to engineer larger molecules with several valencies or specificities and can be easily conjugated to imaging agent or drug delivery systems. Importantly, their high modularity increases further the format possibilities to crease small size antibody-derived molecules for therapy and imaging (Figure 1). In this review, we discuss the potential of nanobodies and nanobody-based engineered molecules for the immunotherapy and immunoimaging of tumors (Figure 2).

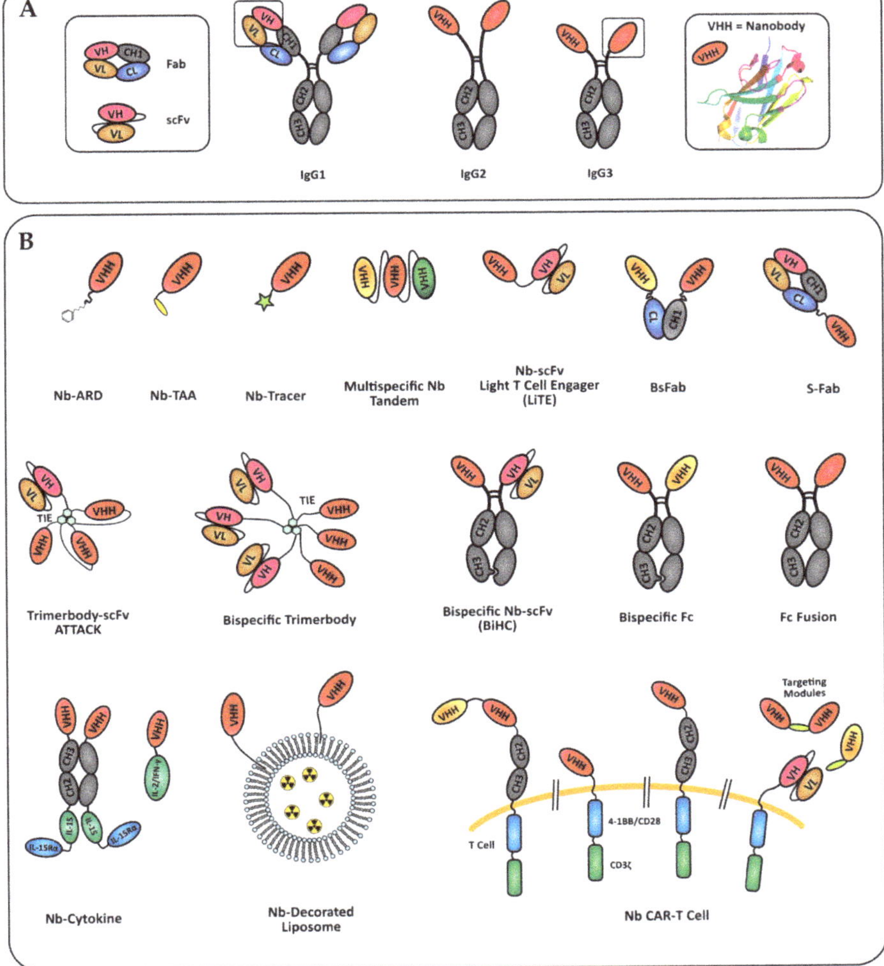


**Figure 1.** Nanobody-based formats in development for tumor immunotherapy and imaging. (**A**) Camelids specificity domains derived of conventional IgG1 or HcAbs (IgG2 and IgG3). The nanobody crystal structure shown is pdb entry 6GZP. (**B**) Formats of nanobody engineered molecules discussed in this review. Nb: nanobody; ARD: antigen recognition domain; TAA: tumor associated antigen.

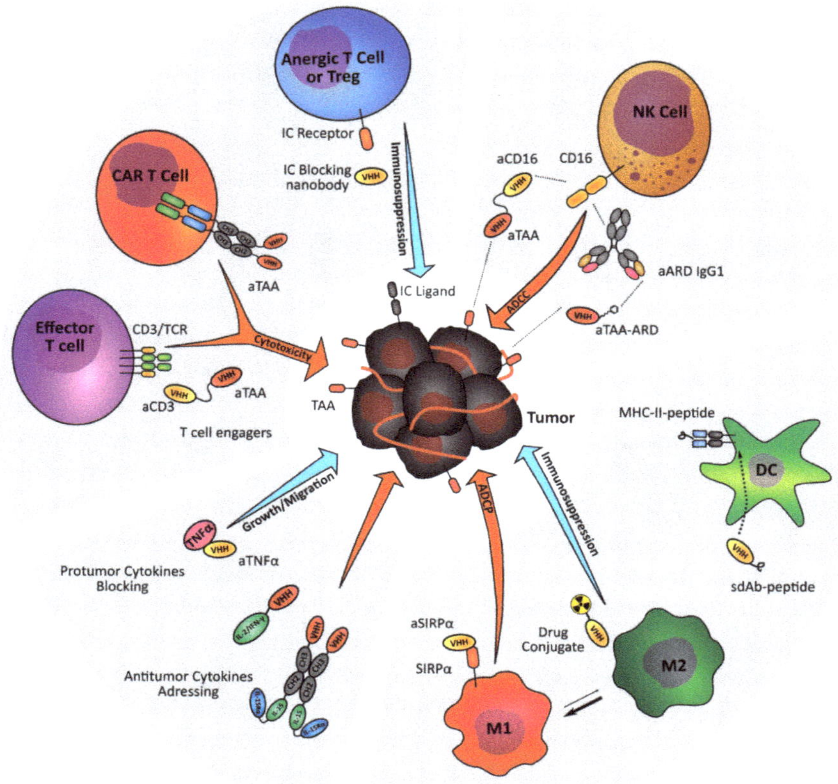


**Figure 2.** Nanobody-based strategies targeting the immune stroma of tumors. Nanobody-derived immunomodulatory molecules are under investigation to increase anti-tumor immunity (orange arrows) and prevent tumor-driven immune suppression (blue arrows). TAA: Tumor associated antigen; IC: Immune checkpoint; ARD: Antibody recruiting domain.

## 2. Targeting T Cell Activation and Cytotoxicity

T cells are adaptive immune cells endowed with a crucial role in anti-tumor cytotoxicity. Following the success of immune checkpoint blockade therapies, different strategies based on monoclonal antibodies aiming at activating the T cell mediated anti-tumor response and overcome immunotherapy resistance are currently under intense investigation [15]. Within this line, the high modularity of nanobodies is being actively exploited to design innovating drug candidates. Current research mostly covers T cell redirection strategies through the generation of bispecific format to recruit and activate cytotoxic or $\gamma\delta$ T cells, and the design of nanobody-derived chimeric antigen receptor (CAR) T cells. Some immune checkpoint blocking nanobodies are also being developed.

### 2.1. CD3 Bispecific Nanobodies: BiTE-Like Formats

CD3, and most specifically its $\zeta$ chain, belongs to the T cell receptor (TCR) complex. Anti-CD3 antibodies are commonly used to activate T cells in vitro. However therapeutic anti-CD3 antibodies can induce systemic inflammation and strong adverse effects. Bispecific antibodies allow a specific recognition and killing of tumor cells by bridging together the tumor and immune cells. Importantly, monovalent CD3 binding in the periphery does not trigger T cell activation and cytokine release syndrome. Bispecific T cell engagers (BiTEs) rely on a bispecific format combining an anti-CD3 single chain Fv (scFv) with a tumor associated antigen (TAA) specific scFv. BiTEs have been developed against several tumor antigens and blinatumomab (Blincyto, Abgen), a CD3 × CD19 BiTE is now approved by the FDA for the treatment of B-cell acute lymphoblastic leukemia [16]. Nanobodies allow for generation of BiTE-like format with a smaller size and higher modularity, and have been used to target several tumor antigens.

Her2 is a member of the epidermal growth factor receptor (EGFR) family, commonly known as a tumor antigen over-expressed in 20 to 30% of breast cancers, but also expressed in many other cancer types [17]. An anti-Her2 antibody-based immunotherapy with two monoclonal antibodies targeting different Her2 epitopes (trastuzumab + pertuzumab) is now first line treatment for patients with metastatic Her2$^+$ breast cancer. Lin et al. designed the S-Fab format, combining the Fab fragment of the well characterized anti-CD3 UTCH1 clone bearing an anti-Her2 nanobody fused to the C-terminus of its light chain [18]. The construction led to encouraging pre-clinical results, including a strong in vitro cytotoxicity and in vivo tumor growth inhibition. An alternative of the CH1-Cκ heterodimerization motif is the knob-into-hole technology, relying on CH3 engineering to prevent homodimerization of two distinct CH3 mutants [19]. The resulting bispecific Fc fragment increases the size of the molecule to ~80 kDa, still significantly smaller than a full-size IgG (~150 kDa). This format retains the serum half-life and effector functions of an IgG while affording a higher tissue penetration. A Her2 × CD3 bispecific antibody using the anti-CD3 UTCH1 scFv and an anti-Her2 nanobody (Nb) was generated with this technology [20]. The construction led to a strong anti-tumor growth effect with 4 doses spaced every 3 days instead of daily doses for 7 days for the previous Fab-Nb format.

Carcinoembryonic antigen (CEA) is a cell surface GPI-anchored glycoprotein commonly overexpressed in solid tumors [21]. It is a popular antigen for targeted therapy designs. A CEA × CD3 S-Fab showed strong T cell-mediated cytotoxicity against CEA$^+$ colon carcinoma cells LS174T and in vivo tumor inhibition [22]. In a more recent study, the authors used PEGylation to increase the in vivo serum half-life of the construct from 3 to 36 h in rats [23]. The anti-tumor effect was not impacted in a daily injection model, showing that PEGylation does not reduce the molecule activity, thus permitting to reach the same efficacy with less frequent injections.

Mølgaard et al. generated a bispecific light T cell engager (LiTE) comprising an anti-CD3 UTCH1 scFv fused in N- or C-terminal with an anti-EGFR nanobody [24]. The very small size of the LiTE format leads to a high tissue penetration but requires continuous injection for therapeutic applications or gene-based delivery. The authors reported similar binding properties and in vitro effects for C and N terminal fusions. The team used the nanobody modularity to build an original new format combining the CD3 scFv and 3 anti-EGFR nanobodies in a trimerbody format [25]. The interest of this format resides in a controlled orientation of the 3 nanobodies via collagen-derived trimerization domains (TIE), causing an increased affinity for EGFR without losing affinity for CD3. Although high affinity to tumor antigens can also lead to off-target effect on healthy cells expressing the receptor at lower levels, this model can be applied to target more tumor-specific antigens or to combine several lower affinity nanobodies. This format was further developed into a bispecific trivalent 4-1BB × EGFR construct combining 3 4-1BB scFv-TIE-EGFR Nb chains [26]. The 4-1BB is a strong co-stimulatory receptor expressed on T and NK cells and induces cytotoxicity, proliferation and cytokine release. However, systemic activation of 4-1BB causes over-activation of T and NK cell responses and sometimes severe toxicities. This trivalent/bispecific format was able to mainly localize into tumors, allowing an efficient anti-tumor effect with significantly reduced off-target cytotoxicity.

### 2.2. *γδT Cell Activation*

γδT cells are a subpopulation representing 1–10% of leukocytes. Their activation is MHC-independent, and leads to cytokine release and cytotoxicity. They are a heterogeneous population with pro-tumor (regulatory $\gamma\delta$T, $\gamma\delta$T17) and anti-tumor (V$\gamma$9V$\delta$2 T) activities. V$\gamma$9V$\delta$2 T cells constitute the main population of $\gamma\delta$T cells and their tumor infiltration have been associated with good prognosis factor in various tumor types [27–29]. Proliferation and activation of anti-tumor V$\gamma$9V$\delta$2 T cells appear as a promising therapeutic strategy [30]. De Bruin et al. screened several nanobodies for their ability to specifically bind to V$\gamma$9V$\delta$2 TCR and activate these $\gamma\delta$ T cells [31]. Using one of the activating nanobodies, they built a bispecific antibody combining anti-EGFR and anti-V$\gamma$9V$\delta$2 TCR nanobodies, separated with a $G_4S$ linker [32]. This antibody was able to activate V$\gamma$9V$\delta$2 T cells and triggered in vitro cytotoxicity against $EGFR^+$ cancer cell lines. In humanized xenografted mice, V$\gamma$9V$\delta$2 T cells injection alone caused improved survival and lower tumor burden and this effect was further increased with co-injection with the bispecific antibody.

### 2.3. Engineering Nanobody-Derived TCR in CAR-T Cell Therapy

Alternatively, T cell retargeting can be achieved through genetic engineering. Chimeric antigen receptor (CAR) T cells are genetically modified T cell expressing a non MHC-restricted antigen receptor. The intracellular domain is a fusion between a CD3$\zeta$ signaling domain and one or two co-stimulatory intracellular domains (CD28, OX-40, 4-1BB), and the extracellular domain is an antigen-binding site derived from an antibody. Great success and FDA-approval of anti-CD19 scFv-based CAR T cells (KYMRIAH) for treatment of B cell ALL has further increased interest in this field for cancer therapies. The main issues with CAR T cells are adverse reactions and the difficulty to control T cell proliferation in patients [33]. Many CAR T cell formats have been described, varying substantially is their antigen binding site, extracellular spacer and intracellular signaling domains.

More recently, CARs have been designed with a nanobody as antigen recognition site. The first nanobody-based CAR T cell used a MUC-1 targeting nanobody [34]. Interestingly, in a second generation of this CAR T cell, the authors added a caspase8-mediated suicide switch to control in vivo T cell proliferation and reduce potential unwanted damages [35]. This switch is triggered by adding a dimerization protein inducing caspase8 cell death signaling.

A novel modular universal CAR platform technology named UniCAR was developed to control the T cell activity and proliferation after injection. The authors used an inert CAR directed against a short peptide (5B9) and a targeting module (TM) directed against a tumor antigen and tagged with 5B9 [36]. The specific anti-tumor activity of these CAR T cells relies on the presence of a targeting module that can be switched during treatment to change antigen specificity, or to stop treatment. Recently, a nanobody-based anti-EGFR CAR-T cell was generated using the UniCAR technology [37]. As expected, these CAR T cells were unable to induce significant lysis of $EGFR^+$ cells and cytokine release in the absence of the anti-EGFR TM. PET imaging showed that the TM was rapidly eliminated via kidneys and that TM-CAR complexes were reversible. The team recently published a nanobody-based bivalent anti-EGFR module with increased affinity for EGFR to be used in the UniCAR system [38]. The increased affinity allows binding on $EGFR^{low}$ cells. The bivalent construct showed increase cytokine release and tumor rejection. This strategy could also be used to produce bispecific CAR T cells.

Bispecific CAR T cells can be designed to reduce the risk of tumor escape by loss of a tumor antigen and increase targeting accuracy. A bispecific Her2 $\times$ CD20 CAR was designed using a dual nanobody tandem as antigen recognition domain [39]. In this study, the dual specific CAR were equally efficient against CD20 or Her2 cells, proving the potential of bispecific CAR T cells to overcome the loss of antigen issue.

Sharifzadeh et al. generated different CAR formats with an anti-TAG72 nanobody to compare different costimulatory domains. They confirmed that CD28-OX40-CD3$\zeta$ CAR was more potent than the shorter CD28-CD3$\zeta$ intracellular domain. Similarly, significant anti-tumor effects in mice was achieved with nanobody-derived CAR targeting Her2 in breast cancers [40], glypycan-2 in neuroblastoma models [41] or CD38 in multiple myeloma [42].

### 2.4. Immune Checkpoint Blockade

Immune checkpoints are a group of inhibitory receptors mainly expressed in T and NK cells. The success of checkpoint therapy led to FDA approval of monoclonal antibodies against PD-1, PD-L1, and CTLA-4 [43]. Nanobodies are being developed as an alternative to monoclonal antibodies for therapy or as tools to investigate immune checkpoint biology. For instance, using an anti-murine CTLA-4 nanobody either as a monovalent domain, as a bivalent dimer or fused to an IgG2a constant region, Ingram et al. demonstrated that blocking CTLA-4 was not sufficient to trigger an anti-tumor activity. Indeed, only a Fc fusion of the nanobody triggered Treg depletion, controlled tumor growth and increased overall survival, suggesting a role of the Fc receptor-mediated immune response through the involvement of NK-mediated antibody dependent cell-mediated cytotoxicity (ADCC) [44]. However, another study showed an anti-tumor effect for a monovalent anti-CTLA-4 nanobody in a similar B16 mice model [45].

A recent study demonstrated a comparable level of $CD4^+$ T cell activation and tumor growth inhibition for a bivalent anti-PD-L1 nanobody-Fc fusion and durvalumab (a FDA-approved anti-PD-L1 IgG1κ) [46]. In addition to T cell activation, an effect was mediated by ADCC against $PD\text{-}L1^+$ tumor cells, consistent with the known involvement of FcγRs in PD-1/PD-L1 therapies [47].

Other checkpoint blocking nanobodies are being investigated, such as an anti-TIM-3 nanobody that demonstrated an anti-proliferative effect against the human leukemia cell line HL-60 [48].

## 3. Enhancing NK Cell-Mediated Antitumor Activity

### 3.1. Anti-CD16 Bispecific Antibodies

NK cells are innate effector immune cells playing a role in eliminating malignant cells and pathogens. They play a role in early cancer killing and immunosurveillance [49]. NK cell infiltration within tumor is a good prognosis in cancer patients [50]. They mediate direct killing of tumor cells independently of antigen presentation. NK cells are also important mediators of the adaptive immune response through cytokines and chemokine release, dendritic cells recruitment and T cell activation [51–53].

CD16 (FcγRIII) is a receptor for IgG1 and IgG3 Fc fragment expressed on NK cells, macrophages and γδT cells and mediates ADCC and phagocytosis of antibody-opsonized cells. NK cell-mediated ADCC is an important factor in the success of anti-tumor antigens antibody therapies and CD16 polymorphisms with higher affinity for Fc fragment are correlated with improved response to these treatments [54–57]. Fc engineering has become a major approach to improve efficiency of therapeutic antibodies [58]. As stated previously, the presence or absence of an Fc fragment can impact the action of nanobody-based constructs [44]. The engagement of CD16 also activates NK cell proliferation and functions independently of the ADCC effect through PI3K/MAPK pathways, without the need of a costimulatory signal [59,60]. The use of CD16 × TAA bispecific antibodies rather than Fc fragments was proposed to avoid binding to other Fc receptors, and sensitivity to CD16 polymorphisms. C21, A nanobody with high affinity for CD16 (10 nM, as determined by Biacore) was able to induce IL-2 and IFNγ secretion by NK cells in vitro after multimerization by sdAb biotinylation and incubation with streptavidin [61]. Different bispecific formats using the C21 nanobody are being developed to achieve this dual effect to improve tumor targeted therapy and allow a better recruitment of NK cells.

*Tandem*: The smallest bispecific nanobody format consists in a tandem of two nanobodies linked head to tail by a short peptide linker. Wang et al. constructed two tandems based on the anti-CD16 nanobody C21 and an anti-MUC-1 nanobody with a 2 amino acid linker (GS) [62] or an anti-CEA nanobody with a larger linker ($(G_4S)_3$) [63]. The major interests of this format are an effective and low-cost production in bacterial systems and a high stability compared to classical scFvs. In both studies, the authors observed a high in vitro cytotoxicity and in vivo tumor growth inhibition in NOD/SCID mice xenografted with $MUC\text{-}1^+/CEA^+$ colon carcinoma line LS174T, humanized by PBMC injection and treated daily with bispecific antibody injections. Therapeutic molecules based on this format would thus require PEGylation or the addition of an anti-albumin binding domain to increase their serum half-life, or the use of infusion pumps to deliver a constant flow rate, such as those used for blinatumomab in the clinic.

*BsFab*: The so-called bsFab format, combining the C21 nanobody with an anti-CEA nanobody through CH1-Cκ heterodimerization motif was developed [64]. The team showed that albeit the bsFab alone could not activate CD16-transfected Jurkat cells, it could induce IL-2 and IFN-γ secretion through CD16 clustering in the presence of CEA$^+$ colon carcinoma cells (LS174T). The bsFab showed potent in vitro NK cytotoxicity against CEA$^+$ cancer cells independently of CD16 polymorphisms, leading to tumor growth inhibition in vivo. Such bispecific formats might constitute a promising approach to specifically activate NK cells within tumor tissues. A similar construct using C21 nanobody with an anti-Her2 nanobody was compared to Trastuzumab [65]. The anti-Her2 bsFab showed increased cytokine release in vitro and similar tumor growth inhibition in vivo on Her2$^+$ breast cancer cell lines SK-BR-3 and BT 474. Interestingly the bsFab also led to NK cell activating and tumor growth inhibition using the trastuzumab-refractive Her2$^{low}$ cell line MCF-7 model.

*BsFc*: A bispecific CD16 × CEA was generated using anti-CD16 and anti-CEA nanobodies linked to two different mutants of an IgG1 to produce a bispecific Fc via the knob-into-hole technology [66]. This construct resulted in tumor growth inhibition in mice models. Unfortunately, the half-life of this molecule was not assessed in this study but can be expected to be higher than bsFab or tandem formats.

*S-Fab*: Wang et al. used the S-Fab format by combining an anti-CD16 Nanobody with the Fab of the anti-Her2 mAb Trastuzumab [67] or Pertuzumab [68]. The authors compared the efficiency of both constructs and showed that the pertuzumab-based construct was efficient at lower doses, but both induced potent tumor cell killing in vitro and reduced tumor growth in vivo. Interestingly the 2 antibodies used together seemed to synergize on Her2$^{med}$ cell line LS174T. An anti-Glypican 3 (GPC3) × anti-CD16 antibody was developed using the same S-Fab format. GPC3 is a tumor antigen overexpressed in hepatocellular carcinomas (HCC) with low normal tissue expression. In this video article, the authors described the process of production of this antibody and showed an effective in vitro cytotoxicity effect on HCC cell lines [69].

### 3.2. Nanobody Coupling to An Antibody Recruiting Domain

Alternatively, recruiting NK cells on target cells and trigger ADCC can be achieved by the coupling of an antibody recruiting domain (ARD) to an antigen binding domain (ABD). One study used an anti-Her2 nanobody as ABD and a dinitrophenol group (DNP) as the ARD [70]. DNP is an environmental chemical contaminant, and around 0.8% of IgG1s circulating in human serum bind DNP. The ABD binds HER2 positive tumor cells, thereby recruiting DNP-specific IgG1 serum antibodies, which trigger NK cells through FcγRIII, resulting in tumor cell killing by ADCC. The modularity of this approach might translate into broad and potent therapies, however patient-to-patient variability in serum reactiveness to DNP could be an issue.

## 4. Modulation of Antigen Presenting Cells

Antigen presenting cells (APC) are key mediators of the adaptive immune response. APC phenotypes are very heterogeneous. M2 macrophages, myeloid derived suppressor cells (MDSCs) and tolerogenic dendritic cells promote an immunosuppressive environment and favor tumor growth and angiogenesis [71]. At the opposite of the spectrum, M1 Macrophages, B cells, and cDC1 dendritic cells mediate the anti-tumor immune response through antigen presentation [72,73]. These last populations are very important to trigger an effective lymphocyte response. Nanobody-based therapeutics can be used to reprogram or eliminate suppressor cells or to enhance anti-tumor APC effector functions.

### 4.1. Innate Immune Checkpoint Blockade

The SIRPα-CD47 axis is one of the most important innate checkpoints. CD47 is a ubiquitous marker of self and acts as a "don't eat me" signal. It prevents phagocytosis by macrophages through interaction with the immune checkpoint receptor SIRP-α. By overexpressing CD47, many cancer types hijack this pathway to protect themselves from immune attacks [74]. Consequently, anti-CD47 nanobodies have been developed to restore macrophage-mediated phagocytosis. For instance, an anti-mouse CD47 nanobody significantly increased antibody-dependent phagocytosis of tumor cells by macrophages in vitro. However in vivo anti-tumor activity was only achieved in combination with anti-PD-L1 treatment [75]. The main issues with CD47 blockade are healthy cells toxicity (anemia) and antigen sink due to the ubiquitous expression of CD47. The challenge of CD47 therapy is thus to address the antibodies to the tumor site to reduce toxicity. Ingram et al. genetically engineered a B16 cell line to express the anti-CD47 nanobody within the tumor microenvironment to simultaneously address the antigen sink and nanobody intrinsic low half-life. This approach showed an effect of local CD47 therapy in combination with anti-TAA mAbs [76].

### 4.2. Nanobody-Based Immunization Strategies

One of the challenges in immunotherapy is to engineer dendritic cells to present TAA and provide effective immunization against the tumor. Adoptive transfer of autologous DCs matured in vitro to present TAA has been extensively studied in phaseI/II trials with promising results for glioma, melanoma, and lung cancers [77,78]. However, feasibility and cost of such strategies can negatively impact their therapeutic development. An easier and cheaper strategy for DC-based vaccine is to target TAA or TAA-derived peptides to the DC in vivo directly into patients. The major interests of nanobodies are their high penetrability within tissues and efficient coupling. Duarte et al. isolated nanobodies against the APC markers CD11b, CD36, and MHC-II combined to immunogenic peptides to determine the best marker for targeted antigen delivery [79]. An anti-MHC-II nanobody fused to an antigen was able to deliver the antigen to all DC and B cells subpopulations and induced a strong B cell response, as measured by serum responsiveness against the antigen. Interestingly, dimerization of the nanobody further increased B cell response, suggesting a role of antibody affinity or receptor clustering, or both, for internalization of the construction. This MHC-II nanobody was used to induce an immune response against the tumor antigen MUC1 [80]. Injection of the nanobody conjugated to MUC1-derived peptide induced strong humoral and CD4 responses. This strategy was applied to a different peptide format. A plant peptide scaffold was used to generate cyclic peptide and graft them onto the anti-MHC-II nanobody. Cyclization makes peptides more stable and less susceptible to proteolytic degradation. The nanobody targeting effectively activated an immunization against cyclical conformation of the peptides [81]. These studies highlight the potential of nanobody-conjugated TAA to induce an effective immune response against different antigen configurations in a non-invasive and cost-effective manner.

### 4.3. Drug Delivery

Macrophage mannose receptor (MMR)-expressing tumor associated macrophages (TAM) are involved in immune suppression and angiogenesis and are a marker of bad prognosis [82,83]. Specific elimination of the MMR$^+$ macrophages is proposed as a therapeutic strategy to increase the M1/M2 ratio towards a more favorable for tumor regression. Nuhn et al. used a nanobody with a high affinity to MMR (~20 nM) and conjugated it to a polymeric nanogel to produce MMR-specific 40 nm particles carrying drugs or imaging agents [84]. In a mouse lung cancer model, they successfully induced fluorescent nanogel internalization by MMR$^+$ TAMs and that phenotype was reverted in MMR KO mice. Drug delivery can be achieved by linking an internalization receptor. A nanobody-drug conjugate Targeting MHC-II on B cell lymphoma models presented a quick internalization and efficient drug delivery [85]. While this shows the potential of monovalent nanobody for MHC-II-dependent internalization and drug delivery, possibly representing a good alternative for therapy of anti-CD20 resistant lymphomas, it cannot be effectively use for specific APC targeting as it will kill pro-tumor as well as anti-tumor APCs.

### 4.4. APC Reprogramming

APC reprogramming from immunosuppressive to inflammatory phenotype has achieved great success in re-establishing an anti-tumor microenvironment in pre-clinical studies [86]. Gefitinib (Gef) and simvastatin (SV) are anti-tumor drugs with an effect on TAM activation [87,88]. In vitro treatment of macrophages with Gef/SV treatment inhibits HIF-2$\alpha$ and vascular endothelial growth factor expression. Recently, Yin et al. generated anti-PD-L1-nanobody-decorated liposomes to deliver Gef/SV treatment to PD-L1-expressing cancer cells and macrophages [89]. The liposome was effectively delivered to NSCLC endothelial cells and M2-like macrophages and the drug combination had potent anti-tumor effect on Gef-resistant and SV-resistant cells. Moreover, the M1/M2 ratio switched from M2-dominant to M1-dominant. This research shows the effective use of a nanobody to target both immune and cancer cells and remodel the tumor environment.

CD1d is a non-classical MHC involved in glycolipid presentation to NKT cells. This interaction activates NKT cell to produce cytokines and enhances IL-12 secretion and stimulatory capabilities of DCs. CD1d engagement potentiates anti-tumor T cell response [90]. Lameris et al. generated 22 anti-CD1d nanobodies with distinct biological activities [91]. Two of these nanobodies were able to induce NKT-independent IL-12 production by monocyte-derived DCs. Other nanobodies could block the NKT TCR-CD1d interaction or induce CD1d$^+$ cell apoptosis, with a potential use for CD1d$^+$ B lymphoblasts or multiple myelomas. This work highlights the small size of nanobodies and their ability to bind non-conventional epitopes as strong assets for the generation of molecules with various and well-defined effects.

Goyvaerts et al. developed a very interesting nanobody-lentivirus approach to address lentiviral vectors to specific cell populations [92]. The modified virus could effectively transduce DCs in a nanobody-dependent manner by targeting a so far unidentified target, with a strong potential for DC immunization. Unfortunately, the approach induced a weaker antigen-specific CD8$^+$ response compared to a broad tropisms lentivirus, possibly due to an infection-related inflammation [93]. However, this works clearly demonstrates the potential of nanobody–lentivirus constructs to transduce cell populations for reprogramming and activation of dendritic cells.

## 5. Targeting the Tumor Environment Cytokines and Chemokines

### 5.1. Pro-Tumor Cytokines Targeting: TNF/G-CSF

Cytokines are key modulators of immune cell states of activation. Inflammatory cytokines and growth factor are involved in cancer progression. Another immunotherapy approach aims at modulating the immune cells or cancer cells activity by targeting cytokines and their receptors.

TNF-$\alpha$ is a pro-inflammatory cytokine inducing many cellular processes, among which cell death, but also proliferation and angiogenesis. TNF-$\alpha$ serum levels are increased in cancer patients and correlate with disease progression [94]. Many tumors evolve to escape TNF-$\alpha$-mediated cytotoxicity and use it as a growth factor helping their survival and migration. Moreover TNF-$\alpha$ acts as an immunosuppressive cytokine by increasing MDSC and Treg proliferation [95,96]. Anti-TNF-$\alpha$ blockade has shown interesting anti-tumor effects [97]. An anti-TNF-$\alpha$ nanobody was able to reduce TNF-mediated proliferation and migration potency of breast cancer cell lines. The nanobody alone did not significantly reduce tumor growth in vivo but greatly inhibited lung metastasis and increased potency of the antimitotic Paclitaxel [98].

An alternative to block a cytokine activity is to target its receptor expressed on tumor cells. Bakherad et al. isolated an anti-granulocyte-colony stimulating factor receptor (G-CSFR) nanobody [99]. G-CSFR is mostly expressed in neutrophils and mediates their proliferation and activity. G-CSFR is expressed on gastric, colon, or lung carcinomas and leads to increased tumor growth and metastasis in a G-CSF-dependent matter [100–102]. In this study, the blocking nanobody successfully inhibited the G-CSF-induced gastric cancer cell proliferation in vitro via the SOCS3 signaling pathway.

*5.2. Modulating the CXCR4/CXCR7/CXCL12 Chemokines Axis in Solid Cancers*

Chemokines are important signals to orient cell migration within the organism, in particular for the immune cell infiltration into the tumor [103]. However, cancer cells can also use these signals to favor their migration into the blood stream. The CXCR4/CXCR7/CXCL12 chemokine axis is physiologically critical for hematopoiesis and T cell homing into inflammation sites. Chemokines CXCR4 and CXCR7 are overexpressed in many solid cancers and involved in metastasis towards CXCL12-rich tissues (such as lung, liver, and bone marrow) [104,105]. CXCR7 can also induce angiogenesis and tumor growth independently of its effect on tumor cell migration. CXCR4 blocking strategies using antibodies [106] or small inhibitors [107,108] have proven to be effective to reduce metastatic burdens and a phase IIb clinical trial for a CXCR4 antagonist in combination with anti-PD1 therapy in advanced pancreatic cancer is ongoing (NCT02907099).

Some nanobodies targeting immune cell chemotaxis are being developed to prevent tumor cell migration. An anti-CXCR7 nanobody was selected to block its interaction with CXCL12. The nanobody reduced tumor growth via its anti-angiogenesis effect [109]. However, blocking CXCR7 alone does not appear sufficient to stop epithelial-mesenchymal transition and metastasis. It would thus be of interest to associate CXCR7 and CXCR4 blockade to reduce CXCL12-mediated migration in addition of the anti-tumor effect of CXCR7 antagonists [110,111]. Several blocking anti-CXCR4 nanobodies have been used to reduce T cell chemotaxis in HIV infections contexts. These nanobodies prevented CXCL12-mediated migration of CXCR4$^+$ cells [112,113]. By fusing these nanobodies to an Fc domain, the authors increased the apparent affinity and the blocking potency against CXCR4, and induced ADCC and complement dependent cytotoxicity of CXCR4-expressing cells [114]. To our knowledge these anti-CXCR4 nanobodies have not been tested in cancer cell migration models but the combination of anti-CXCR4 and CXCR7 blocking nanobodies could be synergizing tools to tackle CXCL12-mediated metastasis.

*5.3. Nanobody as Carriers for Cytokine Delivery*

With the well-demonstrated anti-tumor effect of some cytokines, numerous anti-tumor cytokines have been proposed as cancer treatment. IL-2, a cytokine involved in immune cell activation and proliferation, was the first immunotherapy approved for cancers as early as 1992 [115]. However, while cytokines have great anti-tumor potency, the main limitation is off-target activation and associated toxicity. The main strategies to circumvent this limitation are gene transfer to produce cytokines locally, or antibody-directed cytokines (immunocytokines). Nanobodies can also be used as carrier for cytokines to restrict the activation to the tumor microenvironment. Dougan et al. used an anti-PD-L1 nanobody to deliver the anti-tumor cytokines IL-2 and IFN-$\gamma$ to immunologically impaired pancreatic tumors [116]. The nanobody was able to deliver the cytokines in well vascularized B16 melanoma tumors but also to denser mice pancreatic tumors, whilst the nanobody-Fc fusion did not achieve equally well. Both IL-2-nanobody and IFN-$\gamma$-nanobody synergized with an anti-TAA antibody for tumor growth inhibition. Immune infiltrate analysis showed increased CD8$^+$ counts but also Tregs proliferation with the IL-2 construct. The IFN-$\gamma$-nanobody significantly decreased MDSCs and redirected the DCs towards a MHC-II$^+$ phenotype. A similar approach was developed by Liu et al. with an IL15-Linker-IL-15R$\alpha$ construct fused to the C-terminus of an anti-CEA nanobody-Fc fusion [117]. IL-15 binding to its soluble receptor IL15-R$\alpha$ increases IL-15 effector function [118]. Contrarily to the broad spectrum of IL-2, IL-15 mostly affects NK and CD8$^+$ T cells proliferation and activation. Despite the relatively large size of this construct (~140 kDa), the authors demonstrated a strong anti-tumor effect associated with CD8$^+$ recruitment within the tumor in mouse models.

## 6. Nanobodies as Potent Imaging Tools

### *6.1. Importance of Molecular Imaging for Cancer Diagnostics*

With the ongoing development of in targeted therapies, it has become more and more important to visualize the presence tumor antigens and immune infiltrates to predict responsiveness. Molecular imaging with labeled antibodies has been intensely developed but the difficult tissue penetration and long half-life are strong obstacles to obtain good contrast and cancer detection. Consequently, nanobodies recently emerged as powerful tools for in vivo and in vitro imaging for diagnosis [119]. As opposed to the therapeutic setting, the fast elimination of nanobodies in vivo due to their small size and the absence of Fc fragment avoiding recycling constitutes a strong advantage for imaging, as it reduces background and generates a high contrast rapidly after injection. For in vitro staining, the small size of nanobodies also affords a better tissue penetration and staining compared to full size IgG (Figure 3) [120]. Nanobody-based imaging agents are very promising for an accurate and fast diagnosis in cancer therapy. Immuno-imaging requires the labeling of the targeting agent with an imaging probe. NHS ester-based chemistry targeting lysine side chains is often not ideal for nanobody labeling, because their small size is associated with a relatively high risk of impacting their binding activity. Moreover, it is of crucial importance to control batch-to-batch reproducibility concerning labeling ratio and orientation. Most current strategies are using sortase-mediated coupling. Sortase is a transpeptidase coupling a LPXTG motive to an N-terminal glycine [121,122]. This strategy is a very simple, cost-effective, and versatile way to label biomolecules in a controlled fashion. Moreover, the reaction cleaves the end of the LPXTG motive, allowing the convenient use of a purification tag (such as polyHis), which is ultimately replaced by the probe after coupling. Other site-directed coupling strategies can be used, such as His tag directed coupling using $^{99m}$Tc-tricarbonyl for Single-photon emission computed tomography (SPECT) [123].

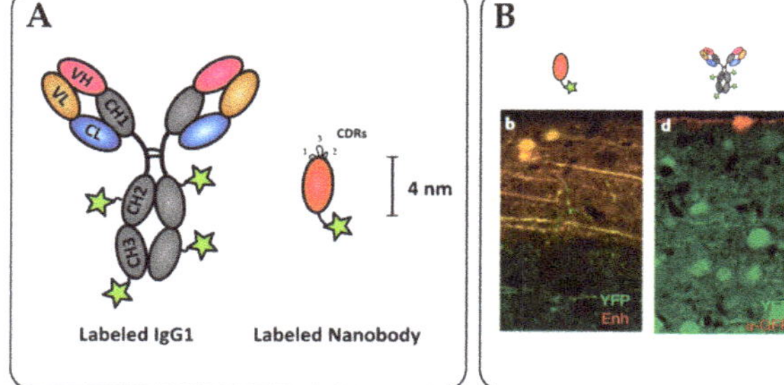


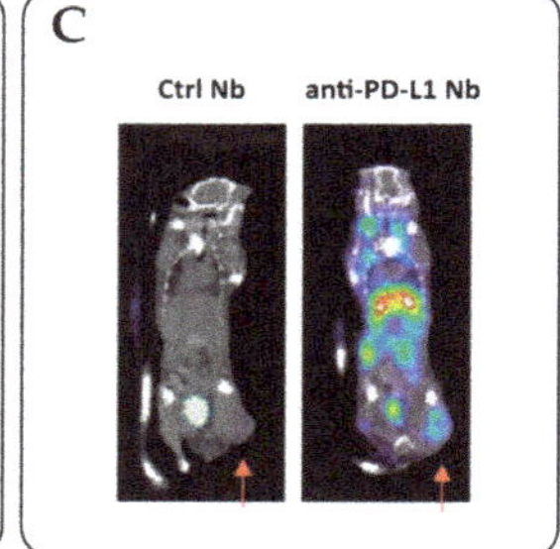


**Figure 3.** Nanobodies as potent tools for tumor immunoimaging. (**A**) Nanobody labelling strategies allow for site-specific and oriented conjugation. (**B**) Penetration of an anti-GFP nanobody (left) or full-size IgG (right) within an YFP-expressing brain tissue in vitro. Adapted from Fang T. et al. [120]. (C) SPECT/CT imaging of PD-L1 positive mouse lung epithelial cell line TC-1 in C57/BL6 mice with radiolabeled $^{99m}$Tc nanobodies 1 h after injection. The arrows indicate the tumor site. Adapted from Broos K. et al. [124].

### *6.2. Cancer Cell Detection*

The most advanced molecule in non-invasive imaging is an anti-Her2 nanobody used to detect Her2 expression in breast cancers via positive electron tomography-computed tomography (PET/CT). A phase I trial demonstrated a quick elimination, allowing measurement at 60–90 min after injection [125]. An ongoing phase II clinical trial investigates the uptake of the radiopharmaceutical 68-GaNOTA-Anti-HER2 nanobody in brain metastasis using PET/CT imaging (NCT03331601). Many nanobodies, including anti-CD20 [126], anti-Dipeptidyl-Peptidase 6 [127], and anti-HER3 nanobodies [128] are currently being studied.

### 6.3. Monitoring of Immune Infiltration

Nanobody-based imaging agents are very promising for an accurate and fast diagnosis in cancer therapy. Importantly, nanobodies are been explored as imaging agents to assess the immune infiltration within tumors prior or during immunotherapy. Indeed, therapeutic responses and patient prognosis are often linked to the nature, density, and activation status of immune cells infiltrated within the tumor microenvironment. Understanding the immune contexture of tumor would help in delivering the right treatment to patients.

#### 6.3.1. Tumor Infiltrating Lymphocyte Monitoring

T cells are a particularly interesting target for molecular imaging as their infiltration and state of activation are crucial to immunotherapy responsiveness. Bannas et al. used an anti-ART2 nanobody as imaging tool for optical imaging of T cells within lymph nodes. While a nanobody-Fc fusion displayed better in vitro staining, the single domain nanobody had more imaging potency in vivo due to reduced background. Good signal/noise ratio was observed 2 h post administration, while a monoclonal antibody required at least 24 h. Interestingly, this nanobody possesses a functional ART2 inhibiting property and could be used for both therapy and immunomodulation [129]. A PEGylated $^{89}$Zr anti-CD8$^{+}$ nanobody was produced to monitor the CD8$^{+}$ T cell infiltration within the tumor by PET/CT. PEGylation with a 20 kDa PEG moiety greatly reduced kidney retention of the antibody and thus gave a more specific staining of lymphoid organs [130]. In an immunized B16 mouse model giving partial regression with anti-CTLA-4 therapy, the authors showed that a homogeneous CD8$^{+}$ infiltration detected with the anti-CD8 nanobody in PET/CT correlated with anti-CTLA-4 responsiveness. Direct assessments of the checkpoint blockade targets have been investigated with a $^{99m}$Tc-laballed anti-PD-L1 nanobody for SPECT imaging. Potent PD-L1$^{+}$ tumor labeling was achieved 1 h post-inoculation and PD-L1 expression correlated with CD8$^{+}$-dependent reduced tumor development [124].

#### 6.3.2. APC Monitoring

Macrophage polarization is another important prognosis marker. The ratio between pro and anti-inflammatory macrophages shapes the tumor microenvironment.

Anti-MMR nanobodies have been evaluated to detect pro-tumor TAMs. Using a $^{99m}$Tc-labeled nanobody, Movahedi et al. managed to detect TAMs in mice [131]. In a following study, the authors compared the use of an anti-MMR nanobody with $^{99m}$Tc or $^{18}$F labeling respectively for SPECT or PET/CT imaging. Their work showed a greatly reduced liver and kidney uptake of the fluorinated nanobody compared to the radiometal-labeled version [132].

Conversely to MMR, MHC-II is a good prognosis marker and is associated with an efficient antigen presentation. An anti-MHC-II nanobody labeled with a near infrared fluorochrome was used to monitor human cell infiltration in NOD/SCID humanized mice. That probe was efficient for organ labeling and flow cytometry analysis, but the authors used a $^{64}$Cu staining for in vivo imaging. The nanobody displayed a good PET/CT signal/noise ratio 2 h after injection, and allowed imaging of MHC-II$^{+}$ cells in the spleen and bones despite kidney and bladder non-specific accumulation [133]. To take advantage of the higher availability of $^{18}$F-2-deoxyfluoroglucose ($^{18}$F-FDG) compared to other $^{18}$F probes, the authors developed an interesting strategy using $^{18}$F-FDG as a labeling probe. The labeled high affinity nanobody could detect MHC-II lymphoid organs more efficiently than the previously described low affinity anti-MHC-II $^{18}$F-nanobody. This approach allowed the detection of early tumors (~1 mm diameter) that cannot be detected with $^{18}$F-FDG due to their low metabolic activity [134]. This strategy combines the feasibility of $^{18}$F-FDG PET/CT imaging and precision of molecular nanobody-based imaging.

## 7. Conclusions

As shown by the multitude of different formats with promising pre-clinical effects reported in the last years, nanobodies emerge as powerful antibody engineering tools to replace scFv fragments as building blocks, and allow the generation of molecules with carefully designed affinities, valencies, and specificities. Nanobodies also rise as very potent imaging agents due to their stability, production and coupling efficiency, tissue penetrability, and fast elimination from the blood stream. We forecast that, in the near future, nanobodies will lead to many innovative and high potential molecules for cancer immunoimaging and immunotherapy.

**Author Contributions:** The manuscript was written by T.C. and P.C.

**Funding:** This work was supported by INSERM, CNRS and Cancéropôle PACA.

**Conflicts of Interest:** The authors declare no conflict of interest.

## References

1. Stambrook, P.J.; Maher, J.; Farzaneh, F. Cancer Immunotherapy: Whence and Whither. *Mol. Cancer Res.* **2017**, *15*, 635–650. [CrossRef] [PubMed]
2. Hahn, A.W.; Gill, D.M.; Pal, S.K.; Agarwal, N. The future of immune checkpoint cancer therapy after PD-1 and CTLA-4. *Immunotherapy* **2017**, *9*, 681–692. [CrossRef] [PubMed]
3. Reck, M.; Rodríguez-Abreu, D.; Robinson, A.G.; Hui, R.; Csőszi, T.; Fülöp, A.; Gottfried, M.; Peled, N.; Tafreshi, A.; Cuffe, S.; et al. Pembrolizumab versus Chemotherapy for PD-L1–Positive Non–Small-Cell Lung Cancer. *N. Engl. J. Med.* **2016**, *375*, 1823–1833. [CrossRef]
4. Larkin, J.; Chiarion-Sileni, V.; Gonzalez, R.; Grob, J.J.; Cowey, C.L.; Lao, C.D.; Schadendorf, D.; Dummer, R.; Smylie, M.; Rutkowski, P.; et al. Combined Nivolumab and Ipilimumab or Monotherapy in Untreated Melanoma. *N. Engl. J. Med.* **2015**, *373*, 23–34. [CrossRef]
5. Thorsson, V.; Gibbs, D.L.; Brown, S.D.; Wolf, D.; Bortone, D.S.; Ou Yang, T.-H.; Porta-Pardo, E.; Gao, G.F.; Plaisier, C.L.; Eddy, J.A.; et al. The Immune Landscape of Cancer. *Immunity* **2018**, *48*, 812–830.e14. [CrossRef] [PubMed]
6. Del Bano, J.; Chames, P.; Baty, D.; Kerfelec, B. Taking up Cancer Immunotherapy Challenges: Bispecific Antibodies, the Path Forward? *Antibodies* **2016**, *5*, 1. [CrossRef]
7. Brinkmann, U.; Kontermann, R.E. The making of bispecific antibodies. *mAbs* **2017**, *9*, 182–212. [CrossRef]
8. Hamers-Casterman, C.; Atarhouch, T.; Muyldermans, S.; Robinson, G.; Hamers, C.; Songa, E.B.; Bendahman, N.; Hamers, R. Naturally occurring antibodies devoid of light chains. *Nature* **1993**, *363*, 446–448. [CrossRef]
9. Vincke, C.; Loris, R.; Saerens, D.; Martinez-Rodriguez, S.; Muyldermans, S.; Conrath, K. General Strategy to Humanize a Camelid Single-domain Antibody and Identification of a Universal Humanized Nanobody Scaffold. *J. Biol. Chem.* **2009**, *284*, 3273–3284. [CrossRef]
10. Duggan, S. Caplacizumab: First Global Approval. *Drugs* **2018**, *78*, 1639–1642. [CrossRef]
11. Genst, E.D.; Silence, K.; Decanniere, K.; Conrath, K.; Loris, R.; Kinne, J.; Muyldermans, S.; Wyns, L. Molecular basis for the preferential cleft recognition by dromedary heavy-chain antibodies. *Proc. Natl. Acad. Sci. USA* **2006**, *103*, 4586–4591. [CrossRef] [PubMed]
12. Mitchell, L.S.; Colwell, L.J. Analysis of nanobody paratopes reveals greater diversity than classical antibodies. *Protein Eng. Des. Sel.* **2018**, *31*, 267–275. [CrossRef] [PubMed]
13. Cartwright, A.N.R.; Griggs, J.; Davis, D.M. The immune synapse clears and excludes molecules above a size threshold. *Nat. Commun.* **2014**, *5*, 5479. [CrossRef] [PubMed]
14. Van Roy, M.; Ververken, C.; Beirnaert, E.; Hoefman, S.; Kolkman, J.; Vierboom, M.; Breedveld, E.; 't Hart, B.; Poelmans, S.; Bontinck, L.; et al. The preclinical pharmacology of the high affinity anti-IL-6R Nanobody®ALX-0061 supports its clinical development in rheumatoid arthritis. *Arthritis Res. Ther.* **2015**, *17*, 135. [CrossRef] [PubMed]
15. O'Donnell, J.S.; Teng, M.W.L.; Smyth, M.J. Cancer immunoediting and resistance to T cell-based immunotherapy. *Nat. Rev. Clin. Oncol.* **2018**. [CrossRef] [PubMed]

16. Huehls, A.M.; Coupet, T.A.; Sentman, C.L. Bispecific T-cell engagers for cancer immunotherapy. *Immunol. Cell Biol.* **2015**, *93*, 290–296. [CrossRef]
17. Martin, V.; Cappuzzo, F.; Mazzucchelli, L.; Frattini, M. HER2 in solid tumors: More than 10 years under the microscope; where are we now? *Future Oncol.* **2014**, *10*, 1469–1486. [CrossRef]
18. Lin, L.; Li, L.; Zhou, C.; Li, J.; Liu, J.; Shu, R.; Dong, B.; Li, Q.; Wang, Z. A HER2 bispecific antibody can be efficiently expressed in Escherichia coli with potent cytotoxicity. *Oncol. Lett.* **2018**, *16*, 1259–1266.
19. Ridgway, J.B.B.; Presta, L.G.; Carter, P. 'Knobs-into-holes' engineering of antibody CH3 domains for heavy chain heterodimerization. *Protein Eng. Des. Sel.* **1996**, *9*, 617–621. [CrossRef]
20. Xing, J.; Lin, L.; Li, J.; Liu, J.; Zhou, C.; Pan, H.; Shu, R.; Dong, B.; Cao, D.; Li, Q.; et al. BiHC, a T-Cell–Engaging Bispecific Recombinant Antibody, Has Potent Cytotoxic Activity Against Her2 Tumor Cells. *Transl. Oncol.* **2017**, *10*, 780–785. [CrossRef]
21. Hammarström, S. The carcinoembryonic antigen (CEA) family: Structures, suggested functions and expression in normal and malignant tissues. *Semin. Cancer Biol.* **1999**, *9*, 67–81. [CrossRef] [PubMed]
22. Li, L.; He, P.; Zhou, C.; Jing, L.; Dong, B.; Chen, S.; Zhang, N.; Liu, Y.; Miao, J.; Wang, Z.; et al. A novel bispecific antibody, S-Fab, induces potent cancer cell killing. *J. Immunother.* **2015**, *38*, 350–356. [CrossRef] [PubMed]
23. Pan, H.; Liu, J.; Deng, W.; Xing, J.; Li, Q.; Wang, Z. Site-specific PEGylation of an anti-CEA/CD3 bispecific antibody improves its antitumor efficacy. *Int. J. Nanomed.* **2018**, *13*, 3189–3201. [CrossRef] [PubMed]
24. Mølgaard, K.; Harwood, S.L.; Compte, M.; Merino, N.; Bonet, J.; Alvarez-Cienfuegos, A.; Mikkelsen, K.; Nuñez-Prado, N.; Alvarez-Mendez, A.; Sanz, L.; et al. Bispecific light T-cell engagers for gene-based immunotherapy of epidermal growth factor receptor (EGFR)-positive malignancies. *Cancer Immunol. Immunother.* **2018**, *67*, 1251–1260. [CrossRef] [PubMed]
25. Harwood, S.L.; Alvarez-Cienfuegos, A.; Nuñez-Prado, N.; Compte, M.; Hernández-Pérez, S.; Merino, N.; Bonet, J.; Navarro, R.; Bergen en Henegouwen, P.M.; Lykkemark, S.; et al. ATTACK, a novel bispecific T cell-recruiting antibody with trivalent EGFR binding and monovalent CD3 binding for cancer immunotherapy. *Oncoimmunology* **2018**, *7*, e1377874. [CrossRef] [PubMed]
26. Compte, M.; Harwood, S.L.; Muñoz, I.G.; Navarro, R.; Zonca, M.; Perez-Chacon, G.; Erce-Llamazares, A.; Merino, N.; Tapia-Galisteo, A.; Cuesta, A.M.; et al. A tumor-targeted trimeric 4-1BB-agonistic antibody induces potent anti-tumor immunity without systemic toxicity. *Nat. Commun.* **2018**, *9*, 4809. [CrossRef] [PubMed]
27. Lozupone, F.; Pende, D.; Burgio, V.L.; Castelli, C.; Spada, M.; Venditti, M.; Luciani, F.; Lugini, L.; Federici, C.; Ramoni, C.; et al. Effect of human natural killer and gammadelta T cells on the growth of human autologous melanoma xenografts in SCID mice. *Cancer Res.* **2004**, *64*, 378–385. [CrossRef]
28. Duault, C.; Betous, D.; Bezombes, C.; Roga, S.; Cayrol, C.; Girard, J.-P.; Fournié, J.-J.; Poupot, M. IL-33-expanded human Vγ9Vδ2 T cells have anti-lymphoma effect in a mouse tumor model. *Eur. J. Immunol.* **2017**, *47*, 2137–2141. [CrossRef]
29. Beck, B.H.; Kim, H.-G.; Kim, H.; Samuel, S.; Liu, Z.; Shrestha, R.; Haines, H.; Zinn, K.; Lopez, R.D. Adoptively-transferred ex vivo expanded γδ-T cells mediate in vivo antitumor activity in preclinical mouse models of breast cancer. *Breast Cancer Res. Treat.* **2010**, *122*, 135–144. [CrossRef]
30. Pauza, C.D.; Liou, M.-L.; Lahusen, T.; Xiao, L.; Lapidus, R.G.; Cairo, C.; Li, H. Gamma Delta T Cell Therapy for Cancer: It Is Good to be Local. *Front. Immunol.* **2018**, *9*, 1305. [CrossRef]
31. de Bruin, R.C.G.; Lougheed, S.M.; van der Kruk, L.; Stam, A.G.; Hooijberg, E.; Roovers, R.C.; van Bergen en Henegouwen, P.M.P.; Verheul, H.M.W.; de Gruijl, T.D.; van der Vliet, H.J. Highly specific and potently activating Vγ9Vδ2-T cell specific nanobodies for diagnostic and therapeutic applications. *Clin. Immunol.* **2016**, *169*, 128–138. [CrossRef] [PubMed]
32. de Bruin, R.C.G.; Veluchamy, J.P.; Lougheed, S.M.; Schneiders, F.L.; Lopez-Lastra, S.; Lameris, R.; Stam, A.G.; Sebestyen, Z.; Kuball, J.; Molthoff, C.F.M.; et al. A bispecific nanobody approach to leverage the potent and widely applicable tumor cytolytic capacity of Vγ9Vδ2-T cells. *OncoImmunology* **2018**, *7*, e1375641. [CrossRef]
33. June, C.H.; O'Connor, R.S.; Kawalekar, O.U.; Ghassemi, S.; Milone, M.C. CAR T cell immunotherapy for human cancer. *Science* **2018**, *359*, 1361–1365. [CrossRef] [PubMed]
34. Iri-Sofla, F.J.; Rahbarizadeh, F.; Ahmadvand, D.; Rasaee, M.J. Nanobody-based chimeric receptor gene integration in Jurkat cells mediated by PhiC31 integrase. *Exp. Cell Res.* **2011**, *317*, 2630–2641. [CrossRef] [PubMed]

35. Khaleghi, S.; Rahbarizadeh, F.; Ahmadvand, D.; Rasaee, M.J.; Pognonec, P. A caspase 8-based suicide switch induces apoptosis in nanobody-directed chimeric receptor expressing T cells. *Int. J. Hematol.* **2012**, *95*, 434–444. [CrossRef] [PubMed]
36. Cartellieri, M.; Feldmann, A.; Koristka, S.; Arndt, C.; Loff, S.; Ehninger, A.; von Bonin, M.; Bejestani, E.P.; Ehninger, G.; Bachmann, M.P. Switching CAR T cells on and off: A novel modular platform for retargeting of T cells to AML blasts. *Blood Cancer J.* **2016**, *6*, e458. [CrossRef] [PubMed]
37. Albert, S.; Arndt, C.; Feldmann, A.; Bergmann, R.; Bachmann, D.; Koristka, S.; Ludwig, F.; Ziller-Walter, P.; Kegler, A.; Gärtner, S.; et al. A novel nanobody-based target module for retargeting of T lymphocytes to EGFR-expressing cancer cells via the modular UniCAR platform. *OncoImmunology* **2017**, *6*, e1287246. [CrossRef]
38. Albert, S.; Arndt, C.; Koristka, S.; Berndt, N.; Bergmann, R.; Feldmann, A.; Schmitz, M.; Pietzsch, J.; Steinbach, J.; Bachmann, M. From mono- to bivalent: Improving theranostic properties of target modules for redirection of UniCAR T cells against EGFR-expressing tumor cells in vitro and in vivo. *Oncotarget* **2018**, *9*, 25597–25616. [CrossRef]
39. De Munter, S.; Ingels, J.; Goetgeluk, G.; Bonte, S.; Pille, M.; Weening, K.; Kerre, T.; Abken, H.; Vandekerckhove, B. Nanobody Based Dual Specific CARs. *Int. J. Mol. Sci.* **2018**, *19*, 403. [CrossRef]
40. Jamnani, F.R.; Rahbarizadeh, F.; Shokrgozar, M.A.; Mahboudi, F.; Ahmadvand, D.; Sharifzadeh, Z.; Parhamifar, L.; Moghimi, S.M. T cells expressing VHH-directed oligoclonal chimeric HER2 antigen receptors: Towards tumor-directed oligoclonal T cell therapy. *Biochim. Biophys. Acta* **2014**, *1840*, 378–386. [CrossRef]
41. Li, N.; Fu, H.; Hewitt, S.M.; Dimitrov, D.S.; Ho, M. Therapeutically targeting glypican-2 via single-domain antibody-based chimeric antigen receptors and immunotoxins in neuroblastoma. *Proc. Natl. Acad. Sci. USA* **2017**, *114*, E6623–E6631. [CrossRef]
42. An, N.; Hou, Y.N.; Zhang, Q.X.; Li, T.; Zhang, Q.L.; Fang, C.; Chen, H.; Lee, H.C.; Zhao, Y.J.; Du, X. Anti-Multiple Myeloma Activity of Nanobody-Based Anti-CD38 Chimeric Antigen Receptor T Cells. *Mol. Pharm.* **2018**, *15*, 4577–4588. [CrossRef]
43. Hargadon, K.M.; Johnson, C.E.; Williams, C.J. Immune checkpoint blockade therapy for cancer: An overview of FDA-approved immune checkpoint inhibitors. *Int. Immunopharmacol.* **2018**, *62*, 29–39. [CrossRef]
44. Ingram, J.R.; Blomberg, O.S.; Rashidian, M.; Ali, L.; Garforth, S.; Fedorov, E.; Fedorov, A.A.; Bonanno, J.B.; Gall, C.L.; Crowley, S.; et al. Anti–CTLA-4 therapy requires an Fc domain for efficacy. *Proc. Natl. Acad. Sci. USA* **2018**, *115*, 3912–3917. [CrossRef]
45. Wan, R.; Liu, A.; Hou, X.; Lai, Z.; Li, J.; Yang, N.; Tan, J.; Mo, F.; Hu, Z.; Yang, X.; et al. Screening and antitumor effect of an anti-CTLA-4 nanobody. *Oncol. Rep.* **2018**, *39*, 511–518. [CrossRef]
46. Zhang, F.; Wei, H.; Wang, X.; Bai, Y.; Wang, P.; Wu, J.; Jiang, X.; Wang, Y.; Cai, H.; Xu, T.; et al. Structural basis of a novel PD-L1 nanobody for immune checkpoint blockade. *Cell Discov.* **2017**, *3*, 17004. [CrossRef]
47. Dahan, R.; Sega, E.; Engelhardt, J.; Selby, M.; Korman, A.J.; Ravetch, J.V. FcγRs Modulate the Anti-tumor Activity of Antibodies Targeting the PD-1/PD-L1 Axis. *Cancer Cell* **2015**, *28*, 285–295. [CrossRef]
48. Homayouni, V.; Ganjalikhani-hakemi, M.; Rezaei, A.; Khanahmad, H.; Behdani, M.; Lomedasht, F.K. Preparation and characterization of a novel nanobody against T-cell immunoglobulin and mucin-3 (TIM-3). *Iran. J. Basic Med. Sci.* **2016**, *19*, 1201–1208.
49. Imai, K.; Matsuyama, S.; Miyake, S.; Suga, K.; Nakachi, K. Natural cytotoxic activity of peripheral-blood lymphocytes and cancer incidence: An 11-year follow-up study of a general population. *Lancet* **2000**, *356*, 1795–1799. [CrossRef]
50. Rusakiewicz, S.; Semeraro, M.; Sarabi, M.; Desbois, M.; Locher, C.; Mendez, R.; Vimond, N.; Concha, A.; Garrido, F.; Isambert, N.; et al. Immune Infiltrates Are Prognostic Factors in Localized Gastrointestinal Stromal Tumors. *Cancer Res.* **2013**, *73*, 3499–3510. [CrossRef]
51. Böttcher, J.P.; Bonavita, E.; Chakravarty, P.; Blees, H.; Cabeza-Cabrerizo, M.; Sammicheli, S.; Rogers, N.C.; Sahai, E.; Zelenay, S.; Sousa, C.R. e NK Cells Stimulate Recruitment of cDC1 into the Tumor Microenvironment Promoting Cancer Immune Control. *Cell* **2018**, *172*, 1022–1037.e14. [CrossRef]
52. Krebs, P.; Barnes, M.J.; Lampe, K.; Whitley, K.; Bahjat, K.S.; Beutler, B.; Janssen, E.; Hoebe, K. NK cell–mediated killing of target cells triggers robust antigen-specific T cell–mediated and humoral responses. *Blood* **2009**, *113*, 6593–6602. [CrossRef]
53. Kelly, J.M.; Darcy, P.K.; Markby, J.L.; Godfrey, D.I.; Takeda, K.; Yagita, H.; Smyth, M.J. Induction of tumor-specific T cell memory by NK cell–mediated tumor rejection. *Nat. Immunol.* **2002**, *3*, 83–90. [CrossRef]

54. Cartron, G.; Dacheux, L.; Salles, G.; Solal-Celigny, P.; Bardos, P.; Colombat, P.; Watier, H. Therapeutic activity of humanized anti-CD20 monoclonal antibody and polymorphism in IgG Fc receptor FcγRIIIa gene. *Blood* **2002**, *99*, 754–758. [CrossRef]
55. Arnould, L.; Gelly, M.; Penault-Llorca, F.; Benoit, L.; Bonnetain, F.; Migeon, C.; Cabaret, V.; Fermeaux, V.; Bertheau, P.; Garnier, J.; et al. Trastuzumab-based treatment of HER2-positive breast cancer: An antibody-dependent cellular cytotoxicity mechanism? *Br. J. Cancer* **2006**, *94*, 259–267. [CrossRef]
56. Maréchal, R.; De Schutter, J.; Nagy, N.; Demetter, P.; Lemmers, A.; Devière, J.; Salmon, I.; Tejpar, S.; Van Laethem, J.-L. Putative contribution of CD56 positive cells in cetuximab treatment efficacy in first-line metastatic colorectal cancer patients. *BMC Cancer* **2010**, *10*, 340. [CrossRef]
57. Veeramani, S.; Wang, S.-Y.; Dahle, C.; Blackwell, S.; Jacobus, L.; Knutson, T.; Button, A.; Link, B.K.; Weiner, G.J. Rituximab infusion induces NK activation in lymphoma patients with the high-affinity CD16 polymorphism. *Blood* **2011**, *118*, 3347–3349. [CrossRef]
58. Sondermann, P.; Szymkowski, D.E. Harnessing Fc receptor biology in the design of therapeutic antibodies. *Curr. Opin. Immunol.* **2016**, *40*, 78–87. [CrossRef]
59. Trotta, R.; Kanakaraj, P.; Perussia, B. Fc gamma R-dependent mitogen-activated protein kinase activation in leukocytes: A common signal transduction event necessary for expression of TNF-alpha and early activation genes. *J. Exp. Med.* **1996**, *184*, 1027–1035. [CrossRef]
60. Lee, H.-R.; Son, C.-H.; Koh, E.-K.; Bae, J.-H.; Kang, C.-D.; Yang, K.; Park, Y.-S. Expansion of cytotoxic natural killer cells using irradiated autologous peripheral blood mononuclear cells and anti-CD16 antibody. *Sci. Rep.* **2017**, *7*, 11075. [CrossRef]
61. Behar, G.; Sibéril, S.; Groulet, A.; Chames, P.; Pugnière, M.; Boix, C.; Sautès-Fridman, C.; Teillaud, J.-L.; Baty, D. Isolation and characterization of anti-FcγRIII (CD16) llama single-domain antibodies that activate natural killer cells. *Protein Eng. Des. Sel.* **2008**, *21*, 1–10. [CrossRef]
62. Li, Y.; Zhou, C.; Li, J.; Liu, J.; Lin, L.; Li, L.; Cao, D.; Li, Q.; Wang, Z. Single domain based bispecific antibody, Muc1-Bi-1, and its humanized form, Muc1-Bi-2, induce potent cancer cell killing in muc1 positive tumor cells. *PLoS ONE* **2018**, *13*, e0191024. [CrossRef]
63. Dong, B.; Zhou, C.; He, P.; Li, J.; Chen, S.; Miao, J.; Li, Q.; Wang, Z. A novel bispecific antibody, BiSS, with potent anti-cancer activities. *Cancer Biol. Ther.* **2016**, *17*, 364–370. [CrossRef]
64. Rozan, C.; Cornillon, A.; Pétiard, C.; Chartier, M.; Behar, G.; Boix, C.; Kerfelec, B.; Robert, B.; Pèlegrin, A.; Chames, P.; et al. Single-Domain Antibody–Based and Linker-Free Bispecific Antibodies Targeting FcγRIII Induce Potent Antitumor Activity without Recruiting Regulatory T Cells. *Mol. Cancer Ther.* **2013**, *12*, 1481–1491. [CrossRef]
65. Turini, M.; Chames, P.; Bruhns, P.; Baty, D.; Kerfelec, B. A FcγRIII-engaging bispecific antibody expands the range of HER2-expressing breast tumors eligible to antibody therapy. *Oncotarget* **2014**, *5*, 5304–5319. [CrossRef]
66. Li, J.; Zhou, C.; Dong, B.; Zhong, H.; Chen, S.; Li, Q.; Wang, Z. Single domain antibody-based bispecific antibody induces potent specific anti-tumor activity. *Cancer Biol. Ther.* **2016**, *17*, 1231–1239. [CrossRef]
67. Li, A.; Xing, J.; Li, L.; Zhou, C.; Dong, B.; He, P.; Li, Q.; Wang, Z. A single-domain antibody-linked Fab bispecific antibody Her2-S-Fab has potent cytotoxicity against Her2-expressing tumor cells. *AMB Express* **2016**, *6*, 32. [CrossRef]
68. Deng, W.; Liu, J.; Pan, H.; Li, L.; Zhou, C.; Wang, X.; Shu, R.; Dong, B.; Cao, D.; Li, Q.; et al. A Bispecific Antibody Based on Pertuzumab Fab Has Potent Antitumor Activity. *J. Immunother.* **2018**, *41*, 1–8. [CrossRef]
69. Wang, Y.; Liu, J.; Pan, H.; Xing, J.; Wu, X.; Li, Q.; Wang, Z. A GPC3-targeting Bispecific Antibody, GPC3-S-Fab, with Potent Cytotoxicity. *J. Vis. Exp.* **2018**, *137*, e57588. [CrossRef]
70. Gray, M.A.; Tao, R.N.; DePorter, S.M.; Spiegel, D.A.; McNaughton, B.R. A Nanobody Activation Immunotherapeutic that Selectively Destroys HER2-Positive Breast Cancer Cells. *Chembiochem* **2016**, *17*, 155–158. [CrossRef]
71. Ostrand-Rosenberg, S.; Fenselau, C. Myeloid-Derived Suppressor Cells: Immune-Suppressive Cells That Impair Antitumor Immunity and Are Sculpted by Their Environment. *J. Immunol.* **2018**, *200*, 422–431. [CrossRef]
72. Engblom, C.; Pfirschke, C.; Pittet, M.J. The role of myeloid cells in cancer therapies. *Nat. Rev. Cancer* **2016**, *16*, 447–462. [CrossRef]

73. Böttcher, J.P.; e Sousa, C.R. The Role of Type 1 Conventional Dendritic Cells in Cancer Immunity. *Trends Cancer* **2018**, *4*, 784–792. [CrossRef]
74. Majeti, R.; Chao, M.P.; Alizadeh, A.A.; Pang, W.W.; Jaiswal, S.; Gibbs, K.D.; van Rooijen, N.; Weissman, I.L. CD47 is an adverse prognostic factor and therapeutic antibody target on human acute myeloid leukemia stem cells. *Cell* **2009**, *138*, 286–299. [CrossRef]
75. Sockolosky, J.T.; Dougan, M.; Ingram, J.R.; Ho, C.C.M.; Kauke, M.J.; Almo, S.C.; Ploegh, H.L.; Garcia, K.C. Durable antitumor responses to CD47 blockade require adaptive immune stimulation. *Proc. Natl. Acad. Sci. USA* **2016**, *113*, E2646–E2654. [CrossRef]
76. Ingram, J.R.; Blomberg, O.S.; Sockolosky, J.T.; Ali, L.; Schmidt, F.I.; Pishesha, N.; Espinosa, C.; Dougan, S.K.; Garcia, K.C.; Ploegh, H.L.; et al. Localized CD47 blockade enhances immunotherapy for murine melanoma. *Proc. Natl. Acad. Sci. USA* **2017**, *114*, 10184–10189. [CrossRef]
77. Tang, C.; Wang, X.; Li, Z.; Yue, Q.; Yang, Z.; Fan, K.; Hoon, D.; Hua, W. A Systemic Review of Clinical Trials on Dendritic-Cells Based Vaccine Against Malignant Glioma. *J. Carcinog. Mutagen* **2015**, *6*. [CrossRef]
78. Garg, A.D.; Perez, M.V.; Schaaf, M.; Agostinis, P.; Zitvogel, L.; Kroemer, G.; Galluzzi, L. Trial watch: Dendritic cell-based anticancer immunotherapy. *OncoImmunology* **2017**, *6*, e1328341. [CrossRef]
79. Duarte, J.N.; Cragnolini, J.J.; Swee, L.K.; Bilate, A.M.; Bader, J.; Ingram, J.R.; Rashidfarrokhi, A.; Fang, T.; Schiepers, A.; Hanke, L.; et al. Generation of Immunity against Pathogens via Single-Domain Antibody–Antigen Constructs. *J. Immunol.* **2016**, *197*, 4838–4847. [CrossRef]
80. Fang, T.; Van Elssen, C.H.M.J.; Duarte, J.N.; Guzman, J.S.; Chahal, J.S.; Ling, J.; Ploegh, H.L. Targeted antigen delivery by an anti-class II MHC VHH elicits focused αMUC1(Tn) immunity †Electronic supplementary information (ESI) available. *Chem. Sci.* **2017**, *8*, 5591–5597. [CrossRef]
81. Kwon, S.; Duarte, J.N.; Li, Z.; Ling, J.J.; Cheneval, O.; Durek, T.; Schroeder, C.I.; Craik, D.J.; Ploegh, H.L. Targeted Delivery of Cyclotides via Conjugation to a Nanobody. *ACS Chem. Biol.* **2018**, *13*, 2973–2980. [CrossRef]
82. Sun, X.; Gao, D.; Gao, L.; Zhang, C.; Yu, X.; Jia, B.; Wang, F.; Liu, Z. Molecular imaging of tumor-infiltrating macrophages in a preclinical mouse model of breast cancer. *Theranostics* **2015**, *5*, 597–608. [CrossRef]
83. Dong, P.; Ma, L.; Liu, L.; Zhao, G.; Zhang, S.; Dong, L.; Xue, R.; Chen, S. CD86+/CD206+, Diametrically Polarized Tumor-Associated Macrophages, Predict Hepatocellular Carcinoma Patient Prognosis. *Int. J. Mol. Sci.* **2016**, *17*, 320. [CrossRef]
84. Nuhn, L.; Bolli, E.; Massa, S.; Vandenberghe, I.; Movahedi, K.; Devreese, B.; Van Ginderachter, J.A.; De Geest, B.G. Targeting Protumoral Tumor-Associated Macrophages with Nanobody-Functionalized Nanogels through Strain Promoted Azide Alkyne Cycloaddition Ligation. *Bioconjug. Chem.* **2018**, *29*, 2394–2405. [CrossRef]
85. Fang, T.; Duarte, J.N.; Ling, J.; Li, Z.; Guzman, J.S.; Ploegh, H.L. Structurally Defined αMHC-II Nanobody–Drug Conjugates: A Therapeutic and Imaging System for B-Cell Lymphoma. *Angew. Chem. Int. Ed.* **2016**, *55*, 2416–2420. [CrossRef]
86. Georgoudaki, A.-M.; Prokopec, K.E.; Boura, V.F.; Hellqvist, E.; Sohn, S.; Östling, J.; Dahan, R.; Harris, R.A.; Rantalainen, M.; Klevebring, D.; et al. Reprogramming Tumor-Associated Macrophages by Antibody Targeting Inhibits Cancer Progression and Metastasis. *Cell Rep.* **2016**, *15*, 2000–2011. [CrossRef]
87. Tariq, M.; Zhang, J.; Liang, G.; He, Q.; Ding, L.; Yang, B. Gefitinib inhibits M2-like polarization of tumor-associated macrophages in Lewis lung cancer by targeting the STAT6 signaling pathway. *Acta Pharmacol. Sin.* **2017**, *38*, 1501–1511. [CrossRef]
88. Alupei, M.C.; Licarete, E.; Patras, L.; Banciu, M. Liposomal simvastatin inhibits tumor growth via targeting tumor-associated macrophages-mediated oxidative stress. *Cancer Lett.* **2015**, *356*, 946–952. [CrossRef]
89. Yin, W.; Yu, X.; Kang, X.; Zhao, Y.; Zhao, P.; Jin, H.; Fu, X.; Wan, Y.; Peng, C.; Huang, Y. Remodeling Tumor-Associated Macrophages and Neovascularization Overcomes EGFRT790M-Associated Drug Resistance by PD-L1 Nanobody-Mediated Codelivery. *Small* **2018**, *14*, 1802372. [CrossRef]
90. McEwen-Smith, R.M.; Salio, M.; Cerundolo, V. The Regulatory Role of Invariant NKT Cells in Tumor Immunity. *Cancer Immunol. Res.* **2015**, *3*, 425–435. [CrossRef]
91. Lameris, R.; de Bruin, R.C.; van Bergen en Henegouwen, P.M.; Verheul, H.M.; Zweegman, S.; de Gruijl, T.D.; van der Vliet, H.J. Generation and characterization of CD1d-specific single-domain antibodies with distinct functional features. *Immunology* **2016**, *149*, 111–121. [CrossRef]

92. Goyvaerts, C.; De Groeve, K.; Dingemans, J.; Van Lint, S.; Robays, L.; Heirman, C.; Reiser, J.; Zhang, X.-Y.; Thielemans, K.; De Baetselier, P.; et al. Development of the Nanobody display technology to target lentiviral vectors to antigen-presenting cells. *Gene Ther.* **2012**, *19*, 1133–1140. [CrossRef]
93. Goyvaerts, C.; De Vlaeminck, Y.; Escors, D.; Lienenklaus, S.; Keyaerts, M.; Raes, G.; Breckpot, K. Antigen-presenting cell-targeted lentiviral vectors do not support the development of productive T-cell effector responses: Implications for in vivo targeted vaccine delivery. *Gene Ther.* **2017**, *24*, 370–375. [CrossRef]
94. Wang, X.; Lin, Y. Tumor necrosis factor and cancer, buddies or foes? *Acta Pharmacol. Sin.* **2008**, *29*, 1275–1288. [CrossRef]
95. Zhao, X.; Rong, L.; Zhao, X.; Li, X.; Liu, X.; Deng, J.; Wu, H.; Xu, X.; Erben, U.; Wu, P.; et al. TNF signaling drives myeloid-derived suppressor cell accumulation. *J. Clin. Investig.* **2012**, *122*, 4094–4104. [CrossRef]
96. Okubo, Y.; Mera, T.; Wang, L.; Faustman, D.L. Homogeneous Expansion of Human T-Regulatory Cells Via Tumor Necrosis Factor Receptor 2. *Sci. Rep.* **2013**, *3*, 3153. [CrossRef]
97. Bertrand, F.; Montfort, A.; Marcheteau, E.; Imbert, C.; Gilhodes, J.; Filleron, T.; Rochaix, P.; Andrieu-Abadie, N.; Levade, T.; Meyer, N.; et al. TNFα blockade overcomes resistance to anti-PD-1 in experimental melanoma. *Nat. Commun.* **2017**, *8*, 2256. [CrossRef]
98. Ji, X.; Peng, Z.; Li, X.; Yan, Z.; Yang, Y.; Qiao, Z.; Liu, Y. Neutralization of TNFα in tumor with a novel nanobody potentiates paclitaxel-therapy and inhibits metastasis in breast cancer. *Cancer Lett.* **2017**, *386*, 24–34. [CrossRef]
99. Bakherad, H.; Gargari, S.L.M.; Sepehrizadeh, Z.; Aghamollaei, H.; Taheri, R.A.; Torshabi, M.; Yazdi, M.T.; Ebrahimizadeh, W.; Setayesh, N. Identification and in vitro characterization of novel nanobodies against human granulocyte colony-stimulating factor receptor to provide inhibition of G-CSF function. *Biomed. Pharmacother.* **2017**, *93*, 245–254. [CrossRef]
100. Fan, Z.; Li, Y.; Zhao, Q.; Fan, L.; Tan, B.; Zuo, J.; Hua, K.; Ji, Q. Highly Expressed Granulocyte Colony-Stimulating Factor (G-CSF) and Granulocyte Colony-Stimulating Factor Receptor (G-CSFR) in Human Gastric Cancer Leads to Poor Survival. *Med. Sci. Monit.* **2018**, *24*, 1701–1711. [CrossRef]
101. Morris, K.T.; Khan, H.; Ahmad, A.; Weston, L.L.; Nofchissey, R.A.; Pinchuk, I.V.; Beswick, E.J. G-CSF and G-CSFR are highly expressed in human gastric and colon cancers and promote carcinoma cell proliferation and migration. *Br. J. Cancer* **2014**, *110*, 1211–1220. [CrossRef]
102. Agarwal, S.; Lakoma, A.; Chen, Z.; Hicks, J.; Metelitsa, L.S.; Kim, E.S.; Shohet, J.M. G-CSF promotes neuroblastoma tumorigenicity and metastasis via STAT3-dependent cancer stem cell activation. *Cancer Res.* **2015**, *75*, 2566–2579. [CrossRef]
103. Sackstein, R.; Schatton, T.; Barthel, S.R. T-lymphocyte homing: An underappreciated yet critical hurdle for successful cancer immunotherapy. *Lab. Investig.* **2017**, *97*, 669–697. [CrossRef]
104. Liao, Y.X.; Zhou, C.H.; Zeng, H.; Zuo, D.Q.; Wang, Z.Y.; Yin, F.; Hua, Y.Q.; Cai, Z.D. The role of the CXCL12-CXCR4/CXCR7 axis in the progression and metastasis of bone sarcomas (Review). *Int. J. Mol. Med.* **2013**, *32*, 1239–1246. [CrossRef]
105. Krikun, G. The CXL12/CXCR4/CXCR7 axis in female reproductive tract disease: Review. *Am. J. Reprod. Immunol.* **2018**, *80*, e13028. [CrossRef]
106. Brennecke, P.; Arlt, M.J.E.; Campanile, C.; Husmann, K.; Gvozdenovic, A.; Apuzzo, T.; Thelen, M.; Born, W.; Fuchs, B. CXCR4 antibody treatment suppresses metastatic spread to the lung of intratibial human osteosarcoma xenografts in mice. *Clin. Exp. Metastasis* **2014**, *31*, 339–349. [CrossRef]
107. Benedicto, A.; Romayor, I.; Arteta, B. CXCR4 receptor blockage reduces the contribution of tumor and stromal cells to the metastatic growth in the liver. *Oncol. Rep.* **2018**, *39*, 2022–2030. [CrossRef]
108. Lefort, S.; Thuleau, A.; Kieffer, Y.; Sirven, P.; Bieche, I.; Marangoni, E.; Vincent-Salomon, A.; Mechta-Grigoriou, F. CXCR4 inhibitors could benefit to HER2 but not to triple-negative breast cancer patients. *Oncogene* **2017**, *36*, 1211–1222. [CrossRef]
109. Maussang, D.; Mujić-Delić, A.; Descamps, F.J.; Stortelers, C.; Vanlandschoot, P.; Walsum, M.S.; Vischer, H.F.; van Roy, M.; Vosjan, M.; Gonzalez-Pajuelo, M.; et al. Llama-derived Single Variable Domains (Nanobodies) Directed against Chemokine Receptor CXCR7 Reduce Head and Neck Cancer Cell Growth In Vivo. *J. Biol. Chem.* **2013**, *288*, 29562–29572. [CrossRef]
110. Zheng, N.; Liu, W.; Chen, J.; Li, B.; Liu, J.; Wang, J.; Gao, Y.; Shao, J.; Jia, L. CXCR7 is not obligatory for CXCL12-CXCR4-induced epithelial-mesenchymal transition in human ovarian cancer. *Mol. Carcinog.* **2019**, *58*, 144–155. [CrossRef]

111. Duda, D.G.; Kozin, S.V.; Kirkpatrick, N.D.; Xu, L.; Fukumura, D.; Jain, R.K. CXCL12 (SDF1α)-CXCR4/CXCR7 Pathway Inhibition: An Emerging Sensitizer for Anticancer Therapies? *Clin. Cancer Res.* **2011**, *17*, 2074–2080. [CrossRef]
112. Van Hout, A.; Klarenbeek, A.; Bobkov, V.; Doijen, J.; Arimont, M.; Zhao, C.; Heukers, R.; Rimkunas, R.; de Graaf, C.; Verrips, T.; et al. CXCR4-targeting nanobodies differentially inhibit CXCR4 function and HIV entry. *Biochem. Pharmacol.* **2018**, *158*, 402–412. [CrossRef]
113. Jähnichen, S.; Blanchetot, C.; Maussang, D.; Gonzalez-Pajuelo, M.; Chow, K.Y.; Bosch, L.; Vrieze, S.D.; Serruys, B.; Ulrichts, H.; Vandevelde, W.; et al. CXCR4 nanobodies (VHH-based single variable domains) potently inhibit chemotaxis and HIV-1 replication and mobilize stem cells. *Proc. Natl. Acad. Sci. USA* **2010**, *107*, 20565–20570. [CrossRef]
114. Bobkov, V.; Zarca, A.M.; Van Hout, A.; Arimont, M.; Doijen, J.; Bialkowska, M.; Toffoli, E.; Klarenbeek, A.; van der Woning, B.; van der Vliet, H.J.; et al. Nanobody-Fc constructs targeting chemokine receptor CXCR4 potently inhibit signaling and CXCR4-mediated HIV-entry and induce antibody effector functions. *Biochem. Pharmacol.* **2018**, *158*, 413–424. [CrossRef]
115. Rosenberg, S.A. IL-2: The First Effective Immunotherapy for Human Cancer. *J. Immunol.* **2014**, *192*, 5451–5458. [CrossRef]
116. Dougan, M.; Ingram, J.R.; Jeong, H.-J.; Mosaheb, M.M.; Bruck, P.T.; Ali, L.; Pishesha, N.; Blomberg, O.; Tyler, P.M.; Servos, M.M.; et al. Targeting Cytokine Therapy to the Pancreatic Tumor Microenvironment Using PD-L1–Specific VHHs. *Cancer Immunol. Res.* **2018**, *6*, 389–401. [CrossRef]
117. Liu, Y.; Wang, Y.; Xing, J.; Li, Y.; Liu, J.; Wang, Z. A novel multifunctional anti-CEA-IL15 molecule displays potent antitumor activities. *Drug Des. Dev. Ther.* **2018**, *12*, 2645–2654. [CrossRef]
118. Rubinstein, M.P.; Kovar, M.; Purton, J.F.; Cho, J.-H.; Boyman, O.; Surh, C.D.; Sprent, J. Converting IL-15 to a superagonist by binding to soluble IL-15Rα. *Proc. Natl. Acad. Sci. USA* **2006**, *103*, 9166–9171. [CrossRef]
119. Schoonooghe, S.; Laoui, D.; Van Ginderachter, J.A.; Devoogdt, N.; Lahoutte, T.; De Baetselier, P.; Raes, G. Novel applications of nanobodies for in vivo bio-imaging of inflamed tissues in inflammatory diseases and cancer. *Immunobiology* **2012**, *217*, 1266–1272. [CrossRef]
120. Fang, T.; Lu, X.; Berger, D.; Gmeiner, C.; Cho, J.; Schalek, R.; Ploegh, H.; Lichtman, J. Nanobody immunostaining for correlated light and electron microscopy with preservation of ultrastructure. *Nat. Methods* **2018**, *15*, 1029–1032. [CrossRef]
121. Witte, M.D.; Wu, T.; Guimaraes, C.P.; Theile, C.S.; Blom, A.E.M.; Ingram, J.R.; Li, Z.; Kundrat, L.; Goldberg, S.D.; Ploegh, H.L. Site-specific protein modification using immobilized sortase in batch and continuous-flow systems. *Nat. Protoc.* **2015**, *10*, 508–516. [CrossRef]
122. Massa, S.; Vikani, N.; Betti, C.; Ballet, S.; Vanderhaegen, S.; Steyaert, J.; Descamps, B.; Vanhove, C.; Bunschoten, A.; van Leeuwen, F.W.B.; et al. Sortase A-mediated site-specific labeling of camelid single-domain antibody-fragments: A versatile strategy for multiple molecular imaging modalities. *Contrast Media Mol. Imaging* **2016**, *11*, 328–339. [CrossRef]
123. Xavier, C.; Devoogdt, N.; Hernot, S.; Vaneycken, I.; D'Huyvetter, M.; De Vos, J.; Massa, S.; Lahoutte, T.; Caveliers, V. Site-Specific Labeling of His-Tagged Nanobodies with 99mTc: A Practical Guide. In *Single Domain Antibodies: Methods and Protocols*; Saerens, D., Muyldermans, S., Eds.; Methods in Molecular Biology; Humana Press: Totowa, NJ, USA, 2012; pp. 485–490, ISBN 978-1-61779-968-6.
124. Broos, K.; Keyaerts, M.; Lecocq, Q.; Renmans, D.; Nguyen, T.; Escors, D.; Liston, A.; Raes, G.; Breckpot, K.; Devoogdt, N.; et al. Non-invasive assessment of murine PD-L1 levels in syngeneic tumor models by nuclear imaging with nanobody tracers. *Oncotarget* **2017**, *8*, 41932–41946. [CrossRef]
125. Keyaerts, M.; Xavier, C.; Heemskerk, J.; Devoogdt, N.; Everaert, H.; Ackaert, C.; Vanhoeij, M.; Duhoux, F.P.; Gevaert, T.; Simon, P.; et al. Phase I Study of 68Ga-HER2-Nanobody for PET/CT Assessment of HER2 Expression in Breast Carcinoma. *J. Nucl. Med.* **2016**, *57*, 27–33. [CrossRef]
126. Krasniqi, A.; D'Huyvetter, M.; Xavier, C.; der Jeught, K.V.; Muyldermans, S.; Heyden, J.V.D.; Lahoutte, T.; Tavernier, J.; Devoogdt, N. Theranostic Radiolabeled Anti-CD20 sdAb for Targeted Radionuclide Therapy of Non-Hodgkin Lymphoma. *Mol. Cancer Ther.* **2017**, *16*, 2828–2839. [CrossRef]
127. Balhuizen, A.; Massa, S.; Mathijs, I.; Turatsinze, J.-V.; Vos, J.D.; Demine, S.; Xavier, C.; Villate, O.; Millard, I.; Egrise, D.; et al. A nanobody-based tracer targeting DPP6 for non-invasive imaging of human pancreatic endocrine cells. *Sci. Rep.* **2017**, *7*, 15130. [CrossRef]

128. Warnders, F.J.; van Scheltinga, A.G.T.T.; Knuehl, C.; van Roy, M.; de Vries, E.F.J.; Kosterink, J.G.W.; de Vries, E.G.E.; Hooge, M.N.L. Human Epidermal Growth Factor Receptor 3–Specific Tumor Uptake and Biodistribution of 89Zr-MSB0010853 Visualized by Real-Time and Noninvasive PET Imaging. *J. Nucl. Med.* **2017**, *58*, 1210–1215. [CrossRef]
129. Bannas, P.; Well, L.; Lenz, A.; Rissiek, B.; Haag, F.; Schmid, J.; Hochgräfe, K.; Trepel, M.; Adam, G.; Ittrich, H.; et al. In vivo near-infrared fluorescence targeting of T cells: Comparison of nanobodies and conventional monoclonal antibodies. *Contrast Media Mol. Imaging* **2014**, *9*, 135–142. [CrossRef]
130. Rashidian, M.; Ingram, J.R.; Dougan, M.; Dongre, A.; Whang, K.A.; LeGall, C.; Cragnolini, J.J.; Bierie, B.; Gostissa, M.; Gorman, J.; et al. Predicting the response to CTLA-4 blockade by longitudinal noninvasive monitoring of CD8 T cells. *J. Exp. Med.* **2017**, *214*, 2243–2255. [CrossRef]
131. Movahedi, K.; Schoonooghe, S.; Laoui, D.; Houbracken, I.; Waelput, W.; Breckpot, K.; Bouwens, L.; Lahoutte, T.; Baetselier, P.D.; Raes, G.; et al. Nanobody-Based Targeting of the Macrophage Mannose Receptor for Effective In Vivo Imaging of Tumor-Associated Macrophages. *Cancer Res.* **2012**, *72*, 4165–4177. [CrossRef]
132. Blykers, A.; Schoonooghe, S.; Xavier, C.; D'hoe, K.; Laoui, D.; D'Huyvetter, M.; Vaneycken, I.; Cleeren, F.; Bormans, G.; Heemskerk, J.; et al. PET Imaging of Macrophage Mannose Receptor–Expressing Macrophages in Tumor Stroma Using 18F-Radiolabeled Camelid Single-Domain Antibody Fragments. *J. Nucl. Med.* **2015**, *56*, 1265–1271. [CrossRef]
133. Van Elssen, C.H.M.J.; Rashidian, M.; Vrbanac, V.; Wucherpfennig, K.W.; el Habre, Z.; Sticht, J.; Freund, C.; Jacobsen, J.T.; Cragnolini, J.; Ingram, J.; et al. Noninvasive Imaging of Human Immune Responses in a Human Xenograft Model of Graft-Versus-Host Disease. *J. Nucl. Med.* **2017**, *58*, 1003–1008. [CrossRef]
134. Rashidian, M.; Keliher, E.J.; Dougan, M.; Juras, P.K.; Cavallari, M.; Wojtkiewicz, G.R.; Jacobsen, J.T.; Edens, J.G.; Tas, J.M.J.; Victora, G.; et al. Use of 18F-2-Fluorodeoxyglucose to Label Antibody Fragments for Immuno-Positron Emission Tomography of Pancreatic Cancer. *ACS Cent. Sci.* **2015**, *1*, 142–147. [CrossRef]

MDPI
St. Alban-Anlage 66
4052 Basel
Switzerland
Tel. +41 61 683 77 34
Fax +41 61 302 89 18
www.mdpi.com

*Antibodies* Editorial Office
E-mail: antibodies@mdpi.com
www.mdpi.com/journal/antibodies

www.ingramcontent.com/pod-product-compliance
Lightning Source LLC
LaVergne TN
LVHW071523170726
843515LV00008B/2045
* 9 7 8 3 0 3 6 5 0 3 7 8 3 *